The Prehistory of Mathematical Structuralism

LOGIC AND COMPUTATION IN PHILOSOPHY

Series Editor

*

Published in the Series:

MATHEMATICS AND MIND
Alexander George

THE LOGIC OF RELIABLE INQUIRY
Kevin T. Kelly

IN THE LIGHT OF LOGIC
Solomon Feferman

THE PROVENANCE OF PURE REASON
Essays in the Philosophy of Mathematics and Its History

THE PREHISTORY OF MATHEMATICAL STRUCTURALISM
Edited by Erich H. Reck and Georg Schiemer

The Prehistory of Mathematical Structuralism

Edited by

ERICH H. RECK AND GEORG SCHIEMER

OXFORD
UNIVERSITY PRESS

OXFORD
UNIVERSITY PRESS

Oxford University Press is a department of the University of Oxford. It furthers the University's objective of excellence in research, scholarship, and education by publishing worldwide. Oxford is a registered trade mark of Oxford University Press in the UK and certain other countries.

Published in the United States of America by Oxford University Press
198 Madison Avenue, New York, NY 10016, United States of America.

Library of Congress Cataloging-in-Publication Data
ISBN 978–0–19–064122–1

1 3 5 7 9 8 6 4 2

Printed by Integrated Books International, United States of America

Contents

Acknowledgments vii
About the Contributors ix

1. Introduction and Overview 1
Erich H. Reck and Georg Schiemer

I. MATHEMATICAL DEVELOPMENTS

2. Grassmann's Concept Structuralism 21
Paola Cantù

3. Dedekind's Mathematical Structuralism: From Galois Theory to Numbers, Sets, and Functions 59
José Ferreirós and Erich H. Reck

4. Pasch's Empiricism as Methodological Structuralism 88
Dirk Schlimm

5. Transfer Principles, Klein's Erlangen Program, and Methodological Structuralism 106
Georg Schiemer

6. The Ways of Hilbert's Axiomatics: Structural and Formal 142
Wilfried Sieg

7. Noether as Mathematical Structuralist 166
Audrey Yap

8. The Functional Role of Structures in Bourbaki 187
Gerhard Heinzmann and Jean Petitot

9. Saunders Mac Lane: From *Principia Mathematica* through Göttingen to the Working Theory of Structures 215
Colin McLarty

II. LOGICAL AND PHILOSOPHICAL REFLECTIONS

10. Logic of Relations and Diagrammatic Reasoning: Structuralist Elements in the Work of Charles Sanders Peirce 241
Jessica Carter

11. Poincaré and the Prehistory of Mathematical Structuralism 273
Janet Folina

12. "If Numbers Are to Be Anything At All, They Must Be Intrinsically Something": Bertrand Russell and Mathematical Structuralism 303
Jeremy Heis

13. Cassirer's Reception of Dedekind and the Structuralist Transformation of Mathematics 329
Erich H. Reck

14. Methodological Frames: Paul Bernays, Mathematical Structuralism, and Proof Theory 352
Wilfried Sieg

15. Carnap's Structuralist Thesis 383
Georg Schiemer

16. Explication as Elimination: W. V. Quine and Mathematical Structuralism 421
Sean Morris

Name Index 443
Subject Index 447

Acknowledgments

We would like to thank all the contributors to this volume not only for their essays, but for their fruitful and enjoyable collaboration more generally.

During the period when we worked on the volume, we interacted with a large number of people whose helpful comments and broader support we would like to acknowledge as well, including: Andy Arana, Bahram Assadian, Steve Awodey, Neil Barton, Francesca Biagioli, André Carus, Günther Eder, Philip Ehrlich, William Ewald, Juliet Floyd, Marco Giovanelli, Eduardo Giovannini, Norbert Gratzl, Mirja Hartimo, Henning Heller, Pierre Keller, Ansten Klev, Johannes Korbmacher, Martin Kusch, Hannes Leitgeb, Øystein Linnebo, Paolo Mancosu, Jean-Pierre Marquis, Chris Menzel, Marco Panza, Inger Bakken Pedersen, Esther Ramharter, Vincenzo de Risi, Andrea Sereni, Stewart Shapiro, Clinton Tolley, John Wigglesworth, and Richard Zach.

In addition, we would like to thank Florian Kolowrat and Henning Heller for their editorial assistance.

This project received funding from the European Research Council (ERC) under the European Union's Horizon 2020 research and innovation program (Grant Agreement No. 715222).

About the Contributors

Paola Cantù is a CNRS tenured researcher affiliated with the Centre Gilles Gaston Granger at Aix-Marseille Université. She has published on the epistemology of the Peano School, on Gödel's philosophy, on Bolzano, on Veronese's infinitesimals, and on argumentation theory.

Jessica Carter is Associate Professor at the Department of Mathematics and Computer Science at the University of Southern Denmark. Her research combines the philosophy of mathematics with the history and contemporary "practice" of mathematics. She has in particular contributed to two areas, namely visual thinking and structuralism in mathematics. She is also among the founders of the Association for the Philosophy of Mathematical Practice.

José Ferreirós is Professor of Logic and Philosophy of Science at the University of Seville in Spain. He is the author of *Labyrinth of Thought: A History of Set Theory and Its Role in Modern Mathematics* (1999) and of *Mathematical Knowledge and the Interplay of Practices* (2015).

Janet Folina is Professor of Philosophy at Macalester College. She is the author of *Poincaré and the Philosophy of Mathematics* (1992). Her current research interests include the philosophy of mathematics of the 19th and 20th centuries, and the concept of proof in mathematics.

Gerhard Heinzmann is Professor of Philosophy at the University of Lorraine in Nancy and founder of the history of science and philosophy laboratory Archives Henri Poincaré. He is an expert on the mathematical and philosophical work of Poincaré and has published extensively on Poincaré, Wittgenstein, and Cavaillès, among others. He is co-editor of *The Age of Alternative Logics: Assessing Philosophy of Logic and Mathematics Today* (2006) and of *Freedom in Mathematics* (2016).

Jeremy Heis is Associate Professor at the Department of Logic and Philosophy Science at the University of California, Irvine. His research focuses on Kant's philosophy of mathematics and logic, neo-Kantianism (especially Ernst Cassirer), and early analytic philosophy. He also has more general interests in the history and philosophy of mathematics and logic.

Colin McLarty is Truman P. Handy Professor of Intellectual Philosophy and Professor of Mathematics at Case Western Reserve University. His research focuses on logic, the history and philosophy of mathematics, and on structural foundations for mathematics in category theory. He is the author of *Elementary Categories, Elementary Toposes* (1992).

Sean Morris is Associate Professor of Philosophy at Metropolitan State University of Denver. His research focuses on the history of analytic philosophy and logic. He is particularly interested in the interaction of technical developments in mathematics and the special sciences with philosophy. He is the author of *Quine, New Foundations, and the Philosophy of Set Theory* (2018).

Jean Petitot was Full Professor at the Mathematical Center of the School for Advanced Studies in Social Sciences (EHESS) in Paris and former director of the Research Center in Applied Epistemology (CREA) at the École Polytechnique. His main research interests concern the epistemology of mathematics and neurocognitive models. He is the author of over 300 research articles and eight books, including *Elements of Neurogeometry: Functional Architectures of Vision* (2017).

Erich H. Reck is Professor of Philosophy at the University of California, Riverside. His main research interests lie in the philosophy of mathematics, the history and philosophy of logic, and the history of 19th/20th-century philosophy. He is the editor of several collections of articles, beginning with *From Frege to Wittgenstein: Perspectives on Early Analytic Philosophy* (2002), and most recently *Logic, Philosophy of Mathematics and their History: Essays in Honor of W.W. Tait* (2018). He has also published a number of articles on Dedekind, Frege, Carnap, Wittgenstein, and Cassirer.

Georg Schiemer is Associate Professor at the Department of Philosophy at the University of Vienna as well as external fellow at the Munich Center for Mathematical Philosophy at Ludwig Maximilian University Munich. He is also principal investigator of the project "The Roots of Mathematical Structuralism" funded by an European Research Council Starting Grant (2017–2022). His research focuses on the history and philosophy of mathematics and early analytic philosophy. He is also interested in the history and philosophy of logic and formal philosophy of science.

Dirk Schlimm is Associate Professor in the Department of Philosophy and Associate Member in the School of Computer Science at McGill University. He is also an External Member of the Munich Center for Mathematical Philosophy (MCMP), Ludwig Maximilian University Munich. His research interests fall in the areas of history and philosophy of mathematics and science, epistemology, and cognitive science. In particular, he is interested in the developments in the 19th and early 20th centuries that led to the emergence of modern mathematics and logic, and in systematic investigations regarding axiomatics, the use of notation, analogical reasoning, concept formation, and theory development.

Wilfried Sieg is the Patrick Suppes Professor of Philosophy at Carnegie Mellon University. He received his Ph.D. from Stanford University in 1977. From 1977 to 1985, he was Assistant and Associate Professor at Columbia University. In 1985, he joined the Carnegie Mellon faculty as a founding member of the University's Philosophy Department and served as its head from 1994 to 2005. He is internationally known for mathematical work in proof theory, historical work on modern logic and mathematics, and philosophical essays on the nature of mathematics. He is the author of *Hilbert's Programs and*

Beyond (2013 and 2019); he edited and co-edited a number of volumes including *David Hilbert's Lectures on the Foundations of Arithmetic and Logic* (2013). Sieg is a Fellow of the American Academy of Arts and Sciences.

Audrey Yap is Associate Professor of Philosophy at the University of Victoria, in British Columbia. Her primary research areas are the history and philosophy of mathematics and logic, as well as feminist epistemology. She is currently on the Association for Symbolic Logic Committee for Logic Education and a board member of the Philosophy of Mathematics Association.

The Prehistory of Mathematical Structuralism

1
Introduction and Overview

Erich H. Reck and Georg Schiemer

1. Structuralism in the Philosophy of Mathematics: A Brief History

The core idea of mathematical structuralism is that mathematical theories, always or at least in many central cases, are meant to characterize abstract structures (as opposed to more concrete, individual objects). Thus, arithmetic characterizes the natural number structure, analysis the real number structure, and traditional geometry the structure of Euclidean space. As such, structuralism is a general position about the subject matter of mathematics, namely abstract structures; but it also includes, or is intimately connected with, views about its methodology, since studying such structures involves distinctive tools and procedures. The goal of the present collection of essays is to discuss mathematical structuralism with respect to both aspects. And this is done by examining contributions by a number of mathematicians and philosophers of mathematics from the second half of the 19th and the early 20th centuries.

In English-speaking philosophy, structuralist ideas have played a role for a while; but the current discussion of structuralism, as a main philosophical position, started in the 1960s. A crucial article often referred to in this context is Paul Benacerraf's "What Numbers Could Not Be" (1965). This article was a reaction against the view, dominant at the time, that numbers and other mathematical objects are all sets. For example, the natural numbers are the finite von Neumann ordinals familiar from Zermelo-Fraenkel set theory; and the real numbers are Dedekind cuts constructed in a set-theoretic way. According to Benacerraf, this kind of position misrepresents mathematics by leaving out its structuralist aspects. Beyond Benacerraf, there were other reactions against such a set-theoretic, foundationalist orthodoxy. For example, in Hilary Putnam's article "Mathematics without Foundations" (1967), a form of if-then-ism for mathematics was suggested instead (more on both subsequently).

It took until the 1980s for the debates about mathematical structuralism to really pick up steam. The main impetus came from a number or writings by Michael Resnik, Stewart Shapiro, Geoffrey Hellman, and Charles Parsons (cf. Resnik 1981, 1997, Shapiro 1983, 1997, Hellman 1989, 1996, Parsons 1990, 2009, among

Erich H. Reck and Georg Schiemer, *Introduction and Overview* In: *The Prehistory of Mathematical Structuralism*. Edited by: Erich H. Reck and Georg Schiemer, Oxford University Press (2020). © Oxford University Press.
DOI:10.1093/oso/9780190641221.003.0001

others). While Benacerraf had suggested thinking of the natural numbers, say, as an "abstract structure," distinct from all set-theoretic systems, he remained noncommittal and somewhat vague about the nature of such structures. Resnik and Shapiro took that notion more seriously, suggesting that we should think of them as abstract "patterns." In Shapiro's hands, especially, the patterns were then conceived of as a novel kind of abstract entity, to be described and studied in a general "structure theory." While more focused on epistemological issues, Resnik resisted reifying the relevant patterns. But for both Resnik and Shapiro, particular mathematical objects, such as specific natural or real numbers, are "positions" in such structures. In addition, Charles Parsons developed a distinctive variant of such a "structuralist view of mathematical objects" (more on the differences soon).

Shapiro characterized his position further in a twofold way: as a form of "realism"; and as "*ante rem* structuralism." What exactly realism amounts to in this context is a difficult, slippery question. But at a minimum, it involves taking mathematical statements, such as 2 + 3 = 5, at face value, in the sense that "2," "3," and "5" are seen as singular terms referring to abstract objects to which we ascribe properties, etc. Shapiro called his position *ante rem* since he took his abstract structures to be "ontologically independent" of their more concrete "instantiations," including set-theoretic ones. Often it is assumed in this context that the *ante rem* aspect directly implies the "realist" one. But as we will see later, this is misleading and wrong in general; the two can be, and have at times been, separated. Moreover, Parsons explicitly distinguishes further metaphysical claims involving "realism" from the basic structuralist conception he accepted. This means that one can be a "pattern structuralist" without being a realist, except in the minimal sense already mentioned. (Both points will matter later.)

In widely adopted terminology, Parsons also distinguished between "eliminative" and "non-eliminative" forms of structuralism (Parsons 1990). According to "non-eliminative structuralism," abstract structures are accepted, or postulated, as *sui generis* objects (different from other kinds of objects, including set-theoretic systems). Shapiro's *ante rem* structuralism is a main example; but Resnik's and Parsons's forms of structuralism are others. A paradigmatic example of "eliminative structuralism" is Geoffrey Hellman's "modal structuralism," intentionally devised to be a "structuralism without structures" (Hellman 1996). Building on Putnam's if-then-ism (and earlier ideas in Russell), Hellman proposed to interpret every mathematical statement as having a modalized if-then form. For example, "2 + 3 = 5" has the form "Necessarily, for all models M of the Dedekind-Peano axioms, $2_M + 3_M = 5_M$," where 2_M, 3_M, and 5_M are what "play the roles" of 2, 3, and 5 in the model M, etc. (cf. Reck and Price 2000 for details). Along such lines, structures seen as abstract objects are "eliminated"; we don't need to assume their existence. In fact, Hellman's position is "eliminativist" in a very strong sense, since reference to abstract objects is avoided altogether. Instead, mathematics becomes the study of certain possibilities (the Dedekind-Peano axioms, say, have to be possible) and necessities (general if-then statements such as the preceding example).

Shapiro and Hellman have worked out their positions in great detail. For both, this includes distinguishing "algebraic" mathematical theories, such as group theory, lattice theory, and topology, from "non-algebraic" ones, such as the theories of the natural numbers, real numbers, and sets. With respect to the latter, we are dealing with categorical (or at least quasi-categorical) theories, which means that all their models are isomorphic (up to the height of the set-theoretic hierarchy in the case of Zermelo-Fraenkel set theory). It is such theories to which their accounts are meant to apply primarily. For the non-categorical ones a more indirect approach is used. Beyond Shapiro's and Hellman's positions, other versions of structuralism have been proposed, and they usually involve a similar distinction. We already mentioned Resnik's and Parsons's positions on the non-eliminative side; Charles Chihara's is another example on the eliminative side (Chihara 2004); and we will encounter more later.

Since the 1980s, the debates about structuralism in mathematics have been extended in other respects too. Three trends stand out especially. First, some comparative studies have been offered (Hellman 2001, 2005; Cole 2010; Shapiro 2012; also Reck and Price 2000, on which we will build). One of their results is that Shapiro's, Hellman's, and similar positions rely, at bottom, on the assumption of a kind of "coherence" for the mathematical theories at issue, besides their categoricity. (In light of Gödel's theorems, this replaces provable consistency as a basic requirement for mathematics.) Somewhat surprisingly, such positions are thus very similar in a basic respect (which, among others, puts the "realism/anti-realism" distinction into a new light). A second development, from around 2000 on, has been to further probe certain features of Shapiro's structuralism especially, but also of other forms of non-eliminative structuralism. One example is that "positions" in structures are taken to be "ontologically dependent" on the whole structure. But how exactly is that to be understood? (Cf. Linnebo 2008, among others.) Another example is that, according to Shapiro's and similar positions, "structurally indistinguishable" objects should be identified. Yet that leads to problems in the case of "nonrigid" structures (with nontrivial automorphisms), such as the system of complex numbers (see, e.g., Keränen 2001; Leitgeb and Ladyman 2008; and Shapiro 2008).

A third main development since the 1980s has been the introduction and promotion of category-theoretic forms of structuralism, by Steve Awodey, Elaine Landry, Jean-Pierre Marquis, Colin McLarty, and others (cf. Awodey 1996, Landry 2009, McLarty 2004, Marquis 2008). While all the versions of structuralism we have discussed work, in one form or another, with first-order logic and set theory (perhaps modified slightly, e.g., in terms of Hellman's modal logic), category theory involves a radical shift away from that framework. This affects the way in which "structuralist" ideas are implemented. Roughly speaking, in categorical language only "structural properties" are expressible (see Landry and Marquis 2005, Korbmacher and Schiemer 2018 for more); and crucial features involving them can be highlighted further, e.g., in terms of "universal mapping properties." This makes the approach "structuralist" in a distinctive, very basic way. But category theory is also taken to be an alternative, significantly different "foundational" framework for mathematics (which has led to debates about the notion, or notions, of foundations involved). For both reasons, "categorical" versions of structuralism are hard to compare with those mentioned earlier. That being said, category theory is in line with an important shift in mathematical methodology that emerged in the 19th and early 20th centuries, and investigating that shift further can help us understand "structuralism" better in general, including its categorical versions (as will become clearer later).

2. The Varieties of Mathematical Structuralism: Extending the Taxonomy

As the discussion in the previous section shows, it is misleading to speak of "structuralism" as if this label attached to a unique, unified position in the philosophy of mathematics. (Occasionally "structuralism" is identified, even more misleadingly, with Shapiro's position, since it is the most prominent one.) Rather, a whole variety of "structuralist" positions have been proposed in the literature. They all share the core idea with which we started this introduction, namely that "mathematical theories characterize abstract structures." But how that slogan is interpreted varies widely. Previously we used a threefold taxonomy so as to introduce some order and clarity. It started with Parsons's distinction between eliminative and non-eliminative forms of structuralism, with Hellman's and Shapiro's positions as paradigm cases. Then we added categorical structuralism as a third alternative, one that is not easy to compare to the others. But actually, the options one should consider are more varied than that; hence, a comprehensive taxonomy for structuralism has to be broader and richer. We will now take some steps in that direction.

One further distinction (largely ignored for long, but related to the difficulties in comparing categorical and other forms of structuralism) is very basic and should be introduced before all the others. It is the distinction between "methodological structuralism," on the one hand, and "metaphysical structuralism," on the other (cf. Awodey 1996; Reck and Price 2000). In related terminology, one can distinguish between "mathematical" and "philosophical structuralism." In what follows, we will treat these two dichotomies as the same (with only a slight difference in what is highlighted). The former term in each case, i.e., "methodological/mathematical structuralism," is meant to capture a distinctive way of doing mathematics, i.e., a certain "methodology," "form of practice," or "mathematical style." (For related discussions, cf. Corry 2004, Carter 2008, and Landry 2018, among others.) Roughly, it consists of doing mathematics by "studying abstract structures"; but this slogan requires again clarification. In addition, the methodology at issue comes with a general assumption on what mathematics is about, or what its subject matter is, namely "abstract structures." Then again, methodological/mathematical structuralism does not include, in itself, claims about what these structures are, i.e., about their "nature," "abstractness," "existence," etc. That is exactly what is added when we move on to "metaphysical/philosophical structuralism." In other words, there is a basic distinction between one kind of structuralism focused on "methodological," or more generally "mathematical," issues, while the other kind adds specific "metaphysical," or more broadly "philosophical," theses to the mix. Hence the labels.

With respect to mathematical practice, or to pursuing mathematical research fruitfully, one typically does not need to consider the specific "metaphysical" or more generally "philosophical" questions just mentioned. In fact, mathematicians often dismiss them as misleading or misguided (with important exceptions, as we will see). In contrast, it is exactly such questions that philosophers of mathematics try to address, including Benacerraf, Resnik, Shapiro, Hellman, and Parsons. Of course, the philosophers' answers should be grounded in mathematical practice, i.e., the goal should be a philosophical position not only compatible with but informed by mathematical practice, thus appropriate for it. Categorical structuralists often try to remain on the methodological/mathematical side alone. Their concern is then how to think through, and develop further, the methodology emerging in the late 19th- and early 20th-century mathematics by category-theoretic means. But sometimes metaphysical/philosophical views are added along the way also along such lines (e.g., when category theory is interpreted in a formalist way).

One main goal of the present collection of essays is to clarify the origins, and with it the nature, of methodological/mathematical structuralism up to the rise of category theory (from Grassmann, Dedekind, and Klein to Noether, Bourbaki, and Mac Lane). This is intended to clarify what it has meant, and still often means, to do mathematics by "studying abstract structures." A second main goal is to illustrate that the emergence of methodological/mathematical structuralism, in that sense, was accompanied, from early on, by reflections that shade over into "metaphysical/philosophical structuralism." And it was not only philosophers who engaged in such reflections, but also mathematicians themselves. (In several cases, the philosopher and mathematician at issue is one and the same person.) To be able to pursue this second goal fruitfully, several further distinctions concerning "structuralism in mathematics" are called for, now especially on the philosophical side.

Parsons's dichotomy between eliminative and non-eliminative forms of structuralism will remain helpful in what follows, so that we will keep using it. But it should be added, right away, that one can find relevant positions in the literature that are "semi eliminativist," unlike Hellman's position, which is "fully eliminativist." This concerns structuralist positions that reject the postulation of structures as distinctive, independent abstract objects, but accept other kinds of abstract objects, e.g., sets (thought of in some nonstructuralist way then). In other words, there are structuralist positions that are eliminativist about structures, but are not nominalist. They still count as forms of eliminative structuralism, but not of eliminativism about abstract objects generally.

One example is what is sometimes called "set-theoretic structuralism" (cf. Reck and Price 2000). According to this position, the natural numbers, say, should not be identified, in any strict or absolute sense, with the finite von Neumann ordinals. Why not? Because, exactly as Benacerraf argued, there are various set-theoretic models of the Dedekind-Peano axioms, indeed infinitely many, and none of them is privileged in a metaphysical sense (as opposed to some weaker pragmatic sense). This becomes a form of structuralism if one adds that "any set-theoretic model will do," so that the intrinsic, nonstructural properties of its elements do not matter. In other words, we can identify "the natural numbers" with the finite von Neumann ordinals, but do so in a pragmatic sense and with the proviso that we could have identified them with, say, the finite Zermelo ordinals too. In John Burgess's words (Burgess 2015), this position involves a "indifference to identify" them with any particular model of the Dedekind-Peano axioms; similarly in other cases. (Strictly speaking, this position is "structuralist" with respect to some objects but not generally, e.g., not for sets.)

One can generalize this approach. Set-theoretic structuralism is a specific version of "relativist structuralism" (see again Reck and Price 2000). This name derives from the fact that the reference of "the natural numbers," and with it the reference of the numerals "1", "2", "3", etc., is relative to an arbitrary, or only pragmatically determined, choice between equivalent models. Other forms of relativist structuralism result then from modifying the basic framework. For example, one can work not just with pure sets, but also allow for "atoms" or "urelements." Along such lines, one can, in fact, let any objects whatsoever occupy any "position" in a given structure; thus Julius Caesar or some beer mug can "be" the number 2. (If at least some abstract objects are included as candidates here, this will again be a semi-eliminative view broadly speaking, but also a form of eliminative structuralism.)

Yet another kind of structuralism, closely related to relativist structuralism, is "universalist structuralism" (cf. Reck and Price 2000). With it, we come back to if-then-ism, i.e., the suggestion that any mathematical sentence should be seen as quantifying over all models of the relevant axiom system and as consisting of a corresponding if-then claim. In other words, we keep the "universalist" side of Hellman's position but leave out its modal aspect. But what about the existence of the models; i.e., what about the so-called non-vacuity problem for the theory at issue? Or what ensures its "coherence"? Here one can again work with axiomatic set theory as the framework; but there are other options as well. (Once more, this makes the position partly but not fully "structuralist.")

Turning to the side of non-eliminative structuralism, there are additional options available too and further distinctions to be drawn. (At this point, we go beyond Reck and Price 2000.) One of them is indicated, implicitly, by Shapiro's label "*ante rem* structuralism." Shapiro's terminology, explicitly inspired by medieval debates about universals, suggests "*in re* structuralism" as an alternative. (Another alternative might be *post rem* structuralism. Parsons's position has been labeled that way, although this terminology is not widespread. We will not pursue it further here.) In fact, two different forms of *in re* structuralism have played a role in the literature already. For the first, consider again the natural numbers within a set-theoretic framework. There are infinitely many models for the Dedekind-Peano axioms, as we have noted. But then, we can identify "the structure" of the natural numbers with the equivalence class (under isomorphism) of all of them. This class is different from all the models in it, while arguably depending on them ontologically (the way in which a class depends on its elements). In that sense, we have arrived at a form of *in re* structuralism. Actually, this is exactly the position one gets if Russell's "principle of abstraction" (cf. Russell [1903] 1996) is applied to the case at hand, as Rudolf Carnap and others noted.

As the appeal to Russell's "principle of abstraction" indicates but as is true more generally, there are certain forms of structuralism that arise from "structuralist abstraction" (cf. Schiemer and Wigglesworth 2018; Reck 2018). That abstraction can, in turn, be reconstructed as a mathematical function, which maps models of a mathematical theory to a corresponding "abstract structure" as their value. Along Russellian and Carnapian lines, that value is the class of all models isomorphic to the given one (or more generally, equivalent in some other way). A different option is to use the following "abstraction function": it maps any given model of a theory to a novel, privileged model of it. (In the case of the Dedekind-Peano axioms, say, the value then deserves to be called "the natural numbers"; similarly for "the real numbers," etc.) Here the new model is again ontologically dependent on the original ones, since it has been introduced "by abstraction" on their basis. In the recent literature, this position has been explored by Øystein

Linnebo and Richard Pettigrew, building on Dedekind. For these authors, the "principle of abstraction" involved is similar to neo-Fregean "abstraction principles" (cf. Linnebo and Pettigrew 2014; Reck 2018).

Last but not least, let us return to eliminative structuralism once more. Yet another option under that label, different from Hellman's and Chihara's, is "concept structuralism," as advocated recently by Dan Isaacson, Solomon Feferman, Tony Martin, and others (cf. Isaacson 2010, Feferman 2014). The guiding idea for them is that what matters in mathematics in the end is "concepts" as opposed to "objects." Thus, there is the concept "model of the Dedekind-Peano axioms" (in Russell's terminology: "progression"; in Dedekind's: "simple infinity"); likewise for other axiom systems that define (higher-order) concepts, including the concept of set. All that is crucial for mathematics, so the suggestion now, is what is provable from those concepts, thus what is true for all models falling under them. Once again, we avoid postulating "abstract structures" as separate objects. We might even say that the structure simply "is" the concept at issue, parallel to its identification with the (closely related) equivalence class given earlier, except that the structure is not an object in this case.

3. The Pre-History: Key Themes and Features

As should be evident by now, a plethora of positions have been introduced under the name of "structuralism in mathematics" since the 1960s, following the initial lead of Benacerraf, Putnam, later Resnik, Shapiro, Hellman, Parsons, and others. For the most part, they are versions of "metaphysical/philosophical" structuralism." But these positions are all inspired by mathematical practice, at least implicitly, thus by methodological/mathematical structuralism. So far we have not said much about what the latter amounts to, except for mentioning category theory as one version, or one outgrowth, of it. However, it is not the only version, much less the original one. To probe this issue in a deeper way, it becomes important, and will prove illuminating, to consider how "structuralist mathematics" arose historically since the middle of the 19th century. Many of the essays in the present volume will, in fact, address that rise in detail, i.e., they are meant to fill that gap. As further preparation for them, we will now offer a brief overview of the themes and features that play a key role.

A number of developments transformed mathematics radically in the 19th century, as is now widely acknowledged, so much so that some commentators have talked about a "second birth" of the discipline (Stein 1988). The result of that transformation was "modern mathematics." In the 20th century, it was then systematized, provided with a set-theoretic foundation, and later reshaped, once again, along category-theoretic lines. The main innovations that played a role in the 19th century are well known (see, e.g., Boyer and Merzbach 1991, chaps. 24–26). They include the radical broadening and rethinking of geometry, by means of introducing various non-Euclidean theories (projective, elliptic and hyperbolic, n-dimensional, etc.); the rigorization and arithmetization of analysis, including better and more explicit characterizations of the number systems involved (from the natural to the complex numbers), and leading to a broadened conception of function as well; the transformation of algebra, from the study of equations to a much more general, abstract conception of it (Galois theory, the introduction of novel number systems and related innovations, e.g., quaternions, vector spaces, etc.); and the rise of set theory and modern logic (transfinite numbers, generalized notions of set and function, quantification theory, and a logical theory of relations, among others).

These broad developments brought with them several important changes that we consider to be "proto-structuralist," i.e., part of the immediate background for the rise of "structuralist mathematics" but not constitutive of it yet. They include: the rejection of the traditional view that mathematics is "the science of number and quantity," by adding parts that cannot be understood thus (complex analysis, group theory, topology, etc.); the expansion and systematization of traditional theories, by introducing "ideal elements" (points at infinity, points with complex coordinates, ideal divisors, transfinite numbers, etc.); later the reconstruction of such objects in set-theoretic terms (Dedekind cuts and ideals, quotient constructions in algebra, etc.); the adoption of the view that many parts of mathematics are not about particular objects and their properties, but are applicable much more widely (group theory, ring theory, topology, etc.); the related suggestion that mathematics is more about the relations between objects than about their intrinsic, non-relational properties (from number systems to groups, rings, etc.); also the emphasis on the "freedom" of mathematics, in the sense that its development should not be constrained by its direct and readily apparent applicability, but should involve the exploration of new "conceptual possibility" (non-Euclidean geometries, transfinite numbers, etc.); and finally, the suggestion that many parts of mathematics, perhaps even all, can be reconstructed systematically within "logic," including a basic theory of sets and functions (thus basing it on "laws of thought" alone).

As the reader will see, many of these changes play important roles in the essays in this volume. In fact, one function of these essays is to document their increasing significance in 19th- and early 20th-century mathematics. But the features we have listed also brought with them, or soon led to, additional innovations that are more properly "structuralist." Prominent among those are the following six, as we want to suggest: First, there is the suggestion to base various parts of mathematics on fundamental, characteristic concepts ("group," "field," "metric space," also "simple infinity," "complete ordered field," "3-dimensional Euclidean space," etc.); and this leads to the modern axiomatic approach (explicitly in Peano, Hilbert, etc.). Second, the relevant concepts typically specify global or "structural" properties (the "denseness" of an ordered system, the "continuity" of a space, also the "infinity" of a set); and this relies on considering whole systems of objects, as opposed to individual objects, especially various "complete infinities" (the systems of the real numbers, Euclidean space, various function spaces, etc.). Third, increasingly important becomes the study of such systems by relating them to each other, especially in terms of morphisms (homomorphisms, isomorphisms, etc.). A case in point, but also a method applicable more generally, is, fourth, the characterization of various systems or kinds of objects via "invariants" (complex-valued functions via their Riemann surfaces, geometries via their groups of transformation, etc.). Fifth, there is the novel practice of "identifying" isomorphic systems, since they are "essentially the same" from a mathematical point of view (e.g., different models of geometric theories, the system of Dedekind cuts and that of equivalence classes of Cauchy sequences, etc.). Sixth, this can all be seen as culminating in the view that what really matters in mathematics is the "structure" captured axiomatically, on the one hand, and preserved under relevant morphisms, on the other hand (two closely related techniques, both important historically).

What makes a mathematical methodology structuralist, in our view, is not the presence of one or two particular items on the list just given; nor do all six have to be present. Rather, what matters is the self-conscious and fruitful use of several of them together. Put differently, we think it is neither promising nor appropriate to try to define structuralist mathematics in terms of a few essential features (necessary and sufficient conditions). Instead, what we are dealing with is a case of "family resemblances," and hence, of "clusters" of these features emerging and playing a central role. In any case, when all of the corresponding tools and techniques were in place, in the late 19th and early 20th centuries, mathematicians began to study the results more systematically. This led to the introduction of several additional fields in mathematics: axiomatic set theory, seen as a "foundation" for all of mathematics (not just as an exploration of the infinite, although this too remained a goal); model theory, proof theory, recursion theory, etc., as ways to study "metamathematical" or "metalogical" features of mathematical theories (consistency, completeness, and categoricity, but also decidability, mutual interpretability, etc.); and somewhat later, category theory, with its generalization of the use of morphisms, invariants, etc. (initially in algebra and topology, then also more widely, and finally as an alternative foundation for mathematics).

During the period when these innovations became accepted widely, a number of philosophically inclined mathematicians and mathematically informed philosophers also began to reflect on their deeper significance, often in conversation with each other. For many of them this included attempts to say more about how to conceive of the nature of the various "structures" that had arisen, or of the underlying notion of "structure." This means, as we will see, that already toward the end of the 19th and early in the 20th century one can find forms of metaphysical/philosophical structuralism in the literature. And as we would like to emphasize, this happened 60–80 years before Benacerraf, Putnam, etc., began to publish on the topic, i.e., long before what is usually seen as the start of the debates about the topic. A central goal of the present collection is both to recover and to make fruitful this "prehistory of mathematical structuralism".

4. Previews of the Essays, Indicating Their Contributions to the Volume

After this condensed survey of structuralist themes and key features that arose in 19th and early 20th century mathematics, the stage is set for the essays in this volume. In this section of the introduction, we will preview the main themes in them, thus also indicating how each of these essays fits into the volume as a whole.

Overall, the volume is divided into two parts. The essays in Part I are concerned primarily with aspects of methodological/mathematical structuralism as they emerged in the 19th and early 20th centuries. Each focuses on a particular mathematician, from Grassmann to Mac Lane. With Part II, the focus shifts to the metaphysical/philosophical side, as well as to contributions by philosophers. However, the division between the two parts is porous, including many cross-references in the essays themselves. Moreover, while most of the essays in Part II focus on figures usually identified as philosophers, such as Peirce, Russell, and Cassirer, some of the people covered in this part, like Poincaré and Bernays, were also mathematicians, perhaps even primarily so. Why are the essays on them then included in Part II? The reason is that the main focus of these essays is on philosophical (and logical) themes. Yet even by that criterion, some placements of essays could have been different.

Among mathematicians, Richard Dedekind is often regarded as the "founding father" of structuralism; second in that regard is David Hilbert (cf. Shapiro 1996); and third probably Nicolas Bourbaki, especially among historians of mathematics (cf. Corry 2004). All three will be quite prominent in our volume, but it reaches back further, thus starting with Grassmann.

More precisely, the volume starts with an essay by Paola Cantú on Hermann Grassmann, the author of *Die Lineare Ausdehungslehre*, a book that influenced various later structuralists strongly. As Cantú documents, Grassmann suggested conceiving of mathematics as a "general theory of forms," and this was related to his introduction of several new systems of "quantities" (hyperspaces, hyperreal numbers, etc.). In fact, Grassmann emerges as an early proponent of concept structuralism, thus of eliminative structuralism.

In contrast, Dedekind has been interpreted as the first "non-eliminative structuralist" in the literature (Reck 2003), although this is not uncontroversial. In the essay co-written by Erich H. Reck and José Ferreirós in the present volume, the main focus is instead on Dedekind's contributions to methodological/mathematical structuralism. That essay starts with an account of important influences on Dedekind, namely Gauss, Dirichlet, and Riemann. Then his structuralist contributions to algebra and algebraic number theory, including Galois theory, are discussed, making evident their close relation to his work on the foundations of arithmetic and set theory.

A mathematician usually not associated with structuralism, nor recognized much as a philosopher of mathematics more generally, is Moritz Pasch. In Dirk Schlimm's essay, Pasch's work, not only on geometry but also on arithmetic, is put in the context of broader developments in 19th-century mathematics. In doing so, structuralist features of his approach are revealed, e.g., concerning the centrality of duality principles, even though a tension remains with the empiricism that dominates his work philosophically. In addition, Schlimm provides an analysis of what should be seen as central to mathematical/methodological structuralism more generally.

In the next essay, by Georg Schiemer, the investigation of 19th-century geometry with respect to the rise of mathematical structuralism is continued. Here it is Felix Klein's use of group theory in reconceptualizing geometry that becomes the focus. Klein was led to rethink the subject matter of different kinds of geometry in terms of what is invariant under relevant groups of transformations. This culminated in his "Erlangen program," in which various geometries are classified by comparing their respective transformation groups. Another influential structuralist idea one can find in Klein, as Schiemer documents, is the suggestion to show the structural equivalence of different geometries in terms of "transfer principles".

In Wilfried Sieg's essay on David Hilbert, two kinds, or uses, of the axiomatic method are distinguished: there is "structural axiomatics," on the one hand, which grew out of Hilbert's early axiomatization of geometry; and there is "formal axiomatics," on the other hand, which involves the metamathematical study of axiomatic systems in Hilbert's later proof theory. With respect to the former, the "conceptual" methodology advocated earlier by Dedekind and others is brought to full fruition, i.e., their suggestion to base various parts of mathematics on "characteristic concepts." The latter constitutes a major, and very influential, example of studying mathematical theories with respect to "foundational" issues, such as consistency and decidability, by using tools from modern logic.

Another mathematician in the early 20th century who built on Dedekind's work explicitly was Emmy Noether. In Audrey Yap's essay, three phases in Noether's mathematical career are distinguished. It is especially the second and third phases that are relevant for our purposes, since they illustrate the shift from a more concrete, calculational way of doing mathematics, still dominant in Noether's first phase, to a more and more abstract approach. Moreover, the latter became a paradigm of methodological/mathematical structuralism later in the 20th century, strongly influencing the work of Bourbaki and the rise of category theory, among other developments.

The name "Nicolas Bourbaki" stands for a group of mathematicians who worked on reshaping and systematizing modern mathematics from the 1930s on, by building on what they found in Dedekind, Hilbert, Noether, and others. According to Gerhard Heinzmann and Jean Petitot's essay, what lies at the core of the methodology that resulted is a "functional conception of structure." Its main purpose was to help mathematicians in reconceptualizing the interrelations of different theories and, especially, in solving hard problems. The latter is illustrated by an extended case study from algebraic geometry, which leads us from Dedekind through André Weil to Alain Connes. Here issues concerning methodological/mathematical structuralism are illustrated by means of a substantive mathematical example, one that still occupies mathematicians today.

Both in the essays on Noether and Bourbaki, and also already in the essay on Klein, close connections between 19th- and early 20th-century mathematics, on the one hand, and category theory, on the other, start to emerge. This theme is deepened in Colin McLarty's essay on Saunders Mac Lane. In that essay, Mac Lane is presented as a mathematician interested in logical and philosophical issues from early on, although he became disillusioned by their treatment in mainstream philosophy. Later he was led back to some of them from within mathematics. As a result Mac Lane adopted, and promoted explicitly, a form of methodological/mathematical structuralism tied to category theory. McLarty characterizes it as "a working theory of structures for mathematicians."

The first essay in Part II of our volume concerns the logician, philosopher, and scientist C. S. Peirce. (Like Part I, this second part is arranged chronologically by the birthdates of the thinkers under discussion.) In the recent literature, Peirce has been interpreted as subscribing to a form of non-eliminative structuralism (Hookway 2010). In Jessica Carter's essay, the focus is instead on Peirce's distinctive, still relatively unknown views about mathematical inquiry and proof, namely in terms of diagrammatic reasoning. Carter finds some aspects of structuralism in Peirce's works, at least in the sense of methodological/mathematical structuralism. But she refrains from interpreting him as a full-fledged structuralist, since this would oversimplify his multifaceted work.

The second essay in Part II, by Janet Folina, concerns Henri Poincaré. This essay, in particular, could have been put into Part I too, since Poincaré made major contributions to structuralism as a mathematician. But Folina is more interested in metaphysical/philosophical ideas and themes, which one can find in Poincaré's writings as well. She argues, in particular, that Poincaré should be seen as a proponent of *ante rem* structuralism. However, in this case one needs to separate the *ante rem* aspect clearly from the realist aspect, as she adds, even though they are often conflated in the current literature on structuralism. In fact, with respect to mathematics Poincaré turns out to be a "constructivist *ante rem* structuralist," surprising as that may sound at first.

As a prototypical logicist, Bertrand Russell tends to be seen as a strong opponent of structuralism. There is justice to this view, although the story is more complicated and more interesting in the end, as Jeremy Heis documents in his essay. Early in his career, during the years 1900–1903, Russell was intensely interested in Dedekind's works, as some of his posthumously published writings show; and he interpreted Dedekind as holding a non-eliminative structuralist position. While attracted to that position initially himself, he then turned against it, for reasons Heis documents in detail. But Russell was an important contributor to the debates about structuralism in another way as well, namely by means of his promotion of a logic of relations. That logic was taken as the background for reconstructing structuralist ideas by several later thinkers, from Ernst Cassirer in the early 20th century to Geoffrey Hellman today.

Cassirer's explicit and detailed defense of structuralism, both in the methodological/mathematical and in the metaphysical/philosophical senses, is the topic of Erich Reck's second essay in this volume. While Cassirer was very knowledgeable about Felix Klein's work and about developments in 19th-century geometry more generally, the focus in this essay is on his positive reception of Dedekind's structuralist views. This included a defense of them against Russellian objections. But Dedekind's contributions are also embedded into a rich account of the history of mathematical science, guided, among others, by Cassirer's distinction between "substance concepts" and "function concepts".

The last three essays in the volume concern Paul Bernays, Rudolf Carnap, and W. V. O. Quine, respectively. In Wilfried Sieg's second contribution, an essay on Bernays, the connection between methodological/mathematical structuralism, in Dedekind's, Hilbert's, and related works, and 20th-century proof theory is thematized. The core concept for Sieg is that of a "methodological frame," as introduced in Bernays's writings. The role of such frames is to allow for a kind of "reductive structuralism," in the sense of investigating mathematical theories in terms of their underlying deductive structures, thus by utilizing the tools of Hilbertian proof theory. Seen as such, Bernays's work constitutes a reflection on mathematical structuralism from the perspective of mathematical logic.

In Georg Schiemer's second contribution to this volume, Rudolf Carnap's work from the 1920s–1930s is investigated in a parallel way, i.e., with respect to its use of logic. As Schiemer documents, Carnap picked up on Russell's "principle of abstraction," both to demystify the notion of "structure" and to study it further logically. This led him to a form of *in re* structuralism according to which "the structure" of a mathematical theory, such as Dedekind-Peano arithmetic, is identified with the equivalence class of models that satisfy the theory.

In the last essay in our volume, Sean Morris discusses Quine's place in the prehistory of structuralism. Several current structuralists, including Resnik, Shapiro, and Parsons, have acknowledged Quine as a strong influence. Usually this involves Quine's later works, in which he proposed a very general form of structuralism (also for physical objects, not just for mathematical ones). However, Quine's relevant views can be traced back to his earliest publications and his dissertation, as Morris documents. He also argues that Quine stands firmly in the tradition of Russell's and Carnap's "scientific philosophy," including the rejection of traditional metaphysics. Hence Quine's structuralism should not be seen as exemplifying any strong form of realism. Instead, it is grounded in the methodology of the mathematical sciences as interpreted by him.

In fact, the latter holds, *mutatis mutandis*, also for every other figure covered in the present volume. What the essays establish as a whole, then, is that there are very strong ties between methodological/mathematical forms of structuralism and more metaphysical/philosophical views. These should not, and ultimately cannot, be understood independently of each other; they are two sides of the same coin.

5. Gaps in this Volume and Two Final Suggestions

While this collection of essays is meant to recover the prehistory of mathematical structuralism in a substantive and inclusive way, we realize that it is far from complete. In other words, we could, and perhaps should, have included a number of other thinkers and developments as well, both on the mathematical and on the philosophical sides.

One mathematician who comes to mind right away is Bernhard Riemann. Riemann is mentioned in several of our essays, but he would undoubtedly have deserved his own treatment. A second, less prominent example is Hermann Hankel. He too comes up in some of our essays along the way, with his view of mathematics as a "theory of forms" that is similar to Grassmann's. A third example is George Boole, as well as other British algebraist in the mid-19th century, who helped to push mathematics in a structuralist direction too.

Actually, some ideas relevant for us can already be found in late 18th- and early 19th-century thinkers. Abel and Gauss are two cases, with their suggestion that mathematics is more about relations, relations of relations, etc., than about objects. And a few such ideas can be traced back even further, e.g., to D'Alembert's work on the calculus (cf. Folina's essay), or to Leibniz's study of the continuity of space (cf. De Risi in progress). But the further back one goes, the more one should speak of "proto-structuralist" rather than "structuralist" ideas, as we believe.

On the side of philosophers there are gaps too. One notable figure mentioned only tangentially, but who would have deserved a separate essay, is Edmund Husserl. As is well known, Husserl started out as a mathematician, including by serving as an assistant of Weierstrass in Berlin. And he was concerned about a "general theory of manifolds" in some of his later works, thereby building explicitly on Grassmann's, Riemann's, and Klein's writings. There are clear connections to methodological/mathematical structuralism in his works; but one can find related metaphysical/philosophical views too, including perhaps another form of *in re* structuralism.

Somewhat later in the 20th century, another interesting philosopher for our purposes is Albert Lautman. While still largely unknown among English-speaking philosophers, he offered detailed reflections on the mathematics of Bourbaki, and with it, on mathematical structuralism. Lautman's views were mathematically and philosophically sophisticated, thus deserving to be reconsidered. Indeed, we had planned to include essays on both Husserl and Lautman; but because of space and time restrictions, they had to be omitted in the end. And beyond Husserl and Lautman, there surely are further philosophers one could have included. Then again, the volume is already very long as it is.

Because of such omissions, the volume is open to complaints that we did not cover this or that figure who would undoubtedly have deserved a separate essay as well. In response, we want to close with two suggestions: First, one thing we hope this volume will do is to inspire more research on the prehistory of structuralism, thus recovering and reinvestigating other relevant mathematicians and philosophers as well. In other words, we suggest viewing the volume only as the start with respect to covering its topic. Having said that, we hope that it is substantive enough to inspire further work.

Second, while the approaches and treatments in our essays are primarily historical, we hope that the volume will be seen as a contribution to mathematical structuralism in a systematic sense too, i.e., as relevant for current philosophy of mathematics. As we see it, combining historical and systematic investigations can only enrich both sides, also in other cases. More generally, a rich topic such as mathematical structuralism will surely benefit from being studied in several different ways.

References

Awodey, Steve. 1996. Structure in Mathematics and Logic: A Categorical Perspective. *Philosophia Mathematica* 4, 209–237.

Benacerraf, Paul. 1965 [1983]. What Numbers Could Not Be. *Philosophical Review* 74, 47–73. Reprinted in *Philosophy of Mathematics: Selected Readings*, 2nd ed., edited by Paul Benacerraf and Hilary Putnam, pp. 272–294. New York: Cambridge University Press.

Boyer, C., and U. Merzbach. 1991. *A History of Mathematics*. 2nd ed. New York: Wiley & Sons.

Burgess, John. 2015. *Rigor and Structure*. New York: Oxford University Press.

Carter, Jessica. 2008. Structuralism as a Philosophy of Mathematical Practice. *Synthese* 163, 119–131.

Chihara, Charles. 2004. *A Structural Account of Mathematics*. Oxford: Oxford University Press.

Cole, Julian. 2010. Mathematical Structuralism Today. *Philosophy Compass* 5, 689–699.

Corry, Leo. 2004. *Modern Algebra and the Rise of Mathematical Structures*. 2nd ed. Basel: Birkhäuser.

De Risi, Vincenzo. In progress. *Continuity of Space and Theories of Intersections from Euclid to Leibniz*. Manuscript.

Feferman, Solomon. 2014. Logic, Mathematics, and Conceptual Structuralism. In *The Metaphysics of Logic*, edited by P. Rush, pp. 72–92. New York: Cambridge University Press.

Hellman, Geoffrey. 1989. *Mathematics without Numbers: Towards a Modal-Structural Interpretation*. New York: Oxford University Press.

Hellman, Geoffrey. 1996. Structuralism without Structures. *Philosophia Mathematica* 4, 100–123.

Hellman, Geoffrey. 2001. Three Varieties of Mathematical Structuralism. *Philosophia Mathematica* 9, 184–211.

Hellman, Geoffrey. 2005. Structuralism. In *Oxford Handbook of Philosophy of Mathematics and Logic*, edited by Stewart Shapiro, pp. 536–562. New York: Oxford University Press.

Hookway, Christopher. 2010. "The Form of a Relation": Peirce and Mathematical Structuralism. In *New Essays on Peirce's Mathematical Philosophy*, edited by M. Moore, pp. 19–40. Chicago: Open Court.

Isaacson, Daniel. 2010. The Reality of Mathematics and the Case of Set Theory. In *Truth, Reference, and Realism*, edited by Z. Novák and A. Simonyi, pp. 118–140. Budapest: Central European University Press.

Keränen, Jukka. 2001. The Identity Problem for Realist Structuralism. *Philosophia Mathematica* 9, 308–330.

Korbmacher, J., and G. Schiemer. 2018. What Are Structural Properties? *Philosophia Mathematica* 26(3), 295–323.

Landry, Elaine. 2009. How to Be a Structuralist All the Way Down. *Synthese* 179, 435–454.

Landry, Elaine, ed. 2018. *Categories for the Working Philosopher*. New York: Oxford University Press.

Landry, Elaine, and Jean-Pierre Marquis. 2005. Categories in Context: Historical, Foundational, and Philosophical. *Philosophia Mathematica* 15, 1–43.

Leitgeb, H., and J. Ladyman. 2008. Criteria of Identity and Structuralist Ontology. *Philosophia Mathematica* 16, 388–396.

Linnebo, Øystein. 2008. Structuralism and the Notion of Dependence. *Philosophical Quarterly* 58, 59–79.

Linnebo, Øystein, and Richard Pettigrew. 2014. Two Types of Abstraction for Mathematical Structuralism. *Philosophical Quarterly* 64, 267–283.

Marquis, Jean-Pierre. 2008. *From a Geometric Point of View: A Study of the History and Philosophy of Category Theory*. New York: Springer.

McLarty, Colin. 2004. Exploring Categorical Structuralism. *Philosophia Mathematica* 12, 37–53.

Parsons, Charles. 1990. The Structuralist View of Mathematical Objects. *Synthese* 84, 303–346.

Parsons, Charles. 2009. *Mathematical Thought and its Objects*. Cambridge, MA: Harvard University Press.

Putnam, Hillary. [1967] 1983. Mathematics without Foundations. *Journal of Philosophy* 64, 5–22. Repr. in *Philosophy of Mathematics: Selected Readings*, 2nd ed., edited by Paul Benacerraf and Hilary Putnam, pp. 295–313. New York: Cambridge University Press.

Reck, Erich. 2003. Dedekind's Structuralism: An Interpretation and Partial Defense. *Synthese* 137, 369–419.

Reck, Erich. 2018. On Reconstructing Dedekind Abstraction Logically. In *Logic, Philosophy of Mathematics and Their History: Essays in Honor of W.W. Tait*, edited by Erich Reck, pp. 113–137. London: College Publications.

Reck, Erich, and Michael P. Price. 2000. Structures and Structuralism in Contemporary Philosophy of Mathematics. *Synthese* 125, 341–383.

Resnik, Michael. 1981. Mathematics as the Science of Patterns: Ontology and Reference. *Noûs* 15, 529–550.

Resnik, Michael. 1997. *Mathematics as the Science of Patters*. New York: Oxford University Press.

Russell, Bertrand. [1903] 1996. *Principles of Mathematics*. New York: Norton.

Schiemer, Georg, and John Wigglesworth. 2018. The Structuralist Thesis Reconsidered. *British Journal for the Philosophy of Science*. Published online January 30, 2018.

Shapiro, Stewart. 1983. Mathematics and Structure. *Philosophy of Science* 50, 523–548.

Shapiro, Stewart. 1997. *Philosophy of Mathematics: Structure and Ontology*. Oxford: Oxford University Press.

Shapiro, Stewart. 2012. Mathematical Structuralism. *Internet Encyclopedia of Philosophy*, www.iep.utm.edu/m-struct/print/.

Stein, Howard. 1988. *Logos*, Logic, and *Logistiké*: Some Philosophical Remarks on Nineteenth Century Transformations of Mathematics. In *History and Philosophy of Mathematics*, edited by William Aspray and Philip Kitcher, pp. 238–259. Minneapolis: University of Minnesota Press.

PART I

MATHEMATICAL DEVELOPMENTS

2
Grassmann's Concept Structuralism

Paola Cantù

1. Introduction

It is hard to determine whether Hermann Grassmann should be considered a mathematically inclined philosopher or a philosophically inclined mathematician, for he was an autodidact in mathematics (he learned mathematics mainly from the books of his father, Justus, and from Legendre's treatise), in Greek philology, and partly also in philosophy. He studied theology at the Berlin University at the end of the 1820s, and attended, among the philosophy courses, only Schleiermacher's lectures on dialectics and Ritter's lectures on the history of philosophy. In any case, his main contributions concern mathematics and linguistics, rather than philosophy. Or rather, he got recognition mainly for his mathematical results and his linguistic achievements, whereas his philosophy of mathematics did not receive similar attention, not even after his death. Yet a large part of Grassmann's mathematical work is specifically devoted to (*a*) the relation between the emergence of a new abstract mathematical theory and the need for a new philosophical frame to understand it, (*b*) the relation between certain applications of this theory and Leibniz's universal characteristics, and (*c*) the characterization of mathematical disciplines by means of a philosophical deduction of their fundamental concepts and of mathematics as the science of particulars generated from a given element.

Notwithstanding the growing number of publications concerning specific aspects of Grassmann's mathematical or philosophical writings,[1] it is still difficult to find a comprehensive treatment of his philosophy of mathematics. There are several reasons for this: (1) Grassmann's philosophy of mathematics varies in different writings, (2) it is difficult to clearly distinguish his conception from that of his brother Robert, (3) where a distinction can be traced, Robert appears to have been the one who was most interested in logic and philosophy of logic

[1] See, for example, Banks (2013); Radu (2013); Petsche et al. (2011); Schubring (2005); Flament (2005); Radu (2003); Darrigol (2003); Schubring (1996a); Dorier (1995); Schreiber (1995); Flament (1994); Boi et al. (1992); Châtelet (1992); Otte (1989); Hestenes (1986); Schlote (1985); Echeverría (1979); Lewis (1977); Heath (1917).

Paola Cantù, *Grassmann's Concept Structuralism* In: *The Prehistory of Mathematical Structuralism*. Edited by: Erich H. Reck and Georg Schiemer, Oxford University Press (2020). © Oxford University Press.
DOI:10.1093/oso/9780190641221.003.0002

(see Peckhaus 2011 and Grattan-Guinness 2011), (4) Grassmann's philosophical style, typical of early 19th-century natural philosophy, cannot easily be read by contemporary philosophers, (5) Grassmann was interpreted in different ways in the second half of the 19th century and at the beginning of the 20th century (see, e.g., Hankel 1867; Cassirer 1910; and Klein 1875). These interpretations exemplify Grassmann's philosophical destiny, which is perhaps less "tragic" than his mathematical fortune, yet not really fortunate, because he was often *used* to corroborate a given conception of mathematics rather than *read* to verify what his own view really was. These interpretations did not adequately emphasize the role of particulars in Grassmann's mathematics (Cassirer), the role of intuition and the reasons for a quasi-axiomatic presentation of extension theory and arithmetic (Klein), the differences between the general theory of forms and a symbolic treatment of mathematical objects as signs whose referent does not matter (Hankel). Yet all these aspects are extremely relevant to grasp Grassmann's understanding of concept formation in mathematics and his contribution to the history of methodological and philosophical structuralism. I will try to reconstruct Grassmann's definition of mathematics as the science of the particular, and to investigate his complex distinction between formal and real, referring to some philosophical interpretations discussed in Lewis (1977), Flament (1994), Banks (2013), and Schlote (1996).

So in the following it will not be sufficient to recall several of Grassmann's mathematical contributions that are relevant for the structuralist transformation of mathematics, such as abstract algebra, linear algebra, and number theory (§2). The most important task will be that of giving a plausible and comprehensive reconstruction of Grassmann's philosophy of mathematics (§3), as it emerges from his own mathematical works, rather than from subsequent influential interpretations, such as those by Hankel, Cassirer, and Klein. As a result, it will emerge that the notions of linear combination, series, and addition are more important to Grassmann than the notions of function, mapping, and order. Mathematics is the science of the particular, and the general theory of forms does not properly belong to it, because it is about underdetermined connections.

The main aim of the chapter will be to analyze Grassmann's contribution to structuralism, discussing differences and similarities between our interpretation and some received views in the literature (§4). In particular, I will try to evaluate Grassmann's work with respect to two different issues that are often mixed up in the literature or, when they are clearly distinguished, are often called by different names or defined in slightly different ways: methodological (or mathematical)[2] and philosophical structuralism.

[2] In the literature, this methodology is often called "mathematical structuralism" rather than "methodological structuralism." I prefer Reck and Price's (2000) terminological choice for two reasons. On the one hand, this choice does better justice to the idea that the structuralist philosophical

By *methodological structuralism* I intend an analysis of the method that is applied by mathematicians when they are doing mathematics and that has evolved in time. Reck and Price have defined methodological structuralism as a methodology that "motivates, explicitly or implicitly, many of the structuralist views in the philosophical literature" (2000, 345). Reck and Schiemer in the introduction to this volume enucleate a list of conditions that should characterize methodological structuralism. Later in this chapter, I broadly follow their suggestion and associate methodological structuralism with questions concerning (1) criticism of mathematics as the science of a given domain of objects (e.g., quantities), concerning objects in isolation rather than relations, (2) the role of intuition and formal deductions, (3) the role of axioms, invariants, and applications, and (4) the relation between alternative ways to frame mathematics (e.g., set theory, category theory). This methodological structuralism tackles deep philosophical questions, which often arise in mathematical practice itself or in historical analysis of the development of mathematical theories.

Philosophical structuralism is used here as a collective name for a large number of different philosophical theories centering on the fundamental question, "What is a structure?" Typical issues concern, for example (1) whether there are objects and operations, and what their relations to structures might be, (2) whether general structures can be distinguished from particular structures and from exemplars, (3) what is that we call "formal" in a structure and what role is played by axiomatics within it. In section 4.3 the analysis of these issues is interconnected with the study of answers given in the contemporary philosophical debate by Shapiro, Parsons, Feferman, Isaacson, and Burgess. A tentative distinction between concept structuralism and object structuralism is used to characterize Grassmann's own perspective with respect to some contemporary approaches.

The objective is certainly not to determine whether Grassmann was a forerunner of a specific philosophical position in the contemporary debate. This would be quite anachronistic, because both mathematics and philosophy have deeply evolved from Grassmann's time. On the one hand several conceptions of structuralism are grounded either in a set-theoretic or in a categorical framework that had not yet been developed at the time; on the other hand the analytic approach to structuralism is based on a new understanding of mathematics and logic introduced, e.g., by Dedekind, Frege, Peano, Russell, and Hilbert, which makes it difficult to separate our common use of certain notions (such as

viewpoint emerges in mathematical practice, and that a study of the mathematical method might already be philosophical in nature. On the other hand, it avoids the mistake of considering the methodological component of structuralism as the only mathematical aspect of it, whereas also the so-called philosophical structuralism might be the result of mathematical self-reflection.

function, concept, equality) from the corresponding use made by Grassmann. Yet, provided that historical differences are spelled out clearly, it is not anachronistic to evaluate Grassmann from the perspective of contemporary philosophy of mathematics, to verify whether he asked questions that challenge certain structuralist views or raised issues that still need to be clarified.

2. Grassmann's Mathematics

Hermann Grassmann's contributions to mathematics and to its applications to physics are numerous; we will recall them very shortly. A clear and detailed presentation of Grassmann's mathematical writings can be found in Schubring (1996b) and Petsche et al. (2011). In the following we will restrict our attention to several contributions that might be relevant for the development of structuralism and that derive mainly from the following works: *Ausdehnungslehre* (both in the 1844 and in the 1862 revised edition), *Geometrische Analyse, Lehrbuch der Arithmetik*, and Robert Grassmann's *Formenlehre*.

2.1. Linear Algebra

Grassmann's extension theory (ET) (*Ausdehnungslehre*) introduces several fundamental concepts of linear algebra: basis, dimension, generator, linear dependence and independence, but there is no axiomatization of the theory (Dorier 1995; Zaddach 1994). Grassmann's vector theory is developed in a purely abstract way (in modern parlance, the vector system is a module over a field), and conceptually distinguished from geometry, which is considered as an applied science (it is the application of ET to three-dimensional space).

Grassmann's theory partially differs from contemporary vector-based systems, such as vector analysis, exterior algebra, and geometric algebra, both from a technical and from a philosophical point of view. Differences concern the closure of the operations, the condition of homogeneity on addition, and the conception of the product (Cantù 2011, 96–98). Besides, an important characteristic of Grassmann's system is that his notions of base and of a system of (independent) generators does not aim at the introduction of a system of coordinates, but rather at expressing the idea that all the magnitudes of the system are characterized by some generating law.

Following Grassmann, who uses a geometrical analogy to make the abstract presentation more intuitive, we will introduce the fundamental notions of ET (element, generating law, simple extensive formation, extensive magnitude) by analogy with geometry (point, movement, bound vector, vector). An extensive formation (*Ausdehnungsgebilde*) is "the collection of all elements into which the generating element is transformed by continuous evolution":[3] geometrically speaking, it is the geometrical figure resulting from the different positions of a point in continuous movement. An elementary (*einfach*) extensive formation "is produced by continuation of the same fundamental evolution" (Grassmann 1844, 48, my trans.): geometrically, it is a straight line that results from the movement of a point in just one direction.

An extensive magnitude is the class of extensive formations that are generated according to the same law by means of equal evolutions (Grassmann 1844, 48–49); that is, the vector defined as an equivalence class of bound vectors having the same direction, the same orientation, and the same size.

Given Grassmann's understanding of equality as an identity whose criterion is substitutivity, one cannot say that two extensive formations (two bound vectors) are equal (in the sense that they are equivalent), but rather that their extensive magnitudes (their corresponding free vectors) are equal (see §2.3.1). An extensive formation is determined by the elements it is composed of. An extensive magnitude, on the contrary, is determined only by direction, size, and orientation; that is, it does not depend on the initial element of the generation (Grassmann 1844, 49).

2.2. Number Theory

2.2.1. Natural Numbers

The theory of natural numbers is presented by Grassmann in the *Lehrbuch der Arithmetik* (1861), which is the result of collaboration with his brother Robert. Here the term "magnitude" (*Grösse*) replaces "form"; mathematics is defined as the science of magnitudes, that is, of anything that should be set equal or unequal to another thing (Grassmann 1861, 1). This general definition of magnitude might apply to any kind of form: arithmetical, extensive, or combinatorial. In any case, arithmetical magnitudes are characterized by a further property, that is, the fact that they are obtained by successive applications of a specific kind of connection (an addition) to a single magnitude taken as given and denoted by the sign *e*. It should be noted that Grassmann does not mention the number 1 as the arithmetic unit. Any magnitude that is taken as initial element to build the arithmetic series, which he calls *Grundreihe*, by successive addition of that initial magnitude

[3] See Grassmann (1844, 48; 1995, 47). Cf. also Lewis (1977, 150). By translating *Ausdehnungsgebilde* by "extensive formation" rather than "extensive structure" (Kannenberg) or "extensive entity" (Lewis), I follow here the French translation by Flament and Bekemeier in Grassmann (1994).

can play the role of a unit. The commutativity and associativity of any arithmetical magnitudes denoted by the symbols *a*, *b*, *c* is not introduced as an axiom, but derived inductively (*inductorisch*) from the commutativity and associativity of $a + e = e + a$ and $(a + b) + e = a + (b + e)$ respectively (Grassmann 1861, 1). This shows the essential role played by the initial element and by the operation of addition in the definition of an arithmetical magnitude, and thus of the notion of series (see §3.1.1).[4]

The *Lehrbuch* has been very influential, because it introduces (1) a clear distinction between the symbols used to denote (*bezeichnen*) the concepts and the concepts themselves, (2) the parallelism between the symbolic development (*Formelentwicklung*) and the conceptual development (*Begriffsentwicklung*) of a proof, (3) a clear separation between primitive and derived propositions, and (4) the use of induction as a method of inference.

2.2.2. Real Numbers

Real numbers are not introduced in arithmetic, but in ET. Grassmann, at least in the first edition of the *Ausdehnungslehre*, defines real numbers as ratios of extensive magnitudes of the same dimension: they are thus introduced as magnitudes of grade zero, that is, as magnitudes that have no dimension. The idea that numbers are themselves magnitudes is familiar in modern linear algebra, where the real number field can itself be represented as a vector system (a module on the field of real numbers). In particular, the fact of having no dimension allows for the product of real numbers to be commutative, even if the product between extensive forms is generally non-commutative. So all properties of the usual arithmetical operations hold for the so-introduced real numbers, which are the only magnitudes whose product commutes (Cantù 2011, 98).

Once real numbers have been introduced according to the operation that generates them (division), they can be used as a tool in the symbolic definition of extensive magnitudes given in the second edition of the *Ausdehnungslehre*. Relying on an analogy with the generation of natural numbers as successive additions of the unity, Grassmann defines several unit magnitudes $e_1, e_2, \ldots$ and then introduces extensive magnitudes as additions of the products of these units by real numbers, as in the following polynomial: $a_1e_1 + a_2e_2 + \ldots$. Yet real numbers, although presupposed in the definition, can still be conceived as extensive magnitudes "if the system consists only of the absolute unity (1) (Grassmann 1862, 12, my trans.).[5]

[4] The notion of power series emerges already in arithmetic, because Grassmann investigates which powers can be transformed into a power series of the form $ax^n + bx^{n-1} + cx^{n-2} + \ldots$ with x as base. The complex relations between the solution of systems of linear equations, analysis, and extension theory is here evident.

[5] This sentence is omitted in Grassmann (2000). Grassmann thus uses the notion of series to express natural numbers, extensive magnitudes, and also real numbers. Besides, he often uses it as a tool

Natural numbers are based on addition of absolute unities, rational numbers are based on division of natural numbers, and real numbers are obtained as the quotient of extensive magnitudes. This approach does not provide a unified notion of number that includes natural, rational, and real. Grassmann does not seem to be bothered by the piecemeal character of the definition. On the contrary, he aims to ground each kind of number in the operation that is used to generate it, and seems to consider as most primitive those notions that are built on the basis of addition alone (see §2.3.1 for a discussion of this algebraic hierarchy between operations). So natural numbers are more primitive than rational numbers because the former are introduced by an operation of addition, whereas the latter need multiplication and division. For the same reason, extensive magnitudes are more primitive than reals, which are magnitudes and not numbers: extensive magnitudes are introduced by addition, whereas reals are obtained as the quotient of extensive magnitudes.

2.3. Algebra and Logic

2.3.1. Abstract Algebra

Under the name of "general theory of forms" (GTF) Grassmann gathers the investigation of equality, difference, and the common properties of some connections that make their appearance in all branches of mathematics. Contrary to the usual treatment in modern algebra, he does not investigate sets of objects endowed with a given operation, but rather considers the connections in a purely formal way, abstracting from the elements they might be applied to. It is true that sometimes he reasons as if in specific mathematical branches one should consider the connections as always holding between certain given magnitudes, and then show that these connections satisfy the requirements that allow one to call them addition and multiplication respectively (Schlote 1996, 168). Yet this can be done only once GTF has been established.[6] This explains why Grassmann claims that

in the solution of problems in different mathematical branches. The ubiquity of the notion of series as well as its capacity to express the generating rule of mathematical forms attests to the foundational role Grassmann attributes to it.

[6] See, e.g., the following passage in the second edition of the *Ausdehnungslehre*: "We therefore also call such a method of conjunction a multiplication, provided only that its multiplicative relation to addition is demonstrated, or in other words, provided only that the equal entry of all the terms of the conjunctive factors into the conjunction is established in the above sense" (Grassmann 2000, 43). In other words, Grassmann defines *in abstracto* what addition, multiplication, and raising to a power are, and then, given a domain closed under an operation, he determines whether it is an addition, a multiplication, or a raise to a power, and specifies its further characteristics.

GTF should precede all other mathematical branches in the exposition: there are both epistemological and didactic grounds, because GTF provides a foundation of all other branches of mathematics—in that it presents as united what should be united, and has the highest degree of generality—and also because it spares useless repetitions of basic concepts in a mathematical treatise (Grassmann 1844, 28).

Two forms are said to be equal when they can be substituted one for the other in any connection they occur in. Equality is transitive—if two forms are equal to a third, then they are equal one to the other and has the following property: forms that are generated in the same way from equal forms are equals (Grassmann 1844, 28).

Forms are determined by their generating law, and are therefore equal if the same law from the same initial element generates them. Grassmann has often been criticized for his adoption of a Leibnizian conception of equality as substitutability *salva veritate* instead of a Euclidean conception of equality as equivalence (Helmholtz 1887, 377n): as I read him, his equality lies midway between Leibniz and Euclid, because he defines it as an identity, and restricts it to some features of concepts rather than defining it between objects themselves.[7]

Given that forms are not given objects but the results of an act of thought that generates them according to a certain law, only the characteristics that depend on the specific way in which forms have been generated will be taken into account in the comparison: the substitutability is thus limited to pertinent contexts.

Grassmann then considers three connections and introduces a four-level distinction based on their decreasing generality.

1. Grassmann believes that the most restrictive conditions to be required from any mathematical connection depend on the number of connections that are introduced and on their reciprocal relations. So he requires from a first connection (*connection of first order*) that it be commutative and associative, and from a second connection related to the first (*connection of second order*) that it satisfy the distributive laws with respect to the first.[8] At this level of generality, the two connections (denoted respectively by $\cap$ and $\cup$) are pre-mathematical operations between concepts.
2. Then there is the formal level, where the conditions are less restrictive and the connections of first and second order are respectively called "formal addition" and "formal multiplication" and denoted by the usual arithmetical symbols $+$ and $\cdot$. A formal addition is a simple synthetic connection with

[7] E.g., two vectors can be considered as equal—in some extended sense—because their directions and lengths are equal, i.e., identical (Grassmann 1844, 28).

[8] The distributive laws are two, because the second connection need not be commutative.

a single-valued analytical operation, whereas a formal multiplication is a connection of second order with respect to the given addition. This is the level of GTF, which is occasioned by an investigation of certain properties of the connections that are common to different mathematical branches.

In modern parlance, one could say that Grassmann's notion of formal addition corresponds to a commutative group, and the notion of formal multiplication corresponds to a ring under two operations (Schlote 1996, 168). Yet it has often been remarked in the literature (e.g., in Lewis 1977, 140, 146, and Flament 1992, 216) that one should not consider the properties of the connections of first and second order, or the properties of the formal addition and formal multiplication as axioms,[9] or as a reductionist kind of foundation. I believe that a comparison with ancient proportion theory might be illuminating, because—as Aristotle himself observed—the theorems of the theory of proportions could be demonstrated not only separately for numbers and for geometrical magnitudes but also in a more general way. Just as the formulation of proportion theory did not imply (at least not until the 18th century) the creation of a new genus of objects (Cantù 2008, §3), the fact that Grassmann assembled a list of propositions that "relate to all branches of mathematics in the same way" (Grassmann 1844, 33; 1995, 33) does not imply the construction of a new branch of mathematics or of a new domain of objects. So formal operations have the properties that are common to real operations, the latter being the operations that generate the mathematical forms in each mathematical discipline. This explains why Grassmann considers addition as being always commutative: he had not encountered any example of a non-commutative additive group in mathematics.

3. Third, there are abstract connections between thought forms, which might have different properties depending on the thought forms they are applied to. For example, at this level we find addition and multiplication between natural numbers, or addition and multiplication between extensive magnitudes. These abstract connections might be different (e.g., the multiplication is commutative in the first case and not commutative in the second case), but only with respect to the properties that were not already contained in their respective "formal" notions. These are what Grassmann calls "real" connections between forms: they are "real" because the law of connection is specified and grounds the generation of the forms.[10] This is

[9] See Radu (2003) for a discordant point of view on this issue.

[10] "So far we have developed the concept of addition in a purely formal manner, since we have defined it from the validity of certain laws of conjunction. This formal concept also remains the only general one. Yet it is not the way we arrive at this concept in the individual branches of mathematics. Rather in them a characteristic method of conjunction is obtained from the generation of the magnitudes itself, which manifests itself as an addition in precisely the general sense given,

the mathematical treatment of connections, as it is developed in its mathematical branch. The idea that in each branch of mathematics, one should verify whether the connections that can be introduced can be called addition or multiplication confirms that there should be a distinction between the level of formal operations and the level of specific mathematical branches, where Grassmann refuses to admit a domain of elements given prior to, or independently from, the generation of the elements themselves (Cantù 2011, 100).[11]

4. Fourth, there is the application of mathematical operations to physical reality, as in the case of the addition of masses or forces, or segments. To this level belongs the investigation of the connections that one finds in geometry.

Mathematics, as we will see in section 3.1, is for Grassmann the science of the particular. GTF, on the contrary, investigates formal operations, which are necessarily underdetermined, because the nature of the forms and their generating law induce the properties of the operation, which might vary relative to the domain of application. Grassmann considers as more "general" the product relative to a variable domain—a domain that is not closed under the operation but rather a result of our carrying out the operation itself. It is more general in the sense that it is underdetermined, because the determination or particularization of the operation depends on further conditions dictated by the nature of mathematical objects and by generating rules. The refusal to admit a domain of elements given prior to, or independently from, the generation of the elements themselves is an idea that Grassmann never abandons, and a basic assumption of his "constructivism" (see §4.2.2 and Cantù 2011, 100).

To resume, Grassmann's GTF, that is, the study of some fundamental relations and operations that occur in all branches of mathematics, is not itself part of mathematics, because it contains formal operations that are underdetermined and that might receive full determination only when they become real operations in mathematical disciplines or in the applications of mathematics.

since those formal laws apply to it" (Grassmann 1844, 40; 1995, 39, trans. slightly modified). "We have therefore formally defined the general concept of this multiplication as well; if the nature of the magnitudes to be conjoined is given, then to this formal concept must correspond a real concept that expresses the method of generation of the product by the factor." (Grassmann 1844, 44; 1995, 42, trans. slightly modified).

[11] It is interesting to note that different notions of product might occur in the same mathematical branch: for example, in the case of ET, there are a real addition between homogenous magnitudes (e.g., between segments) and a formal addition between non-homogeneous magnitudes (e.g., the addition between a point and a segment gives a point, because the symbol of addition has to be interpreted as a movement from one point to another point rather than as a concatenation of magnitudes) (Cantù 2011, 97).

2.3.2. Logic

Hermann Grassmann's contributions to logic concern some reflections (1) on the notion of primitive proposition—and in particular on the idea that in formal sciences there are only definitions, and no axioms, because mathematics concerns abstract concepts and not given objects—and (2) on a general logical law (*law of progression*).

The nature of primitive propositions varies according to the kind of science in question: formal sciences start from definitions, while real sciences start from axioms.

> Now if truth is in general based on the correspondence of the thought with the existent, then in particular in the formal sciences it is based on the correspondence of the second thought process with that existent established by the first, that is, on the correspondence of the two thought processes. . . . Consequently, the formal sciences cannot begin with axioms [*Grundsätze*], as do the real; rather, definitions comprise their foundation. (Grassmann 1844, 22; 1995, 23, trans. slightly modified)

Unlike Kant's formal criterion of truth, which concerns only the form and not the content of knowledge, and thus cannot say anything on the eventual contradiction between knowledge and its object (Kant 1787, 82), Grassmann's condition on the consistency of two thought processes is a condition on the consistency between an object of thought (the result of the first thought process) and a thought, that is, between two concepts.

Grassmann mentions a law of progression (*Fortschreitungsgesetz*) that he considers a general logical law:[12] it guarantees that "anything that is asserted about a series of things in the sense that it should hold for each individual of the series can actually be asserted about each individual belonging to the series" (Grassmann 1844, 65, my trans.).[13]

[12] Here "logical law" should be intended as a law of thought, rather than as a law of propositional or predicate logic. This interpretation is supported by what Grassmann's brother Robert explicitly claimed, i.e., that mathematical proof is independent of any given natural language, and of any logical theory (syllogistic logic in particular). Even if Robert's conception differs from that of Hermann inasmuch as he treated GTF as being itself mathematics, and logic as being one of its branches (Grassmann 1872, 17–18), I think the two brothers would agree that the notion of series is more pervasive in science (and thus more fundamental or more general) than universal instantiation in predicate logic: Hermann does not mention the latter at all, neither in GTF nor in mathematical branches such as arithmetic and extension theory, whereas Robert considers universal instantiation as occurring in a specific mathematical branch: logic.

[13] On the rule of progression cf. Cassirer (1910, 20–21). I translate "law of progression" rather than "procedural principle," as does Kannenberg (Grassmann 1995, 62), to underline the relationship to the concept used by Cassirer in a passage of *Substanzbegriff und Funktionsbegriff* where he discusses Grassmann's ideas ("In all these cases, we are not concerned in analyzing a given 'whole' into parts similar to it, or in compounding it again out of these, but the general problem is to combine any conditions of progression in a series in general into a unified result" [Cassirer 1910, 127; 1923, 96]).

This law, which we could understand as universal instantiation for series (rather than a law explaining the universal quantification of a predicate), just makes explicit what Grassmann means by general proposition, and should not be assumed as an axiom of mathematics. It is rather what allows us to demonstrate mathematical results, because proofs are considered concatenations (*Aneinanderketten*) of definitions, that is, themselves series of thoughts.

3. Grassmann's Philosophy of Mathematics

3.1. Mathematics as the Science of the Particular

Defining mathematics as the science of (thought) forms, Grassmann claims that mathematics is about concepts, which are considered particulars generated by means of an act from some initial element.

1. Thought forms are particulars that have "come to be through thought" (Grassmann 1844, Intro., §§2–3, 22–23).
2. Thought forms might come to be by different types of generation and different ways to relate them to the initial element of the generation (Grassmann 1844, Intro., §5, 25).

But he also characterizes mathematics by contrasting its peculiar method to the method followed in philosophy.

3. The mathematical method goes from the particular to the general (Grassmann 1844, Intro., §13, 30).

According to both characterizations of mathematics, which are not mutually exclusive but rather complementary, mathematics is conceived as the science of the particular.

3.1.1. Mathematical Thought Forms as Particulars

Mathematical thought forms (*Denkformen*)—or simply forms—are determined by their generating law: any form is a particular being that has come to be by some act of thought (it is the result of a particular act).

> Pure mathematics is therefore the science of the *particular* existent that *has come to be* by thought. The particular existent, viewed in this sense, we call a thought form or simply a *form*: thus pure mathematics is the *theory of forms*. (Grassmann 1995, 24)

Table 1 The Partition of Mathematics

Elements	Discrete generation	Continuous generation
Equal	Arithmetic (natural numbers)	Analysis (intensive magnitudes)
Different	Combinatorics (permutations)	Extension theory (extensive magnitudes)

Forms are abstract concepts that result from a generating thought from an initial element, which is itself a particular concept. Mathematics is thus the science of the particular that is posited by thought and not the science of the general laws of thinking (logic).[14]

The generation of the forms is so intrinsic to their nature that it also explains the partition of mathematical disciplines: depending on the relation between the elements (equal or different) and on the kind of generating law (discrete or continuous) that is applied to an initial element, one obtains a partition of mathematics into four branches: arithmetic, analysis, combinatorics, and extension Theory (see Table 1).

The partition of mathematics is based on the generating law and on the relation of the generated element to the initial element; that is, it is based on operations and relations, but also on the determinateness of the initial element. One reason why Grassmann introduced the term "form" in the definition of mathematics is that it contains a reference to formation, that is, to the way mathematical particulars are generated by a certain law, which alone guarantees their determinateness.[15]

> Since what is different from something given [*von einem Gegebenen*] may be different in infinite ways, the difference would get lost completely in the indeterminate, were it not constrained by a fixed law. (Grassmann 1995, 29, trans. modified)

[14] "The formal sciences treat either the general laws of thought or the particular as established by means of thought, the former being the dialectic (logic), the latter pure mathematics" (Grassmann 1995, 23–24).

[15] Other reasons might be the influence of Leibniz as well as dissatisfaction with the usual term "magnitude" (§4.2.1).

Grassmann is a conceptualist and a constructivist: what mathematics is about are concepts determined by construction, according to a law, from a particular that is considered as given (even if it is itself a concept).[16]

The relation of difference and equality between elements is introduced as a difference or equality with something given, that is, a first element from which the forms are successively generated. The importance of the individuation of a first element in the series becomes particularly relevant in arithmetic, where it grounds Grassmann's conception of numerical induction, but it plays a fundamental role also in ET, where it grounds the notion of generator, and in applications (e.g., in the geometrical calculus the initial element is the point, whereas in the barycentric calculus it is the point magnitude). Yet this characteristic of Grassmann's philosophy of mathematics has often been neglected or underestimated in the literature.

However, it plays an essential role in the notion of equality (e.g., two extensive magnitudes are equal if they are generated in the same way by equal elements), in the definition of thought form, and in the clarification of the specificity of the mathematical method (§3.1.2). Besides, it is relevant in the distinction between operations that are defined on a fixed domain and operations whose domains depend on the forms they are applied to: see, for example, the regressive or applied product (Grassmann 1844, 206), which is relative to the system that two magnitudes have in common (Cantù 2011, 94). Finally, it constitutes an essential aspect of Grassmann's understanding of concept formation, which is better represented by the notion of series than by the notion of function. Whereas the notion of function is introduced in modern mathematics by a correspondence between two given sets of elements, a series is always determined by an initial element and a law of development.[17]

One of the most acute interpreters of Grassmann's notion of series was Ernst Cassirer. Yet he underestimated the role of the initial element that is taken as given in order to go from the particular to the general, as well as the additive group of elements required by the notion of a series, insisting rather on what he calls the "concrete universality" of mathematical functions, and on the order relation between the members of a series. Cassirer understood Grassmann's claims about the initial element in ET by analogy with arithmetic rather than with geometry, and tried to go beyond the limit to "definite kinds of transformations" by highlighting the fact that "the element . . . is . . . only a pure particular grasped as different from other particulars," whereby no "specific content" is assumed

[16] By concept I mean what Grassmann calls "a thought representing an existent"; the existent might be given independently from thought, as in real sciences, or be itself a thought, as in formal sciences.

[17] In modern parlance we could say that a function could be intended as a logical notion, whereas a series is usually considered a mathematical notion.

(Cassirer 1923, 97–98).[18] But Grassmann explicitly adds that a further determination and distinction is guaranteed by the generation law, which might generate the forms in a discrete or in a continuous way. This is needed in order to let a real concept of operation correspond to the formal concept of operation.

Cassirer rightly underlined what he considered the "most general function of the mathematical concept: . . . giving some qualitatively definite and unitary rule that determines the form of the transition between the members of a series" (1923, 98). But then he concluded by inscribing Grassmann among the authors who considered mathematics as the science of relations:

> The considerations by which Grassmann introduces his work thus create a general logical schema under which the various forms of calculus, which have evolved independently of the *Ausdehnungslehre*, can also be subsumed; for they only show from a new angle that the real elements of mathematical calculus are not magnitudes but relations. (Cassirer 1923, 99).

So, even if Cassirer's reading of Grassmann is partially accurate and faithful, he inscribes Grassmann in a tradition to which Grassmann does not properly belong, especially if one acknowledges that Grassmann's GTF, which actually deals only with relations and operations, does not really belong to mathematics.[19]

[18] "First, in place of the point, that is, of the particular position (locus), we here substitute the *element*, by which we mean a mere particular, conceived as distinguished from other particulars; and indeed we attribute to the element in the abstract science absolutely no other content. There can therefore be no question as to what sort of particular it properly is—for it is the particular *per se*, devoid of any real content—, or in what sense this one is distinguished from the others—for it is merely defined as the distinct *per se*, without establishing any real content that might account for its distinctness. Our science has this concept of an element in common with combination theory, and thus the designation of elements by different letters is also common to both. The difference consists only in the way forms are obtained from the elements in the two sciences; that is, in combination theory by simple conjunction and thus discretely, but here by continuous generation. The different elements can now also be regarded as different states of the same generating element, and this difference of stages corresponds to differences in position (locus)" (Grassmann 1844, 47; 1995, 46, trans. modifified). Cf. also Grassmann (1994, 12).

[19] Erich Reck rightly noticed that Cassirer's notion of function cannot be reduced to the concept of a mapping between two domains. I agree that the notion of series grounds what Cassirer says about conceptual understanding: an intuitive multiplicity can be understood conceptually only if its elements can be seen as the elements of a series (Cassirer 1910, 19–20). Yet the notion of function is distinct from the notion of series: it is "some general law of arrangement through which a thoroughgoing rule of succession is established. . . . it is the rule of progression, which remains the same, no matter in which member it is represented" (Cassirer 1910, 20–21; 1923, 17). In another passage, Cassirer again distinguishes between "a series which has a first member, and for which a certain law of progress has been established, of such a sort that to every member there belongs an immediate successor with which it is connected by an unambiguous transitive and asymmetrical relation, that remains throughout the whole series" and the "progression" (*Fortschritt*) in which "we have already grasped the real fundamental type with which arithmetic is concerned" (Cassirer 1910, 49; 1923, 38). On the one hand, the notion of function is influenced by Dedekind's notion of *Abbildung* (Cassirer 1910, 50); on the other hand, Cassirer apparently agrees with Russell that arithmetic is a formal study of relations (Cassirer 1910, 48).

More specifically, Cassirer's move from series to relations is too quick. Grassmann's philosophy of mathematics is rooted in the notions of series and additive group rather than in the notions of function and order: the former and not the latter are relevant in concept formation.

In the introduction to the second edition of the *Ausdehnungslehre*, Grassmann explains the different role played by the notion of linear combination, which is essential to defining the notion of elementary extensive magnitude, and by the notion of function, which is not an expression between signs: its value is itself a magnitude, and precisely a composite magnitude (Grassmann 1862, 7; 2000, xvi, trans. modified). The notion of function occurs only in the second part of the book and presupposes the notion of magnitude, which is defined by means of the notion of linear combination.

> Definition. When a magnitude *u* depends on one or several magnitudes *x*, *y*, . . . in such a way that, whenever *x*, *y*, . . . assume determinate values, then *u* also assumes a determinate (univocal) value, then we call *u* a function of *z*, *y*, ... (Grassmann 1862, 224, my trans.)

So, according to Grassmann, neither linear combinations nor functions are notational abbreviations (Grassmann 1862, 5; 2000, xiv): they are rather essential tools for concept formation. It is only because a new autonomous concept has been formed by addition and multiplication that inverse operations can arise and the concept of a negative quantity can be introduced. In particular, Grassmann's notion of function has nothing in common with the notion of a mapping between two domains, because the latter does not include any privileged reference to operations. And even in the case of numbers, the operation of successor is considered a generating law that builds numbers by addition (the simplest example of linear combination, characterized by a single unit: the absolute unit). Thus, Grassmann highlights the similarities between numbers and extensive magnitudes, which are generated by similar concept formation tools, rather than their respective differences (the kind of order, and the commutative property of the product).

3.1.2. Mathematical (and Scientific) Method: From the Particular to the General (and Back)

The notion of a particular is at the basis of the mathematical method too. Mathematics and philosophy are characterized by an opposite movement: philosophy starts from the general and arrives at the particular with an analytical process of decomposition of a complex concept in more simple concepts; mathematics proceeds in a synthetic way and links several particulars to get a new particular, that is, links several concepts to get a new concept. The philosophical (dialectical/logical) method and the mathematical method can be better understood by analogy with the two different moments of the Platonic dialectic process: reduction and division. Philosophy starts from an overview of the totality, which is successively articulated and ramified, whereas mathematics starts by connecting particulars and aims at their unity (Grassmann 1844, 30).[20]

It is still controversial whether Schleiermacher's *Dialektik* had a decisive influence on Grassmann's understanding of mathematics.[21] Without pretending to give historical support to the claim, I would like to recall two issues of Schleiermacher's *Dialectic* that might clarify my interpretation of the difference between formal and real operations as a difference in determinateness, and offer a key to understanding the importance of the initial element in Grassmann's generation of mathematical objects.[22]

Schleiermacher distinguishes between the construction of the one (the initial element) and the combination of the one with another one (generation law starting from an initial element and determining the other elements). Even if mathematics mainly deals with combination, it does not ignore that each particular is in turn the result of a thought process, and in particular cannot ignore that the initial element is the result of a construction, which is relevant in the determination of the outcome of the combination. Production (or construction) and combination condition each other reciprocally (Schleiermacher 1839 [1986], 179).

The knowledge of a single concept obtained by a process of concept formation is an incomplete knowledge that needs to be further determined by the connection of that concept with other concepts. The mathematical method concerns specifically the determination of connections, but these cannot be separated from the particulars they should connect, and from the initial element whose knowledge should be completed by further determinations.[23] The initial element

[20] For a convincing interpretation of Schleiermacher's influence on Grassmann's distinction between dialectic and mathematics see Lewis (1977). See especially the distinction between construction (concept formation) and combination (connection of particular concepts).

[21] See, for example, Engel (1911), Lewis (1977), and Petsche (2004), who claim it did, and Schubring (2008), who denies it—or at least restricts the influence to other domains than mathematics, as, for example, philology.

[22] Both issues are actually mentioned in Lewis (1977), but are not related to Grassmann's *Ausdehnungslehre*, nor specifically to Grassmann's ideas on concept formation and indeterminacy. Schleiermacher's influence is rather recognized by Lewis in Grassmann's deduction of mathematical branches by opposition of the fundamental concepts (equal, unequal, discrete, continuous). See also Schubring (2008, 63). I thank Jamie Tappenden for calling my attention to this review.

[23] "Even if a concept is formed it can never by itself completely represent the existent since as concept it only contains what is based in the particularity of this existent and not what is posited in it as a consequence of its associations [with other existents]. . . . Thus each given thought contains the requirement of finding another new thought and of determining that which is left undetermined. The first is the extensive direction within combination and the latter the intensive, and it will be in the oscillation between the two that we must progress. The method in the first direction—to find from a given knowledge a new one—we call the heuristic; that in the other—to connect the dispersed

of the series allows the construction of the successive element, which in turn adds determinateness to the preceding element.

The scientific method, which is for Grassmann the method of presentation of a science in a treatise, that is, a pedagogical method (Flament 1992, 215), should sum up in itself both the specificity of the philosophical method (which proceeds from the general to the particular) and the specificity of the mathematical method (which proceeds from the particular to the general). So any scientific exposition should combine two aspects, which Grassmann calls respectively content and form: the content (*Inhalt*) of a science is the development that goes from one individual truth to another individual truth in demonstrations, whereas the form (*Form*) is a guiding idea, which is either a presentiment of the searched truth or a conjectural analogy with other well-known branches of knowledge (Grassmann 1844, 31).

3.2. Formal and Real

There is an ambiguous use of the terms "formal," "form," and *Formel* in Grassmann, which explain the difficulty in understanding and appreciating his philosophy of mathematics. These expressions occur in different contexts with different meanings.

1. Forms are thought forms, which are opposed to what exists independently of thought ("das Sein als das dem Denken selbständig gegenübertretende"), to what is given and cannot be itself generated by thought (e.g., space) (Grassmann 1844, Intro., §1, p. 22). The former notion is connected to the distinction between "formal sciences" and "real sciences," whereby formal sciences concern the purely conceptual (*rein begrifflich*), whereas real sciences concern what is given outside thought (e.g., spatial [*räumlicher Natur*] notions) (Grassmann 1844, 22).
2. *Formeln* are symbolic expressions, as opposed to their denotations: concepts (Grassmann 1861, 6).
3. "Formal operations" in GTF are the formal addition and the formal multiplication that are opposed to "real operations" (e.g., addition between numbers in arithmetic or between extensive magnitudes in ET) (Grassmann 1844, 41, 42n).
4. The form of a science is its treatment or exposition, as opposed to its content (concatenations of truths) (Grassmann 1844, 31).

and isolated given material—the architectonic" (Schleiermacher [1839] 1986, 179–180; trans. in Lewis 1977).

In the following, I will discuss in more detail the first three occurrences, having already discussed the last one in the method of scientific exposition (§3.1.2).

3.2.1. Purely Conceptual versus Spatially Intuitive

We have already mentioned that thought forms are the objects of mathematics. "Formal" is also used to characterize several sciences in contrast to real sciences: the main differences between them concern their respective objects (abstract vs. intuitive), their primitive propositions (definitions vs. axioms), and their criterion of truth (correspondence between two acts of thought versus correspondence with some external thing). Independence from intuition is clearly stated at the end of Grassmann's *Geometrische Analyse*, where abstract as purely conceptual is opposed to real as associated with spatial intuition.

> Now, in fact, as is demonstrated throughout Grassmann's *Ausdehnungslehre*, all concepts and laws of the new analysis can be developed completely independently of spatial intuition [*unabhängig von der räumlichen Anschauung*], since they can be tied only to the abstract concept of a continuous transformation; and, once one has grasped the idea of this purely conceptually conceived [*rein begrifflich gefassten*] continuous transformation, it is easy to see that also the laws developed in this essay can be conceived as freed from spatial intuition [*von der räumlichen Anschauung gelösten*]. (Grassmann 1995, 384, trans. modified)

The main point is not to do away with intuition, but to give it its own role. The analogy with geometry is essential in the method of exposition of the *Ausdehnungslehre* and as a heuristic guide in the search for theorems. One thing is to consider pure concepts as independent from intuition and another thing is to assume that Grassmann calculates with signs devoid of meaning.

3.2.2. Symbolic versus Conceptual

Formel occurs in expressions such as *Formelentwicklung*, where it might be considered synonymous with what we now call symbolic. In the *Treatise on Arithmetic*, formal as symbolic is mentioned in the inferential development of arithmetical truths, which are expressed by symbols but denote concepts. There is thus no idea of symbols or "signs or letters the referent of which did not matter" in Grassmann (Darrigol 2003, 522). This formalistic interpretation of Grassmann's number theory has been encouraged by Hankel, who maintained a distinction between actual and formal numbers, but identified formal numbers with signs (Hankel 1867, 36). Quoting Grassmann as a fundamental source of his work (Hankel 1867, 16), Hankel indirectly suggested that Grassmann shared his point of view. Yet Grassmann claimed that "the symbolic development [*Formelentwicklung*] and the conceptual development [*Begriffsentwicklung*] should go hand in hand. . . . The whole treatment will proceed along a conceptual development, whereas the formulas added at each step symbolically represent the conceptual advancement." (1861, 6, my trans.)

The mentioned definition of function, as well as the refusal to consider linear combination as a notational abbreviation, also supports a non-formalistic reading of Grassmann.[24]

3.2.3. Formal versus Real: Two Levels of Abstractness

The adjective *formell* occurs in GTF as a way to distinguish formal addition and multiplication from real addition and multiplication. Here the formal concerns an underdetermination of the concept of a connection, which gets embodied and becomes real only in each specific mathematical discipline. Both the formal connections and the real connections are thus abstract and purely conceptual (*rein begrifflich*), and thus opposed to intuitive or spatial notions that can be found in applied mathematics (e.g., in a real science as geometry). The ambiguity of the terminology is here evident and explains why it is difficult to understand Grassmann's philosophy of mathematics. The notion of real connection, opposed to that of formal connection, is not to be found in real sciences, but in formal sciences! It is thus abstract and opposed as such to what is real in the sense of concrete, as connections between geometrical figures.

Are the formal operations of GTF merely expressed by signs devoid of reference, as several authors took them to be? Here Grassmann's idealistic philosophy explains why this is not the case.

> As the general sign for conjunction we take the symbol $\cap$; now if a and b are the factors, with a the prefactor, b the postfactor, then we indicate the product of their conjunction as $(a \cap b)$, where the parentheses here express that the conjunction indicates that the factors are no longer separate, but that their concepts are unified. (Grassmann 1995, 34)

It is certainly true that the level of generality and the abstraction from specific features of the real operations suggest that the formal connections do not refer in the same way, because they concur in the formation of concepts once applied

[24] For a different reading see Darrigol (2003, 522), but also Klein, who encouraged a formalistic reading of Grassmann, as he praised his ingenious algorithms (Klein 1979, 166–167). Klein's reading suggests that ET contains algorithms that refer to geometry, thereby ignoring Grassmann's abstract level that is situated between the formal and the real concrete level (§2.4).

to some concept. Just as proportion theory (from which GTF inherits the analysis of equality and of addition) needs to be applied to specific mathematical branches, so does Grassmann's GTF. The following passage supports this interpretation, according to which formal and real should be understood as disembodied and embodied respectively:

> Incidentally, it lies in the nature of things that the conceptual determination of these connections is here purely formal, whereas only in the single sciences it can be embodied by means of real definitions. (Grassmann 1844, 42n, my trans.)

The insistence on the separation of the different mathematical branches and on the purity of proofs in each domain is incompatible both with the idea that GTF has as its object an abstract structure, and with the view that it constitutes a symbolic calculus devoid of reference or meaning.

3.2.4. Hankel's Three-Level Distinction

Hankel rightly distinguishes the first and the second notions of formal and real, individuating three levels in the *Ausdehnungslehre*: formal, real abstract, and real concrete, which correspond to the laws of GTF, of extensive magnitudes and of geometrical figures respectively (Hankel 1867, 16–17).

Hankel thus traces a distinction between (1) the level of formal laws, (2) the level of abstract content, and (3) the level of real content. There are at least two distinct interpretations of this tripartition in the literature on Grassmann: (*a*) universal algebra, different algebras, and physical instantiations of such algebras, (*b*) abstract algebra, linear algebra, and geometry. A critical discussion of these two interpretations is useful to understand Grassmann's contributions to universal algebra, abstract algebra, and non-Euclidean geometry, and therefore to the transformation of mathematics into a science of structures, but it is also useful to compare contemporary philosophical structuralism with Grassmann's peculiar understanding of mathematical objects and structures.

The second interpretation of the tripartition holds for what Grassmann does in the *Ausdehnungslehre*, provided that one also remarks that of the three levels, only the second properly pertains to pure mathematics, whereas the former is not mathematics, and the latter is applied mathematics, and provided that one recalls the differences between Grassmann's approach and modern algebra. The main question here is whether GTF (1) does not belong to mathematics yet, because it has not been sufficiently developed or because it cannot be part of mathematics, given that it is only formally abstract and not really abstract, or (2) cannot belong to mathematics, given its too general nature. In support of interpretation (1), it should be noted that Grassmann himself declares that "such a general branch is not yet available" and that he has developed it only as far as it is needed for ET, thereby neglecting a third possible connection: raising to a power (Grassmann 1844, 33, 42n). In support of interpretation (2) there is the fact that, according to Grassmann's conception of the mathematical method, GTF cannot belong to mathematics, because it does not go from the particular to the general. And in fact the connections are considered independently from their application to a first element.

My claim is that GTF should be considered as something that has to do with the scientific method, which, as we have seen, has to incorporate both the philosophical and the mathematical method, going both from the unity of the idea to the multiplicity of particulars and back. This interpretation seems to be confirmed by the role Robert Grassmann assigns to it in the *Formenlehre*. Robert develops what Grassmann calls GTF as a theory of magnitudes (*Grösenlehre*),[25] including the general definitions and theorems that make a rigorous scientific thought possible, teaching us how to make scientific inferences (*wissenschaftlich beweisen*) (Grassmann 1872, 1), and characterizes it as the general part in opposition to special disciplines.[26]

Now, if one considers not only the *Ausdehnungslehre* but more generally the totality of Grassmann's writings, then one might have an argument for the first interpretation of the tripartition already mentioned. The level of formal laws might correspond to a certain way of doing universal algebra; the level of abstract content might correspond to different algebras developed by Grassmann, among which is vector space theory, but also some non-commutative algebras that he developed in his essay on different kinds of multiplication (Grassmann 1855); and the level of real content would correspond to geometry, to the barycentric calculus, and to other physical instantiations of such algebras.

[25] Yet Robert Grassmann uses a different terminology and inverts the presentation: he begins by raising to a power (*Anreihung*, which is not commutative), then introduces multiplication (*Einigung*, which is associative), and finally introduces addition (*Vertauschung*, which is commutative) (Grassmann 1872, 15–24). He then considers direct (*Trennung* or *trennbare Knüpfung*) and inverse operations (*Lösung* or *untrennbare Knüpfung*), where the former are univocal and the latter are not univocal.

[26] "*Grösenlehre*, the first or most general discipline of *Formenlehre*, teaches us to recognize those connections between magnitudes that are common to all disciplines of *Formenlehre*. It develops the laws of equality, addition or *Fügung*, multiplication or *Webung*, and exponentiation or *Höchung*. The four special disciplines of *Formenlehre* emerge from *Grösenlehre* through the introduction of new conditions" (Grassmann 1872, 11–12, my trans.).

4. Is Grassmann a Structuralist?

4.1. Mathematical Contributions

Grassmann's contributions to mathematics already tell us something about his relation to methodological structuralism (see §1). If the mathematical structuralist methodology is the result of several important innovations such as abstract algebra, axiomatic method, set theory, and Bourbaki's structuralism (Reck and Price 2000, 346), Grassmann did explicitly contribute to the first factor, thanks to his contribution to vector space theory, which clearly favored the development of abstract algebra. Vector space theory is also an example of a non-Euclidean geometry, because the vector space has dimension n, with $n \in \mathrm{N}$, and thus includes the investigations of abstract spaces with dimension > 3. Grassmann's distinction between real and formal sciences thus contributed to the liberation of abstract geometry (linear algebra) from physical space.

Even if Grassmann's presentation of extensive forms is not strictly axiomatic, Grassmann contributed to the development of axiomatics for at least three reasons:

1. He gave an axiomatic presentation of natural numbers in 1861, where, thanks also to the collaboration with his brother Robert, specific attention was given not only to the propositions chosen as axioms but also to demonstrative inferences (and in particular to which propositions are used in each step of the derivation).
2. He developed a purely abstract treatment of linear magnitudes that is completely independent from concrete intuition.
3. He was the source of inspiration of Giuseppe Peano, who published an explicitly axiomatic presentation of vector theory in 1888 and of arithmetic in 1889.[27]

Grassmann contributed to the investigation of the abstract structure of a system of extensive magnitudes. He highlighted similarities and differences concerning the operations of different systems of mathematical forms. On top of that, he favored a comparison of the abstract structures of numbers with the abstract structures of magnitudes, individuating their main difference in dimensionality and in the commutativity of the product. Grassmann thus clearly contributed to

[27] Shapiro himself, literally quoting Nagel (1939), acknowledges the contribution of Grassmann's theory of extension as a prefiguration of "the method of implicit definition" (Shapiro 1997, 147). On the relation between Peano's axiomatic vector theory and Grassmann's extension theory see in particular Dorier (1995, 247) and Cantù (2003, 331–338).

the development of abstract algebra, and, if universal algebra is conceived as a comparative investigation of different algebras—either to see what they have in common (Grätzer 1968, 7) or in connection with their interpretations in order to find a generalized notion of space that might serve as a uniform method of interpretation of the various algebras (Whitehead [1898], 1960, v, 29)—then he contributed to the development of universal algebra too. In particular, Grassmann's comparison of different structures was motivated by a foundational effort to distinguish different branches of mathematics according to the structural relations of their elements (§3.1.1). Yet one should remember that Grassmann did not understand algebraic systems as sets of given entities closed with respect to certain operations, and did not investigate classes of algebras, but only the different possible properties of operations.

If an essential condition for the development of methodological structuralism was the "transition from geometry as the study of physical or perceived space to geometry as the study of freestanding structures," a transition that was accomplished through the development of analytical geometry, projective geometry and non-Euclidean geometry (Shapiro 1997, 14), Grassmann's distinction between what we now call linear algebra (vector space theory in n dimensions) and geometry (the three-dimensional application of the former to physical space) (Grassmann 1845, 297) was certainly a relevant step, even if, as is often said, Grassmann did not bridge the gap between the discrete and the continuous, at least in the sense that he never considered real numbers as an extension of the system of rational numbers. On the contrary, he defined real numbers as being themselves magnitudes, thereby emphasizing the difference between discrete natural numbers and continuous real numbers, and grounding ET independently from arithmetic.

4.2. Methodological Structuralism

4.2.1. Mathematics Is Not the "Science of Quantity"

Methodological structuralism is often associated with a criticism of the definition of mathematics as science of quantity and number. Grassmann criticizes the traditional definition of mathematics as "science of quantity or magnitudes" (*Grössenlehre*) for two reasons. First, the word *Größe* refers only to continuous magnitudes and thus does not apply to the whole of mathematics.

> The name "theory of magnitude" is inappropriate for all mathematics, since one finds no use for magnitude in a substantial branch of it, namely combination

> theory, and even in arithmetic only in an incidental sense. (Grassmann 1995, 24)

That *Größe* refers only to continuous quantities is proved linguistically: in the German language *vermehren* and *vermindern* are connected to number, while *vergrössern* and *verkleinern* are connected to continuous quantities. Distinguishing what Wolff in *Mathematisches Lexikon* had not explicitly separated (Wolff translated both Latin terms *magnitudo* and *quantitas* by the same German word, *Grösse* [Cantù 2008]), Grassmann refuses to admit the reduction of geometry to algebra and to subsume continuous geometrical figures and real numbers under a single genus. Grassmann considers natural numbers as discrete quantities generated by repetition of a unit. Therefore the language rightly distinguishes numbers that increase or decrease from continuous magnitudes (including real numbers) that become bigger or smaller.

Second, the word *Größe* fails to express the main characteristic of mathematical objects, that is, that they are not given but generated according to a rule (§3.1.1). It is only in this second sense that Grassmann's remarks might be interpreted as having some relationship to structuralist approaches.

4.2.2. Mathematics Is Not about "Objects" but about Relations

A second fundamental feature of methodological structuralism is that mathematics is not about objects but about relations, or at least about objects only inasmuch as they are positions in a structure. Recalling what we have said about mathematics as the science of the particular, and especially about the role of the initial element in the "real" generation of mathematical abstract forms, it seems implausible to associate Grassmann with the conception of mathematics as the science of relations, notwithstanding Cassirer's and Hankel's tendency to do it. A further argument against this assimilation might come from some remarks by Banks, who insists on Grassmann's belonging to a German tradition that was interested in the development of a physical monadology in a Leibnizian sense (Banks 2013, 20–21), or the investigations by Brigaglia, who considers Grassmann to be the inspiring source for Segre's generalization of the notion of point (Brigaglia 1996, 159–160).

Yet there might be reasons to claim that, even if Grassmann's mathematics cannot be considered, *sensu stricto*, a science of relations, it might be an intermediate step between the traditional conception of mathematics and a structuralist approach. Such reasons are his constructivism and his consideration of operations as corresponding to pre-mathematical operations that can be applied to any kind of domain. Grassmann's constructivism is based on the idea that forms are the results of processes of connection, which construct or generate them, so that, in a dialectical perspective inherited from Schleiermacher, forms cannot really be distinguished from the process of their construction, and thus from the operations that occur in their concept formation and that determine their relations to other forms.

4.2.3. Mathematics Is the Study of "Relational Systems"

A third feature of methodological structuralism is the idea that mathematics investigates different "relational" systems, such as number systems, geometrical manifolds, various algebras, and so on. Again, Grassmann's separation between GTF, the specific branches of pure mathematics, and applied mathematics make it difficult to compare this approach to methodological structuralism. Certainly, he did not deeply investigate order relations, and he had a quite intuitive notion of continuous transformation. On the other hand, the effort to introduce a partition of mathematics that is based on different properties of the operations—an effort that became systematic especially in Robert's *Formenlehre*—or Grassmann's abstract analysis of different kinds of formal multiplications and their possible "realizations" in mathematical theories (Grassmann 1855, 216–217) can be considered a step toward the development of the project of a systematic investigation of relational systems.

4.2.4. Mathematics Is Not "Directly about the World"

There is at least one feature of methodological structuralism that Grassmann entirely subscribed to: it is the separation between pure and applied mathematics, which implies that mathematics is about abstract forms, and thus is not about the external world, or, in Grassmann's parlance, is not about a given that is not itself constructed by thought.

4.2.5. Mathematical Inferences Are "Formal"

A further feature of methodological structuralism is based on the idea that deductions are merely formal. Grassmann explicitly recognized that mathematical inferences are independent of intuition in the sense that primitive propositions can be conceived purely conceptually and that the only general logic law is the law of progression. Even if he admitted a relevant role of intuition as a heuristic tool, he never conceded that it should play a role in deductions. Yet Engel criticized Grassmann exactly because he did not manage to fulfill his project, maintaining an intuitive and unclear notion of continuous transformation (Grassmann 1844, 405). But this again is a controversial issue: Lawvere claims on the contrary that Grassmann's continuous transformation is unclear if wrongly conceived as a spatial translation, but that it becomes philosophically clear if it is understood as an action of the additive monoid of time.[28]

[28] "Grassmann philosophically motivated a notion of a 'simple law of change,' but his editors in the

4.2.6. Mathematics Goes toward Set Theory or Category Theory

If either set theory or category theory is a necessary condition for the development of methodological structuralism (an exception made, maybe, for Hellman's modal structuralism), then one should note that Grassmann did not contribute to the development of a theory of sets. On the contrary, his notion of product is incompatible with the modern understanding of a function, and his constructivism is incompatible with a set-theoretic perspective, where operations are defined on previously given sets of individuals (Cantù 2011, 2016). Grassmann did not contribute to the theory of category either, from a strict mathematical point of view, but Lawvere considers that category theory makes it "possible to recover some of Grassmann's insights and to express these in ways comprehensible to the modern geometer," and claims that Grassmann can be considered as precursor of category theory (Lawvere 1996, 255–256).

4.2.7. Mathematics Is Based on Some Kind of Axioms

We have seen that Grassmann did not present ET in an axiomatic way, at least not in a Hilbertian sense. But he considered mathematics to be based on concepts, because he deduced the differences between mathematical disciplines and mathematical forms by means of four fundamental concepts: equal, different, discrete, and continuous. And there is another sense in which GTF is based on some general conditions upon which a real operation might be called an addition or a multiplication (see §2.3.1). Finally, mathematics, being a formal science, does not have axioms in a traditional sense, but definitions. Yet the treatment of ET is not presented axiomatically, and GTF rather describes the operational features common to all operations that can be found in known mathematical disciplines, rather than an axiomatic description of algebraic structures.

4.2.8. Mathematics Studies Invariants

Structuralism is often associated to the investigation of invariant properties of different systems. GTF can be interpreted as a unifying perspective that studies what is invariant in different mathematical operations. Yet, as we have repeated many times, it is not a branch of mathematics. Some authors have tried to show that, even if Grassmann did not himself develop a comparison and classification of different mathematical systems by means of groups, his remarks on affine geometry influenced Klein's Erlangen program (see Engel 1911, 312, quoted in Toebies 1996, 120–122). Yet this is controversial, given that it has been argued

1890's found this notion incoherent and decided he must have meant mere translations. However, translations are insufficient for the foundational task of deciding when two formal products are geometrically equal axial vectors. But if 'law of change' is understood as an action of the additive monoid of time, 'simple' turns out to mean that the action is internal to the category *A* [of affine-linear spaces and maps] at hand" (Lawvere 1996, 255).

that Klein took his inspiration from Riemann and from projective geometry rather than from Grassmann and affine geometry (Rowe 2010, 142). Besides, Kannenberg has claimed that even if Grassmann individuated the group of circular and linear transformations starting from an analysis of different side conditions for the multiplication of extensive magnitudes, "the relation between groups and 'species of multiplication' is not reciprocal" (Grassmann 1995, 469).

To resume, Grassmann's GTF, which corresponds to the level of formal laws, is not part of mathematics, because it is underdetermined: it does not speak about a specific structure or a class of structures, but rather about the possible ingredients of a structure, being thus more similar to a metatheory of abstract structures. Grassmann's arithmetic contains just one kind of multiplication, whereas Grassmann's ET (the level of abstract content) concerns several kinds of multiplication, which can be fully determined only by further side conditions.

4.3. Philosophical Structuralism

In an earlier section of my chapter (see §3), I have tried to reconstruct Grassmann's conception from a perspective internal to his writings and to the spirit of his time. Now I will try to look at Grassmann's philosophy of mathematics from the present perspective, and thus look at some of the questions raised by Grassmann in the light of contemporary philosophical structuralism.

4.3.1. Grassmann's Claims on Structures

Even if Grassmann never used the term "structure" himself, I suggest that he might have agreed that (1) mathematical objects are characterized by structural properties; (2) structures are not given axiomatically; (3) general structures are distinguished from particular structures and from exemplars; (4) there is an interdependence between a structure and its objects, and (5) pre-mathematical operations between concepts are distinguished from operations in structures.

1. Mathematical objects are characterized by means of their relations and operations and by the relations between such operations (i.e., by structural properties). Structures are not themselves mathematical forms, because mathematical structures are universals, whereas mathematical forms, although being themselves concepts, are particulars.
2. Structures are not given axiomatically (construction versus postulation), and certainly not defined as in model theory by means of a domain and some relations and operations on it.
3. GTF is the study of relations and operations and concerns what we now call general structures (monoids, groups, rings). Mathematics is the study

of particular thought forms and concerns what we would now call particular structures. Applications of mathematics study particulars considered as given independently from thought and concern the investigation of exemplars of particular structures.

4. There is a dialectic between general structures, particular structures, and their exemplars, which allows further determination of mathematical forms as well as their relational properties. General structures do not exist independently from particular structures. Particular structures both determine and are determined by their objects. The distinction between GTF and mathematics concerns the question of the interdependence between relation and relata. In formal operations the operation might stand without its factors, whereas in mathematics it cannot. GTF is a sort of metatheoretical discourse on mathematical operations rather than itself a theory having mathematical structures as its objects.
5. There are some pre-mathematical relations and operations (equality, connection, and separation) that express some general operations of composition of concepts. They have some very general properties, such as substitutivity, commutativity and distributivity respectively. They are underdetermined with respect to mathematical operations, which have further properties: for example, the properties of the additive operation in an abelian group, or of the additive and multiplicative operations in a ring.

4.3.2. Grassmann's Claims Evaluated from the Perspective of Contemporary Philosophical Structuralism

If one evaluates the previous claims from the perspective of contemporary philosophical structuralism, one might remark that (1) there are no structures as universal objects in Grassmann; (2) there is no set-theoretic approach in Grassmann, (3) Isaacson's distinction between general and particular structures might apply to Grassmann's distinction between ET and arithmetic, (4) Grassmann's epistemology might be fruitfully compared to Parsons's non-eliminative structuralism as well as to (5) Feferman's conceptualism.

1. No universals: Grassmann's epistemology suggests the need for a constructivist alternative to the *ante rem* / *in re* ontology, an alternative that might speak about structures without considering them as mathematical objects, and especially not as universals.

Grassmann certainly had an ontological perspective, at least in the sense that he was an idealist and a constructivist: mathematical forms have an objective nature. All products of thought processes become objective in the moment of their construction, and can thus be successively taken as given (Grassmann 1844, 22). For this objective nature of thought forms, it is certainly not easy to associate Grassmann with an eliminative (nominalistic) position à la Hellmann (1990). Nor does it seem possible to consider thought forms as *in re* universals (as in Shapiro's account of eliminative structuralism [1997, 9]).

Yet the question about Grassmann's structuralism can be asked once more at another level, that is, at the level of GTF. Grassmann could be associated with an eliminative approach at this level, because there are no such things as the objects of a general structure. General structures are not the genus of which spaces, number systems, and so forth are species (Burgess 2015, 107–108), but are based on underdetermined concepts that get their full determination once applied to particulars, and it is this very application that makes further side conditions explicit and allows for the determinateness that is needed to treat something as an object of mathematics.

2. No set-theoretic notion of structure: a preliminary objection might concern the anachronism of applying a philosophical perspective that is grounded on different notions of function, object, and concept. According to Grassmann, operations are not closed on a domain, either because the domain might be considered variable (in mathematics) or because the operations might be considered independently from their factors or from a domain on which the factors should vary.

The main problem in the case of Grassmann is to exactly determine what he might mean by "structure." Whereas the model-theoretic notion of structure is based on a domain (a set) to which the operation is applied (and the definition of the structure concerns this domain, at least inasmuch as it has properties of closure with respect to operations), there is not even the possibility of determining closure properties in Grassmann's consideration of formal operations.

3. General and particular structures (Isaacson and Shapiro): a comparison with Isaacson's structuralism is interesting in order to appreciate another aspect of Grassmann's structuralism: the distinction between formal and real operations. Isaacson's structuralism is antithetic to Grassmann's perspective, at least inasmuch as it defends the existence of structures but not of mathematical objects (structures themselves are not mathematical objects), and he centers his perspective on axiomatic postulation rather than on construction.

Isaacson distinguishes between general and particular structures. The distinction is derived from the way we linguistically refer to them, either by the determinate article (the structure of natural numbers) or by the indeterminate article (a group) (Isaacson 2011, 2–3, 18). Isaacson remarks that Bourbaki believed that the mathematical interest was mainly on the side of general structures, and Grassmann might agree on that point.

Yet, according to Isaacson, the philosophical interest is all on the side of particular structures, because structuralist realism concerns the existence of particular structures. This might be related, I think, to Grassmann's choice to consider GTF as not properly belonging to mathematics: it concerns general and not particular structures. Besides, Isaacson notes that particular structures can themselves be classified into abstract and concrete structures (type and tokens), being in the relation one-many. This might correspond to Grassmann's distinction between vector space theory and 3-dimensional geometry, or between the abstract real level and the level of applications.

Shapiro had introduced a distinction between "algebraic" and "non-algebraic" fields of mathematics, that is, between mathematical subdisciplines that concern a class of structures or a single structure respectively.[29] Even if one might claim that algebraic fields are about general structures, whereas non-algebraic fields are about particular structures, the use of Shapiro's distinction is problematic, because it does not do justice to Grassmann's idea that all mathematical fields are about particular structures. It is only GTF that concerns general structures. This is an important aspect of what we might call Grassmann's *concept structuralism*, as opposed to an *object structuralism*, which requires a complete determinateness of the objects and therefore an identity criterion.[30] And this might explain why Grassmann would probably disagree with the idea (shared, e.g., by Isaacson

[29] See Shapiro (1997, 40–41). The distinction made by Grassmann between arithmetic and ET can be compared with Shapiro's distinction between non-algebraic (e.g., arithmetic and analysis) and algebraic fields (e.g., group theory, field theory, or topology, which are about a class of related structures).

[30] More should be said on this notion of "concept structuralism," but this would require a new article. For the sake of the understanding of Grassmann's perspective, it might suffice to say what concept structuralism is not, and how it is related to a dynamic process of mathematical determination of pre-theoretical notions. (1) Concept structuralism is not a historical tradition like "conceptual mathematics" (see, e.g., Stein 1988 and Ferreirós 2007). (2) Concept structuralism is not necessarily characterized as a weak form of Platonism (see, e.g., Ferreirós's effort to define conceptual structuralism as a version of weak Platonism, suggesting that structures exist as abstract entities but are not necessarily independent from the mathematician). Structures are conceptual tools that describe general properties of the operations among particular entities. In a proper sense, only the particulars can be said to exist as fully determined objects. (3) Concept structuralism is based on the idea that mathematics is a dynamic process that tries to further determine some pre-theoretic notions, e.g., by considering the algebraic closure of an underdetermined operation, so that mathematical objectivity is ultimately grounded in processes of concept formation. I would like to thank José Ferreirós for the rich discussion we had on the topic, and for the useful insights I got from the reading of his essay on mathematical practices (Ferreirós 2016), and his unpublished manuscript on Feferman's conceptualism (Ferreirós 2018).

and Shapiro) that the philosophical problem consists in accounting only for the existence of particular structures.

Another interesting point is Isaacson's remark that there cannot be objects without particularity and without an identity criterion: therefore Shapiro has a problem when he pretends to speak about the objects of a structure (as is proved by Keranen's objections). I take Isaacson's remark to suggest that whenever structures are introduced axiomatically (or by postulation), then one cannot talk about mathematical objects of these structures, because no identity criteria are available. Grassmann avoids introducing vector space systems by postulation, exactly because he believes that they concern mathematical forms whose construction is determined by their generating laws, which also allow for identity criteria. Construction rather than postulation has for Grassmann a foundational value. This position is again antithetic to the position of Isaacson, who believes that only postulation has foundational value, and that construction was fundamental only in the logicist perspective, because the construction should prove the logical nature of mathematical concepts.

4. Parsons's non-eliminative structuralism: Grassmann's approach can be interestingly compared with Parsons's version of non-eliminative structuralism.[31] Mathematical objects are taken to be particular forms (e.g., numbers, extensive magnitudes, etc.). Neither formal operations nor structures themselves seem to be considered mathematical objects, because they appear in GTF as underdetermined, devoid of an identity criterion, which on the contrary seems to be a necessary condition for something to be an object (Isaacson 2011). Talk about formal operations is rather metatheoretic, and general structures (in Grassmann's sense) are not even deficient-property objects (Burgess 2015), because they are not structures whose elements have no individual nature, but rather a bunch of operations considered independently from their "application" to particulars. There is a dialectic between particular structures and their exemplars, as in the case of the geometric analogy that guides the development of Grassmann's ET. Similarly, Parsons considers structures to be not

[31] A general classification of all kinds of contemporary variants of structuralism is not available, and various terminologies conflict one with the other. I will adopt Parsons's terminology, and distinguish *eliminative* from *non-eliminative* structuralism: "Eliminative structuralism . . . proposes some procedure for paraphrasing the language that refers to the objects we are concerned with, usually either the numbers of one of the number systems, or sets, so that commitment to the objects concerned, even the conception of them as a distinctive kind of object, disappears. . . . [Non-eliminative structuralism] takes the ideas behind structuralism not as the basis for a program for eliminating numbers, sets and other pure mathematical objects, but rather as the basis for an account of them as objects, as the objects which theories of numbers and sets talk about when taken more or less naïvely" (Parsons 2004, 57).

free-standing but connected to instantiations developed in mathematical practice. Grassmann's vector space theory is presented in a purely abstract way in the first edition of the *Ausdehnungslehre*, but a geometric analogy guides the development of ET. This dialectic between the particular structure and one of its exemplars suggests a comparison with Parsons's claim that talking about mathematical objects is legitimate in structuralism, even if their identity criteria cannot be established exclusively by means of structural properties, but require some reference to extra-structural properties.

Grassmann similarly believes that it is possible to talk both about operations that are only partially determined and about operations that are fully determined in some particular structure or in an exemplar of it. This is legitimate, because, according to Parsons, structures are not free-standing but are somehow connected to instantiations developed in mathematical practice.

With Parsons, Grassmann would agree that mathematical objects such as natural numbers are usually given in a realization of the structure, and that "some mutual dependence in understanding what the objects of a domain are and what their most important properties and relations are" need not be circular (Parsons 2004, 73). I suggest that Grassmann would understand in a dialectical way the relation between the so-called intended model and the axiomatic formulation of a structure.

Grassmann's perspective cannot be compared with Parsons's Quinean approach, according to which "speaking of objects just is using the linguistic devices of singular terms, predication, identity, and quantification to make serious statements" (Parsons 1982, 497). Yet I think Grassmann shares what I take to be a presupposition of Parsons's structuralism: the possibility of talking both about objects that are only partially determined (e.g., determined only by their structural properties, even when this does not allow us to distinguish objects in the structure, as might be the case for i and $-i$ in the structure of the complex numbers) and about objects that are fully determined in some instantiation of the structure (where one might have identity criteria or knowledge of specific relations between the objects).

5. Feferman's conceptualism: Feferman's conceptual structuralism is based on the belief that the general ideas of order, succession, collection, relation, rule, and operation are pre-mathematical. Likewise, Grassmann's conceptual constructivism distinguishes pre-mathematical operations between concepts (some general notions of composition) from mathematical operations.

According to Feferman, the basic objects of mathematical thought exist only as abstract mental conceptions resulting from processes that are independent of the concrete objects to which they are applied, and based on pre-mathematical concepts such as relations, rules. and operations (Feferman 2009). Grassmann might substantially agree on several of Feferman's 10 theses that characterize his version of conceptual structuralism.[32]

As in Feferman's version of conceptual structuralism, mathematics does not concern only universal or relational concepts, but also particular concepts considered as autonomous thought forms. The focus is on the procedures of concept formation.

4.3.3. Grassmann's Challenges to Contemporary Structuralism

The comparison between Grassmann's epistemology and contemporary philosophical structuralism can be used both to better understand Grassmann's philosophy and to consider whether new challenges might derive from his "obsolete" perspective.

Grassmann certainly contributed to the development of methodological structuralism. He criticized the traditional definition of mathematics as a science of magnitudes, and even if he still associated it with particular thought forms, he considered the latter to be determined by their generating law applied to an initial element. Grassmann clearly separated pure from applied mathematics, and developed a formal analysis of certain properties of connections that can be found in all mathematical branches. Even if, *sensu stricto*, he did not axiomatize mathematics, he individuated certain side conditions of the general connections that can be considered as invariant under specific kinds of transformations.

From a philosophical perspective, Grassmann's general theory of forms and the general definition of multiplication that occurs in ET can be interestingly compared to a non-eliminative structuralism associated with a constructivist ontology, as for example Parsons's or Feferman's structuralism. With the latter Grassmann would share the idea that the basic objects of mathematical thought exist only as abstract mental conceptions resulting from processes that are independent of the concrete objects to which they are applied, and based on pre-mathematical concepts such as relations, rules, and operations (Feferman 2009). With the former Grassmann would share the idea that mathematical forms (including numbers and extensive magnitudes) are the objects that mathematics talks about (Parsons 2004, 57).

Even if most questions related to the development of philosophical structuralism, such as Benacerraf's dilemma on natural numbers, cannot really be

[32] See in particular Feferman's theses 1, 2, 3, 8, and 9 (2009, 3).

compared with Grassmann's pre-set-theoretic approach, the epistemological challenge is taken into account in his constructivism. So Grassmann's most interesting contributions to contemporary structuralism might be seen in several challenges: (*a*) find a constructivist alternative to the *ante rem* / *in re* ontology, (*b*) verify whether a form of conceptualism might explain how mathematicians talk about structures without wholly abstracting from their instantiations, (*c*) consider the foundational role of series in mathematical and scientific thought, (*d*) develop an investigation of the differences between what we have called concept structuralism and object structuralism.

References

Banks, Erik C. 2013. Extension and Measurement: A Constructivist Program from Leibniz to Grassmann. *Studies in History and Philosophy of Science Part A* 44(1), 20–31.

Boi, Luciano, Dominique Flament, and Jean-Michel Salanskis, eds. 1992. *1830–1930: A Century of Geometry. Epistemology, History and Mathematics*. New York: Springer.

Brigaglia, Aldo. 1996. The Influence of Grassmann on Italian Projective n-Dimensional Geometry. In Schubring 1996a, pp. 155–164.

Burgess, John P. 2015. *Rigor and Structure*. New York: Oxford University Press.

Cantù, Paola. 2003. "La matematica da scienza delle grandezze a teoria delle forme: L'*Ausdehnungslehre* di H. Grassmann." PhD thesis, University of Genoa, January.

Cantù, Paola. 2008. Aristotle's Prohibition Rule on Kind-Crossing and the Definition of Mathematics as a Science of Quantities. *Synthese* 174(2), 225–235.

Cantù, Paola. 2011. Grassmann's Epistemology: Multiplication and Constructivism. In *From Past to Future: Graßmann's Work in Context*, edited by Hans-Joachim Petsche, pp. 91–100. New York: Springer.

Cantù, Paola. 2016. Peano and Gödel. In *Kurt Gödel: Philosopher-Scientist*, edited by Gabriella Crocco and Eva-Maria Engelen, pp. 107–126. Aix-en-Provence: Presses Universitaires de Provence.

Cassirer, Ernst. 1910. *Substanzbegriff und Funktionbegriff*. Berlin: Cassirer Verlag.

Cassirer, Ernst. 1923. *Substance and Function and Einstein's Theory of Relativity*. Chicago: Open Court.

Châtelet, Gilles. 1992. La capture de l'extension comme dialectique géométrique: Dimension et puissance selon l'*Ausdehnungslehre* de Grassmann (1844). In Boi et al. 1992, pp. 222–244.

Darrigol, Olivier. 2003. Number and Measure: Hermann von Helmholtz at the Crossroads of Mathematics, Physics, and Psychology. *Studies in History and Philosophy of Science* 34(3), 515–573.

Dorier, Jean-Luc. 1995. A General Outline of the Genesis of Vector Space Theory. *Historia Mathematica* 22, 227–261.

Echeverría, Javier. *1979*. L'analyse géométrique de Grassmann et ses rapports avec la *Caractéristique Géométrique* de Leibniz. *Studia Leibnitiana* 11(2), 223–273.

Engel, Friedrich. 1911. *Graßmanns Leben*. In Grassman 1894-1911, vol. 3.2.

Feferman, Solomon. 2009. Conceptions of the Continuum. *Intellectica* 51, 169–189.

Ferreirós, José. 2007. *Labyrinth of Thought: A History of Set Theory and Its Role in Modern Mathematics*. Basel: Birkhäuser.

Ferreirós, José. 2016. *Mathematical Knowledge and the Interplay of Practices*. Princeton, NJ: Princeton University Press.

Ferreirós, José. 2018. On Feferman's Conceptual Structuralism, and Objectivity. Manuscript.,

Flament, Dominique. 1992. *La lineale Ausdehungslehre* (1844) de Hermann Günther Grassmann. In Boi et al. 1992, pp. 205–221

Flament, Dominique. 1994. Préface. Hermann Günther Grassmann: L'homme et l'oeuvre. In Grassmann 1994, pp. 7–50.

Flament, Dominique. 2005. H. G. Grassmann et l'introduction d'une nouvelle discipline mathematique: L'*Ausdehnungslehre*. *Philosophia Scientiæ* 5(CS), 81–141.

Grassmann, Hermann Günther. 1844. *Die Wissenschaft der extensiven Grösse oder die Ausdehnungslehre, eine neue mathematische Disciplin dargestellt und durch Anwendungen erläutert. Erster Theil. Die lineale Ausdehnungslehre ein neuer Zweig der Mathematik dargestellt und durch Anwendungen auf die übrigen Zweige der Mathematik, wie auch auf die Statik, Mechanik, die Lehre von Magnetismus und die Krystallonomie erläutert*. Leipzig: Wigand. Reprinted in Grassmann 1894–1911, vol. 1.1.

Grassmann, Hermann Günther. 1845. Kurze Uebersicht über das Wesen der Ausdehnungslehre. *Grunert Archiv* 6, 337-350. Reprinted in Grassmann 1894–1911, vol. 1.1, pp. 297–312. Translated by W. W. Beman as A Brief Account of the Essential Features of Grassmann's Extensive Algebra, *Analyst* pp. 114–124, 1881.

Grassmann, Hermann Günther. 1855. Sur les différents genres de multiplication. *Crelle's Journal* 49, 123–141.

Grassmann, Hermann Günther. 1861. *Lehrbuch der Arithmetik*. Berlin: Enslin. Partially reprinted in Grassmann 1894–1911, vol. 2.1, pp. 295–349.

Grassmann, Hermann Günther. 1862. *Die Ausdehnungslehre, vollständig und in strenger Form bearbeitet*. Leipzig: Enslien. Reprinted in Grassmann 1894–1911, vol 1.2.

Grassmann, Hermann Günther. 1894–1911. *Gesammelte mathematische und physikalische Werke*. Edited by Friedrich Engel. 3 vols. Leipzig: Teubner.

Grassmann, Hermann Günther. 1994. *La science de la grandeur extensive*. Translated by D. Flament and B. Bekemeier, revised by E. Knobloch. Paris: Blanchard.

Grassmann, Hermann Günther. 1995. *A New Branch of Mathematics: The "Ausdehnungslehre" of 1844 and Other Works*. Translated by Lloyd C. Kannenberg. Chicago: Open Court.

Grassmann, Hermann Günther. 2000. *Extension Theory*. Translated by Lloyd C. Kannenberg. Providence: American Mathematical Society.

Grassmann, Robert. 1872. *Die Formenlehre oder Mathematik*. Stettin: Grassmann.

Grattan-Guinness, Ivor. 2011. Discovering Robert Grassmann (1815–1901). In Petsche et al. 2011, pp. 19–36.

Grätzer, George. 1968. *Universal Algebra*. Princeton, NJ: Van Nostrand.

Hankel, Hermann. 1867. *Vorlesungen über die complexen Zahlen und ihre Functionen*. Leipzig: Voss.

Heath, A. 1917. E. The Geometrical Analysis of Grassmann and Its Connection with Leibniz's Characteristic. *The Monist* 27, 36–56.

Hellman, Geoffrey. 1990. Modal-Structural Mathematics. In *Physicalism in Mathematics*, edited by D. Irvine, pp. 307–330. Dordrecht: Kluwer.

Helmholtz, Hermann von. 1887. Zählen und Messen, erkenntnistheoretisch betrachtet. In *Philosophische Aufsätze, Eduard Zeller zu seinem fünfzigjährigen Doctorjubiläum* gewidmet, pp. 11–52. Leipzig: Fues.

Hestenes, David. 1986. *New Foundations for Classical Mechanics*. Dordrecht: Kluwer.

Isaacson, Daniel. 2011. The Reality of Mathematics and the Case of Set Theory. In *Truth, Reference and Realism*, pp. 1–76. New York: Central European University Press.

Kant, Immanuel. 1787. *Kritik der reinen Vernunft*. 2nd ed. In *Gesammelte Schriften/ Akademieausgabe*, volume Abt.*1*, Werke, Bd.3, Berlin: Reimer, 1904. Translated by P. Guyer and A. Wood as *Critique of Pure Reason*. Cambridge: Cambridge University Press, 1988.

Klein, Felix. 1875. Review of Victor Schlegel, *System der Raumlehre, Erster Teil. Jahrbuch über die Fortschritte der Mathematik*, pp. 231–235.

Klein, Felix. 1979. *Development of Mathematics in the 19th Century*. Translated by M. Ackerman. Brookline, MA: Math Sci Press.

Lawvere, William F. 1996. Grassmann's Dialectics and Category Theory. In Schubring 1996a, pp. 255–264.

Lewis, Albert C. 1977. H. Grassmann's 1844 *Ausdehnungslehre* and Schleiermacher's *Dialektik*. *Annals of Science* 34(2), 103–162.

Nagel, Ernst. 1939. The Formation of Modern Conceptions of Formal Logic in the Development of Geometry. *Osiris* 7, 142–224.

Otte, Michael. 1989. The Ideas of Hermann Grassmann in the Context of the Mathematical and Philosophical Tradition since Leibniz. *Historia Mathematica* 16(1), 1–35.

Parsons, Charles. 1982. Objects and Logic. *The Monist* 65(4), 491–516.

Parsons, Charles. 2004. Structuralism and Metaphysics. *Philosophical Quarterly* 54(214), 56–77.

Peckhaus, Volker. 2011. Robert and Hermann Grassmann's Influence on the History of Formal Logic. In Petsche et al. 2011, pp. 217–227.

Petsche, Hans-Joachim. 2004. *Graßmann*. Basel: Springer.

Petsche, Hans-Joachim, A. C. Lewis, J. Liesen, and S. Russ, eds. 2011. *From Past to Future: Graßmann's Work in Context. Graßmann Bicentennial Conference, September 2009*. Berlin: Springer.

Radu, Mircea. 2003. A Debate about the Axiomatization of Arithmetic: Otto Hölder against Robert Grassmann. *Historia Mathematica* 30(3), 341–377.

Radu, Mircea. 2013. Otto Hölder's 1892 "Review of Robert Graßmann's 1891 Theory of Number." Introductory Note. *Philosophia Scientiae* 17(1), 53–56.

Reck, Erich, and Michael Price. 2000. Structures and Structuralism in Contemporary Philosophy of Mathematics. *Synthese* 125(3), 341-383.

Rowe, David E. 2010. Debating Grassmann's Mathematics: Schlegel versus Klein. *Mathematical Intelligencer* 32(1), 41–48.

Schleiermacher, Friedrich D. [1839] 1986. *Dialektik (1811)*. Berlin: Reimer. Reprint, Hamburg: Meiner.

Schlote, Karl-Heinz. 1985. H. Graßmanns Beitrag zur Algebrentheorie. *Janus* 72, 225–255.

Schlote, Karl-Heinz. 1996. Hermann Günther Grassmann and the Theory of Hypercomplex Number Systems. In Schubring 1996a, pp. 165–174.

Schreiber, Peter. 1995. Hermann Grassmann: Werk und Wirkung. In *Internationale Fachtagung anläßlich des 150. Jahrestages des ersten Erscheinens der "linearen Ausdehnungslehre (Lieschow/Rügen, 23.-28.5.1994)*. Greifswald: Universität Greifswald.

Schubring, Gert, ed. 1996a. *Hermann Günther Grassmann (1809–1877): Visionary Mathematician, Scientist and Neohumanist Scholar: Papers from a Sesquicentennial Conference*. Dordrecht: Kluwer.

Schubring, Gert. 1996b. The Cooperation between Hermann and Robert Grassmann on the Foundations of Mathematics. In Schubring 1996a, pp. 59–70.

Schubring, Gert. 2005. *Conflicts between Generalization, Rigor, and Intuition: Number Concepts Underlying the Development of Analysis in 17th–19th Century France and Germany*. New York: Springer.

Schubring, Gert. 2008. Graßmann (Vita mathematica). [Review]. *Mathematical Intelligencer* 30, 62–64.

Shapiro, Stewart. 1997. *Philosophy of Mathematics: Structure and Ontology*. New York: Oxford University Press.

Stein, H. 1988. Logos, Logic and *Logistiké*: Some Philosophical Remarks on Nineteenth Century Transformations in Mathematics. In *History and Philosophy of Modern Mathematics*, edited by W. Aspray and Philip Kitcher, pp. 238–259. Minneapolis: University of Minnesota Press.

Toebies, Renate. 1996. The Reception of Graßmann's Mathematical Achievements by A. Clebsch and His School. In Schubring 1996a, pp. 117–130.

Whitehead, Alfred North. [1898] 1960. *A Treatise on Universal Algebra, with Applications*. Cambridge: Cambridge University Press. Reprint, New York: Hafner.

Zaddach, Arno. 1994. *Grassmanns Algebra in der Geometrie*. Mannheim: Wissenschaft-Verlag.

3

Dedekind's Mathematical Structuralism: From Galois Theory to Numbers, Sets, and Functions

José Ferreirós and Erich H. Reck

In this essay, "mathematical structuralism" will be understood mainly as a style of work, a methodology for mathematics—but methodological choices can hardly be made without concern for the subject matter. Richard Dedekind's case was no exception to this rule. Thus his mathematical structuralism, which will be our main concern, was supplemented by a philosophical conception of mathematical objects.[1]

What is meant by "structure" in this context? Roughly, a structure is a relational system, a framework (*Fachwerk*, truss) of relations between elements—where the emphasis is on the relations (and relations of relations, etc.), in the sense that the same structure can be instanced by different kinds of elements. This rough sketch can be elaborated in a number of different ways, both mathematically and philosophically.

What, then, do we mean by "mathematical structuralism"? It is a style of work that takes results in a given branch of mathematics to emerge from the nature of relevant structures (exemplified therein), and often typically, from certain interrelations between structures of different kinds. A clear and paradigmatic example, also for Dedekind, is Galois theory, as we will see.

The essay will proceed as follows: After some background on Dedekind's main forerunners (§1), we will consider structuralist themes in his approach to Galois theory and algebraic number theory (§2). Then we will turn to his rethinking of the real numbers (§3) and the natural numbers (§4), within a general framework of sets and functions. The essay will end with a brief summary and conclusion (§5).

[1] The way in which we use "mathematical structuralism" in this essay makes it closely related to "methodological structuralism" in Reck and Price (2000); cf. the editorial introduction to this volume. We also use "style" in a methodological and epistemological sense, as opposed to a personal, national, or merely aesthetic one; cf. Mancosu (2017) for a general discussion.

José Ferreirós and Erich H. Reck, *Dedekind's Mathematical Structuralism* In: *The Prehistory of Mathematical Structuralism.* Edited by: Erich H. Reck and Georg Schiemer, Oxford University Press (2020). © Oxford University Press.
DOI:10.1093/oso/9780190641221.003.0003

1. Forerunners: Gauss, Dirichlet, and Riemann

As just indicated, a core ingredient of mathematical structuralism is the emphasis on relations, as opposed to objects standing in those relations. Henri Poincaré is well known for having written in *Science and Hypothesis*: "Mathematicians do not study objects, but the relations between objects; to them it is a matter of indifference if these objects are replaced by others, provided that the relations do not change" (Poincaré [1902] 2011, 20). Less well known is the fact that he said this as a preparation for explaining Richard Dedekind's account of the real numbers as defined by cuts.[2] Yet this point of view has deeper roots, also reaching further back than Dedekind.

By 1900, a structuralist approach was natural for many mathematicians, especially those, like Poincaré, used to working with group theory; similarly for Hilbert and mathematicians influenced by his application of the axiomatic method to geometry. But already in the 1820s, C. F. Gauss had argued that "mathematics is, in the most general sense, the science of relations" (Gauss [1917] 1981, 396).[3] This is so since "the mathematician abstracts entirely from the nature of the objects and the content of their relations; he is concerned solely with counting and comparison of the relations among themselves" (Gauss [1831] 1863, 176).[4] In another pregnant remark, he wrote that some mathematical results should be obtained "from notions [i.e., concepts], not from notations" (quoted in Dedekind 1895, 54). At the same time, Gauss's style of doing mathematics was still mostly classical; and while he took care to reformulate some existing theories in terms of pregnant "notions" (such as the congruence relation, ≡, in number theory), his writings often seem more calculational than structural.

Around 1850, several German mathematicians insisted that one ought to "put thoughts in the place of calculations", as Dirichlet wrote in his obituary of Jacobi. In other words, they adopted the principle—later attributed by Hermann Minkowski to Dirichlet himself—of obtaining mathematical results with a "minimum of blind calculation, a maximum of clear-seeing thought" (quoted in Stein 1988, 241). And by the end of the 19th century it had become customary to speak of a *conceptual approach* to mathematics in this connection, as opposed to more calculational approaches.[5] Riemann and Dedekind,

[2] See the essay on Poincaré in the present volume for more.

[3] Our translation; in the original German: "Die Mathematik ist so im allgemeinsten Sinne die Wissenschaft der Verhältnisse [in der] man von allem Inhalt der Verhältnisse abstrahiert" (Gauss [1917] 1981, 396).

[4] Our translation; in the original German: "Der Mathematiker abstrahirt gänzlich von der Beschaffenheit der Gegenstände und dem Inhalt ihrer Relationen; er hat es bloss mit der Abzählung und Vergleichung der Relationen unter sich zu thun" (Gauss 1831, 176).

[5] For more on the opposition between a "conceptual" and a more "computational" approach to mathematics, cf. Stein (1988), Laugwitz (2008), also Tappenden (2006), Reck (2016).

two young mathematicians influenced directly by Dirichlet, adopted this principle wholeheartedly. They also gave it a particularly abstract twist, or as one might say, a philosophical bent.

The initial model in this connection was Dirichlet's work from the 1830s, specifically his contributions to analytic number theory and the theory of trigonometric series. Gustav Lejeune Dirichlet is not as well known today as he deserves; but his mathematical results were "jewels" (Gauss in an 1845 letter to Humboldt)[6] that greatly influenced the development of mathematics. Moreover, his lectures—recorded, edited, and published by Dedekind—were highly influential and celebrated for their conceptual clarity. When he proved a theorem, one would never get lost in a jungle of calculations; instead, one would come away with clear insight into the chain of reasons, into the crucial steps that make the result possible. In addition, in Dirichlet's work on Fourier series (1829) he promoted analysis with more rigor than Cauchy. He was able to prove the existence of a Fourier-series representation for any function that is continuous and does not oscillate too often. Crucially, this result *necessitated* a "conceptual approach," since the goal was to establish the existence of a series representation merely from some very general traits of functions.

Dirichlet's application of methods from analysis to pure number theory (1837) was also greeted as an impressive novelty. The first example was his theorem that there are infinitely many primes of the form $a + n \cdot b$, with a and b coprime. The key point here is that recourse to certain functions in analysis (called L-series) was presented as indispensable; thus a result about finite numbers could only be obtained via a detour through the infinitesimal calculus. This stimulated much thought about the foundations of mathematics, especially by Kronecker and Dedekind. Dirichlet's own conclusion seems to have been that *pure mathematics is just arithmetic*, i.e., that all of analysis and algebra is nothing but a heavily developed number theory. Thus, as Dedekind later recalled, in the 1850s he often heard Dirichlet say that any result of algebra or analysis, no matter how complex or apparently remote, could be reformulated purely as a theorem about the natural numbers ([1888a] 1963a, 35). This would, among others, justify the application of analytic methods to number theory in a deep way, implying that there is nothing "impure" in it.

One more aspect of these contributions by Dirichlet is crucial for our purposes. It is his conceptual approach to mathematics that led him to emphasize the idea of an *arbitrary function*. Up to then, a "function" was supposed to be given explicitly by means of a formula, say polynomial or a concrete infinite

6 And "one does not weigh jewels on a grocer's scales" (Biermann 1977, 88).

series. However, Dirichlet defined a function to be a "law" according to which "to any x there corresponds a single finite y", i.e., an arbitrary correspondence of numerical values (Ferreirós 1999, 148). A function f is then *continuous* if small variations of x correspond to small variations of $f(x)$. Assuming now that, within an interval, the function f is bounded, is continuous except in finitely many points, and has finitely many maxima and minima, Dirichlet established that there is a Fourier-series representation for it.

A general way to understand this result is that the notion of function representable by a Fourier series, which makes it "calculational", is tantamount to a notion defined more abstractly or *conceptually*, namely that of a piecewise continuous, piecewise monotone function f. This is how Bernhard Riemann presented the matter in the introduction to his PhD thesis on the theory of analytic (complex-valued) functions. As such, Dirichlet's approach constitutes a substantial triumph for the conceptual style of thinking. Riemann then made it his programmatic goal to base the theory of *complex* functions on a similarly conceptual starting point, leaving the development of explicit "forms of representation" for the very end of the treatment. Here is how he characterized the resulting methodological perspective:

> Previous methods of treating these functions were always based on an expression for the function, taken as its definition, which determined its value for *each* value of the argument. Our investigation has shown that, as a consequence of the general characteristics of [analytic] functions of a complex variable, in such a definition a part of the determining elements follows from the rest; and the extension of those determining elements has been reduced to what is strictly necessary. This simplifies their treatment considerably. To give an example, in order to establish the equality of two different expressions for the same function, it was necessary to transform one into the other, that is, to show that they coincided for each value of the variable magnitude; now it is sufficient to show their coincidence in a far more restricted domain.
>
> A theory of such functions in accordance with the foundations established here would determine the configuration of the function (that is, its value for each value of the argument) independently of forms of determination by means of operations; to the general concept of a[n analytic] function of a complex magnitude, one would only add the necessary traits for determining the function, and only afterwards would one move on to the different expressions which the function admits. What is common to a species of functions that have been expressed in a similar way by means of operations would then be represented by means of boundary and discontinuity conditions. (Riemann [1851] 1876, §20, 38–39; our trans.)

The basis for Riemann's approach was his definition of an *analytic function* via the Cauchy-Riemann conditions,[7] together with his study of functions by means of their associated Riemann surfaces, plus some additional conditions regarding points of discontinuity (poles and singularities) and boundary conditions.[8]

The association of a *Riemann surface*—a geometric or, better, topological object—with each analytic function was a very fruitful move too, but one that remained somewhat mysterious at the time. In retrospect it can be regarded as another step toward mathematical structuralism: the study of one kind of object (a complex function) by associating it with an object of a different kind (a surface in n-dimensional space). The price paid by Riemann and his followers was foundational worries concerning the nature of these novel objects, which required the development of n-dimensional geometry and topology in order to be fully resolved. Finally, applying the same kind of methodology to the study of Euclidean space, Riemann subsumed the latter under the much richer and quite abstract concept of *continuous (and differentiable) manifold*, endowed with a certain metric.[9] The idea here was to look for further conditions so as to gradually narrow the scope of spaces falling under this general concept, thereby clarifying the nature of the assumptions behind Euclidean geometry and its links to other recently developed geometries, like the non-Euclidean one of Lobatchevsky-Bolyai, or even more generally, to geometries in spaces of variable curvature (cf. Ferreirós 2006).

In Riemann's work, the conceptual style of doing mathematics became very explicit and exclusive. As a consequence, it was criticized by other mathematicians—most importantly by Weierstrass and his Berlin school—who wanted to remain closer to the previous concrete and constructive style of mathematics.[10] In particular, Weierstrass gave preference to explicit representations of functions by means of power series. He argued, among others, that the class of differentiable functions had not been characterized completely yet (constructively, as one should add); and along such lines, the definition of analytic functions given by the Cauchy-Riemann conditions was not entirely satisfactory. For Dedekind, in contrast, the example of Riemann's style of mathematics became the model to emulate. Thus, when Dedekind makes his most committed

[7] These conditions say, in essence, that a function is *analytic* or holomorphic if and only if it is differentiable (in the complex domain); they state: $\frac{\partial u}{\partial x} = \frac{\partial v}{\partial y}$ and $\frac{\partial u}{\partial y} = -\frac{\partial v}{\partial x}$ for $u + vi = f(x + yi)$.

[8] Thus, if the function has discontinuities only in isolated points, and they consist in its "becoming infinite with finite order," then the function "is necessarily algebraic" and vice versa.

[9] Cf. Scholz (1999). An n-dimensional manifold is currently defined as a topological space that, locally, behaves like Euclidean space—but globally it won't in general be like $\mathbb{R}^n$. Riemann introduced the idea in connection with his reflections about n-dimensional geometry: they generalized the idea of a 2-dimensional surface to three and more dimensions. The Riemann surfaces are 2-dimensional manifolds, and it may be impossible to embed them in Euclidean space.

[10] Cf. Bottazzini and Gray (2013), 320–324, and Tappenden (2006), 108–122.

statements about mathematical method, there is typically a reference to Riemann involved, as in the following example:

> In these last words, if they are taken in their most general sense, we find the expression of a great scientific thought: the decision for the inner in contrast to the outer. This contrast also comes up in almost all fields of mathematics. One only has to think of function theory, of Riemann's definition of functions by means of characteristic inner properties, from which the outer forms of representation arise with necessity. But in the much more limited and simple field of ideal theory too, both directions have their validity. (Dedekind 1895, 54–55)[11]

Similarly, in the preface to his 1871 book on algebraic number theory, which contains his first presentation of ideal theory, Dedekind expressed his "hope that the effort to obtain characteristic basic concepts [*das Streben nach characteristischen Grundbegriffen*], which has been crowned with such beautiful success in other areas of mathematics, may not have eluded me completely" (Dedekind 1930–32, 3:396–397, our trans.). The same outlook is presented in a letter to Lipschitz from 1876, again with reference to Riemann.[12]

Dedekind was exposed to Dirichlet's and Riemann's conceptual style of thought during his time as privatdozent at the University of Göttingen. This proved to be a crucial experience for him. Not only did he later repeat Dirichlet's view that all of algebra and analysis is an extended form of arithmetic, as already noted; he also adopted his general notion of function (with consequences we explore further subsequently). And whenever it came to expressing his most deeply cherished methodological preferences, he wrote that his aim (in algebra, in number theory, etc.), like Riemann's in his theory of functions, was to base his results on "characteristic concepts," while letting concrete "forms of representation" emerge only as end products.

Actually, in Dedekind's hands the consistent promotion of such goals led even further—to a form of *mathematical structuralism*. This Dedekindian move brought with it novel set- and map-theoretic methods. But before we turn to those, the historical roots of yet another core ingredient of his mathematical

[11] Our translation; in the original: "In diesen letzten Worten liegt, wenn sie im allgemeinsten Sinn genommen werden, der Ausspruch eines großen wissenschaftlichen Gedankens, die Entscheidung für das Innerliche im Gegensatz zu dem Äußerlichen. Dieser Gegensatz wiederholt sich auch in der Mathematik auf fast allen Gebieten; man denke nur an die Funktionentheorie, an Riemanns Definition der Funktionen durch innerliche charakteristische Eigenschaften, aus welchen die äußerlichen Darstellungsformen mit Notwendigkeit entspringen. Aber auch auf dem bei weitem enger begrenzten und einfacheren Gebiet der Idealtheorie kommen beide Richtungen zur Geltung."

[12] From the letter to Litschitz (in our trans.): "My efforts in the theory of numbers are directed ... —though this comparison may sound pretentious—to attain in this field something similar to what Riemann did in the field of function theory" (Dedekind 1930–32, 3:468, our trans.). For additional remarks, compare, e.g., p. 296.

structuralism should be made explicit, namely: the systematic exploitation of relations, not just between particular mathematical objects, but between whole systems of objects as exemplified by Galois theory.

2. The Algebraic Context: From Galois Theory to Algebraic Number Theory

B. L. van der Waerden, the author of the classic textbook *Moderne Algebra* (1930), stated: "Galois and Dedekind are those who gave modern algebra its structure—the supporting skeleton of this structure comes from them" (1964, vii). Let us consider what he meant by that, including how Galois' and Dedekind's approaches are related. After that, we will turn to Dedekind's closely related use of Galois theory in his work on algebraic number theory.

After having finished his dissertation under Gauss in 1852, Dedekind remained in Göttingen as a privatdozent for six more years. He hesitated about what to do next. He also attended several of Dirichlet's and Riemann's classes so as to broaden and deepen his knowledge of mathematics. In 1855, he found his first great field of work: the contributions of Abel and Galois to higher algebra, into which he immersed himself, and especially, Galois's theory (first published, posthumously, in 1846 and quite difficult to understand at the time). In 1856–57 and 1857–58, Dedekind gave the first university courses in Germany on Galois theory. And it is here that he started to develop "the structural and conceptual methodology that will be characteristic of his whole mathematical work" (Scharlau 1981, 336, our trans.).

As is well known, algebra had been understood as the general theory of the symbolic resolution of equations for centuries; or as Isaac Newton put it, it was a kind of "universal arithmetic" that worked with a symbolic or literal calculus (in German: *Buchstabenrechnung*) instead of ordinary arithmetical calculations. It was primarily Galois' work in the early 19th century that led to a novel, much more abstract understanding of algebra—later often called "modern algebra"—in which the resolution of equations is relegated to the level of applications, while issues involving general theories of groups and fields come to the forefront (see Corry 2004, chap. 1). Dedekind played a crucial role in that development.

A central algebraic problem, from the 15th to the 19th century, was to find general methods for solving polynomials of any degree by means of radicals—just as the second-degree equation $ax^2 + bx + c = 0$ is solved by taking $x = (-b \pm \sqrt{(b^2 - 4ac)})/2a)$. Analogous resolutions were found for equations of degree 3 and 4 in the 16th century. But around 1800 mathematicians were convinced that a general solution, for all degrees n, is impossible to obtain. Lagrange, Ruffini, and Abel provided increasingly fine-grained analyses of this question, leading to Abel's proof that equations of degree 5 are in general not solvable (by radicals). This line of mathematics emphasized analyzing *permutations* of the roots of the equation at issue, and some expressions that remain *invariant* under such permutations. The very young Galois picked up on that approach, noting that all the permutations together form a *group*—a very innovative and rather abstract concept. This led him to associate with each equation its "Galois group" G, and then to investigate subgroups of G with particular attention to what later (by Heinrich Weber) would be called "normal" subgroups, which proved to be crucial.

Rather quickly, Dedekind obtained remarkable clarity in rethinking Galois' crucial innovation. Here is one passage in which he explains the path he took:

> During my first in-depth study of [the Gaussian theory of] cyclotomy[13] during the Pentecost holidays of 1855, I had, while well understanding all the details, to fight long and hard until I found the crucial principle in irreducibility; I only had to direct simple, natural questions at it so as to be led, with necessity, to all the details. Through a careful study of the algebraic investigations of *Abel* and, especially, *Galois*, and by my discovery, in early December of the same year, of the most general relation between any two irreducible equations, these thoughts were brought to a certain conclusion. Later I employed the method I had found also in the two winter courses on cyclotomy and higher algebra [given at Göttingen] in 1856–58. (Dedekind 1930–32, 3:414–415)[14]

Dedekind's lecture notes from these courses were only published, by Wilfried Scharlau, in the 1980s. In Scharlau's evaluation, his presentation of Galois theory—with its group-theoretic and field-theoretic foundations (see below in this section)—was far ahead of his time, even satisfying 20th-century expectations (Scharlau 1981, 341).[15] A similar level would only be achieved again in

[13] Cyclotomy is the study of roots of equations of the form $x^m = 1$, with m a positive integer. These roots are points on the unit circle (and thus cut it, "cyclotomy").

[14] Our translation; in the original German: "Bei meinem ersten gründlichen Studium der Kreisteilung in den Pfingstferien 1855 hatte ich, obgleich ich das Einzelne wohl verstand, doch lange zu kämpfen, bis ich in der Irreduktibilität das Prinzip erkannte, an welches ich nur einfache, naturgemäße Fragen zu richten brauchte, um zu allen Einzelheiten mit Notwendigkeit getrieben zu werden. Nachdem diese Gedanken durch eine eingehende Beschäftigung mit den algebraischen Untersuchungen von *Abel* und namentlich von *Galois* vervollständigt und durch die im Anfang Dezember desselben Jahres gelungene Auffindung der allgemeinsten Beziehungen zwischen irgend zwei irreduktiblen Gleichungen zu einem gewissen Abschluß gekommen waren, habe ich später in meinen beiden Wintervorlesungen über Kreisteilung und höhere Algebra 1956–1958 die damals gewonnene Methode befolgt."

[15] Dedekind's version of Galois theory was also much superior to contemporary ones, e.g., those by Betti or Serret (or Galois himself). It is comparable to the (often celebrated) Jordan (1870), but may be again superior to it as a presentation of the theory as a whole.

Heinrich Weber's *Lehrbuch der Algebra* (1895) and in Dedekind's Supplement XI to Dirichlet's *Vorlesungen über Zahlentheorie* (1894).[16]

Two ingredients of Galois theory and Dedekind's reception of it are of special importance for our purposes: the group-theoretic aspect of Galois's original contribution, developed further by Dedekind; and the introduction of the concept of a field. The latter was only implicit, thus still obscure, in Galois's writings, while Dedekind made it explicit and very central. Concerning the former, in his 1857–58 lectures Dedekind presents very clearly a theory of finite groups, which he already understands in a general, abstract way. Thus he writes:

> The following investigations are based solely on the two fundamental theorems just proven,[17] together with the fact that the number of substitutions is finite. Hence its results will be valid equally for *any domain* with a finite number of *elements, things, concepts* $\theta, \theta', \theta'', \ldots$ that admits of a *composition* $\theta\theta'$, from θ and θ', which is defined arbitrarily but so that $\theta\theta'$ is again a member of that domain and this kind of composition obeys the *laws* expressed in both fundamental theorems. In many parts of mathematics, but especially in number theory and algebra, we repeatedly find examples of this theory; and the same methods of proof are valid there too. (Scharlau 1981, 63, emphasis added)[18]

The structuralist flavor of this passage is undeniable. It is also not hard to see that the two theorems or laws mentioned suffice to axiomatize finite group theory. Dedekind then adds the idea of partitioning a group by a normal subgroup, with an induced law of composition. All of this is quite remarkable for the 1850s.

Dedekind introduces the notion of a field initially under the label "rational domain" (*rationales Gebiet*). The insight that, when studying an algebraic equation, one has to pay attention to the domain of numbers in which its coefficients

[16] One of the students attending the courses, Paul Bachmann, remarked about Dedekind: "In his calmly flowing, never halting presentation, [he was able to] present the theory with such exceptional clarity and simplicity that it was not hard for me to comprehend the material, then still quite foreign to me, despite its abstractness—the concept of group played a big role" (our trans.). In the original German: Dedekind was able "in ruhig fliessendem, niemals stockenden Vortrage die Theorien mit so ausnehmender Klarheit und Einfachheit [vorzutragen], dass es mir nicht schwer wurde, den mir damals noch ganz fremden Gegenstand trotz seiner Abstraktheit—der Gruppenbegriff spielte eine grosse Rolle—verständnisvoll zu erfassen" (quoted in Landau 1917, 53).

[17] The theorems in question state the associativity of the product, and a law of simplification: from any two of the three equations $\phi = \theta$, $\phi' = \theta'$, $\phi\phi' = \theta\theta'$, the third follows.

[18] In the original German: "Die nun folgenden Untersuchungen beruhen lediglich auf den beiden soeben bewiesenen Fundamentalsätzen und darauf, dass die Anzahl der Substitutionen endlich ist: Die Resultate derselben werden deshalb genau ebenso für ein Gebiet von einer endlichen Anzahl von Elementen, Dingen, Begriffen $\theta, \theta', \theta'', \ldots$ gelten, die eine irgendwie definierte Composition $\theta\theta'$ aus θ und θ' zulassen, in der Weise, dass $\theta\theta'$ wieder ein Glied dieses Gebietes ist, and dass diese Art der Composition den Gesetzen gehorcht, welche in den beiden Fundamentalsätzen ausgesprochen sind. In vielen Theilen der Mathematik, namentlich aber in der Zahlentheorie und Algebra, finden sich fortwährend Beispiele zu dieser Theorie; dieselben Methoden der Beweise gelten hier wie dort."

live, together with the domain containing its roots (regarded as different from the first), was due to Galois. The way in which he introduced them was by considering rational functions of given quantities supposed to be "known *a priori*"; as he writes: "We shall call *rational* any quantity which can be expressed as a rational function of the coefficients of the equation and of a certain number of *adjoined* quantities arbitrarily agreed upon" (quoted in Toti Rigatelli 1996, 119). Galois was not more explicit than that—but it was now relatively easy for mathematicians like Dedekind, or Kronecker, to go further and explicitly define fields. This can be done in different ways, and it is instructive to compare the contrasting styles involved.

We mentioned earlier that Weierstrass wanted to remain close to a "premodern", concrete, and calculational style of mathematics. The same applies, all the more, to Kronecker. He essentially followed Galois in defining a "domain of rationality" (*Rationalitätsbereich*) as the totality of quantities that are rational functions of some given quantities $r', r'', r''', \ldots$. Kronecker was explicit in preferring this kind of (constructivist) approach, via explicit expressions, to its more abstract alternative. Dedekind, in contrast, chose to emphasize the link between the notion of a "field" (*Körper*)—as he came to call it around 1870—and the "simplest arithmetic principles" (Dedekind 1930–32, 3:400). Thus, he defined a field as a set of numbers "closed in itself" under addition, subtraction, multiplication, and division. In doing so, he was *directly avoiding* any reliance on explicit expressions for numbers, since this would "spoil" (*verunzieren*) the presentation.

These two definitions are closely related but not exactly equivalent. Kronecker's "domains of rationality" are always engendered by *finitely* many elements $r', r'', r''', \ldots$, while Dedekind's "rational domains" or "fields" do not face such a restriction. As a consequence, the totality of *all* algebraic numbers is a Dedekindian field, but not a Kroneckerian domain of rationality; similarly for the field $\mathbb{R}$ of all real numbers, which was not accepted by Kronecker at all. Moreover, in Dedekind's treatment of what he called a "finite field", i.e., a finite extension of $\mathbb{Q}$, he was not happy with the definition that it is the extension of $\mathbb{Q}$ obtained by adjoining a number α, i.e., the set $\mathbb{Q}[\alpha]$ of all numbers $x_0 + x_1\alpha + x_2\alpha^2 + \ldots + x_{n-1}\alpha^{n-1}$ with coefficients $x_i \in \mathbb{Q}$. Instead, he preferred to call K a "finite field" over $\mathbb{Q}$ when there are only a finite number of subfields K' such that $\mathbb{Q} \subseteq K' \subseteq K$. This is again a *conceptual* definition. It is also one that directly points to an invariant property, as Dedekind was well aware (see Ferreirós 1999, 94). And again, explicit equations or "forms of representation" are relegated to being auxiliary means.

The contrast between the very different methodologies involved—Kronecker's constructivist approach and the conceptual/structural approach of Dedekind—became even clearer and more explicit in their divergent ways of dealing with ideal theory (or the "theory of divisors" in Kronecker's terminology). We will say more about the latter soon. But the style of Dedekind's work is already visible in general traits of his approach to Galois theory. Note, in addition, that Dedekind focuses on the basic foundations of the *whole theory*, i.e., on what we would call its structural underpinnings. In doing so, he relegated the study of concrete solutions of equations to a secondary role, thereby also departing from Galois.[19] What he was mainly concerned about was a general understanding of the *existence* and *nature* of such solutions, not concrete processes of solution.

It remains to highlight one further aspect of the shift from Galois to Dedekind. From today's point of view, the key moves in Galois theory are the following: (i) we associate with a given equation its Galois group G, so as to investigate its subgroups; (ii) we note that there is a correspondence between the subgroups of G and intermediate fields K (intermediate between the base field B, where the coefficients of the equation lie, and its extension E, containing all the roots of the equation); and (iii) we investigate the conditions for obtaining the splitting field E (as a finite extension of B) by studying the properties of the subgroups of G.[20] Galois introduced aspect (i), while (ii) and (iii) were added, and well understood, by Dedekind already in the 1850s. They also illustrate an element of mathematical structuralism we take to be central. Namely, a structuralist methodology often involves addressing *problems about certain structures* by studying their *interrelations with other structures*, perhaps of a *different kind*; and these structural correspondences may require the introduction of *novel objects* along the way.[21] We would like to highlight this aspect especially, since it is often ignored or at least underemphasized by philosophers of mathematics in discussing structuralism.

During the 1860s, a period in which Dedekind moved from Göttingen to Zürich for his first salaried position and then back to his hometown of Braunschweig as professor, he came to view the concept of a number field as the central object of study for algebra. This was consistent with the arithmetizing orientation he had encountered in Dirichlet's work, which guided his research on pure mathematics from early on (like that of several other mathematicians at the time: Weierstrass, Cantor, etc.). To provide outsiders at least with a rough sketch of this conception of algebra, he wrote in 1873 that it deals with the "algebraic [family] relations between numbers" or, better, that it is "the science of [family] relations between fields" (Dedekind 1930–32, 3:409).[22] In particular, the

[19] Fragments of Galois's writings that were oriented more toward this question included details not given by Dedekind (e.g., about irreducible equations of prime degree); cf. Scharlau (1981, 107).

[20] For a classic presentation of Galois theory along such lines, cf. Artin (1942).

[21] Concerning the latter, cf. the introduction of Riemann surfaces. Concerning the former, this amounts to studying relevant morphisms and functors (in category-theoretic language).

[22] Our translation; in the original German: The new algebra deals "von den algebraischen Verwandtschaften der Zahlen"; it is "die Wissenschaft von der Verwandtschaft der Körper."

properties of equations studied both traditionally and in Galois theory can be reconceived as properties of fields and their interrelations (base field, splitting field), as previously noted.

As Scharlau remarked (1981, 106), Dedekind was close to publishing the first textbook of "modern algebra", with a careful redaction of his 1856–58 notes on Galois theory. He failed to do so only because he found "an even more interesting" field of work in algebraic number theory, to which he then directed most of his energies. The exact date of the redaction at issue is not fully clear, but it seems safe to assume that it must have been finished by 1860, if not earlier. In any case, the structure of Dedekind's carefully written notes is distinctive and instructive. Its first section contains an investigation of the group-theoretic results needed in Galois-theoretic algebra; the concept of a (finite) group is isolated and investigated separately; and both are given an abstract, fully general presentation.

What is characteristic here, and a constant in Dedekind's subsequent writings, is this: while investigating a given area of mathematics, he was always on the lookout for *new concepts* that might be useful; and when he became convinced that a certain new idea was needed, he would isolate it and develop its *general theory* separately. As another example, his 1877 presentation of ideal theory begins with a section entitled "Auxiliary Theorems from the Theory of Modules" (in which he introduces an antecedent of the more general 20th-century concept of R-module, where R is a ring);[23] and in all later presentations, the theory of modules forms a section of its own, rising to a rather central role in his 1894 version of ideal theory.

Galois theory remained important for Dedekind's work in algebraic number theory. His first approach to the latter was in terms of a combination of the principles of Galois with a theory of "higher congruences" (Dedekind 1930–32, 3:397).[24] Algebraic numbers are those numbers (real or complex) that are roots of a polynomial with rational coefficients, e.g., $\sqrt{-3}$ (root of $x^2 + 3$) or $\sqrt{1+\sqrt{5}}$ (root of $x^4 - 2x^2 - 4$). Now, in certain simple cases it was clear at the time which numbers should be regarded as algebraic *integers* in such contexts, e.g., numbers of the form $a + b\sqrt{3}$, with a, b integers.[25] But in general the situation was not so clear. Both Dedekind and Kronecker considered this issue; and each of them was helped by previous acquaintance with the concept of a field or "rational domain". Each realized that one has to go to the relevant field first, so as then to isolate the ring of integers in it (to use current terminology). As a consequence both hit on the right definition of an algebraic integer, namely a number (real or complex)

[23] Dedekind's "modules" are in fact $\mathbb{Z}$-modules, where $\mathbb{Z}$ is the usual ring of integers.

[24] Meant are polynomial congruences modulo a prime; cf. Haubrich (1992, chap. 8).

[25] Adjoining $\sqrt{3}$ to $\mathbb{Q}$, we obtain a number-field, denoted $\mathbb{Q}[\sqrt{3}]$, that is a finite extension of $\mathbb{Q}$. The numbers specified are the integers corresponding to that field.

that is the root of a *monic* polynomial with integral coefficients ($\sqrt{1+\sqrt{5}}$ is an example).[26]

When studying the number theory of certain algebraic integers and building on the cases treated by Gauss earlier, Ernst Kummer had found the following problem: one often ends up in a situation in which *prime* integers do not conform to our expectations. Dedekind later gives this simple example: In the domain of integers $\mathbb{Z}[\sqrt{-5}]$, the numbers 2, 3, $1 + \sqrt{-5}$ and $1 - \sqrt{-5}$ are indecomposable, i.e., they are not the product of two other integers of this kind. However, they do not behave like regular primes, for $2 \cdot 3 = (1 + \sqrt{-5}) \cdot (1 - \sqrt{-5}) = 6$, i.e., *unique decomposability* of integers into prime factors *fails*.[27] Kummer then had the brilliant idea of introducing "ideal numbers," objects that do not exist in the given domain of integers, but that, once assumed, allow us to recover the principle of unique decomposition.

The main issue in algebraic number theory on which Dedekind was working, from the 1860s on, was to develop an analysis of all the domains of algebraic integers in which the fundamental principle of unique decomposition holds. The core question became how to define Kummer's "ideal numbers" in a way that could be applied to any ring of integers and that was rigorous, e.g., by allowing us to introduce the product operation on them carefully and explicitly. Around 1860 he worked with a theory based on "higher congruences," as already mentioned, which led him close to that goal. However, he was not fully satisfied with this approach, both since it was not completely general and since it was not sufficiently conceptual. The key to his eventual success, 10 years later, was the *extensionalization* of the whole problem, in the sense of its analysis in *set-theoretic* terms. As he put it himself:

> I did not arrive at a general theory . . . until I abandoned the old, more formal approach completely and replaced it by another, one that departs from the simplest basic conception and fixes the eyes directly on the end. In that approach, new creations are not needed any more, like those of Kummer's ideal number. It is entirely sufficient to consider systems of really existing numbers, which I call ideals. The power of this notion rests on its extreme simplicity. (Dedekind 1877, 268, our trans.)[28]

[26] Monic means that the lead coefficient of the polynomial is 1, as happens in the case of $x^4 - 2x^2 - 4$.

[27] This refers to the *Fundamental Theorem* of number theory, due to Gauss, which holds for the regular integers (in **Z**) as well as for the Gaussian integers $a + bi$ (with $i = \sqrt{-1}$).

[28] Our translation; in the original French: "Je ne suit parvenu à la théorie générale . . . qu'après avoir entièrement abandonné l'ancienne marche plus formelle, et l'avoir remplacée par une autre partant de la conception fondamentale la plus simple, et fixant le regard immédiatement sur le but. Dans cette marche, je n'ai plus besoin d'aucune création nouvelle, comme celle du nombre idéal du Kummer, et il suffit complétement de la considération de ce système de nombres réellement existants, que j'appelle un idéal. La puissance de ce concept reposant sur son extrême simplicité."

That is to say, instead of considering an ideal number *p* in Kummer's sense, which was only a fiction introduced formally, Dedekind considers the totality of algebraic integers in the given ring divisible by *p*—which forms an infinite set. This set is called an *ideal A*. In some cases (those of principal ideals), it corresponds to a number in the ring that divides all the elements of *A*, but not in other cases. For Dedekind the task now became to find a simple definition of such ideals *A*; and he found that two conditions suffice: (i) sums and differences of elements of *A* are again elements of it; (ii) the products of elements of *A* with any integers in the ring are again in *A*. His new definition worked fully generally; and he proceeded to treat ideals (infinite sets) as if they were simple numbers, operating on them as "new arithmetical elements". Doing so allowed him to define the product of ideals; it also made possible the proof of the fundamental theorem for any ring of algebraic integers.

We will not go into further details concerning Dedekind's theory of ideals, since it has been analyzed extensively elsewhere.[29] But two general observations are worth adding in our context. First, the downside of Dedekind's success with his conceptual, set-theoretic, and structuralist techniques was that others at the time were puzzled by his very "abstract" moves. Those moves were natural for him, but foreign to most mathematicians of that generation. Consequently, his ideal theory was not accepted until the 1890s; and even then, David Hilbert, Adolf Hurwitz, and others preferred more formal approaches.[30] As late as 1917, Edmund Landau would remark that in a "modern lecture" aiming to prove the main results, without gaps but briefly, one would prefer Hurwitz's approach, and "Dedekind's definition of an ideal is not used as basic any more [*wird kaum noch zu Grunde gelegt*]" (Landau 1917, 59).

The merit of Hurwitz's more formalistic way was that it avoided "the long chain of classical concepts and theorems of Dedekind's, about field permutations [automorphisms], modules, modules of rang *n*, etc." (Landau 1917, 59). Dedekind published a paper on methodology (1895) in which he explained why his self-contained approach was to be preferred to the Hilbert-Hurwitz way of relying on established algebraic theories. But his structuralist methodology, exemplified by his contributions to Galois theory and algebraic number theory, only came into vogue in the 1920s and later, with works by Emmy Noether, Emil Artin, B. L. van der Waerden, etc.[31] Thus the "modern algebra" of the 1920s would take Dedekind's side—whence Noether's well-known phrase, "It's all in Dedekind already."

[29] Cf. Avigad (2006), earlier Edwards (1980) and Ferreirós (1999, 95–107).

[30] Hurvitz took inspiration from Kronecker's use of polynomial rings and the "method of indeterminates" (*Methode der Unbestimmten*). Hilbert followed that style in his famous *Zahlbericht*, which made it much less structuralist than Dedekind's work (see the introduction to Hilbert [1897] 1998).

[31] Cf. Corry (2004), as well as the essays on Noether, Bourbaki, and Mac Lane in this volume.

The second general observation is that similar "abstract" moves, which again elicited negative reactions, characterize Dedekind's contributions to more foundational issues, as we will see next. The latter also led him to a kind of logicism.

3. The Real Numbers: From Arithmetization to Dedekindian Logicism

Early in the 20th century, Charles Sanders Peirce called Dedekind, very aptly, a "philosophical mathematician." Or to quote him more fully: "The philosophical mathematician, Dr. Richard Dedekind, holds mathematics to be a branch of logic" (Peirce [1902] 2010, 32). Dedekind's logicism was developed in the context of reconceptualizing first the real and then the natural numbers. But it is illuminating to go back further, to Dedekind's earliest foundational reflections.

Dedekind's is a singular case in the history of mathematics, in our judgment, because of the intensity and the success with which he devoted himself to reshaping his discipline. Indeed, he worked on a systematic reshaping of all the "pure mathematics" of his time—arithmetic, algebra, analysis—a fact that has not been recognized enough so far.[32] In doing so, he set the stage for various 20th-century developments—by being a key precursor of Hilbert, Bourbaki, and, above all, "modern algebra". From the beginning of his career, Dedekind was deeply concerned about foundational issues in mathematics as well. In fact, foundational and more mainstream issues were intimately intertwined for him.

Dedekind's interest in foundations is already apparent in his habilitation lecture, whose topic was "the introduction of new functions in mathematics" (Dedekind 1854). In this lecture, he proposed a genetic viewpoint on the number systems, one according to which "the requirement of the unrestricted possibility of carrying through the indirect or inverse operations [subtraction, division, etc.] leads with necessity to the creation of new classes of numbers" (quoted in Ferreirós 1999, 218). However, the set-theoretic considerations typical of his later writings were not present in this discussion yet, which focused on how to redefine such operations rigorously and non-arbitrarily in expanded domains (e.g., how to extend the arithmetic operations from the positive and negative integers to the rational numbers). On the other hand, it is noteworthy that Dedekind speaks of new kinds of numbers as our "creations" already in this context. He also believed that the main difficulties in systematizing arithmetic begin with the imaginary numbers (Dedekind 1930–32, 3:434).

[32] For the rise of "pure mathematics" in this sense, including Gauss's role, cf. Ferreirós (2007).

Interestingly, the latter is an issue to which he would never contribute. The reason seems to be that he found a completely satisfactory solution just a few years later, while reading W. R. Hamilton.[33] Here the ordered pair <*a*, *b*> is not yet conceived as a set—but we are moving in that direction. In all likelihood, Dedekind was completely satisfied with this *reduction* of complex arithmetic to the arithmetic of the real numbers. And the same move became then a central part of his foundational project: to reduce expanded number-domains, together with their operations and laws, to simpler ones. The quintessential example—and a key advancement for the foundations of mathematics—can be found already in 1858, with Dedekind's new approach to the real numbers. But its results were only published in 1872, in his well-known essay *Stetigkeit und irrationale Zahlen*.[34]

As the details of this episode are again well known (or easy to find in the literature),[35] we will only highlight the core ideas. Dedekind starts by assuming that the arithmetic of the rational numbers $\mathbb{Q}$ (an ordered field) has been satisfactorily developed. His goal is to introduce "new arithmetic elements"—the irrationals—*in one step*, as a whole system. By only presupposing $\mathbb{Q}$, he thus reduces the newly created numbers (and their operations) to the rational numbers. In particular, Dedekind proves all the fundamental properties of the new number domain $\mathbb{R}$ based on the operations on and properties of the rationals: his 1872 essay contains a proof that $\mathbb{R}$ is an *ordered* field with the (topological) property of *continuity* or, in later terminology, *line-completeness*. An essential proviso, however, is this: Dedekind needs to regard as unproblematic that we can work set-theoretically with the totality of rational numbers—the *reduction* of $\mathbb{R}$ to $\mathbb{Q}$ is *by set-theoretic means*.

The key in Dedekind's approach to the real numbers is his concept of a *cut*: a Dedekind-cut $<A_1, A_2>$ on $\mathbb{Q}$ is a pair of (non-empty) sets A_1, A_2 such that each element of A_1 is *less than* any element of A_2, i.e., $\forall x \in A_1\ \forall y \in A_2\ (x < y)$. Crucially for him, cuts on the system of rational numbers are a "purely arithmetical phenomenon" (Dedekind [1888a] 1963, 35–36, 40). By presupposing as given also the *totality* of all Dedekind-cuts for the number-system $\mathbb{Q}$, we have

[33] Cf. Ferreirós (1999, 220–221). Hamilton, in the introduction to *Lectures on Quaternions* (1853), defined the complex numbers $a + bi$ as ordered pairs of real numbers <*a*, *b*>, including corresponding operations. In manuscripts by Dedekind from the 1860s, perhaps earlier, he defines the integers as pairs of natural numbers and the rationals as pairs of integers; cf. Sieg and Schlimm (2005).

[34] Dedekind started teaching the calculus at the University of Zürich in 1858. It is in that context that he came up with his theory of cuts; cf. Dedekind (1872, 1), and Dedekind (1888a, 36). (In the English translation of the latter, 1853 is wrongly given as the relevant year).

[35] Besides the original Dedekind (1872), see, e.g., Courant and Robbins (1996, 71–72), Ebbinghaus et al. (1983, 30–31), or earlier Landau (1930, chap. 3). Dedekind was not the only mathematician working on this topic at the time, as mentioned by these writers; but his approach to it was quite original.

essentially introduced *the real number system* in its entirety—some cuts will correspond to rational numbers, while others will not, e.g., the cut $A_1 = \{x: x^2 < 3\}$, $A_2 = \{x: x^2 > 3\}$. It is by means of the latter that the irrationals numbers are introduced.

Dedekind's other central contribution in this context is his masterful definition of *continuity*: a set of elements S endowed with an ordering $<$ is continuous if and only if, given a corresponding cut of its elements into two (non-empty) classes C_1 and C_2 (as defined previously), there exists *one and only one* element c_0 of S that "produces" it. (This definition presupposes implicitly that S is a *densely ordered* set, a point that gave rise to some debate and misunderstandings at the time. Also relevant is the fact that Dedekind continuity implies the Archimedean property.)[36] A straight line in geometry, with an ordering of its points left-and-right, intuitively has the mentioned property: for any cut, there is a point that produces it.[37] As Dedekind established explicitly, the system of all cuts on $\mathbb{Q}$ has the property too.

Using the concept of a *field isomorphism*—present already in Dirichlet (1871), a year before the publication of *Stetigkeit und irrationale Zahlen*[38]—his procedure for introducing the system of real numbers can then be described as follows: $\mathbb{R}$ is defined as a *novel* number system isomorphic to the system of cuts on $\mathbb{Q}$. More specifically, we "create new numbers" corresponding to all the cuts, including those not produced by rational numbers, and together these form the system $\mathbb{R}$. The arithmetic properties of the real numbers, thus introduced, are derived rigorously from the arithmetic of the rational numbers; similarly for a linear ordering on $\mathbb{R}$, induced by that of $\mathbb{Q}$. And $\mathbb{R}$ can now be shown to be *continuous* in the precise sense introduced earlier (just like the system of cuts on $\mathbb{Q}$).

As emphasized already, (infinitary) set theory is functioning as a key background assumption in Dedekind's foundational work (also in his work in algebra and algebraic number theory). But how did Dedekind understand its status? Consider again his view that cuts are a "purely arithmetic phenomenon." Underlying it is the assumption that set theory is *pure logic*; and hence, set-theoretic constructions on $\mathbb{Q}$ are *pure arithmetic*, since we are allowed to employ all of logic's resources in it. It is on this basis that the phenomenon of cuts

36 This says that any positive number r, multiplied by itself n times, will be greater than any other number s. The Archimedean property excludes *infinitesimal* numbers.

37 In the introduction to Dedekind (1888a) he points out, however, that we can conceive of a geometric space that does not have this property, e.g., $\mathbf{A}^3$ where $\mathbf{A}$ is the set of algebraic numbers. This is relevant for evaluating Euclid's traditional approach to geometry.

38 The label "isomophism" is not Dedekind's, however. In 1871, he spoke of a field *substitution* (*Substitution*) instead. The term "isomorphism" was employed early on in crystallography; it was also used in Jordan (1870, 56) for groups. Compare http://jeff560.tripod.com/i.html.

appears "in its logical purity" according to him ([1888a] 1963, 40). Notice also, once again, that together with the set $\mathbb{Q}$ of rational numbers Dedekind assumes as given the totality of all cuts on $\mathbb{Q}$—a strong assumption equivalent to (an application of) Zermelo's power set axiom.

While controversial today, the idea that the concept of set is purely logical was common during Dedekind's time, e.g., in the tradition of the algebra of logic from Boole onward (cf. Ferreirós 1996; 1999, 47–53). Dedekind adopted this view early on, it seems, and it formed a key ingredient in his promotion of an early form of *logicism*. Already in a manuscript drafted in 1872, the same year in which his essay on the real numbers was published, he introduced sets in general as follows: "A *thing* is any object of our thought. . . . A *system* or *collection* [*Inbegriff*] *S* of things is determined when for any thing it is possible to judge whether it belongs to the system or not" (Dugac 1976, 293, our trans.). And in 1887, while preparing the final version of his essay on the natural numbers, he noted that the theory of sets, or "systems of elements," is "logic" (quoted in Ferreirós 1999, 225).

Because Dedekind regarded set theory as pure logic, the fact that the theory of the real numbers can be reduced to the arithmetic of the rational numbers by set-theoretic means implied for him that the notion of the continuum *does not* have to be seen as grounded in perception or geometric intuition. As he puts it, the number concept is "entirely independent of the intuitions of space and time" (Dedekind [1888a], 1963, 31); and the creation of the "pure, continuous number domain" ($\mathbb{R}$) is not dependent on the notion of magnitude. Instead, its creation takes the form of "a finite system of simple steps of thought" (340), and we get a "purely logical construction" (*Aufbau*) of arithmetic—in the broad sense, from $\mathbb{N}$ to $\mathbb{R}$, or even to the field $\mathbb{C}$ of complex numbers.

Clearly the set-theoretic reduction of the irrationals to more elementary number systems was a crucial step for Dedekind. It also seems that he was the first mathematician to consciously avoid reliance on the traditional notion of magnitude in this context (cf. Epple 2003). A further reason for this avoidance was a *requirement of purity*. As he wrote: "I demand that arithmetic shall be developed out of itself" (Dedekind [1872] 1963, 10) and, more particularly, "without any admixture of foreign ideas (such as that of measurable magnitudes)" ([1888a] 1963 35, trans. modified). Again, Dedekind's initial goal—delineated already in 1854, clarified while reading Hamilton, and encouraged by Dirichlet's approach—was to develop the complex number system starting from the natural numbers. Other contributors to "arithmetization," like Weierstrass, shared this goal; but unlike them, Dedekind realized this could be done with the help of set theory alone. Arithmetic is thus shown to be an outgrowth of the "pure laws of thought" (Dedekind [1882] 1963, 31).

Dedekind's version of logicism was highly influential during the 1890s—much more so than Frege's—by affecting authors such as Schröder and Hilbert.[39] The Peirce quotation given at the beginning of this section reflects this state of affairs. On the other hand, Dedekind's talk of "creation" has often been taken to throw doubts on the alleged *logical* nature of his point of view. And it has to be conceded that his way of expressing things sometimes runs the risk of conflating logic and psychology.[40] Was he then guilty of a problematic form of psychologism (as later criticized by Frege and Husserl)?[41] Dedekind was always convinced that mathematical objects and concepts are our "creations"—in his eyes, the prototype objects are numbers, and these are "free creations [*freie Schöpfungen*] of the human mind" ([1888a] 1963, 35; [1872] 1963, 4; also 1854). This was perhaps his most persistent philosophical conviction, from 1854 until his death.[42] Yet such talk about "the human mind" does not have to be understood in a subjectivist sense, as psychologistic thinkers are usually assumed to do. Instead, it can be interpreted in a Kantian or neo-Kantian way; it can thus be seen as a reference to our collective "mind" and its products, thus to human cognition and culture.[43] And as we will see in the next section, by 1888 the "creation" of the natural numbers consists merely in a step of *abstraction* from a more concrete "simply infinite set," so that strictly logico-mathematical results *determine* every single aspect of arithmetic.[44]

One final observation concerning the real numbers: how Dedekind proceeds in this context is closely related to his approach to *ideal theory*—methodologically the two are *of a piece*. Indeed, in a French essay of 1877 he explicitly compares the two cases (Dedekind 1877, 268–269). In both, we introduce new "arithmetical elements" in the progressive expansion of the number systems (although Dedekind does not "create" new objects corresponding to his set-theoretic ideals). And in both he is guided by the following desiderata: (1) "Arithmetic ought to be developed out of itself" ([1872] 1963, 10, trans. modified), thus avoiding any "foreign elements" and "auxiliary means" (magnitudes in the case of the reals, polynomials or other specific representations in the case of ideals).

[39] Cf. Ferreirós (2009), later also Reck (2013a).

[40] The same happens in Schröder's logical writings. And traces of it are still visible in Hilbert, e.g., when he writes: "We think [*wir denken*] of three sets [*Systeme*] of things" (Hilbert 1930, 2); similarly in his paper on the real numbers: "We think of a set of things [*Wir denken ein System von Dingen*]" (Hilbert 1900, 181). Notice the use of Dedekind's terminology in both cases.

[41] Cf. Reck (2013b) for related charges, as well as Dedekind's more general reception.

[42] In a letter to Weber of 1888, he wrote that we have the right to claim for ourselves such a creative power: "We are of divine lineage and there is no doubt that we possess creative power, not only in material things (railways, telegraphs), but quite specially in mental things" (Dedekind 1888b).

[43] Cf. the use of *Geisteswissenschaften* in German, later often translated as "cultural sciences." Many 19th-century philosophers were intensely concerned about them.

[44] Note also that, despite his frequent talk of "construction," Dedekind's basic tendency is not at all constructivistic (in the technical sense). As his theory of the real numbers shows, it is classical and objectivistic, just like Frege's. More on the underlying set theory in the next section.

(2) When new elements are introduced, they must be defined in terms of operations and laws found in the previously given domains (the arithmetic of $\mathbb{C}$ in the case of ideals).[45] (3) The new definitions must be completely general, applying "invariantly" to all relevant cases (we should not define some irrationals as roots, others as logarithms, etc.; we should not employ different means when determining ideal factors in various cases, as Kummer had done). (4) The definitions must offer a solid foundation for the deductive structure of the whole theory; they ought to be not just sound definitions, but the basis for rigorous proofs for all relevant theorems.

These four desiderata are closely related to Dedekind's mathematical structuralism, especially (3) and (4). Moreover, they guide his approach to the natural numbers too, as we will see in the next section.

4. Natural Numbers, Sets, and Functions: Logicism Systematized

While working on Galois theory and algebraic number theory, Dedekind distills out the core concepts of group and field, so as then to investigate them further abstractly and generally (similarly for the concepts of ideal, module, and, in later work, lattice). When developing his theory of the real numbers, his approach is similarly *conceptual*. The concept of field is again crucial in this context, but also that of continuity, defined in terms of cuts. Importantly, these concepts all involve global properties, which affects entire systems of objects—they are "structural" in that sense. We noted earlier that mathematical structuralism typically also involves the study of interrelations between such systems. This too is true for Dedekind's approach to the reals. Not only is an *isomorphism* (for ordered fields) between the system of cuts and that of the real numbers involved, at least implicitly;[46] his domain extension from $\mathbb{Q}$ to $\mathbb{R}$ also brings with it a corresponding *homomorphism*, as he is well aware. And while more heuristic than formally rigorous, his comparison of the reals with the intuitive geometric line involves such an interrelation too.

Dedekind's approach to the natural numbers in his 1888 essay displays the same general features; but there are also some noteworthy changes. In his approach

[45] This requirement was particularly critical at the time. Today we usually treat number systems axiomatically, but this is done (explicitly or implicitly) within the framework of set theory.

[46] Similarly, Dedekind acknowledges an isomorphism between his system of cuts and the reals constructed via (equivalence classes of) Cauchy sequences, as Cantor, Méray, etc. proposed. This is implicit in his remark (letter to Lipschitz, July 27, 1876) that Cantor and Heine have achieved the same goals as himself (reduction to the rational numbers, establishment of the continuity property), and that their expositions are different "only externally". See also Sieg and Schlimm (2017).

to algebra, algebraic number theory, and analysis, Dedekind always deals with subsets of the complex numbers (and related operations and functions). When dealing with the natural numbers, in contrast, he starts to consider sets (*Systeme*) of objects in *complete generality*. As he writes: "It very frequently happens that different things . . . can be considered from a common point of view, can be associated in the mind, and we say that they form a *system S*". Moreover, the concept of thing involved here is very inclusive: "I understand by *thing* every object of our thought" (Dedekind [1888a] 1963, 44). The other crucial aspect about sets *S* is that their identity is now understood *extensionally*—all that matters is that "it is determined with respect to every thing whether it is an element of *S* or not" ([1988a] 1963, 45). In a footnote Dedekind adds that a decision procedure is not required in this connection, thereby distancing himself from Kronecker. Clearly his notion of set is classical, not constructivist.

Parallel to this generalized notion of set, Dedekind introduces a generalized notion of function—or "mapping" (*Abbildung*). In his own words again: "By a *mapping* Φ of a system *S* we understand a law according to which to every determinate element *s* of *S* there *belongs* a determinate thing called the image of *s* and denoted $\Phi(s)$" (Dedekind [1888a] 1963, 50, trans. modified).[47] As Dedekind's use of the term "law" in this passage indicates, he is consciously building on Dirichlet's notion of function, while also broadening it even further (from an arbitrary functional correlation between sets of numbers to one between any two sets of objects). And unlike in axiomatic set theory, he does not reduce functions to sets of tuples; for him the notions of set and function are equally basic. Indeed, both belong to *pure logic*, in line with our earlier discussion. At a few points, Dedekind even seems to suggest that the notion of function or mapping is the really basic one.[48]

What Dedekind proposes in his 1888 essay is, thus, a general *logicist framework* in which to reconstruct arithmetic (from $\mathbb{C}$ all the way down). However, he does not formulate basic laws or axioms for it (as Frege was quick to point out).[49] Instead, he applies it in his reconstruction of the natural number sequence, i.e., in reducing the latter to logic. The core concept here is that of a *simply infinite system* (*einfach unendliches System*) which involves the concept of *infinity* for sets. Famously, a set *S* is (Dedekind-)infinite if it can be mapped 1-1 onto a proper subset of itself (Dedekind [188a] 1963, 63). A set *N* is simply infinite if,

[47] In W. W. Beman's translation (1963) of Dedekind (1888), *Abbildung* is rendered as "transformation," which seems awkward and is less appropriate than "mapping".

[48] As Dedekind writes, he was led to it by scrutinizing counting and numbers. It constitutes "an ability without which no thinking is possible"; and in particular, the entire science of numbers is built "upon this unique and in any event absolutely indispensable foundation" (Dedekind 1963, 32). He does not write anything as strong about the notion of set; compare Ferreirós (2017).

[49] Cf. Reck (2019), also for Dedekind's relation to Frege more generally.

in effect, there is an element *a* in *N* and a 1-1 function *f* on *N* such that $N = \{a, f(a), f(f(a)), \ldots\}$. More rigorously and formally, Dedekind's definition of being simply infinite involves four conditions, including one that uses the abstract concept of a "chain" to express a minimality condition on the set *N*, thus guaranteeing induction for simply infinite systems. [50] It is not hard to see that these four conditions constitute a (more abstract) variant of the Peano axioms—or better, the *Dedekind-Peano axioms*.

Dedekind's reconstruction of the natural numbers is again well known, so that we will only survey some highlights here (cf. Reck 2003). Important for him is to establish both the existence of a simply infinite system and (what we would call) the categoricity of that notion—the fact that any two simple infinities are isomorphic. In a well-known letter to Keferstein (Dedekind 1890), he clarifies that the former is meant to ensure the consistency of the notion of simple infinity. And with his categoricity theorem, Dedekind makes explicit an aspect not present yet in his earlier treatment of the reals. (Any two continuous ordered fields are isomorphic too, but this was not proved in 1872.) In addition, categoricity implies, as noted in passing, that exactly the same theorems hold for all simply infinite systems: i.e., the Dedekind-Peano axioms are semantically complete.[51] Finally, a careful justification for proofs by mathematical induction and for definitions by recursion is provided.

There are two controversial parts of Dedekind's 1888 essay. First, his (attempted) proof for the existence of a simply infinite system, which proceeds *via* arguing that an infinite system exists, relies on a universal set, which makes it fall prey to Russell's antinomy.[52] Second, Dedekind includes the following additional step not mentioned so far: start with a simply infinite system (any of them will do, since they are all isomorphic); then "neglect the special character of the elements, simply retaining their distinguishability and taking into account only the relations to one another" ([1888a] 1963, 68) It is exactly at this point in his essay that Dedekind adds: "With reference to this freeing the elements from every other content (*abstraction*) we are justified in calling numbers a *free creation of the human mind*" (68, emphasis added). However, it is not obvious how to interpret Dedekind's appeal to "abstraction" and "free creation," especially in a non-psychologistic way.

[50] Modernizing his notation slightly, the four conditions are the following: Consider a set *S* and a subset *N* of *S* (possibly equal to *S*). *N* is said to be *simply infinite* if there exists a function *f* on *S* and an element *a* in *N* such that (i) *f* maps *N* into itself; (ii) *N* is the minimal closure of $\{a\}$ under *f* in *S*; (iii) *a* is not in the image of *N* under *f*; and (iv) *f* is a 1-1 function. (Dedekind uses the notion of "chain" in (ii), to capture what it means to be the minimal closure of a set under a function.)

[51] Compare Awodey and Reck (2002a), also for a discussion of the history of these notions.

[52] Dedekind appeals to "the totality of things that can be objects of my thought" (1888a, 64). This may again sound psychologistic, but is meant objectively; cf. Klev (2018).

In correspondence from the same year, Dedekind makes clear that he takes his introduction of "the natural numbers" in his 1888 essay to be exactly parallel to his introduction of "the real numbers" in 1872, although the "abstraction" aspect has now been made more explicit (cf. Dedekind 1888b). Also, in both cases all resulting theorems are determined—entirely objectively—by the basic concepts involved, in the sense that it is determined what holds for any system of objects falling under them.[53] Beyond that, there are two interpretations of "Dedekind abstraction" that have been proposed in the literature. According to the first, a *novel* simply infinite system is introduced by it, a system isomorphic to but not identical with the one we started with and, in addition, determined "purely structurally." According to the second interpretation, such abstraction merely amounts to treating the *original* simple infinity in a certain way, namely by identifying it pragmatically as "the natural numbers," with the proviso that any other simple infinity could play the same role. [54] The case of the reals, or of continuous ordered fields, is parallel.

This essay is not the place to decide which interpretation of "Dedekind abstraction" is more defensible.[55] But with either one of them, we have arrived at a structuralist conception of *mathematical objects* that complements mathematical structuralism in the *methodological* sense; the latter leads to the former in Dedekind's writings, i.e., mathematical structuralism to philosophical structuralism. Turning our attention back to mathematical structuralism, note that, besides Dedekind's continued "conceptualism", the consideration of structure-preserving mappings (morphisms) between different systems of objects has become central in his foundational writings. This is most explicit in the categoricity theorem from his 1888 essay, which involves isomorphisms between any two simply infinite systems. A more implicit case is the treatment of recursive definitions and proofs by induction in it, which relates the natural number sequence to other recursively generated systems in terms of corresponding homomorphisms.[56]

By 1888, Dedekind has come to rely on a general framework of sets and functions for his mathematical structuralism. But as already noted, he does not formulate basic laws or axioms for it. There are some indications that implicitly

[53] As Dedekind writes: "The relations or laws, which are derived entirely from the conditions α, β, γ, δ in (71) are therefore always the same in all ordered simply infinite systems" (1963, 68). (For those conditions, see note 50.)

[54] The first interpretation amounts to reading Dedekind as a "non-eliminative structuralist," while the second amounts to reading him in an "eliminative" way; cf. Reck and Price (2000).

[55] A decision based on Dedekind (1888a) alone may be impossible; both sides can appeal to evidence in it. For the first reading, cf. Reck (2003); for the second, Sieg and Morris (2018). Dedekind may also have moved from one position to the other, i.e., changed his mind in this connection.

[56] From the perspective of category theory, Dedekind's procedure points toward thinking of **N** in terms of a corresponding universal mapping property; cf. McLarty (1993).

he works with a naive comprehension principle for sets.[57] Because of Russell's and related antinomies, this is no longer attractive to us. What one can still do is to carefully reconstruct which more restricted set-formation principles Dedekind actually needs for his overall project. This seems, in fact, to be exactly what Zermelo did while formulating his axiomatization for set theory in 1908. In retrospect, what Dedekind needs is the following: the power set axiom and an axiom of infinity;[58] principles for set-theoretic unions, intersections, or subsets more generally (an axiom of separation); some way of introducing or reconstructing n tuples; and less obviously, the axiom of choice and the axiom of replacement (missed by Zermelo originally).

As Dedekind's work brings out the importance of all these axioms, it makes sense that Zermelo, who knew the history well, considered modern set theory to have been "created by Cantor and Dedekind" (quoted in Ferreirós 1999, xii and 320). Today set theory is no longer considered to be "logic," however, among others because in its axiomatic form it is a specific mathematical theory.

5. Concluding Remarks

Our main concern in this essay has been Dedekind's mathematical structuralism, understood as a methodology or a style of doing mathematics. We can now summarize our main results briefly. From his teachers and mentors in Göttingen, especially Dirichlet and Riemann, Dedekind inherited a *conceptual* way of doing mathematics. This involves replacing complicated calculations by more transparent deductions from basic concepts. Both Dedekind's mainstream work in mathematics, such as his celebrated ideal theory, and his more foundational writings reflect that influence. Thus, he distilled out as central the concepts of group, field, continuity, infinity, and simple infinity. A related and constant aspect in his work is the attempt to characterize whole systems of objects through global properties.

From early on, Dedekind also pursued the program of the *arithmetization* of analysis—in the broad sense, from the complex numbers all the way down to the naturals. A decisive triumph came in 1858, with Dedekind's reductive treatment of the real numbers. From the 1870s on, he added a reduction of the natural numbers to a general theory of sets and mappings. This led to an early form of *logicism*, since he conceived of set theory as a central part of logic; i.e., the

[57] Or equivalently, he might work with a "dichotomy conception" where any division of the universe of objects into two parts creates corresponding sets; cf. Ferreirós (2017).

[58] Zermelo's axiom of infinity was modeled on Dedekind's controversial "proof"; he even called it "Dedekind's axiom." Its standard descendant, modified by von Neumann, still shows this origin.

reduction was ultimately to "the laws of thought." Moreover, in Dedekind's works there is a resolute reliance on the actual infinite—cuts, ideals, etc. are infinite sets. And while problematic in some respects, his attempt to execute a logicist program had a decisive effect on the rise of axiomatic set theory in the 20th century.

Its conceptualist and set-theoretic aspects are central ingredients in Dedekind's *mathematical structuralism*. But we emphasized another characteristic aspect that goes beyond both. This is the method of studying systems or structures with respect to their interrelations with other kinds of structures, and in particular, corresponding *morphisms*. A historically significant example, particularly for Dedekind, was Galois theory. As reconceived by him, in Galois theory we associate equations with certain field extensions, and we then study how to obtain those extensions in terms of the associated Galois group (introduced as a group of morphisms from the field to itself, i.e., automorphisms). Dedekind's more foundational works provide further examples, especially in terms of isomorphisms, such as his celebrated theorem that the Dedekind-Peano axioms are categorical, but also various *homomorphism* results involving the natural and real numbers.

As we saw, Dedekind connected his mathematical or methodological structuralism with a structuralist conception of mathematical objects, i.e., a form of philosophical structuralism (and the latter too involves categoricity results crucially). Central here was Dedekind's long-held view that mathematical objects, and paradigmatically numbers, are "free creations of the human mind," obtained by a kind of "abstraction" from more concrete systems of objects. With respect to Dedekind's logicism and his philosophical structuralism we acknowledged some controversial features. More can, and should, be said about both of them in the end. But we would like to conclude this essay with an observation of a different kind.

Dedekind's methodology was *not static*—it kept evolving. In fact, starting in the 1880s one can discern a subtle shift in his works, from focusing primarily on sets and set-theoretic constructions to taking functions and map-theoretic constructions as more fundamental (cf. Ferreirós 2017). However, there are only some hints to this effect in his writings, and officially both sets and functions remain basic. In addition, it was the aspects of his mathematical structuralism that we highlighted earlier with which he was most influential—on figures from Hilbert and Noether to Zermelo and Bourbaki. Finally, these aspects remain largely intact if one pushes mathematics further in a morphism-theoretic direction, as evidenced by 20th-century category theory and related developments.[59]

[59] Cf. Corry (2004), Awodey and Reck (2002b), and the essay on Mac Lane in this volume.

Acknowledgment

We would like to thank Georg Schiemer for helpful comments on an earlier version of this essay.

References

Artin, Emil. [1942] 1997. *Galois Theory*. 2nd ed. New York: Dover.

Avigad, Jeremy. 2006. Methodology and Metaphysics in the Development of Dedekind's Theory of Ideas. In *The Architecture of Modern Mathematics*, edited by José Ferreirós and Jeremy J. Gray, pp. 159–186. New York: Oxford University Press.

Awodey, Steve, and Erich Reck. 2002a. Categoricity and Completeness, Part I: 19th-Century Axiomatics to 20th-Century Metalogic. *History and Philosophy of Logic 23*(1), 1–30.

Awodey, Steve, and Erich Reck. 2002b. Categoricity and Completeness, Part II: 20th-Century Metalogic to 21st-Century Semantics. *History and Philosophy of Logic 23*(2), 77–94.

Biermann, K.-R., ed. 1977. *Briefwechsel zwischen Alexander von Humboldt und Carl Friedrich Gauß*. New ed. Berlin: Akademie Verlag.

Bottazzini, Umberto, and Jeremy J. Gray. 2013. *Hidden Harmony, Geometric Fantasies: The Rise of Complex Function Theory*. New York: Springer.

Corry, Leo. 2004. *Modern Algebra and the Rise of Mathematical Structures*. 2nd ed. Boston: Birkhäuser.

Courant, Richard, and Herbert Robbins. 1996. *What Is Mathematics? An Elementary Approach to Ideas and Methods*. 2nd ed. Revised by Ian Stewart. New York: Oxford University Press.

Dedekind, Richard. 1854. Über die Einführung neuer Funktionen in der Mathematik. Reprinted in Dedekind 1930–32, vol. 3, pp. 428–438.

Dedekind, Richard. 1872. *Stetigkeit und irrationale Zahlen*. Braunschweig: Vieweg. Reprinted in Dedekind 1930–32, vol. 3, pp. 315–334. English translation, *Continuity and Irrational Numbers*. In Dedekind 1963, pp. 1–22.

Dedekind, Richard. 1877. *Sur la théorie des nombres entiers algébriques*. Paris: Darboux. Reprinted in Dedekind 1930–32, vol. 3, pp. 262–296. English translation, *Theory of Algebraic Integers*. Edited and translated by J. Stillwell. New York: Cambridge University Press, 1996.

Dedekind, Richard. 1888a. *Was sind und was sollen die Zahlen?* Braunschweig: Vieweg. Reprinted in Dedekind 1930–32, vol. 3, pp. 335–391. English translation, *The Nature and Meaning of Numbers*. In Dedekind 1963, pp. 29–115.

Dedekind, Richard. 1888b. Brief an Weber. In Dedekind 1930–32, vol. *3*, pp. 488–490.

Dedekind, Richard. 1890. Letter to Keferstein. Reprinted in *From Frege to Gödel*, edited by Jean v. Heijenoort, pp. 98–113. Cambridge, Ma: Harvard University Press.

Dedekind, Richard. 1895. Über die Begründung der Idealtheorie. In Dedekind 1930–32, Vol. *2*, pp. 50–58.

Dedekind, Richard. 1930–32. *Gesammelte Mathematische Werke*. 3 vols. Edited by Robert Fricke, Emmy Noether, and Øystein Ore. Braunschweig: Vieweg.

Dedekind, Richard. 1963. *Essays on the Theory of Numbers*. Edited and translated by W. W. Beman. New York: Dover. Originally published Chicago: Open Court, 1901.

Dedekind, Richard. 1964. *Über die Theorie der ganzen algebraischen Zahlen: Nachdruck des elften Supplements*. Braunschweig: Vieweg.

Dirichlet, Gustav Lejeune. 1829. Sur la convergence des séries trigonométriques qui servent à représenter une fonction arbitraire entre des limites données. *Journal für reine Mathematik 4*, 157–169. Reprinted in Dirichlet 1889, pp. 283–306.

Dirichlet, Gustav Lejeune. 1837. Über die Darstellung ganz willkürlicher Functionen durch Sinus- und Cosinusreihen. *Repertorium der Physik 1*, 152–174. Reprinted in Dirichlet 1889, pp. 133–160.

Dirichlet, Gustav Lejeune. 1871. *Vorlesungen über Zahlentheorie*, 2nd ed. Edited, with supplements, by Richard Dedekind. Braunschweig: Vieweg.

Dirichlet, Gustav Lejeune. 1889. *G. Lejeune Dirichlet's Werke*. Vol. 1. Edited by Leopold Kronecker. Berlin: Reimer.

Dirichlet, Gustav Lejeune. 1894. *Vorlesungen über Zahlentheorie*. 4th ed. Edited, with additional supplements, by Richard. Dedekind. Braunschweig: Vieweg. Reprinted New York: Chelsea, 1968.

Dugac, Pierre. 1976. *Richard Dedekind et les Fondements des Mathématiques*. Paris: Vrin.

Ebbinghaus, H.-D., et al. 1983. *Zahlen*. Springer: Berlin.

Edwards, Harold. 1980. The Genesis of Ideal Theory. *Archive for History of Exact Sciences* 23, 321–378.

Epple, Moritz. 2003. The End of the Science of Quantity: Foundations of Analysis, 1860–1910. In *A History of Analysis*, edited by H. Jahnke, pp. 291–324. Providence, RI: American Mathematical Society.

Ewald, William, ed. 1996. *From Kant to Hilbert: A Source Book in the Foundations of Mathematics*. 2 vols. New York: Oxford University Press.

Ferreirós, José. 1996. Traditional Logic and the Early History of Set Theory, 1854–1908. *Archive for History of Exact Sciences* 50, 5–71.

Ferreirós, José. 1999. *Labyrinth of Thought: A History of Set Theory and Its Role in Modern Mathematics*. Basel: Birkhäuser.

Ferreirós, José. 2006. Riemann's Habilitationsvortrag at the Crossroads of Mathematics, Physics, and Philosophy. In *The Architecture of Modern Mathematics*, edited by José Ferreirós and Jeremy J. Gray, pp. 67–96. New York: Oxford University Press.

Ferreirós, José. 2007. The Rise of Pure Mathematics, as Arithmetic in Gauss. In *The Shaping of Arithmetic after C.F. Gauss' Disquisitiones Arithmeticae* edited by C. Goldstein, N. Schappacher, and J. Schwermer. Berlin: Springer, pp. 235-268.

Ferreirós, José. 2009. Hilbert, Logicism, and Mathematical Existence. *Synthese* 170, 33–70.

Ferreirós, José. 2016. The Early Development of Set Theory. *Stanford Encyclopedia of Philosophy*. Originally online 2007.

Ferreirós, José. 2017. Dedekind's Map-Theoretic Period. *Philosophia Mathematica* 25(3), 318–340. First online July 15, 2016.

Gauss, Carl Friedrich. 1831. *Theoria residuorum biquadraticorum, Commentatio secunda*. In *Nachrichten der Königlichen Gesellschaft der Wissenschaften zu Göttingen*. Reprinted in Gauss's *Werke*, vol. 2, pp. 169–178. Göttingen: Königliche Gesellschaften der Wissenschaften, 1863. English trans. in Ewald 1996, vol. 1, pp. 306–313.

Gauss, Carl Friedrich. [1917] 1981. *Werke*. Vol. 10.1, *Nachträge zur reinen Mathematik*. Edited by Königliche Gesellschaft der Wissenschaften zu Göttingen. Leipzig: Teubner; reprinted Hildesheim: Olms.

Haubrich, Ralf. 1992. Zur Entstehung der algebraischen Zahlentheorie Richard Dedekinds. Dissertation, University of Göttingen.

Hilbert, David. [1897] 1998. *The Theory of Algebraic Number Fields* [*Zahlbericht*]. Reprinted, with an introduction by F. Lemmermeyer and N. Schappacher, Berlin: Springer.

Hilbert, David. 1900. Über den Zahlbegriff. *Jahresbericht der Deutschen Mathematischen Vereinigung* 8, 180–194. Appendix to Hilbert 1930, pp. 241–246. English trans., On the Concept of Number. In Ewald 1996, vol. 2, pp. 1089–1095.

Hilbert, David. 1930. *Grundlagen der Geometrie*. 7th ed. Leipzig: Teubner. First edition 1899.

Jordan, Camille. 1870. *Traité des substitutions et des équation algébriques*. Paris: Gauthier-Villars.

Klev, Ansten. 2018. A Road Map to Dedekind's Theorem 66. *HOPOS: The Journal of the International Society of the History of Science* 8, 241–277. Available online August 17, 2018.

Landau, Edmund. 1917. Richard Dedekind. *Nachrichten der Königlichen Gesellschaft der Wissenschaften zu Göttingen*, 50–70.

Landau, Edmund. 1930. *Grundlagen der Analysis*. Leipzig: Akademische Verlagsgesellschaft.

Laugwitz, Detlef. 2008. *Bernhard Riemann: Turning Points in the Conception of Mathematics*. Basel: Birkhäuser.

Mancosu, Paolo. 2017. Mathematical Style. *Stanford Encyclopedia of Philosophy*. Originally online 2009.

McLarty, Colin. 1993. Numbers Can Be Just What They Have to Be. *Noûs* 27, 487–498.

Peirce, Charles Sanders. [1902] 2010. The Simplest Mathematics. In *Philosophy of Mathematics: Selected Writings*, edited by M. Moore, pp. 23–36. Bloomington: Indiana University Press.

Poincaré, Henri. [1902] 2011. *Science and Hypothesis*. New York: Dover.

Reck, Erich. 2003. Dedekind's Structuralism: An Interpretation and Partial Defense. *Synthese* 137, 369–419.

Reck, Erich. 2013a. Frege, Dedekind, and the Origins of Logicism. *History and Philosophy of Logic* 34, 242–265.

Reck, Erich. 2013b. Frege or Dedekind? Towards a Reevaluation of their Legacies. In *The Historical Turn in Analytic Philosophy*, edited by Erich Reck, pp. 139–170. London: Palgrave.

Reck, Erich. 2016. Dedekind's Contributions to the Foundations of Mathematics. *Stanford Encyclopedia of Philosophy*. Originally online 2008.

Reck, Erich. 2019. Frege's Relation to Dedekind: *Basic Laws* and Beyond. In *Essays on Frege's Basic Laws of Arithmetic*, edited by P. Ebert and M. Rossberg. New York: Oxford University Press, pp. 264–284.

Reck, Erich, and Michael P. Price. 2000. Structures and Structuralism in Contemporary Philosophy of Mathematics. *Synthese* 125, 341–383.

Riemann, Bernhard. [1851] 1876. *Grundlagen für eine allgemeine Theorie der Functionen einer veränderlichen complexen Grösse*. In *Gesammelte Mathematische Werke*, pp. 3–45. Leipzig: Teubner.

Scharlau, Wilfried, ed. 1981. *Richard Dedekind, 1831–1981: Eine Würdigung zu seinem 150. Geburtstag*. Braunschweig: Vieweg.

Scholz, E. 1999. The Concept of Manifold, 1850–1950. In *History of Topology*, edited by I. M. James, pp. 25–64. Amsterdam: Elsevier.

Sieg, Wilfried, and R. Morris. 2018. Dedekind's Structuralism: Creating Concepts and Deriving Theorems. In *Logic, Philosophy of Mathematics, and Their History: Essays in Honor of W. W. Tait*, edited by Erich Reck. London: College Publications, pp. 251–301.

Sieg, Wilfried, and Dirk Schlimm. 2005. Dedekind's Analysis of Number: System and Axioms. *Synthese* 147, 121–170.

Sieg, Wilfried, and Dirk Schlimm. 2017. Dedekind's Abstract Concepts: Models and Mappings. *Philosophia Mathematica* 25(3), 292–317.

Stein, Howard. 1988. *Logos*, Logic, and *Logistiké*: Some Philosophical Remarks on Nineteenth Century Transformations of Mathematics. In *History and Philosophy of Mathematics*, edited by W. Aspray and P. Kitcher, pp. 238–259. Minneapolis: University of Minnesota Press.

Tappenden, Jamie. 2006. The Riemannian Background to Frege's Philosophy. In *The Architecture of Modern Mathematics*, edited by José Ferreirós and Jeremy J. Gray, pp. 97–132. New York: Oxford University Press.

Toti Rigatelli, Laura. 1996. *Evariste Galois 1811–1832 (Vita Mathematica)*. Basel: Birkhäuser.

Van der Waerden, B. L. 1930. *Moderne Algebra*. Vol. 1. English trans., *Algebra, Vol.* I. New York: Springer, 2003.

Van der Waerden, B. L. 1964. Preface to *Über die Theorie der ganzen algebraischen Zahlen. Nachdruck des elften Supplements*, by Richard Dedekind. Braunschweig: Vieweg.

Weber, Heinrich. 1895. *Lehrbuch der Algebra*. Braunschweig: Vieweg.

4
Pasch's Empiricism as Methodological Structuralism

Dirk Schlimm

1. Introduction

A fundamental tension in philosophy of mathematics, one that goes back at least to Plato's *Meno*, is that between a view of mathematical entities as being abstract in nature and a view of knowledge as being of concrete (or causal) origin. Considered independently, each view can be quite appealing, but their combination raises the serious difficulty of giving a coherent account of mathematical knowledge. Abandoning, or at least substantially weakening, one of these views is a common move to resolve this dilemma. One thinker who resisted the urge to give up his conviction of the empirical origins of human knowledge was Moritz Pasch (1848–1930).

Throughout his life Pasch referred to his own philosophical outlook as *empiricist*. When he proclaimed in the introduction to his *Vorlesungen über neuere Geometrie* that "geometry is seen as nothing else but a part of natural science" (Pasch 1882b, 3), he meant this to be understood literally, in the sense that mathematical theories are based on empirical concepts. Geometric points, for example, are introduced at the beginning of his book as those physical objects that cannot be divided any further within the limits of what we can observe. Likewise, Pasch rejects the common demand that geometric lines should be imagined as being infinitely extended, since this precludes them from being (at least in principle) perceptible objects. Instead, he considers finitely extended line segments to be among the primitive objects of geometry (Pasch 1882b, 4). For him, all mathematical propositions, i.e., not only those of geometry, are ultimately formulated on the basis of observations of physical objects, and he maintains that we can understand the basic mathematical terms only by indicating appropriate objects ("den Hinweis auf geeignete Naturobjecte," Pasch 1882b, 16).[1] The further development of mathematics then proceeds through the deduction of propositions

[1] In this context, one also speaks of "ostensive definitions" of the primitive terms.

Dirk Schlimm, *Pasch's Empiricism as Methodological Structuralism* In: *The Prehistory of Mathematical Structuralism*. Edited by: Erich H. Reck and Georg Schiemer, Oxford University Press (2020). © Oxford University Press.
DOI:10.1093/oso/9780190641221.003.0004

and the definition of new concepts, and both of these processes sustain the epistemological status of their starting points.

Pasch's empiricist view of mathematics appears to be at odds with one of the basic tenets of structuralism, according to which mathematics is about purely abstract structures. Nevertheless, I will argue in the remainder of this chapter that Pasch's mathematical work drove him to adopt an approach that can justly be called "structuralist," in spite of the fact that his deeply held philosophical convictions seem to be incompatible with it. In order to do this, I will begin with discussing the notion of "methodological structuralism" (Reck 2003) and propose two minimal conditions that an approach has to satisfy to qualify as being structuralist (section 2). I will then look in more detail at Pasch's work in geometry (section 3) and the foundations of arithmetic (section 4) to ascertain that it does indeed satisfy the proposed conditions for minimal methodological structuralism. Thus, I conclude that Pasch's approach has its rightful place in an account of the prehistory of mathematical structuralism.

2. Minimal Methodological Structuralism

The notion of "methodological structuralism" was introduced by Reck (2003, 371) in order to distinguish the more ontologically oriented views on the nature of mathematics, like those expressed by Resnik (1997) and Shapiro (1997), from a certain way of practicing mathematics that is (in principle) independent of one's particular ontological commitments.[2] To assess whether a structuralist methodology can be found in the investigations of Pasch and others, it will be useful to identify some of its characteristic features.

A paradigmatic example of a structuralist methodology is the work in modern abstract algebra as presented by van der Waerden (1930). Reck describes this as follows:

> What modern algebraists do is to study various *systems of objects*, of both mathematical and physical natures (the latter at least indirectly), which satisfy certain general conditions: the defining axioms for groups, rings, modules, fields, etc. More precisely, they study such systems *as* satisfying these conditions, i.e., as groups, rings, etc. (2003, 371)

Thus, while an algebraist might explicitly discuss the field of complex numbers in her work, only those properties that are formulated in the field axioms and those

[2] See also the editorial introduction to this volume by Reck and Schiemer.

that follow from them are considered. That only the relations that are specified by the general conditions that define these systems are taken into account, but not any other properties that these objects might have as individuals, is what makes this approach *structural*. Here is how Reck formulates this idea:

> a methodological structuralist will not be concerned about the further identity or nature of the objects in the various systems studied. He or she will simply say: Wherever they come from, whatever their identities and natures, in particular whatever further "non-structural" properties these objects may have, insofar as a system containing them satisfies the axioms . . . , the following is true of it: . . . This is the sense in which methodological structuralism involves a kind of *abstraction*. Here abstraction concerns simply the question of which aspects of a given system are studied and which are ignored when working along such lines. (Reck 2003, 371)

Notice that for a methodological structuralist "abstraction" is not necessarily understood as a process that yields some kind of new abstract entities, but rather as an attitude of restricting oneself to taking into account only some features of the systems under investigation, while disregarding others. In sum, methodological structuralism can be described as the study of systems of objects that are characterized, or defined, axiomatically, with an exclusive focus on the relations that hold between these objects, while ignoring further questions about the nature of the objects. Dedekind's *Was sind und was sollen die Zahlen?* (1888) is a perfect example of an approach that falls under this definition (see Ferreirós and Reck in this volume). However, the insistence on axiomatic definitions seems to be too strong, and Reck himself adds the qualification that methodological structuralism is only "*typically* tied to presenting mathematics in a formal axiomatic way" (Reck 2003, 371; my emphasis). We should also note that the second condition formulated previously (i.e., the *focus* on relations) leaves open the possibility of pursuing structuralist investigations at one time and working along other, nonstructuralist lines at other times. Thus, methodological structuralism can be one particular approach among others pursued by the same mathematician; an approach that can be taken in certain investigations, but that can be ignored in others. It is the result of an attitude about how to conduct certain investigations that can be independent of one's philosophical conceptions of mathematics and the nature of mathematical objects.

With the refined understanding of methodological structuralism given in the previous paragraph we must confront the problem of triviality: Is anything at all excluded by the characterization or has now every mathematician become a methodological structuralist? For example, Euclid can be interpreted as having studied a system of points and lines in his planar geometry, taking into consideration only those relations between them that were licensed by his axioms. This suggests a further criterion to distinguish an approach that is explicitly intended to be structural from one in which axioms are used to describe a single system that is being studied.[3] On the one hand, Euclid investigated only one particular system, which consisted of idealized points and lines, and it seems fair to say that he did not envisage other systems of objects to satisfy the same relational properties. For Dedekind, on the other hand, it was clear that the natural numbers were only one particular instance of a simply infinite system and that there were others as well, like the system of his potential thoughts (*Gedankenwelt*). Similarly, in modern algebra groups and fields can be instantiated by many different systems, like numbers, rotations, etc. Based on these reflections, I propose the following two conditions that must be satisfied by investigations to count as being along the lines of a minimal version of methodological structuralism:

(1) Focus on *relational features* of systems of objects.
(2) The possibility of *multiple* systems that share these relational features must be envisaged.

With these two conditions in hand, we can now look at the works of particular authors and assess whether they qualify as being structuralist in methodology.[4]

3. Empiricist Structuralism in Geometry

Various aspects of Pasch's work in geometry appear to be congenial to methodological structuralism. Pasch presented in his *Vorlesungen über neuere Geometrie* (1882b) the first axiomatization of projective geometry in a way that is considered to be rigorous by contemporary standards. Indeed, Hilbert's axiomatization of Euclidean geometry, *Grundlagen der Geometrie* (1899), can be readily interpreted along structuralist lines (see Sieg's article on Hilbert in this volume) and was heavily influenced by Pasch. Moreover, Pasch famously also gave a characterization of the nature of deduction that emphasizes the relational features of the systems under investigation and which is worth quoting in full:

> In fact, if geometry is genuinely deductive, the process of deducing must be in all respects independent of the *sense* of the geometrical concepts, just as it must

[3] For a discussion of various roles and functions of axioms, see Schlimm (2013a), in particular 49–52 for their descriptive function.

[4] As far as I can tell, the approaches of the authors presented in this volume all satisfy the conditions for minimal methodological structuralism.

> be independent of figures; only the *relations* set out between the geometrical concepts used in the propositions (respectively definitions) concerned ought to be taken into account. (Pasch 1882b, 98)[5]

Pasch's insistence that, in order to be rigorous, deductions must be independent of the meanings of terms and instead rely only on their relational connections, as opposed to the particular meanings of the concepts, forms the cornerstone of his *deductivism*, which he himself referred to as "formalism" (Pasch 1914, 121). This approach meets the first condition for minimal methodological structuralism and thus appears to point to a general structuralist understanding of mathematics. In fact, his axiomatic standpoint has been interpreted as foreshadowing the idea that a system of axioms implicitly defines an abstract structure (Tamari 2007, 6 and 96). According to these indications, it seems straightforward to consider Pasch a methodological structuralist. However, Pasch also held an *empiricist* philosophy of mathematics, a brief sketch of which was given in the introduction to this chapter, which stands in stark contrast to the interpretation of axioms as implicit definitions and requires us to take a closer look at his works and adopt a more nuanced position.

The empiricist standpoint, according to which the fundamental concepts and propositions of mathematics are empirical in nature, is the background for most of Pasch's works, not only in geometry, but also in analysis. For Pasch, the basic, or "core" (*Kern*), propositions that form the starting points of a deductive presentation of mathematics are "directly based on observations" (1882b, 17) and "obtained through experience" (1914, 3). The content of a mathematical discipline like analysis, Pasch maintains, is constituted by facts; these can be derived from basic facts, which are themselves expressed by the basic propositions (1914, 3). However, despite his insistence on the empirical foundation of mathematics, Pasch quickly realized that a deductive development of mathematics cannot be carried out on the basis of empirical facts alone. This led him to distinguish between a mathematical set of axioms called a "stem" (*Stamm*) and a philosophically grounded, empirical set of axioms (first called "basic principles" and later a "core").[6] One of the reasons for this distinction was the observation that the axiomatic presentation of a mathematical theory does not necessarily determine the meanings of its primitive terms in a unique way. This insight was not based on some considerations of first-order logic or nonstandard models as we might be inclined to think nowadays, but on the duality of projective geometry, which was identified in the 1820s by Poncelet and Gergonne (Pasch 1914, 142). Duality

[5] All translations are by the author; translations of Pasch (1920b) and Pasch (1921) are based on those of Pollard (2010).

[6] See Schlimm (2010).

is the curious mathematical phenomenon in which, if the primitive terms (say of "point" and "line," and the relations "lying on" and "contains") of a theorem of projective geometry are interchanged, the result is again a theorem of projective geometry. In Pasch's words, the stem propositions for this discipline form a collection of propositions that is "transformed into itself" if the stem concepts of point and line are interchanged.[7]

> This fact, which is the source of duality, provides the proof that the group of projective stem propositions may not be considered as a definition of the projective stem concepts. Rather, it shows how the relations that are expressed by the projective stem propositions can be satisfied in more than one way. (Pasch 1914, 143)

Thus, the form of the axioms does not determine whether the term "point" indeed refers to points or to lines and, because the axioms of projective geometry cannot fix the meanings of the terms themselves, they cannot be regarded as their definitions.[8]

While some concepts may be defined by the propositions in which they occur, Pasch observes that it is not possible that all concepts could be defined in this way, because this would allow the possibility "that definitions can generate mathematical concepts out of nothing" (1914, 143). He elaborates:

> If one would want to claim that a totality of relations σ between concepts β, e.g., the basic propositions of arithmetic, could constitute a definition of the totality of concepts β, then one would have to be certain that the relations σ could not be satisfied in any other way than by the concepts β, excluding also the case where the concepts β are permuted. (1914, 143)

What Pasch explicitly rejects here is the understanding of a set of axioms (which govern the relations σ) as defining the primitives occurring in them (which refer to the concepts β), which is commonly referred to as an implicit definition. In fact, in reference to the first edition of Schlick's *Allgemeine Erkenntnislehre* (1918), which discusses Hilbert's approach to definitions by axioms, Pasch writes that

> the expression "implicit definition" has a different meaning when used by Mr. Schlick (definition by axioms). I have presented in §72 of *Veränderliche und*

[7] See Eder and Schiemer (2018).

[8] A similar argument is made by Frege in his correspondence with Hilbert (Frege 1976, 58–80).

Funktion [i.e., Pasch 1914, 142–143] the concerns that speak against a definition by axioms. (1920a, 145)

Readers should note that Pasch himself uses the term "implicit definition" in his writings, but in a different sense, namely in the sense of contextual definition, not in relation to axioms.[9] In particular, Pasch does not allow implicit definitions for the basic terms, but only for the introduction of new terms using basic or already defined terms. An implicit definition, in Pasch's sense, tells us how to replace an expression that contains a new term by an expression that does not contain it.[10] Pasch contrasts them with explicit definitions, whereby something that belongs to a genus is defined by specific marks (Pasch 1914, 20). Thus, Pasch's understanding of definitions, which is rooted in his empiricism, is clearly at odds with interpreting him as understanding axioms as implicit definitions of the class of their models or of an abstract structure.

In light of the preceding considerations, we can see how Pasch's move of distinguishing basic concepts and propositions from stem concepts and propositions allowed him to keep a deductivist view of mathematics (according to which a mathematical discipline is developed deductively from its stem), while at the same time retaining his convictions about empiricist foundations for mathematics (for the core). For Pasch, to demonstrate the viability of his empiricist philosophy of mathematics in general, each stem had to be connected to a core. For the case of projective geometry, Pasch showed in his *Vorlesungen* how the stem concepts and propositions can be linked to their empirical basis; I have referred to this project as "Pasch's Programme" (Schlimm 2010). In addition, because in a purely deductive development of a theory the stem propositions of a discipline can play the role of basic propositions (Pasch 1914, 121), Pasch can accommodate the observation that mathematicians can disagree on their philosophical views on the nature of mathematics while at the same time agreeing on the validity of proofs and theorems. After all, from a purely logical point of view, a theory can be developed from either a core or a stem, as long as they are consistent (Pasch 1924, 232).

One reason for Pasch's insistence on an empirical foundation of mathematics is his concern for its use in scientific and everyday applications.

> To apply mathematics, the basic concepts must refer to something that is present in the world of experience and for which the content of the basic propositions is meaningful and valid. We acknowledge this connection with experience as soon as we consider analysis to be something else than . . . an

[9] See Gabriel (1978) and Pollard (2010, 36–39).
[10] His introduction of the term *Menge* (set) is an example (Pasch 1914, 19).

> internally consistent construction [*einen Bau von innerer Folgerichtigkeit*]. (Pasch 1914, 138)

Thus, although Pasch allows for the possibility of working with meaningless terms in mathematics (if the stem is left uninterpreted), or with terms that refer to something other than empirical concepts and relations, he believes that a complete picture of mathematics should include an account of its applicability and that this is best given by empiricism. In addition, the latter removes any doubt about the arbitrariness of mathematics.

> The traditional view renders the mathematical point as a concept that does not refer to something real; I would like to call it a *hypothetical concept*. . . . Now, if hypothetical concepts and the assumed relations between them (hypothetical propositions, hypotheses) are applied to objects of nature, at first to drawn figures, then this remains something arbitrary as long as we do not formulate the laws that govern this application; hereby one has to put up with the imprecision that inheres in the application. It then becomes necessary to make *two different kinds of hypotheses*. Hypotheses of the first kind, which are those already mentioned, only relate the hypothetical concepts with each other, not with empirical ones; hypotheses of the second kind are to establish a bridge between hypothetical and empirical concepts. Compared to the empiricist way of proceeding this is nothing but a detour. (Pasch 1914, 139)

In short, the need for additional hypotheses that connect the mathematical stem concepts to their empirical counterparts when mathematics is applied is used as an argument in favor of using empirical concepts from the start. Notice how Pasch anticipates the need for connecting the scientific terms of a hypothetico-deductive theory with empirical referents; without any reference to Pasch, later philosophers of science referred to his hypotheses of the second kind as "coordinating definitions" (Reichenbach 1928, 31), "bridge laws" (Nagel 1961, chap. 11, sec. 2.3), or "bridge principles" (Hempel 1966, 72). Considerations of parsimony lead Pasch to skip the bridge laws and use empirical hypotheses directly.

Now, where does this discussion leave Pasch with regard to structuralism? He certainly would disagree that mathematics is about abstract structures. However, he would allow us to hold this view if we wanted to, but at the cost of having to explain how these structures can be applied to the world. Pasch himself clearly prefers an empiricist account of mathematics for which the problem of application does not arise. Nevertheless, his mathematical practice satisfies the two conditions for methodological structuralism laid out at the end of section 2: The focus on the relations that are expressed by the axioms of a mathematical discipline, not on the nature of its elements, is what guarantees the rigor of mathematical deductions. Mathematicians can develop their theories on the basis of stem concepts and propositions, which need not have a determinate reference, but can have multiple realizations instead; the paradigmatic example of such a theory is projective geometry. In fact, it was the duality of projective geometry that led Pasch to the distinction between a philosophically meaningful axiomatic foundation (consisting of core concepts and core propositions) and a mathematically sufficient axiomatic basis (consisting of stem concepts and stem propositions).

4. Empiricist Structuralism in Arithmetic

We have seen in the previous section how Pasch's insistence on rigorous deductions together with the surprising fact of the duality of projective geometry pushed him toward a minimal version of methodological structuralism, which he was able to combine with his empiricism about mathematics by separating purely mathematical axioms from philosophically grounded ones. In the present section I want to look at Pasch's work in a different mathematical discipline, namely arithmetic, in order to illustrate that the previous considerations were not unique to geometry, but arose also in other disciplines. This suggests that one ingredient for the emergence of structuralist views of mathematics was a particular attitude towards rigorous deduction that was developed in the 19th century.[11]

In addition to geometry, Pasch also worked on the foundations of arithmetic throughout his entire career; e.g., see Pasch (1882a, 1909, 1914, 1921, and 1924). The development of mathematical concepts on the basis of empirical ones through definitions and deductions, which Pasch presented for geometry, is the same approach he adopted for establishing the foundations of number theory and analysis. Here, too, Pasch aims at reducing the discipline to a core from which everything else can be derived. For him, such a reduction serves to present mathematics as a deductive discipline, justifies confidence in its consistency, allows us to assess its certainty, and forms the basis for any philosophical reflection about mathematics, such as the question of its relation to experience (Pasch 1921, 155).

Pasch disagrees with "the standard practice of putting a more or less finished notion of number at the beginning" of one's mathematical investigations (Pasch 1921, 155).[12] Instead, in his account "the natural numbers do not appear all of a

[11] See Gray (1992) and Detlefsen (1996) for a general overview of these developments.

[12] Contrast this with, for instance, the famous saying attributed to Kronecker that "God created the whole numbers, everything else is the work of man" (Weber 1893, 15).

sudden: they stand at the end of a long and difficult path" (Pasch 1920b, 4). Let us now briefly examine Pasch's account of natural numbers, as presented in his work on the origin of the concept of number.[13] Pasch's empiricist outlook is formulated clearly in the very first paragraph:

> The sort of thought process to be exhibited here might arise in any person who, first, considers only the *things* he himself perceives and distinguishes one from another and who, second, credits himself with eternal life and unlimited memory. Among the things observed by this person are his own actions. (Pasch 1920b, 1)[14]

We notice immediately that Pasch goes beyond assuming what is humanly possible, but instead posits an ideal agent with human-like cognitions, perceptions, and actions, but endowed with "eternal life and unlimited memory." While this move might seem striking at first, it has been popular among empiricists, who would otherwise have to restrict themselves to a finite (and in fact rather small) number of experiences; for example, a very similar starting point of a contemporary empiricist account of mathematical knowledge is Kitcher's *ideal subject* (Kitcher 1983, 109–111). As a careful systematizer, Pasch singles out 11 core concepts to describe the actions of the ideal agent: (1) *things*, (2) *proper names*, and (3) *collective names*, which are themselves things; the actions of (4) *specifying a thing*, (5) *assigning a proper name*, and (6) *assigning a collective name*—collective names can only be assigned to collections of things that were previously specified or assigned a proper name by the agent; any such action is (7) an *event*, which can be temporally related to other events by the relations (8) *earlier*, (9) *later*, and (10) *immediate successor*; finally, an ordered sequence of events forms (11) a *chain of events*. By considering names and events (both of which he considers to be things) in addition to physical objects Pasch frees the ideal agent from being restricted to what is physically present, and by considering experienced events the ideal agent is able to introduce order:

> I assume that I have experienced some events on which I confer the collective name *A*. By experiencing these events, I have registered observations about succession and immediate succession, about precedence and immediate precedence. But the events *A* also produce in me a comprehensive concept that

[13] Pasch's "Der Ursprung des Zahlbegriffs" was completed in 1916 and appeared in print in two parts, Pasch (1920b) and (1921), which were reprinted together in Pasch (1930a). The approach is based on the account given in Pasch (1909), but contains several modifications.

[14] The English translations in this section are taken from Pollard (2010).

> combines them into a whole, into a thing that I call *the chain of the events A* or, more briefly, $\mathfrak{A}$. (Pasch 1920b, 17)

So, while we may think of the events *A* as something like a finite set, $\{s_1, s_2, \ldots, s_n\}$, the chain $\mathfrak{A}$ of events *A* is more like a finite ordered set: $< s_1, s_2, \ldots, s_n >$. Given that Pasch allows the same thing to be given different names and be specified multiple times, he introduces the notion of a *line* for those chains whose elements are all specifications of different things.[15] The *members* of a line are those things that are specified by the elements of the line. Using these notions, Pasch introduces the concept of *number* as follows: to determine the number of a given collection *N*, first an arbitrary larger line $\mathfrak{Z}$ is obtained, whose members have the collective name *z* and whose first member is called *e*. Then,

> from among the members *z* that follow *e* I can specify one and only one member *n* such that the segment of $\mathfrak{Z}$ reaching as far as *n* is equivalent to the collection *N*.
>
> In addition to *N*, all and only the collections that are equivalent to *N* yield this member of the line $\mathfrak{Z}$.
>
> The thing *n* is called the *number* drawn from the line $\mathfrak{Z}$ for the collection *N*.
>
> Any *z* other than *e* can serve as "numbers." (Pasch 1921, 149)

After extending the use of the term "number" also to the member *e* of $\mathfrak{Z}$, and introducing the names "one," "two," "three," etc., for the members of $\mathfrak{Z}$, Pasch concludes:

> Now all the members of the line $\mathfrak{Z}$ have become numbers. Notches in a stick can serve as members of such a line. One notch must be singled out as the first, with all the remaining notches appearing to one side of it. The next member of the line is always the next notch over. (Pasch 1921, 150)

On the one hand, Pasch's example of notches on a stick nicely illustrates the empirical character that the natural numbers have for him; on the other hand, it also illustrates that for him the numbers are not one single, particular system of objects. In fact, it is compatible with this account that Julius Caesar is one of

[15] In (1909) Pasch used the terms *Folge* and *Reihe* (sequence and series), but he changed them in (1920b) to *Kette* and *Rotte* to avoid imbuing terms that already have multiple mathematical meanings with new meanings (Pasch 1920b, 17 and 19). While *Kette* translates straightforwardly as "chain," the term *Rotte* is less familiar and thus more difficult to translate. In a military formation, a *Rotte* consists of those soldiers or planes that are side by side; in this case an individual is called a *Glied*. Accordingly, Pollard (2010, 68) translates *Rotte* as "line" (and *Glied* as "member"), which we follow here, despite the fact that Pasch wanted to use a term that does not already have a mathematical meaning.

the members of 𝔍, and thus a number.[16] If a collection is empty, then there is no member of 𝔍 that can serve as the number of this collection. For this case, Pasch introduces the name "zero" as if it were the name of a thing (using an implicit definition, in Pasch's sense).

Pasch continues his account by introducing the figures "0", "1", . . . , "9", together with rules for obtaining greater numerals (technically, these are chains of specifications of figures) as distinct names. In this way only simple combinatorial processes are required to generate a potentially infinite list of names for numbers.

> For each number drawn from 𝔍, the figures yield a sign [*Zeichen*], and the sign yields a name. So figure-chains will satisfy our need for numerical signs in every case. . . . Conversely, any figure-chain one cares to construct can serve as a numerical sign, as long as I pick a sufficiently "large" 𝔍. (Pasch 1921, 152)

Thus, the system of numerals is a systematically obtained sequence of names that can be used to refer to the members of any chain of things that one decides to use as numbers. In the first exposition of this way of proceeding, Pasch leaves it at that, switching effortlessly and without much ado from numbers as things to their names (e.g., "If a number, (i.e., its name) consists only of nines, . . ." (Pasch 1909, 35); he understands a calculation to be the determination of a fixed name (e.g., in the decimal system) of a number that is given by an arithmetical expression (Pasch 1909, 53). Five years later, in 1914, Pasch is more careful and gives more explicit explanations. After noting that the construction of decimal place-value numerals yields names for each desired number, he notes:

> Once this is achieved, one can disregard which things and which chain of these things were originally used; one only needs to hold fixed *the names of these things, of the numbers*. . . . The decimal place-value name of an absolute whole number counts as a *fixed name*. (Pasch 1914, 33–34)

On the relevance of the decimal place-value system for the development of arithmetic and for everyday life, Pasch approvingly quotes at length a passage from Kronecker's "Über den Zahlbegriff" (Kronecker 1887, 355).[17] In 1921 Pasch reiterates the importance of numerals and the difference between numbers as things and their names, and here a more structuralist perspective emerges. He writes:

[16] Frege famously considered this to be a problem for a definition of numbers (Frege 1884, §55).

[17] Pasch spent two semesters in 1865–866 in Berlin, attending lectures by Kronecker and Weierstrass. He later mentions these as having exerted a great influence on his thinking about the foundations of mathematics (see Pasch 1930b, 7 and Schlimm 2013b, 189). As far I know, Pasch never expressed any explicit criticism of Kronecker's views of the natural numbers (but, see note 12).

> As we moved along, our starting point, the line $\mathfrak{Z}$, receded entirely into the background. We were no longer concerned with our original choice of *things* to serve as members of the line and, so, as numbers—nor did we care what things were added to the line to accommodate larger and larger numbers. We focused entirely on our need for names and signs for numbers of every size.
>
> Indeed, once the nomenclature for the natural numbers is secured, we can quite disregard whatever things might have gotten us to this point. We need only retain the *names* of these things to perform the task for which the natural numbers were intended: determining whether a collection is equal to another or is greater than it or less. (Pasch 1921, 153)

Although for Pasch the natural numbers continue to be a system of things, this system is not a specific, fixed one, nor does it matter which things we choose. It is tempting to speak in this context of an arbitrary choice of representatives, but that would be misleading: the chosen things do not represent numbers for Pasch, they *are* numbers. Nevertheless, we can see here a form of abstraction from the individual nature of the elements, which is characteristic of a structuralist approach. What matters is only the sequential arrangement, or the structure, of these things, their relations among each other. In addition, it is clear that multiple systems of things can instantiate the natural number structure, which is characterized by the line $\mathfrak{Z}$.

We have seen above that, in his more mature writings, Pasch clearly separates the numbers (which he conceives of as things) from their names, e.g., the decimal place-value numerals. While acknowledging that we can get by with a system of numerals, he does not go so far as identifying the numbers with the numerals themselves, in contrast to some of his contemporaries (e.g., Heine and Thomae, who advocated "formal" theories of arithmetic and were severely criticized by Frege).[18] In order to understand Pasch's account better, it will be useful to compare it to those of two contemporaries that he comments on, namely Alfred Pringsheim and David Hilbert.

In his lectures on number theory (1916), Pringsheim introduces numbers as an infinite "ordered system of signs [*Zeichen*] that satisfies certain rules for their combination" (Pringsheim 1916, vii), mentioning Heine and Helmholtz as other proponents of this view.[19] The simplest such system would be a tally system based on a single primitive sign, "|", but for reasons of practicality Pringsheim decides to use the decimal place-value system as the canonical system of natural numbers (Pringsheim 1916, 7).[20] Thus, Pringsheim does not consider the

[18] For a discussion of criticisms (including those by Frege) of this view, which Detlefsen calls "empiricist formalism," see Detlefsen (2005).

[19] See Heine (1872, 173) and von Helmholtz (1887, 21).

[20] For a critical review, see Hahn (1919).

system of natural numbers to be unique (because different systems of numerals would do), but determined only insofar as it obeys certain rules. Pasch mentions Pringsheim's use of decimal numerals approvingly, but he maintains that his own development of them is "completely different in its nature" (Pasch 1921, 153). How so? For Pasch numbers are not signs (numerals), but those things that the numerals refer to. He also does not want to put the numerals at the beginning of arithmetic, but presents the combinatorial concepts and propositions that underlie the use of numerals. Ultimately, Pasch's interests lie deeper, at the level of the combinatorial origins of numbers.

A few years later, Hilbert also put forward an account of arithmetic based on sequences of signs in his *Neubegründung der Mathematik. Erste Mitteilung* (1922). Soon afterward, Pasch gave a reconstruction of Hilbert's approach to arithmetic in light of his own (Pasch 1924). While he argues that formulas that look like Hilbert's axioms could be derived from his core propositions, Pasch objects to Hilbert's conception of the nature of mathematical objects. Hilbert proclaimed his philosophical standpoint on the foundation of pure mathematics as "at the beginning is the sign [*Zeichen*]", listing as his first definition that "The sign 1 is a number" (Hilbert 1922, 163). First, Pasch disagrees with Hilbert's conception of signs. Hilbert seems to consider signs (and in particular numbers) to be types of inscriptions themselves, whereas for Pasch a sign is an inscription type that denotes a thing. The connection between Hilbert's inscriptions and Pasch's view of numbers is that the former could be considered to be marks, just like the notches on a stick, that could serve as the members of the line $\mathfrak{Z}$ (Pasch 1924, 238). Second, Pasch replaces Hilbert's signs "1" and "+" by the aliases (*Decknamen*) "*e*" and "*u*", such that Hilbert's axioms would correspond to stem propositions, obtained from the core propositions by the process of formalization, i.e., the replacement of meaningful terms by meaningless ones (Pasch 1924, 237, 239–240). In other words, while Hilbert presents a *particular* instance of inscription types as numbers, in Pasch's account it is explicitly recognized that these are just one of many possible instantiations. Thus, by building the possibility of multiple instantiations into his account, Pasch's attitude is clearly more structuralist than Hilbert's, because it also satisfies the second criterion for methodological structuralism, namely envisaging multiple realizations, laid out in section 2.

5. Conclusion

In this chapter two conditions were put forward for a minimal version of methodological structuralism, namely (a) the focus on relational features of systems of objects and (b) envisaging the possibility of having multiple systems that share these relational features. Various factors pushed Pasch toward these two aspects of mathematics. In his work on geometry the quest for rigorous deductions led him to focus on the primitives and relations that are expressed by the axioms and to neglect any other properties that mathematical objects might have. The dualism of projective geometry forced him to accept the possibility that the axioms (stem propositions) can be satisfied by different systems of objects. In Pasch's work on the foundations of arithmetic a structuralist perspective emerged from the fact that the canonical names for numbers, namely the decimal numerals, could refer to any appropriate system of objects. Thus, despite the fact that Pasch maintained an empiricist standpoint, according to which all mathematical knowledge is grounded on experiences of physical objects, he nevertheless came to adopt a methodological structuralism that satisfies both conditions (a) and (b). The further development of structuralism toward a more ontologically oriented position regarding the nature of mathematics went well beyond anything that Pasch would have found acceptable. As Dehn remarks, "The fondness for operating with symbols that have gone far beyond what is intuitable has a mythical-revolutionary character; this was completely foreign to Pasch" (Engel and Dehn 1934, 128). What we see in Pasch's work is that methodological structuralism need not be driven by considerations of abstract structures like those found frequently in modern algebra and that it can be combined successfully with an empiricist philosophy of mathematics.

Acknowledgments

The author would like to thank the participants of the workshops "Origins of Mathematical Structuralism" and "Structuralism in 19th Century Mathematics," held in September 2015 and July 2017 in Vienna for discussions that helped to shape this chapter, and Daniel Lovsted and Georg Schiemer for comments on an earlier version. Work on this chapter was funded by Social Sciences and Humanities Research Council of Canada (SSHRC).

References

Dedekind, Richard. 1888. *Was sind und was sollen die Zahlen?* Braunschweig: Vieweg. Reprinted in Dedekind 1930–32, vol. 3, pp. 335–391. English translation by Wooster W. Beman, revised by William Ewald, in Ewald 1996, pp. 787–833.

Dedekind, Richard. 1930–32. *Gesammelte mathematische Werke*. 3 vols. Edited by Robert Fricke, Emmy Noether, and Öystein Ore. Braunschweig: Vieweg.

Detlefsen, Michael. 1996. Philosophy of Mathematics in the Twentieth Century. In *Philosophy of Science, Logic, and Mathematics*, edited by Stuart. G. Shanker, vol. 9 of *Routledge History of Philosophy*, pp. 50–123. New York: Routledge.

Detlefsen, Michael. 2005. Formalism. In *Oxford Handbook of Philosophy of Mathematics and Logic*, edited by Steward Shapiro, pp. 236–317. New York: Oxford University Press.

Eder, Günther, and Georg Schiemer. 2018. Hilbert, Duality, and the Geometrical Roots of Model Theory. *Review of Symbolic Logic* 11(1), 48–86.

Engel, Friedrich, and Max Dehn. 1931. *Moritz Pasch: Zwei Gedenkreden, gehalten am 24. Januar 1931*. Giessen: Töpelmann.

Engel, Friedrich, and Max Dehn. 1934. Moritz Pasch. *Jahresbericht der Deutschen Mathematiker Vereinigung (JDMV)* 44(5/8), 120–142. Reprint, with changes and additions, of Engel and Dehn 1931.

Ewald, William. 1996. *From Kant to Hilbert: A Source Book in Mathematics*. 2 vols. Oxford: Clarendon Press.

Frege, Gottlob. 1884. *Die Grundlagen der Arithmetik*. Breslau: Verlag von Wilhelm Koebner. English translation by J. L. Austin, *The Foundations of Arithmetic*. Oxford: Basil Blackwell, 1953.

Frege, Gottlob. 1976. *Nachgelassene Schriften und Wissenschaftlicher Briefwechsel*. Vol. 2: *Wissenschaftlicher Briefwechsel*. Edited by Gottfried Gabriel, Hans Hermes, Friedrich Kambartel, Christian Thiel, and Albert Veraart. Hamburg: Felix Meiner.

Gabriel, Gottfried. 1978. Implizite Definitionen: Eine Verwechslungsgeschichte. *Annals of Science* 35, 419–423.

Gray, Jeremy J. 1992. The Nineteenth-Century Revolution in Mathematical Ontology. In *Revolutions in Mathematics*, edited by D. Gillies, pp. 226–248. Oxford: Clarendon Press.

Hahn, Hans. 1919. Besprechung von Afred Pringsheim: Vorlesungen über Zahlen- und Funktionenlehre. *Göttingsche gelehrte Anzeigen* 9–10, 321–347. English translation in Hahn 1980, 51–72.

Hahn, Hans. 1980. *Empiricism, Logic, and Mathematics: Philosophical Papers*. Edited by Brian McGuinness. Dordrecht: D. Reidel.

Hallett, Michael, and Ulrich Majer, eds. 2004. *David Hilbert's Lectures on the Foundations of Geometry, 1891–1902*. New York: Springer.

Heine, E. 1872. Die Elemente der Functionenlehre. *Journal für die reine und angewandte Mathematik* 74, 172–188.

Hempel, Carl G. 1966. *Philosophy of Natural Science*. Englewood Cliffs, NJ: Prentice-Hall.

Hilbert, David. 1899. *Grundlagen der Geometrie*. Leipzig: Teubner. Reprinted in Hallett and Majer (2004), pp. 436–525. Translated as *The Foundations of Geometry* by E. J. Townsend. Chicago: Open Court, 1902. Translated as *The Foundations of Geometry* by L. Under. La Salle, IL: Open Court, 1971.

Hilbert, David. 1922. Neubegründung der Mathematik. Erste Mitteilung. *Abhandlungen aus dem Seminar der Hamburgischen Universität 1*, 157–177. Reprinted in Hilbert 1935, 157–177. English translation *The New Grounding of Mathematics: First Report*, by William Ewald in Ewald 1996, 1115–1134.

Hilbert, David. 1935. *Gesammelte Abhandlungen*, Vol. 3. Berlin: Springer.

Kitcher, Philip. 1983. *The Nature of Mathematical Knowledge*. New York: Oxford University Press.

Kronecker, Leopold. 1887. Über den Zahlbegriff. *Journal für die reine und angewandte Mathematik* 101, 337–355. English translation *On the Concept of Number*, in Ewald 1996, pp. 947–955.

Nagel, Ernest. 1961. *The Structure of Science: Problems in the Logic of Scientific Explanation*. New York: Harcourt, Brace and World.

Pasch, Moritz. 1882a. *Einleitung in die Differential- und Integralrechnung*. Leipzig: Teubner.

Pasch, Moritz. 1882b. *Vorlesungen über Neuere Geometrie*. Leipzig: Teubner.

Pasch, Moritz. 1909. *Grundlagen der Analysis*. Leipzig: Teubner.

Pasch, Moritz. 1914. *Veränderliche und Funktion*. Leipzig and Berlin: Teubner.

Pasch, Moritz. 1920a. Die Begründung der Mathematik und die implizite Definition. Ein Zusammenhang mit der Lehre vom Als-Ob. *Annalen der Philosophie* 2(2), 144–162.

Pasch, Moritz. 1920b. Der Ursprung des Zahlbegriffs. *Archiv der Mathematik und Physik* 28, 17–33.

Pasch, Moritz. 1921. Der Ursprung des Zahlbegriffs. *Mathematische Zeitschrift* 11, 124–156.

Pasch, Moritz. 1924. Betrachtungen zur Begründung der Mathematik. *Mathematische Zeitschrift* 20, 231–240.

Pasch, Moritz. 1930a. *Der Ursprung des Zahlbegriffs*. Berlin: Springer.

Pasch, Moritz. 1930b. *Eine Selbstschilderung*. Gießen: von Münchowsche Universitäts Druckerei.

Pollard, Stephen, ed. 2010. *Essays on the Foundations of Mathematics by Moritz Pasch*, New York: Springer.

Pringsheim, Alfred. 1916. *Vorlesungen über Zahlen- und Funktionenlehre*. Vol. 1: *Zahlenlehre, Abteilung 1: Reelle Zahlen und Zahlenfolgen*. Leipzig: Teubner.

Reck, Erich. 2003. Dedekind's structuralism: An interpretation and partial defense. *Synthese* 137, 369–419.

Reichenbach, Hans. 1928. *Philosophie der Raum-Zeit Lehre*. Berlin: Walter de Gruyter. Reprinted in Reichenbach 1977.

Reichenbach, Hans. 1977. *Philosophie der Raum-Zeit Lehre*. Vol. 2 of *Gesammelte Werke*. Edited by Andreas Kamlah and Maria Reichenbach. Braunschweig: Vieweg.

Resnik, Michael D. 1997. *Mathematics as a Science of Patterns*. Oxford: Clarendon Press.

Schlick, Moritz. 1918. *Allgemeine Erkenntnislehre*. Berlin: Springer.

Schlimm, Dirk. 2010. Pasch's Philosophy of Mathematics. *Review of Symbolic Logic* 3(1), 93–118.

Schlimm, Dirk. 2013a. Axioms in Mathematical Practice. *Philosophia Mathematica (Series III)* 21(1), 37–92.

Schlimm, Dirk. 2013b. The Correspondence between Moritz Pasch and Felix Klein. *Historia Mathematica* 40(2), 183–202.

Shapiro, Stewart. 1997. *Philosophy of Mathematics: Structure and Ontology*. New York: Oxford University Press.

Tamari, Dov. 2007. *Moritz Pasch (1843–1930). Vater der modernen Axiomatik. Seine Zeit mit Klein und Hilbert und seine Nachwelt. Eine Richtigstellung*. Aachen: Shaker Verlag.

van der Waerden, B. L. 1930. *Moderne Algebra*. 2 vols. Berlin: Springer.

von Helmholtz, Hermann. 1887. Zählen und Messen, erkenntnistheoretisch betrachtet. In *Philosophische Aufsätze Eduard Zeller zu seinem fünfzigjährigen Doktorjubiläum gewidmet*, pp. 7–52. Leipzig: Fues' Verlag. Reprinted in *Wissenschaftliche Abhandlungen*, vol. 3, pp. 356–391. English translation *Numbering and Measuring from an Epistemological Viewpoint* in von Helmholtz 1977, pp. 75–114.

von Helmholtz, Hermann. 1977. *Epistemological Writings*. Edited by Paul Hertz and Moritz Schlick. Dordrecht: D. Reidel. English translation of *Schriften zur Erkenntnistheorie*. Berlin: Julius Springer, 1921.

Weber, Heinrich. 1893. Leopold Kronecker. *Mathematische Annalen* 43(1), 1–25.

5

Transfer Principles, Klein's Erlangen Program, and Methodological Structuralism

Georg Schiemer

1. Introduction

Structuralism in the philosophy of mathematics holds that mathematics is the science of abstract structures. An alternative characterization of the position does not assume structures as the subject matter of mathematics, but rather holds that mathematical theories study only the structural properties of their objects.[1] The focus on such properties is closely related to criteria of *structural identity* of mathematical objects. Specifically, it is often held that objects that share the same structural properties should be identified. For instance, in the context of non-eliminative structuralism, this view figures prominently in recent debates on the identity of structurally indiscernible positions in a pure structure.[2]

As the present volume shows, there exists a rich and multifaceted mathematical prehistory of these philosophical debates. In particular, one can identify a number of methods and styles of reasoning in 19th-century mathematics that eventually led to a "structural turn" in the discipline.[3] The present article will focus on one important strand in the mathematical roots of structuralism, namely Felix Klein's group-theoretic approach to geometry outlined in his *Erlangen program* of 1872. Klein's program is generally acknowledged today as one of the milestone contributions in 19th-century geometry. Moreover, there is a consensus that his novel algebraic approach in geometry—that is, the study and classification of geometries in terms of transformation groups—had a

[1] These are usually characterized as properties not concerning the intrinsic nature of objects but rather their interrelations with other objects in a system. Compare, for instance, Benacerraf (1965) and Linnebo and Pettigrew (2014).

[2] See, for example, Keränen (2001) and Shapiro (2008). Compare Leitgeb and Ladyman (2008) for a critical discussion of this view.

[3] See the editorial introduction as well as Reck and Price (2000) for a general overview of relevant methodological developments in 19th-century mathematics.

Georg Schiemer, *Transfer Principles, Klein's Erlangen Program, and Methodological Structuralism* In: *The Prehistory of Mathematical Structuralism.* Edited by: Erich H. Reck and Georg Schiemer, Oxford University Press (2020). © Oxford University Press.
DOI:10.1093/oso/9780190641221.003.0005

significant impact on the gradual development of geometry into a science of abstract structures.[4]

Despite the wealth of research on Klein's program and its significance for subsequent developments in geometry, no close study has so far been dedicated to its specific structuralist underpinnings. In particular, Klein's work has not yet been discussed through the lens of modern structuralism.[5] In the present chapter, I want to fill this gap. In particular, I will address the following questions: how, precisely, did Klein contribute to the development of the structural turn in mathematics? In what sense was his group-theoretic approach to geometry structuralist in character? Finally, in what sense did Klein's account anticipate the philosophical debates in structuralism mentioned above?

The aim in this chapter is twofold. The first aim is historical in nature and concerns the geometrical background of Klein's program. In particular, my focus will be on work on duality phenomena in 19th-century projective geometry. The chapter will survey different attempts to justify the principle of duality and then describe two ways in which the principle was generalized in analytic geometry, namely Julius Plücker's contributions to "general reciprocity" and Otto Hesse's so-called transfer principles. Roughly speaking, transfer principles were conceived at the time as mappings between geometrical domains that allow one to translate theorems about configurations of the one domain into corresponding theorems about the second domain. As I will argue, Klein's group-theoretic account in the Erlangen program can be understood as a generalization of this work on reciprocity and transfer principles.

The second aim is more philosophical in character. This is to analyze in closer detail Klein's structuralist account of geometrical knowledge. I will argue here that his group-theoretic approach is best characterized as a kind of "methodological structuralism" regarding geometry (see Reck and Price 2000). Moreover, one can identify at least two aspects of the Erlangen program that connect his approach with present philosophical debates, namely (i) the idea to specify structural properties and structural identity conditions for geometrical figures in terms of transformation groups and (ii) an account of the structural equivalence of geometries in terms of transfer principles. Both ideas clearly present "structural methods" in the sense specified in Reck and Price (2000).

The article is organized as follows. Section 2 will discuss the geometrical background of Klein's program. Specifically, different ways to justify the principle of duality in projective geometry are outlined in section 2.1. In section 2.2, I discuss the use of transfer principles in analytic geometry. Section 3 will then turn

[4] See, e.g., Tobies (1981), Wussing (2007), and Gray (2008).

[5] See, however, Biagioli (2018) for a recent study of the Klein's structuralism underlying his work on non-Euclidean geometry.

to Klein's approach: section 3.1 focuses on his group-theoretic study of geometries in terms of invariants. In section 3.2, I present Klein's method of "transfer by mapping." Section 4 will then discuss several structuralist themes underlying Klein's conception of geometry. Section 4.1 will focus on Klein's account of geometrical properties and congruence specified relative to a group of transformations. In section 4.2, I discuss how Klein's use of transfer principles to identify geometries can be generalized to a notion of structural equivalence in category-theoretic terms. Section 5 contains a short summary.

2. Duality and Transfer Principles

The mathematical background of the Erlangen program is known to be rich and multifaceted.[6] Klein's group-theoretic approach in geometry has different roots, including algebraic work on permutations groups by Camille Jordan and Évariste Galois, Arthur Cayley's invariant-theoretic approach in geometry, as well as Sophus Lie's parallel work on geometry, to name just a few. A different influence on Klein's program concerns the development of projective geometry in the 19th century. Particularly relevant here are, as we will see, different contributions to the principle of duality as well as its generalization in work by Plücker and Hesse. In the present section, I will survey these methodological developments in projective geometry and Klein's reception of them.

2.1. The Principle of Duality in Projective Geometry

Projective geometry, as developed by Jean-Victor Poncelet, Gaspard Monge, Joseph Diez Gergonne, Karl G. C. von Staudt, and Moritz Pasch (among many others), can be characterized as the study of those geometrical properties of figures that remain invariant under certain projective transformations.[7] This approach with its focus on projective invariants was certainly relevant for Klein's subsequent characterization of geometries in terms of their transformation groups. More generally, the development of projective geometry brought with it a certain flexibilization of what count as the primitive elements in a geometry and, in turn, a new focus on geometrical form that clearly stimulated Klein's approach.

[6] See, in particular, Wussing (2007), Rowe (1989, 1992), and Gray (2008) for detailed studies of Klein's program and its mathematical background.

[7] See Torretti (1978) and Gray (2005) on the historical development of projective geometry.

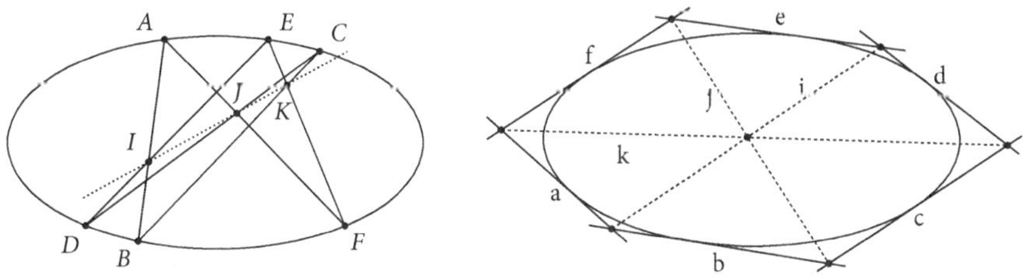

Figure 1 Pascal's and Brianchon's theorem

A central innovation in work by Poncelet, Gergonne, and others was the discovery of the *principle of duality* for theorems in projective geometry.[8] In the case of plane geometry, this principle expresses the fact that for every theorem concerning certain projective properties of configurations in the plane, one can formulate a second theorem with a *dual* (or *reciprocal*) structure based on the method of dualization, that is, by interchanging the words "point" and "line" as well as the relational expressions of "lying on a line" and "meeting in a point."[9] In order to illustrate this principle, consider the following pair of well-known dual theorems, namely Pascal's theorem and Brianchon's theorem.[10] The former theorem expresses the following geometrical fact:

Theorem 1: *Let A, B, C, D, E, F be six points on a conic that form a hexagon. Then the intersection points of the sides* $\overline{AB}$ and $\overline{DE}$, $\overline{FA}$ and $\overline{CD}$, *and* $\overline{BC}$ *and* $\overline{EF}$ *of the hexagon will lie on a line.* (See Fig. 1, left diagram.)

Brianchon's theorem, in turn, states a closely related geometrical fact:

Theorem 2: *Let a, b, c, d, e, f be six lines that form a hexagon circumscribing a conic. Then the principal diagonals i, j, and k of the hexagon meet in a single point.* (See Fig. 5.1, right diagram.)

The two theorems express symmetric facts about the projective structure of hexagons relative to a conic section. That is, any concrete incidence relation between points and lines specified relative to one conic can be shown to correspond to a dualized relation between lines and points specified relative to the second conic. Accordingly, the theorems form an instance of the general principle of projective duality: one can deduce Brianchon's result from Pascal's result (and vice versa) by the previously mentioned technique of dualization, that is, by

[8] The present subsection will closely follow Eder and Schiemer (2018) and Schiemer (2018) in the discussion of the principle of projective duality.

[9] A corresponding principle of duality for solid geometry states that for any theorem of solid projective geometry we get another theorem by interchanging the words 'point' for 'plane' and 'plane' for 'point' (as well as of the primitive incidence relations).

[10] See again Schiemer (2018) for a more detailed discussion of this example. I would like to thank Günther Eder for his permission to use the two diagrams in figure 1 in the present chapter.

interchanging the primitive terms "point" for "line" as well as all the concepts defined in terms of them.

Much work in 19th-century projective geometry was dedicated to the analysis of the principle of duality. Klein's *Vorlesungen über Nicht-Euklidische Geometrie* of 1928 contains an interesting retrospective survey of the different approaches to a general mathematical explanation of duality phenomena. In particular, he distinguishes between three accounts in the geometrical literature from the time (see Klein 1928, 38–39). One approach, which Klein labels the "axiomatic justification of the principle of duality," is ascribed to the works of Gergonne and Pasch. Duality is explained here purely syntactically, in terms of the strictly symmetrical character of the axiom systems describing the projective plane and projective space.

The second approach is more interesting for our discussion and was first formulated in Poncelet's *Traité* of 1822.[11] Duality (or reciprocity) is specified here based on Poncelet's theory of poles and polars and in terms of so-called polar transformations. Roughly speaking, polar transformations are dual correlations between figures that can be constructed relative to a given conic section. Based on a given conic, such a correlation will map every point in the plane to a certain line, its polar, and every line to a single point, its corresponding pole.[12] The central geometrical property of such transformations is that they preserve the incidence relations between points and lines in a given plane. Following Poncelet, this is usually called the *reciprocity* between poles and polars: if a point lies on a line, then the pole of the line will also lie on the polar line corresponding to the point (and vice versa).

According to Poncelet, the principle of duality in projective geometry can be directly explained in terms of the theory of poles and polars. More specifically, in the second volume of the book, Poncelet introduces a general method of constructing new configurations from existing ones based on polar transformations. Given the fact that a polar mapping preserves the incidence properties (up to duality) of the original configurations, it follows that the newly constructed figures have a reciprocal structure. Thus, polar transformations induce a dual translation of theorems about one figure into theorems about its reciprocal figure.

As will be shown in the next section, dual transformations such as those described in Poncelet's polar theory are explicitly discussed in Klein (1872). Moreover, Klein's subsequent writings on geometry, for instance his second volume of *Elementarmathematik vom höheren Standpunkte aus* (1925), also

[11] See again Eder and Schiemer (2018) and Schiemer (2018) for closer discussions of Poncelet's transformation-based account of duality.

[12] See, e.g., Coxeter (1987) for a modern textbook presentation of polar theory.

contain detailed discussions of "transformations with a change of the spatial element" (Klein 1926, 117). However, in contrast to Poncelet's original account of 1822, dual transformations are not understood synthetically here, but analytically in terms of coordinate transformations. This brings us to the third way to think about projective duality mentioned in Klein (1928).

The third approach to justify the principle of duality mentioned in Klein's book is arguably the most relevant one for his Erlangen program. The so-called analytic justification of duality was first formulated by Julius Plücker (1801–1868) in his work on analytic geometry between the late 1820s and the 1840s. Briefly put, Plücker's approach is based on the analytic representation of geometric concepts in terms of equations.[13] Duality (or reciprocity) is discussed most extensively in the second volume of his *Analytisch-geometrische Entwicklungen* (Plücker 1931). The principle is explained here in terms of the *reinterpretation* of symmetric equations expressing geometrical configurations.

To illustrate his account, consider the linear equation presenting the concept of straight lines in the plane:

$$ux + vy + 1 = 0.$$

In the standard interpretation of this equation, u,v are treated as constants that determine a collection of points on a line. Plücker's basic insight was to treat the coefficients u,v instead as "line coordinates" similarly to the point coordinates x,y. Consequently, if x,y are treated as constants and u,v as variables, then the equation determines a collection of lines going through point (x,y). Put differently, whereas the equation $f(x,y) = 0$ in its usual interpretation presents a collection of points (or a point curve) on a line, the reinterpreted equation $f(u,v) = 0$ presents a collection of lines or a line curve. Projective duality is explained by Plücker in terms of the possibility of reinterpreting equations in this sense. More specifically, it is a result of the particular form of this and related *bilinear* equations, that is, of the symmetrical role of the point and line coordinates occurring in them.

Plücker's geometrical work from the time is known for the introduction of a number of different coordinate systems, including triangle coordinates (in Plücker 1830), homogeneous line coordinates for the plane (introduced in volume 2 of *Entwicklungen* of 1831), homogeneous plane coordinates, and line

[13] See Nagel (1939) and Plump (2014) for closer studies of Plücker's work. See Lorenat (2015) for a recent study of the priority dispute on the discovery of duality between Poncelet, Gergonne, and Plücker.

coordinates in space (introduced in Plücker 1846).[14] A central mathematical motivation for this generalization of the concept of coordinates was to be able to reinterpret analytic equations representing geometrical concepts relative to different coordinate systems. As we saw, precisely this method is also used for the justification of projective duality. Compare Plücker on this purely analytic approach:

> Every proof that can be drawn through the connection of general symbols corresponds to two such sentences connected to each other by the principle of reciprocity in case we refer with these symbols to point coordinates at one point and to line coordinates at another point. (Plücker 1931, viii–ix)

According to Plücker, there is thus a direct connection between the reinterpretation of an equation presenting an incidence relation in different coordinates systems and the general idea of "reciprocity" (or "Gergonne-Poncelet duality").

This generalization of the concept of coordinates also brought with it a certain flexibilization of what counts as the "basic elements of space" in a geometry. The main idea underlying Plücker's account of duality is to consider other elements than points as the primitive or basic elements in space. We saw that the line equation stated earlier can be interpreted in two ways, namely as presenting lines as collections of points or points as collections of lines. In the first reading, the points are taken as primitive objects and lines are determined as sets of points. In the second reading, lines are the primitive objects, and points are determined as classes of lines.

Plücker's insight that different objects can serve as the primitive elements of a geometry exercised a strong influence on Klein's subsequent geometrical work.[15] This is documented in several of Klein's later writings on the topic, which contain detailed discussions of the analytic justification of duality. For instance, Klein comments on Plücker's approach in the second volume of *Elementarmathematik* in the following way:

> Now it is Plücker's conception to look upon these u and v as the *"coordinates of the line" and as having equal status with the point coordinates x and y, and as being considered, at times, as variable instead of them. . . . Now the principle of duality resides in the fact, that every equation in x and y, on one hand, and in u and v on the other hand, is completely symmetrical. Everything that we said above*

[14] See Wussing (2007, 28–30) and Plump (2014) for further details on Plücker's work on different coordinate systems.

[15] Klein was a student and assistant of Plücker at the University of Bonn until Plücker's death in 1868. See Rowe (1989) for further details.

> *concerning the duality that is inherent in the axioms of connection resides in this property.* (Klein 2016, 70)[16]

As Klein emphasizes here and in related writings, this insight presupposes the generalized concept of coordinates previously mentioned as well as what he calls "Plücker's general principle of considering any configuration as a space element and its constants as coordinates." (Klein 2016, 72)

Compare the following remark in Klein (1926):

> With this idea of the *arbitrary "element of space"* that can be chosen as the starting point of geometry, a complete clarification of the Poncelet-Gergonnian principle of duality is given: since the equation for the incidence of point and straight line (in the space of point and plane) is symmetrical in the two elements, one can interchange the two words in all sentences that are based on the mere connection of the two elements. (124)

Thus, given this new concept of coordinates, any type of geometrical configuration can serve as the basic elements in geometry, including conic sections, lines, planes, and spheres (among other objects). As we will see in the next section, this insight also led Plücker and other geometers to generalize the original version of Gergonne-Poncelet duality.

2.2. Reciprocity and Transfer Principles

According to the analytic account, the projective duality between points and lines in the plane (as well as between points and planes in space) can be explained in terms of the analytic presentation of the incidence relations between these geometrical concepts. Compare again Plücker on this point in *System der Geometrie des Raumes* of 1846:

> Every geometrical relation is to be viewed as the pictorial representation of an analytic relation, which, irrespective of every interpretation, has its independent validity. Consequently, the principle of reciprocity properly belongs to analysis, and only because we are accustomed to ... express the matter in geometrical language, does it seems to be an exclusively geometrical principle. ...

[16] A similar discussion is given in Klein's *Vorlesungen über die Entwicklung der Mathematik im 19. Jahrhundert* (Klein 1926), now for the related case of the analytic equation presenting straight lines: $u_1x_1 + u_2x_2 + u_2x_2 = 0$. Here again, it is the case that the coefficients u_1, u_2, u_3 and the coordinates x_1, x_2, x_3 have a strictly symmetrical role in the equation. One can therefore interpret the former as "line coordinates" and the equation as expressing a point determined through a bundle of lines.

> Understood purely analytically, the principle of reciprocity is naturally also not bound to the dimensions of space or restricted to them. (Plücker 1846, 322)

The account of reciprocity formulated here is thus not just an analytic reformulation of Poncelet's treatment of duality in synthetic projective geometry, but rather an independent justification with a more general applicability in geometry. Moreover, as Plücker mentions in the preceding passage, the principle is not limited to a particular dimension of the space to be investigated analytically. This insight led him to formulate several alternative generalized notions of reciprocity in his work from the 1820s and 1830s that extend classical Gergonne-Poncelet duality in different ways.[17]

One such extension concerns the introduction of dualities between other geometrical concepts than points, lines, and planes. For instance, Plücker's "Über ein neues Coordinatensystem" (1930) contains a discussion of Poncelet's theory of reciprocity based on the analytic treatment of the concepts of poles and polars. Based on this, Plücker also presents a "generalization" of this theory that applies to higher-order curves. Generally speaking, it is shown here as well as in other publications that one can extend duality to any pairs of geometrical objects with the same dimension-number. Thus, any two geometrical concepts whose analytic representation is based on the same number of independent variables can be shown to have dual properties.[18]

A second extension of Gergonne-Poncelet duality also introduced in Plücker's work is usually called "linear reciprocity." A detailed treatment of it can be found in his *System der analytischen Geometrie* of 1935.[19] The discussion given here concerns the dual correlation between two configurations, where duality is understood in the usual sense that each point and each line in the first figure is mapped to a line and a point in the second, reciprocal figure. Unlike in Poncelet's account of point-line duality, however, this correspondence is not specified within a single geometrical system, that is, within a given projective plane. Instead, reciprocity is specified here with respect to the interpretation in two coordinates systems, one based on point coordinates and the other based on line coordinates.

Plücker calls two such coordinate systems connected by a polar mapping "reciprocal systems" and describes them as follows:

[17] See Nagel (1939) and Plump (2014) for detailed studies of Plücker's generalized notions of reciprocity.

[18] Compare, in particular, Nagel (1939) for a study of this generalized notion of reciprocity in Plücker's work. Klein's *Elementarmathematik* also contains a detailed discussion of Plücker's notion of reciprocity between different higher-order curves based on the "Plückerian principle" to use arbitrary configurations as the primitive elements of a given space.

[19] See, in particular, Plump (2014) for a detailed discussion of this approach.

> We see from this in which sense the relation between the two systems is indeed a mutual one. We call such two systems reciprocally related or reciprocal ones and the principle resulting from this kind of relationship, by which the relations of one of two reciprocal systems can be transferred to the other one, the principle of reciprocity. (Plücker 1835, 74)

The principle stated here clearly presents an extension of the kind of inner-system duality introduced earlier by Poncelet. Duality is now expressed analytically as a correlation between geometrical figures in different coordinate systems with different primitive spatial elements and not between figures within a given system.[20]

A further generalization of classical duality closely related to Plücker's principle of linear reciprocity concerns so-called *transfer principles* in geometry. Roughly speaking, these are analytically defined mappings between different geometrical domains that preserve the relevant projective properties of the configurations in question. Interestingly, the term "transfer principle" first occurs in Plücker's own work in the context of his discussion of reciprocity. In his *System der analytischen Geometrie* (1935), Plücker argues that his concept of general coordinates implies different transfer principles (*Übertragungs-Principe*) based on the (re)interpretation of a given analytic equation in different systems. A transfer is described here as a mapping between the elements of different coordinate systems that allows one to construct, based on a given figure, a corresponding figure in another system (see Plücker 1835, vii).

This account of geometrical transfer principles was further developed in subsequent work on analytic geometry, in particular by Ludwig Otto Hesse (1811–1874). Hesse introduced a particular transfer principle in projective geometry in his article "Über ein Übertragungsprinzip" (1866a).[21] The principle is based on a mapping between points of the complex projective plane and pairs of points on the complex projective line that preserves the projective structure of these two domains. Hesse informally characterizes his approach as follows:

> If one makes to correspond in a univocal way to each point in the plane a pair of points on the straight line and, vice versa, to each pair of points on the straight line a point in the plane, one has a transfer principle that reduces the geometry of the plane to the geometry of the straight line and vice versa. (Hesse 1866a, 15)

The relevant transfer mapping is presented analytically in the following way: Hesse introduces a function from points $P = (x, y)$ in the projective plane to

[20] Compare again Klein (2016, 71–72) for a discussion of this notion of linear reciprocity.

[21] See Hawkins (1984) for a closer discussion of Hesse's transfer principles.

pairs of points $p = \{\lambda_1, \lambda_2\}$ on the projective line (i.e., the fundamental line) specified by the quadratic equation:[22]

$$\phi(\lambda, x, y) = A\lambda^2 + B\lambda + C = 0,$$

where A, B, C are linear functions of coordinates x, y.

This mapping between the plane and the fundamental line is structure-preserving in the sense that it preserves the primitive projective "relations between figures" (*Figurenverhältnisse*) in the two systems. This is established by Hesse in terms of a number of "fundamental theorems" (*Fundamentalsätze*) that show how primitive projective properties of the objects in the first system correspond to properties of pairs of points on the fundamental line. One such theorem concerns the correspondence between the collinearity of points in the plane and the involution between point pairs on the projective line: any three collinear points P_1, P_2, P_3 correspond to three pairs of points p_1, p_2, p_3 on the projective line that are in involution (and vice versa).[23]

As a consequence of this and other fundamental principles, it follows that any projective theorem about the configurations of the one domain can be translated into a theorem about the configurations the other domain and vice versa. As in the case of duality, the method of transfer is thus primarily a method of unification in geometry. It allows one to reapply proven results about a given field to the objects of a different field. Or, as Hesse puts it:

> The principle of transfer developed here gives the opportunity to discover a large number of new theorems from the geometry of the straight line. It presents a recommendable task . . . to prove these theorems not directly in isolation, but to invent proof methods that let the theorems appear as evident in combination. (Hesse 1866a, 20–21)

Hesse's method of transfer used for this identification of the projective geometry of the plane with that of the fundamental line is closely related to Plücker's approach to linear reciprocity. In fact, in his *Vier Vorlesungen aus der analytischen Geometrie* (1866b), Hesse explicitly mentions Plücker's method of reinterpreting equations by the substitution of point coordinates by line coordinates. This

22 The points on the fundamental line are determined in terms of their distance λ from a given point on the line.

23 A second result states that all double points on the fundamental line correspond to the points lying on a given conic in plane and vice versa (Hesse 1866a, 17–20).

method provides a duplication of dual theorems based on the reinterpretation of all formulas used in the proof of a theorem. However, Hesse argues:

> This is a very cumbersome approach, however, to reach from a given theorem to its corresponding one. Geometry therefore replaces the mediating formulas by transfer principles, through which one can immediately deduce the corresponding theorem from a given theorem. In our case this principle is the well-known law of reciprocity. (Hesse 1866b, 32)

This passage clearly indicates the close connection between Hesse's understanding of transfer principles and Plücker reciprocity. Whenever a given equation representing a mathematical concept can be reinterpreted in Plücker's sense, one can also construct a transfer principle that directly maps the objects of the first domain to those of the second domain. In the case of a dual transformations (such as Poncelet's polar transformations), this transfer principle is the principle of reciprocity in Plücker's sense. However, Hesse points out, the method is more general than reciprocity and applies also to non-dual mappings, such as the one previously described. Hesse specifies the general principle as follows:

> In all cases where two geometrical theorems result from different geometrical interpretations of the same analytic formula, a transfer principle can be discovered that replaces the proving formulas in a large number of cases. (1866b, 32)

Thus, according to him, the possible reinterpretation of a given analytic expression in different coordinate systems indicates the existence of a structure-preserving mapping between them that can also be defined analytically. The fact that theorems about different geometrical objects can be proven from the "the same analytic source" shows that one can construct a mapping between these domains that induces a direct translation between the theorems.

Before turning to a closer discussion of Klein's Erlangen program in the next section, let me quickly take stock here. Given the methodological developments in projective geometry already surveyed, one can identify two general structuralist ideas implicit in the work of Poncelet, Plücker, and Hesse. The first one concerns a deliberate indifference with respect to the nature of the primitive spatial elements used for the construction of geometrical configurations and instead a focus on their "invariant form." The second one concerns the emphasis on structure-preserving mappings that allow one to transfer the structure of one geometrical system to a different system. As will be shown in the following section, Klein's work presents a group-theoretical reformulation and further generalization of both ideas.

3. Klein's Erlangen Program

Klein's program was first outlined in his "Vergleichende Betrachtungen über neuere geometrische Forschungen" (1872), a programmatic pamphlet distributed during his inauguration speech at the University of Erlangen.[24] Klein presents here a novel method to study and to classify different geometries in terms of their corresponding transformation groups. While there is scholarly debate on the actual impact of Klein's article for subsequent research in geometry, it is clear that the Erlangen program contributed significantly to a new understanding of the subject matter of geometrical theories.[25] In the following, I will restrict my attention to the presentation of some of the key concepts developed in 1872 (as well as in related writings) and discuss how they are related to the developments in projective geometry sketched above.

3.1. A Group-Theoretic Approach

Klein's approach is motivated by a number of seemingly disconnected fields and methods in 19th-century geometry. Geometry, he writes, "which is after all one in substance, has been only too much broken up in the course of its recent rapid development into a series of almost distinct theories, which are advancing in comparative independence of each other" (1872, 216).[26] Klein's aim in 1872 was therefore to formulate a "general principle" that allows for the comparison and classification of these different geometrical fields. This was, roughly put, the methodological idea that each geometry should be identified with a space and a group of transformations acting on it that leave the relevant geometrical properties invariant.

This algebraic approach to studying the properties of figures clearly brought with it a more abstract conception of the subject matter of geometrical theories. Two issues are noteworthy here. The first point concerns Klein's specific understanding of a geometrical space. It is clear from Klein (1872) as well as from

[24] A revised version of the article was published in *Mathematische Annalen* in 1893 and then again in 1921 in the first volume of Klein's collected works (Klein [1921–23] 1973). In the following, I quote from the English translation by Haskell published in 1892/1893.

[25] Compare Rowe (1989) on this point. See Wussing (2007) for a study of the influence of Klein's approach for the subsequent development of abstract group theory. Compare, in particular, Hawkins (1984) and Birkhoff and Bennett (1988) for partly conflicting assessments of the relevance of Klein's article for subsequent geometrical research.

[26] This is true despite the fact that projective geometry has developed into a fundamental geometrical theory in work by Cayley and Klein in the sense that it not only characterizes the non-metrical properties of configurations but can also be used to represent the metrics of both Euclidean and non-Euclidean geometries. See, in particular, Biagioli (2016) for a discussion of Cayley's work and Klein's projective model of non-Euclidean geometry.

related writings that space is not primarily meant to be physical or intuitive in his account. Rather, geometries study the configurations in formal "manifolds" of arbitrary dimensions "that have been developed from geometry by making abstraction from the geometric spatial image, which is not essential for purely mathematical investigations" (Klein 1872, 216). Klein gives an explicit characterization of the notion in his article "Über die sogenannte Nicht-Euklidische Geometrie (2. Aufsatz)" (1873), which was also written in 1872:

> If n variables $x_1, x_2, ..., x_n$ are given, the infinity to the nth value systems we obtain if we let the variables x independently take the real values from $-\infty$ to $+\infty$, constitute what we shall call, in agreement with usual terminology, a *manifold of n dimensions*. *Each particular value* $(x_1, x_2, ..., x_n)$ is called an *element* of the manifold. (Klein 1873, 116)

The basic spatial elements of a geometry are therefore not genuine geometrical objects such as points or lines, but rather tuples of numbers assigned to the variables in question.[27] Klein's approach is in line here with the purely analytic approach in geometry of Plücker and Hesse discussed in the previous section. As Klein points out in 1872, the reference to genuinely spatial concepts or spatial representation is to be used only for pedagogical purposes. In his own terms, given this purely analytic approach of manifolds, "space-perception has then only the value of illustration" (Klein 1872, 244).

The second issue to be mentioned here concerns Klein's understanding of the notion of geometrical transformations. In his view, one can take "the totality of configurations in space as simultaneously affected by the transformations, and speak therefore of transformations of space" (Klein 1872, 217). Transformations in this sense can include those between spatial elements of the same kind (such as transformations between points), but also those with a change of spatial elements (such as dual mappings).

While Klein gives only an informal description of such spatial transformations and of the geometrical properties preserved by them, his focus on numerical manifolds suggests that they are also treated analytically. In fact, while Klein remains silent on this issue in 1872, he gives a detailed discussion of the analytic representation of various transformations in related writings. For instance, in his 1873 paper, transformations of manifolds are described analytically in the following sense:

[27] In his discussion of manifolds of arbitrary dimensions in 1872, Klein refers both to Hermann Grassmann's *Ausdehnungslehre* as well as to Bernhard Riemann's theory of general manifolds. See Scholz (1980) for a historical survey of the development of the concept.

> A transformation of a manifold into itself is understood as the process that leads from every element to one corresponding element (or several). One may want to specify the transformation in terms of *n* equations, in which the corresponding element depends on the respective original one. The type of equations and their respective relation is at first irrelevant for the concept. In the following, we will always presuppose, however, that they are invertible. The inverted equation presents what should be called inverted transformation. (Klein 1873, 117)

Transformations of a space are thus represented as transformations of coordinates within one or between distinct coordinate systems, specified in terms of a number of analytic or algebraic equations describing the functions between the coordinates.[28]

Klein's work after 1872 also contains an extensive discussion of the geometrical transformations first mentioned in the Erlangen program. Consider his monograph *Elementarmathematik vom höheren Standpunkte aus* of 1908. The "analytic presentation" is described here as follows:

> The analytic expression of a point transformation is what analysis calls the *introduction of new variables* x', y', z':
>
> $$\begin{cases} x' = \varphi(x, y, z) \\ y' = \xi(x, y, z) \\ z' = \psi(x, y, z) \end{cases}$$
>
> We can interpret such a system of equations geometrically in two ways, I might say actively and passively. Passively, it represents a change in the coordinate system, i.e., the new coordinates x', y', z' are assigned to the point with the given coordinates x, y, z. . . . In contrast with this, the active interpretation holds the coordinate system fixed and changes space. To every point x, y, z, the point x', y', z' is made to correspond, so that there is, in fact, a transformation of the points in space. It is with this conception that we shall be concerned in what follows. (Klein 2016, 81–82)

[28] Sophus Lie's *Theorie der Transformationsgruppen* presents the first systematic treatment of the notion of a spatial transformation (Lie 1893). Compare Hawkins (2000) for a detailed study of Lie's work.

Klein's distinction between an "active" and a "passive" interpretation of the equations presenting a transformation is interesting here. The latter account seems similar to Plücker's account of linear reciprocity, and more specifically, to Hesse's analytic presentation of transfer principles between different geometrical fields. The former, active account specifies transformations relative to a given coordinate system as a permutation of all points that also induces a transformation of all configurations in the manifold.

Returning to Klein's 1872 article, it is plausible to assume that this understanding of analytically defined coordinate transformations within a fixed coordinate system also forms the background of his Erlangen program. Klein argues here that one can view different geometrical fields such as Euclidean or projective geometry as determined by a class of relevant transformations. These are the class of isometries in the first case and the projections (including collineations and dual transformations) in the second case. Moreover, given that the transformations of such a class always have inverses and that any two of them can be merged into a new composed transformation, it follows that these classes—equipped with a suitable composition operator—also form *groups* in the algebraic sense of the term. Compare Klein on this point:

> The most essential idea required in the following discussion is that of a group of space-transformations. The combination of any number of transformations of space is always equivalent to a single transformation. If now a given system of transformations has the property that any transformation obtained by combining any transformations of the system belongs to that system, it shall be called a group of transformations. (Klein 1872, 217)[29]

Klein mentions a number of geometrical transformations that form a group in this sense: the class of all movements in a given space; the class of rotations relative to a given point; the class of collineations; as well as the group consisting of all linear substitutions that leave the metric properties unchanged. Klein calls the latter group the "principal group" (*Hauptgruppe*) of a space and the corresponding geometrical discipline "elementary geometry." Dual transformations in the sense specified in the previous section are also mentioned by Klein in this context. In particular, he argues that while such transformations do not form a

[29] It should be noted that Klein does not state the modern axiomatic conditions for abstract groups here (including the associativity of the group operations and the existence of a neutral element). His specification of the concept of groups of transformations in terms of a closure condition for the composition of transformations is directly based on Jordan's theory of permutation groups given in his *Traité*. Compare Wussing (2007, 186) for a detailed survey of Klein concept of groups and his mathematical background.

group by themselves, the class of collineations and dual mappings does form a group (Klein 1872, 217).

Given this conceptual framework, Klein showed in 1872 that groups of transformations allow one to specify the notion of geometrical properties of configurations in a given manifold. More specifically, his proposal was to characterize the relevant properties of a given geometry in terms of an invariance condition specified relative to a group. Thus, given a geometry X with a transformation group G_X, properties of figures are specified as geometrically relevant if they are preserved under the transformations of group G_X. This approach is first characterized informally with respect to the invariance relative to the "principal group":

> Geometric properties are not changed by the transformations of the principal group. And, conversely, geometric properties are characterized by their remaining invariant under the transformations of the principal group. For if we regard space for the moment as immovable, etc., as a rigid manifoldness, then every figure has an individual character; of all the properties possessed by it as an individual, only the properly geometric ones are preserved in the transformations of the principal group. (Klein 1872, 218)

As Klein points out, this invariance-based method not only applies to "elementary geometry" of three-dimensional space, but more generally to any geometry of a formal manifold of arbitrary dimensions that can be characterized in terms of a group of transformations.

This shift of attention from concrete figures to manifolds leads to a "generalization of geometry" that is significant in at least two respects. First, Klein's approach led to the new situation that different (and partly conflicting) geometrical fields were to be treated on equal footing, that is, as equally justified. Or, as Klein puts it, "There is no longer, as there is in space, one group distinguished above the rest by its signification; each group is of equal importance with every other" (Klein 1872, 218). Second, the group-theoretic method implies a radically new conception of the nature of a geometrical theory. A geometry is now conceived as a tuple consisting of a manifold (of a given dimensionality) and a group of transformations acting on this manifold. Consequently, the general task of a geometer is to study those properties of geometrical configurations that are preserved under the transformations in question. Put differently, given this new framework, geometry turns into an invariant theory for the given group:

> Given a manifold and a group of transformations of the same: to investigate the configurations belonging to the manifold with regard to such properties as are not altered by the transformations of the group. . . . Given a manifold

> and a group of transformations of the same: to develop the theory of invariants relating to that group. (Klein 1872, 218–219)[30]

As we saw previously, the transformations in question are generally understood as coordinate transformations expressed by a number of analytic equations. Consequently, geometrical invariants also have to be specified analytically, namely in terms of equations between coordinates and constants representing a geometrical concept that remain preserved under the transformations of a given group.

While Klein does not give a more detailed discussion of the invariant theory related to his group-theoretic approach in 1872, it is developed in his subsequent work.[31] For instance, Klein's *Elementarmathematik* contains a section titled "Group Theory as a Geometrical Principle of Classification" where the analytic invariant theory of various geometries is discussed in further detail. Klein shows here that elementary or "metrical" geometry is characterized by the group of certain special linear substitutions corresponding to the principal group specified in 1872. Geometrical invariants are then given by analytic expressions that remain unaltered by such substitutions. In Klein's terms, "the geometry is thus the invariant theory of these linear substitutions" (Klein 2016, 153).

3.2. Transfer by Mapping

Klein's main focus in 1872 was not the study of particular geometries in isolation but rather the comparison of different theories in terms of their transformation groups. Thus, group theory was to provide a unifying approach that allowed for the classification of different geometrical systems studied at the time. More specifically, Klein's idea to introduce an order of generality between different geometries is based on a relation between their transformation groups. Recall that geometries are conceived in Klein's program as consisting of a manifold and a group of transformations acting on it. Given two such geometries, say A = $\langle M, A \rangle$ and B = $\langle M, B \rangle$, geometry B can be characterized as a *subgeometry* of A if transformation group B forms a *subgroup* of A. It follows from this that every invariant property studied in A (i.e., relative to the transformations in A) is

30 As is shown in Wussing (2007), Klein's use of the notion of invariants can be seen as a concession to the earlier invariant-theoretic approach in geometry, e.g., in work by Cayley and Clebsch, that strongly influenced Klein's own group-theoretic approach.

31 Lie's *Theorie der Transformationsgruppen* contains a detailed presentation of invariants of transformation groups (Lie 1893). Compare also Fano (1907) for a study of the invariants of different transformations groups discussed by Klein and others.

also an invariant in B but not vice versa. Moreover, all theorems of A turn out to be theorems of B.

Klein discusses a number of geometrical theories in 1872 that can be ordered in this way in terms of the relation of subgroups or group extensions. His general approach is to construct subgroups of a given transformation group by restricting the latter to transformations that leave invariant a given spatial element or a given configuration (such as a conic section). The main example in this respect concerns "elementary geometry," specified by the principal group of geometrical transformations. It is shown that the group of projective transformations forms an extension of this group. It follows from this that every property of projective geometry is also a property of elementary geometry but not vice versa. Compare Klein on this point:

> We inquire what properties of the configurations of space remain unaltered by a group of transformations that contains the principal group as a part of itself. Every property found by an investigation of this kind is a geometric property of the configuration itself; but the converse is not true. (Klein 1872, 220)

Thus, while the projective properties—including metrical properties such as the cross-ratio for a given set of points—are also invariant under the transformations of the principal group, properties such as sameness of lengths of segments are not invariant in the projective setting.[32]

A second approach to interrelate different geometries in Klein (1872) concerns so-called transfer principles. Such principles are introduced by Klein as a general method to show the equivalence of geometries in section 4, titled "Übertragung durch Abbildung." The method of "transfer by mapping" is informally characterized here as follows:

> Suppose a manifoldness A has been investigated with reference to a group B. If, by any transformation whatever, A be then converted into a second manifoldnessA', the group B of transformations, which transformed A into itself, will become a group B', whose transformations are performed upon A'. It is then a self-evident principle that the method of treating A with reference to B at once furnishes the method of treating A' with reference to B', i.e., every property of a configuration contained in A obtained by means of the group B furnishes a property of the corresponding configuration in A' to be obtained by the group B'. (Klein 1872, 223)

[32] By the same method, Klein shows that various non-Euclidean geometries form subgeometries of projective geometry. See, in particular, Biagioli (2016) and Torretti (1978) on Klein's discussion of non-Euclidean geometries and the relation to Arthur Cayley's work a generalized metric. Compare Brannan et al. (2011) for a modern presentation of the hierarchy of Kleinian geometries.

To paraphrase Klein's approach in modernized terms: consider two geometries, both understood as tuples consisting of a manifold and a group of transformations acting on it, that is, $G = <A, B>$ and $G' = <A', B'>$, respectively. A transfer principle between G and G' is a mapping between the manifolds $f: A \rightarrow A'$ that induces an isomorphism between the corresponding groups B and B' acting on them. It follows that every invariant property of configuration in A determined with respect to the transformations in B can be mapped to a corresponding invariant property of configurations in A' with respect to B'. Moreover, the transfer principle allows one to translate every theorem of geometry G into a corresponding theorem of geometry G'.

Given Klein's account of transfers by representation, several points of commentary are in order. First, principles of this form play a crucial role in his general program to classify different geometrical fields investigated at the time. He discusses a number of concrete examples of such principles that connect different theories in his 1872 article. This includes a transfer principle between the "theory of binary forms" given by the group of "∞^3 linear transformations" of a straight line and the "projective geometry of systems of points systems on a conic" in the plane (determined by the linear transformations of the conic into itself). The transfer principle in question, Klein argues, preserves the relevant properties of configurations in the two domains. As a consequence, the two geometries are shown to be equivalent:

> The theory of binary forms and the projective geometry of systems of points on a conic are one and the same, i.e., to every proposition concerning binary forms corresponds a proposition concerning such systems of points, and vice versa. (Klein 1872, 223)[33]

This account of transfer principles presented in 1872 is strongly influenced by preceding geometrical research.[34] In particular, Klein explicitly refers to Lie's work as well as to his own article "Über Liniengeometrie und metrische Geometrie" (1872a) for a discussion of another transfer principle connecting line geometry with the metric geometry in four variables. As is shown there, this mapping allows one to "transfer the complete content of metrical geometry to line geometry" and thus induces a "translation into the language of line geometry" (Klein 1872a).

Moreover, the discussion of transfer principles in Klein's 1872 paper was strongly influenced by the developments in projective geometry surveyed in the

[33] A second, analogous example concerns the elementary geometry of the plane and the projective geometry of a quadratic surface with a given fixed point (Klein 1872, 224).

[34] See, in particular, Rowe (1989) for a detailed study of Klein's work on transfer principles and its mathematical background.

previous section, in particular, by Plücker's and Hesse's work on generalized reciprocity and transfer principles. Interestingly, in Klein's article, the very notion of "transfer" is first mentioned in the context of his discussion of the development of projective geometry:

> Every space-transformation not belonging to the principal group can be used to transfer the properties of known configurations to new ones. Thus we apply the results of plane geometry to the geometry of surfaces that can be represented upon a plane; in this way long before the origin of a true projective geometry the properties of figures derived by projection from a given figure were inferred from those of the given figure. (Klein 1872, 220–221)[35]

How are the transfer principles developed in projective geometry related to Klein's own use of "transfers by mapping"? As we saw, transfers were introduced in Plücker's and Hesse's work as mappings between different coordinate systems that induce a translation of the theorems about the projective properties of figures. Klein's method generalizes such principles in the sense that the structure preserved by them is now expressed group-theoretically, that is, in terms of an isomorphism relation between the groups of transformations associated with two manifolds.[36]

A third point to mention here also concerns the projective background of Klein's concept of transfers. Section 5 of the article, titled "On the Arbitrariness in the Choice of the Space Element," shows that such principles can be used to connect geometries describing manifolds with different spatial elements (*Raumelemente*) such as points, lines, higher-order curves, etc. Compare Klein on this point:

> As element of the straight line, of the plane, of space, or of any manifoldness to be investigated, we may use instead of the point any configuration contained in the manifoldness, a group of points, a curve or surface, etc. As there is nothing at all determined at the outset about the number of arbitrary parameters upon which these configurations shall depend, the number of dimensions of our line,

35 As he points out, the transfer of geometrical properties is then generalized in work by Poncelet and others in terms of the introduction of dual transformations, i.e., those based on a change of the elements of space that preserve several symmetrical incidence properties (Klein 1872, 221).

36 Klein, in his 1872 article, does not explicitly use the notion of group isomorphism. However, it is clear from his related writings from the time that this notion or, in his terms, the "similarity" between groups of transformations was assumed in the background of his discussion of transfer principles. See his definition of this notion given in 1873: "Two transformation groups are said to be *similar* if we can associate the transformations of the one group to the transformations of the other group such that composition of corresponding transformations yields corresponding transformations" (118).

> plane, space, etc., may be anything we like, according to our choice of the element. (Klein 1872, 224)

The indifference to the basic nature of geometrical objects expressed here clearly echoes Plücker's idea of a generalized concept of coordinates and the flexibilization of the basic elements of space that comes with it. As we saw in the previous section, Plücker thought of the dimensionality of a space as determined by the number of independent variables needed to present the basic spatial elements in analytic terms. Thus, for instance, a plane is two-dimensional if points are assumed as the basic elements; it is five-dimensional if conic sections are taken as the basic elements. This is precisely the idea also underlying Klein's discussion of manifolds in 1872.[37]

Given this Plückerian account of "spatial elements," Klein's central observation is that the choice of the basic elements and thus of the dimensionality of a given manifold is of secondary importance for the investigation of geometries. What is relevant, from a mathematical point of view, are their group of transformations and the algebraic relations between them. Compare again Klein on this central "structuralist" insight:

> But so long as we base our geometrical investigation on the same group of transformations, the geometrical content [*Inhalt der Geometrie*] remains unchanged. That is, every theorem resulting from one choice of space element will also be a theorem under any other choice; only the arrangement and correlation of the theorems will be changed. The essential thing is thus the group of transformations; the number of dimensions to be assigned to a manifold is only of secondary importance. (Klein 1872, 224–225)

A number of concrete examples of geometries of manifolds with different spatial elements are mentioned by Klein whose equivalence can be established in terms of transfer principles. One such example concerns a mapping between the system of pairs of points on a conic and the plane with straight lines as the basic elements. This mapping assigns to each pair of points (λ_1, λ_2) on a conic the line that intersects the conic at points (λ_1, λ_2) (and vice versa).[38] It thus induces an isomorphism between the group of linear transformations of the conic in itself and the group of linear transformations of the lines in the plane that leave the conic invariant. Interestingly, in the discussion of this and several related

[37] In fact, in a corresponding note in his article, Klein explicitly refers to Plücker's work on "how to regard actual space as a manifoldness of any number of dimensions by introducing as space-element a configuration depending on any number of parameters, a curve, surface, etc." (Klein 1872, 245).

[38] See Fano (1907, 358–359) for a detailed analytic presentation of this mapping and the resulting equivalence theorem.

results, Klein explicitly mentions Hesse's work: "The correlation here explained between the geometry of the plane, of space, or of a manifoldness of any number of dimensions is essentially identical with the principle of transference proposed by Hesse (Borchardt's Journal, vol. 66)" (Klein 1872, 225).

4. Structuralist Themes

The geometrical research surveyed in the last two sections strongly contributed to a general structural turn in 19th-century mathematics. In particular, the systematic use of transformations and transfer principles both in projective geometry and in Klein's program brought with it a new conception of the subject matter of geometry: geometry was no longer understood as the study of concrete figures in intuitive space, but rather as a theory of abstract forms or invariant properties and thus as a branch of pure mathematics. Klein's group-theoretic classification of different geometrical fields in terms of transformation groups in 1872 is often considered a culmination point of this development.[39]

How is the group-theoretic approach in geometry related to modern debates on structuralism? It seems natural to describe Klein's account as a kind of "methodological structuralism," a position first introduced by Reck with respect to Dedekind's foundational work on analysis and arithmetic.[40] This account differs from other philosophical theories of structuralism in the sense that it is more concerned with mathematical methodology than with metaphysical issues concerning the nature of structures. As Reck points out, structural methods in modern mathematics usually imply some form of *abstraction* from the subject matter or the particular nature of the objects described by a mathematical theory (Reck 2003, 371).[41]

Regarding Klein's work, one can identify two different types of structural abstraction in his approach to geometry. The first type is specified relative to a given geometry and concerns the abstraction from particular configurations in order to study their invariant properties. The second type is related to Klein's use of transfer principles. It concerns the abstraction from particular manifolds and

[39] Compare, for instance, Tobies who writes that Klein's Erlangen program "formed a decisive turning point for the geometry of the 19th century. Klein's use of the group concept supported approaches to structural mathematical thinking formed at the end of the 19th century. (Tobies 1981, 36–37, my trans.) See, in particular, Biagioli (2018) for a recent study of Klein's geometrical structuralism.

[40] See, in particular, Reck (2003) as well as Reck and Price (2000) for a more general discussion of the position.

[41] Thus, methodological structuralism can be viewed as the philosophical analysis of styles of reasoning introduced in modern mathematics that allow the mathematician to abstract from particular representations of objects in a system by highlighting their purely structural features or properties.

their basic spatial elements in order to identify the structural content shared by different geometries. In the remaining part of the chapter, I will analyze these two structuralist ideas in Klein's work.

4.1. Invariance and Structural Indiscernibility

A central "structuralist" idea underlying the geometrical developments previously sketched concerns the emphasis on invariant properties. Projective geometry in Poncelet's *Traité* and in subsequent work was viewed as the study of properties of spatial configurations that remain invariant under different types of projections. Generally speaking, invariance criteria were used as a method to carve out those properties that are geometrically relevant. A second and related idea concerns the notion of the geometrical identity (or congruence) of figures. In Euclidean geometry, two figures are usually taken to be distinct if there exist some metrical properties that allow one to discriminate between them. From a projective point of view, however, the same two figures will be treated as indistinguishable in case there exists a projective transformation between them. Thus, the identity of figures is determined here in terms of a primitive concept of structure-preserving transformations.

Obviously, these two ideas in projective geometry formed an important background for Klein's own group-theoretic approach. In fact, in his 1872 paper, the issue of projective identity is explicitly mentioned in his discussion of the extension of the "principal group" by projective transformations. As Klein puts it:

> But projective geometry only arose as it became customary to regard the original figure as *essentially identical* with all those deducible from it by projection, and to enunciate the properties transferred in the process of projection in such a way as to put in evidence their independence of the change due to the projection. (1872, 221)

As was mentioned in section 2, the notion of projective identity discussed here was further generalized in work on duality and general reciprocity. Dual mappings between figures based on Poncelet's theory of poles and polars allow one to identify symmetric incidence relations in a figure that are preserved by such transformations. Moreover, dual figures that share reciprocal properties are usually treated as identical. Compare again Klein on this point:

> From the modern point of view two reciprocal figures are not to be regarded as two distinct figures, but as essentially one and the same. (1872, 221)

Thus, in cases of dual figures, geometers abstract also from the particular nature of the basic elements of geometrical figures (e.g., points or lines in the case of plane geometry).

Arguably, the most systematic expression of these structuralist insights regarding the role of invariants and the nature of geometrical identity is developed in Klein's program. As we saw, both notions are specified here relative to a given group of transformations. Thus, the "elementary" metrical properties of a figure in a given manifold are specified relative to the principal group, its projective properties are specified relative to the extended group of projections and so on. Related to this, a criterion of structural identity is given based on the transformations of a given group.[42]

Expressed more formally in set-theoretic terms, Klein's account can be brought into the following form: let M be a manifold and G a group of transformations $f: M \rightarrow M$ acting on M:

Definition 1 (G-property): A property P of figures in M is a G-property if is it invariant relative to G, i.e., for any $F_1 \subseteq M$: if $P(F_1)$ then for all $f \in G: P(f(F_1))$.

Geometrical properties are conceived extensionally here as classes of configurations of a given manifold. A definition of geometrical identity or congruence of figures can be given within the same framework:

Definition 2 (G-congruence): Two figures F_1, $F_2 \subseteq M$ are G-congruent if there exists a transformation $f \in G$ such that $f(F_1) = F_2$.

This notion of G-congruence can be viewed as an expression of the structural identity of figures: two congruent figures are identical with respect to their structural content or in terms of sharing the same geometrical properties. Similarly, the notion of a G-property can be taken to express the structural properties of a given geometry in terms of an invariance condition.[43]

[42] In a recent analysis of the Erlangen program by Marquis, these two ideas are also emphasized as the philosophically relevant aspects of Klein's approach: "(Transformation groups) constitute in a precise sense the algebraic encoding of a criterion of identity for geometric objects, or to be more precise for geometric object-types. Second, the same transformation groups also encode a definite criterion of meaningfulness for geometric predicates, or, equivalently, a definite criterion for geometric properties" (Marquis 2009, 12).

[43] Notice that, in both definitions, the notion of geometrical structure assumed here is strongly *context-relative*. What counts as a structural property of the figures of a manifold depends critically on the particular transformation group associated with a geometry. Analogously, the congruence conditions for figures within a manifold are also specified in a given geometrical context. Thus, for instance, congruence in affine geometry is specified relative to the group of affine transformations; in Euclidean geometry, it is specified relative to the group of isometries, and so on.

How is Klein's view related to modern structuralism? Given the preceding discussion, several points come to mind here. First, Klein's work on invariants under transformation groups seems closely connected to the structuralists' focus on structural properties of mathematical objects. As mentioned in the introduction, one way to characterize the structuralist thesis is to say mathematical theories describe only structural properties of the objects of their subject domain.[44] For instance, Benacerraf's "What Numbers Could Not Be" (Benacerraf 1965) first emphasized that Peano arithmetic is concerned only with the relations between numbers in ω-sequences and not with particular set-theoretic presentations of them. Klein's approach is similar to Benacerraf's emphasis on purely structural properties. In fact, the former's proposal to think of geometrical properties of figures as invariants relative to a transformation group can be viewed as an early attempt at a mathematically precise characterization of the notion in the context of geometry.

A second point to mention here concerns Klein's understanding of the congruence of geometrical configurations. His account is similar in several respects to recent philosophical work on structuralist identity criteria. We saw that two figures can be identified, according to Klein, in case there exists a transformation of the elements of a space that maps one figure to the other one. One can think of such "internal" identity criteria specified relative to transformation groups in two ways, either (i) as expressing the sameness of figures in a manifold *with respect to* their structural properties or (ii) as expressing the identity of the abstract form shared by these figures.[45]

The first reading connects Klein's account with recent debates on the identity of *structurally indiscernible* objects mentioned in the introduction.[46] Briefly put, this debate concerns the question whether a version of Leibniz's principle of the identity of indiscernible objects presents an adequate identity criterion for structural mathematics. The principle in question holds that two mathematical objects are identical in case that they share the same structural properties. More formally, for any two objects X, Y and structural properties P:

$$X = Y \Leftrightarrow \forall P: \ (P(X) \Leftrightarrow P(Y)). \qquad \text{(PII)}$$

Different versions of (PII) have been discussed in mathematical structuralism. For instance, it has been considered as a criterion of the identity of places in structures in Shapiro's *ante rem* structuralism.

[44] Compare Korbmacher and Schiemer (2017) for a detailed study of the notion of structural properties in mathematics and its possible explications.

[45] Compare again Marquis (2009) for a more detailed discussion of this.

[46] See note 2 for references.

A related discussion can be found in recent work on a structuralist account of mathematics based on homotopy type theory. Awodey (2014) emphasizes that in mathematical practice, isomorphic objects—that is, objects that share the same invariant properties—are usually not distinguished from each other. He takes the idea of treating isomorphic objects as identical to be a general "principle of structuralism" that should be reflected in any philosophical study of modern mathematics.[47] Given Klein's own remarks on the identity of figures stated previously, his approach seems well captured by Awodey's understanding of mathematical identity. The identity of mathematical objects is thus not treated as a primitive notion but as a form of mathematical equivalence defined relative to transformation groups.

The second way to interpret Klein's remarks on congruence, namely as the identity of the abstract shapes of configurations, is also related to non-eliminative structuralism.[48] To see this, compare Marquis's insightful discussion of Klein's notion of identity based on a distinction between "types" and "tokens":

> One aspect of this criterion of identity has to be emphasized immediately: what we are characterizing with its help are *types* of geometric figures, not *tokens* of these figures. . . . Thus, a transformation group specifies the types that are admissible in a geometric space, it determines what there "is" or what can be in a space in an essential way. (Marquis 2009, 20–21)

Thus, according to Marquis, the congruence of figures given by a transformation group induces an identity condition for *types* of figures. For instance, the study of dual transformations between the figures of a given manifold gives a notion of identity for the duality types of figures. Consequently, one can think of the subject matter of geometry not only in terms of the invariant properties, but also in terms of these congruence types of figures.

This philosophical interpretation of Klein's approach presents a particular version of structuralism discussed in the recent literature, namely *in re* structuralism.[49] This is, roughly put, the view that mathematical theories describe abstract structures as their subject matter but that these structures do not exist independently of concrete representations instantiating them. One way of thinking about this dependence relation between a structure and its concrete instantiations is again based on the notion of structural abstraction. Thus,

[47] Structural properties are characterized here in terms of the notion of isomorphism invariance as well as in terms of the definability in a type theoretic language (Awodey 2014).

[48] See Reck and Price (2000) for a general overview of different structure theories.

[49] Compare Shapiro (1997) for a closer discussion of *in re* as opposed to *ante rem* structuralism.

abstract structures are said to be gained from concrete mathematical systems by abstracting away all non-relevant properties of the objects in question.[50]

The Kleinian account of figure types can be understood as a version of *in re* structuralism concerning the subject matter of a particular geometry. As we saw, the study of a space relative to a group of transformations G allows one to treat the concrete configurations in the manifold as instances (or tokens) of more general figure types. A figure type can be instantiated or exemplified by all figures occurring in the manifold that are congruent relative to G. However, the abstract types do not exist independently of their concrete representations but are functionally dependent on them.[51] Moreover, one can think of this dependence relation between types and concrete figures in terms of a notion of abstraction. As Marquis puts this: "A transformation group is a way to *abstract* types from specific tokens" (2009, 21). Given the set-theoretical reconstruction of his approach, one can characterize this notion of Kleinian abstraction more formally in terms of the following abstraction principle:

Definition 3 (Kleinian abstraction): Given a geometry $<M,G>$ and the corresponding congruence relation $\sim_G$, for any two figures $F_1, F_2 \in M$ we have

$$\text{Type}(F_1) = \text{Type}(F_2) \Leftrightarrow F_1 \sim_G F_2.$$

Thus, the types of two figures in a manifold are identical in case that they are congruent relative to the transformation group G.[52]

4.2. Transfer Principles and Structural Equivalence

The second type of structural abstraction developed in Klein's program is related to his use of transfer principles. As we saw, his method of transfer by mapping is closely motivated by previous work on the generalization of Poncelet-Gergonne duality by Plücker and Hesse. In Klein's work, the equivalence of two geometries

[50] See, in particular, Linnebo and Pettigrew (2014) for a recent systematic study of a form of abstraction based structuralism.

[51] Compare again Marquis on this point: "Working with the transformations amounts to working with types instead of working with tokens. Notice, though, that the transformations are applied to tokens of these types and clearly the existence of the latter depends directly on the existence, or should we say the presence, of the former. Thus, a transformation group indicates the presence of geometric types whose existence depends on the existence of geometric tokens" (Marquis 2009, 21).

[52] Notice that this definition of abstraction is again relative to a given choice of a group of transformations. Thus, what counts as an abstract type of a figure differs relative to different groups. To give a simple example: ellipses, parabola, and hyperbola are figure types relative to Euclidean geometry and the group of isometries. In contrast, in projective geometry, these types are reduced to the single, more general type 'conic', given the fact that ellipses, parabola, and hyperbola are equivalent in the projective setting.

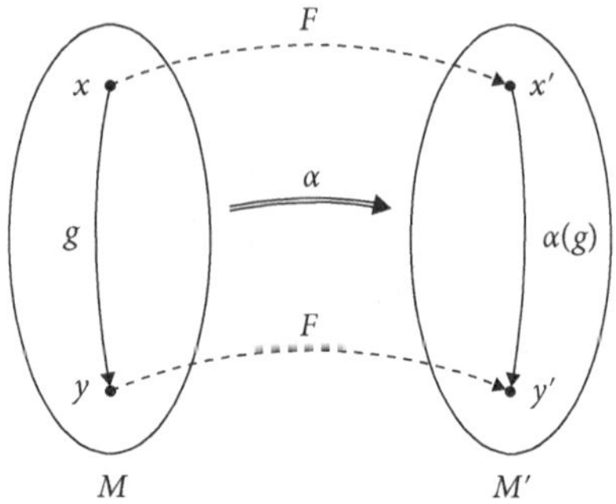

Figure 2 A transfer principle between manifolds M and M'

is formulated in a group-theoretic framework: a transfer is a structure-preserving mapping between two manifolds that induces an isomorphism between the group of transformations acting on the manifolds. As Klein shows, this fact induces a translation between the theorems of the two geometries in question. While his own discussion of transfer principles remains rather schematic in his 1872 article, one can give the following reconstruction of his approach:

Definition 4 (Equivalent geometries): Two geometries $<M, G>$ and $<M', G'>$ are equivalent if there exists a bijection $F: M \rightarrow M'$ and a group isomorphism $\alpha: G \rightarrow G'$ induced by F such that for all $x \in M$ and for all $g \in G$: $F(g(x)) = (\alpha(g))(F(x))$.

A transfer principle in this group-theoretic sense is thus a mapping between two manifolds that allows one to construct an isomorphism between two transformations groups that preserves the group actions on the respective manifolds (see Figure 2).[53]

Given Klein's approach, two points of commentary are in order here. First, notice that by identifying geometries based on their isomorphic transformation groups, one clearly abstracts from the particular nature of the basic objects of a geometry and instead focuses on its general invariant form. The abstraction involved here is more general, however, than the one described in the previous section. It concerns not the specific character of particular figures in a given manifold, but rather the manifolds themselves. In order to grasp the "real content" of a given geometry, Klein argues, the specific character of the spatial elements in the domain is irrelevant. What is relevant is the structural content of a geometry characterized by its transformations group.[54]

[53] Given that α is a group isomorphism, also the composition of transformations as well as the inverse function on transformations are preserved.

[54] Compare Marquis (2009) for a similar assessment of Klein's approach.

Moreover, given Klein's indifference to the basic ontology of geometrical objects, his account of transfer principles can be viewed as a general criterion for the structural equivalence of geometries. To use one of his own examples, the theory of binary forms and the projective geometry of points on a conic are taken to be equivalent in the sense that they share the same structural content, independent of their particular geometrical domains. This sameness of structure is expressed by the fact that their corresponding groups or transformations are isomorphic (or, in Klein's terms, "similar").

How is the structuralism implicit in Klein's account of transfer principles related to contemporary philosophy of mathematics? Surprisingly, there is still yet little discussion on possible criteria of the structural equivalence of mathematical theories in the present debate. As we saw in the previous section, structuralists are mainly concerned with questions regarding the nature of abstract structures and, to a lesser degree, with the question of when two structures should be taken to be equivalent.[55] Nevertheless, there is a close connection between Klein's approach and subsequent developments in category theory. In fact, category theory is often considered as a "conceptual extension" or "generalization" of Klein's program. Consider, for instance, the following well-known passage from Eilenberg and Mac Lane's article "General Theory of Natural Equivalences" of 1945:

> This may be regarded as a continuation of the Klein Erlanger Programm, in the sense that a geometrical space with its group of transformations is generalized to a category with its algebra of mappings. (237)

The relation between the study of categories and Klein's program expressed here seems to be this: in Klein's account, the structure of a geometry is expressed in terms of the group of transformations acting on a given manifold. Similarly, category theory can be understood as the study of particular categories in terms of their objects and structure preserving mappings.[56] As in Klein's account, the category-theoretic study of objects such as graphs or monoids can be understood as the study of the invariant properties expressible in terms of structure-preserving mappings between these objects.

I cannot develop any further here the question in what sense category theory can be viewed as a generalization of Klein's group-theoretic approach in geometry.[57] However, it will be interesting to point to two connections between Klein's conceptual approach and an account of mathematical structuralism motivated

[55] See, in particular, Resnik (1997) and Shapiro (1997) on the characterization of the equivalence of mathematical structures based on the notion of definitional equivalence.

[56] See Awodey (2010) for a textbook presentation of category theory.

[57] See, in particular, Marquis (2009) for an extensive study of this question and the historical development of category theory more generally.

by category theory.[58] A first point of contact between Klein's account and categorical structuralism concerns the indifference with respect to the nature of mathematical objects considered. Categorical structuralists explicitly share Klein's view that what matters in mathematics are not the particular mathematical objects or their set-theoretic representations but rather their "invariant form." Thus, the objects in a particular category are not supposed to have any properties other than those specifiable in terms of mappings between them. Compare Awodey on this structuralist conception of objects:

> This lack of specificity or determination [of particular objects] is not an accidental feature of mathematics. . . . Rather it is characteristic of mathematical statements that the particular nature of the entities involved plays no role, but rather their relations, operations, etc.—the "structures" that they bear—are related, connected, and described in the statements and proofs of theorems. (2004, 59)

The second point of contact concerns the notion of the structural equivalence of theories. We saw that Klein's motivation for his Erlangen program was not to study geometries in isolation but to compare different geometries investigated at the time in terms of their transformation groups. Similarly, research in category theory is usually not confined to the isolated study of particular mathematical categories but mainly concerns the study of relations between different categories. The central concept used for this task is that of a *functor*, i.e., a structure-preserving mapping between categories:

Definition 5 (Functor): A functor between categories **C** and **D** is a mapping F: **C** → **D** of objects to objects and arrows to arrows such that

(a) $F(f: A \to B) = F(f): F(A) \to F(B)$

(b) $F(1_A) = 1_{F(A)}$

(c) $F(g \circ f) = F(g) \circ F(f)$

A functor is a mapping between two categories that leaves invariant the domain and codomains of mappings, the identity mappings, and the composition of mappings. Consequently, each categorical property specifiable in the

[58] See, for instance, Awodey (1996) and McLarty (2004) for different versions of categorical structuralism.

one category will be transferred by the functor into a categorical property of the objects in the second category (see Awodey 2010, 8–9).

It seems natural to think of functors as a mathematical generalization of Klein's notion of transfers. We saw earlier that Klein's Erlangen program gives an account of the "essential sameness" of geometries in terms of transfer principles. A plausible category-theoretic reconstruction of this *Kleinian* notion of inter-theoretic equivalence can be given in terms of the concept of categorical equivalence:

Definition 6 (Equivalence of categories): An equivalence of categories **C** and **D** consists of a pair of functors E: **C** → **D** and F: **D** → **C** such that there are natural isomorphisms:[59]

$$E \circ F \cong 1_D$$

$$F \circ E \cong 1_C$$

Given the conceptual similarity between Klein's program and category theory as a general framework for structural mathematics, one can consider this notion of categorical equivalence as a generalization of Klein's notion of structural equivalence.[60] In both cases, the structure of a given theory is determined by the algebraic properties of mappings or transformations. Moreover, two theories are considered to be identical on a structural level in case there exists a mapping that allows one to transfer the algebraic structure of one theory to the other theory.[61]

5. Conclusion

Klein's Erlangen program of 1872 presents a landmark contribution to algebraic reasoning in geometry and, more generally, to the gradual implementation of a structural approach in modern mathematics. The aim in this chapter was to further substantiate these claims and to specify Klein's particular version of geometrical structuralism. As we saw, his account is based on the systematic use of

[59] Notice that this notion is more general than the isomorphism of categories: functors E and F are not required to be inverses of each other, but only "pseudo-inverses." This means that for any $D \in$ **D**: $E \circ F(D) \cong D$, not necessarily $E \circ F(D) = D$. See Awodey (2010).

[60] See again Marquis (2009) for a closer discussion of the relation between Klein's work and modern category theoretic concepts.

[61] See Barrett and Halvorson (2016) for a recent proposal to explicate the equivalence of scientific theories in terms of the notion of categorical equivalence.

transformation groups in order to specify the invariants of configurations in a manifold as well as the structural content of geometries.

The chapter focused on two thematic points: the first one was an important strand of the mathematical background of Klein's program, namely different proposals to generalize the principle of duality in 19th-century geometry. This included Plücker's purely analytic study of dualities between geometrical configurations of any dimension. It was shown how his approach led to the formulation of different transfer principles in projective geometry. Moreover, Klein developed his own account of geometry in direct continuation with these "structuralist" methods of Plücker and Hesse. Specifically, his approach presents a generalization by group-theoretic means of two ideas first developed in preceding geometrical research, namely (i) the use of structure-preserving mappings in reciprocity and transfer principles and (ii) the focus on *invariant* form in the analytic presentation of geometrical figures and their properties.

The second aim in this chapter was to connect Klein's conception of geometry with current debates on structuralism. As we saw, there are at least two points of contact between his ideas and more recent philosophical work. The first concerns Klein's approach to specify geometrical properties and the notion of congruence (or equivalence) of configurations relative to a given group of transformations. This approach clearly mirrors recent work on structural properties and structural identity conditions for mathematical objects in non-eliminative structuralism. More specifically, building on recent work by Marquis, we saw that Klein's approach can be interpreted as a version of *in re* structuralism for geometry, according to which the real subject matter of a geometry consists of abstract figure *types* specifiable in terms of a congruence relation.

The second point of contact concerns Klein's proposal to specify the structural equivalence of two geometries based on transfer principles. This approach is closely related to later attempts to think about mathematical objects (and the equivalence of theories) in category-theoretic terms. In particular, a natural generalization of Klein's "transfer by mapping" approach can be given in terms of the notion of categorical equivalence of categories of theories. This analogy with modern category theory also suggests to treat Klein's specific geometrical structuralism as a precursor of more recent accounts of categorical structuralism, that is, attempts by Awodey and others to capture the philosophers' talk about mathematical structures in the language of category theory.

Acknowledgments

This project has received funding from the European Research Council (ERC) under the European Union's Horizon 2020 research and innovation program (grant agreement No. 715222). Research related to this article was presented at the conference "The Emergence of Structuralism and Formalism" in Prague in 2016, at the workshop "Roots of Mathematical Structuralism" at University of California, Riverside in 2016, at the workshop "Mathematics and Its Philosophy in the 19th Century" at the University of Amsterdam in 2018, as well as at the HOPOS 2018 meeting at the University of Groningen. I would like to thank the respective audiences for their comments. I would especially like to thank Erich Reck, Francesca Biagioli, Eduardo Giovannini, Günther Eder, and Henning Heller for helpful discussions on the topics addressed in this chapter.

References

Awodey, Steve. 1996. Structure in Mathematics and Logic: A Categorical Perspective. *Philosophia Mathematica* 4, 209–237.

Awodey, Steve. 2004. An Answer to Hellman's Question: Does Category Theory Provide a Framework for Mathematical Structuralism? *Philosophia Mathematica* 12(1), 54–64.

Awodey, Steve. 2010. *Category Theory*. Oxford: Oxford University Press.

Awodey, Steve. 2014. Structuralism, Invariance, and Univalence. *Philosophia Mathematica* 22(1), 1–11.

Barrett, Thomas, and Hans Halvorson. 2016. Glymour and Quine on Theoretical Equivalence. *Journal of Philosophical Logic* 45(5), 467–483.

Benacerraf, Paul. 1965. What Numbers Could Not Be. *Philosophical Review* 74(1), 47–73.

Biagioli, Francesca. 2016. *Space, Number, and Geometry from Helmholtz to Cassirer*. New York: Springer.

Biagioli, Francesca. 2018. Structuralism and Mathematical Practice in Felix Klein's Work on Non-Euclidean geometry. Unpublished manuscript.

Birkhoff, Garrett, and M. K. Bennett 1988. Felix Klein and His "Erlanger Programm." In *History and Philosophy of Modern Mathematics*, edited by William Aspray and Philip Kitcher, pp. 145–176. Minneapolis: University of Minnesota Press.

Brannan, David A., Matthew F. Esplen, and Jeremy J. Gray. 2011. *Geometry*. 2nd ed. Cambridge: Cambridge University Press.

Coxeter, H. S. M. 1974/1987. *Projective Geometry*. 2nd ed. Berlin: Springer.

Eder, Günther, and Georg Schiemer. 2018. Hilbert, Duality, and the Geometrical Roots of Model Theory. *Review of Symbolic Logic* 11(1), 48–86.

Eilenberg, Samuel, and Saunders Mac Lane. 1945. General Theory of Natural Equivalences. *Transactions of the American Mathematical Society* 58, 231–294.

Fano, Gino. 1907. Kontinuierliche geometrische Gruppen: Die Gruppentheorie als geometrisches Einteilungsprinzip. *Encyklopädie der Mathematischen Wissenschaften*, vol. 3, 4B, 289–388.

Gray, Jeremy J. 2005. Felix Klein's Erlangen Program, "Comparative Considerations of Recent Geometrical Researches" (1872). In *Landmark Writings in Western Mathematics, 1640–1940*, edited by Ivor Grattan-Guiness, pp. 544–552. Amsterdam: Elsevier.

Gray, Jeremy J. 2008. *Plato's Ghost: The Modernist Transformation of Mathematics*. Princeton, NJ: Princeton University Press.

Hawkins, Thomas. 1984. The *Erlanger Programm* of Felix Klein: Reflections on Its Place in the History of Mathematics. *Historia Mathematica* 11, 442–470.
Hawkins, Thomas. 2000. *Emergence of the Theory of Lie Groups: An Essay in the History of Mathematics. 1869–1926.* New York: Springer.
Hesse, Ludwig O. 1866a. Ein Übertragungsprincip. *Journal für die reine und angewandte Mathematik* 66, 15–21.
Hawkins, Thomas. 1866b. Vier Vorlesungen aus der analytischen Geometrie. *Zeitschrift für Mathematik und Physik* 11, 369–425.
Keränen, Jukka. 2001. The Identity Problem for Realist Structuralism. *Philosophia Mathematica* 9(3), 308–330.
Klein, Felix. 1872a. Über Liniengeometrie und metrische Geometrie. In *Gesammelte Mathematische Abhandlungen*, vol. 1, pp. 106–126. Berlin: Springer.
Klein, Felix. 1872. *Vergleichende Betrachtungen über neuere geometrische Forschungen.* Erlangen: Deichert.
Klein, Felix. 1873. Über die sogenannte Nicht-Euklidische Geometrie (Zweiter Aufsatz). *Mathematische Annalen* 6, 112–145.
Klein, Felix. [1921–23] 1973. *Gesammelte Mathematische Abhandlungen.* 3 vols. Berlin: Springer.
Klein, Felix. 1926. *Vorlesungen über die Entwicklung der Mathematik im 19. Jahrhundert*, vol. 1. Berlin: Springer.
Klein, Felix. 1928. *Elementarmathematik vom höheren Standpunkte aus.* Vol. 3. Berlin: Springer.
Klein, Felix. 2016. *Elementary Mathematics from an Advanced Standpoint.* Vol. 2, *Geometry*. Translated by E. R. Hedrick and C. A. Noble. Berlin: Springer.
Korbmacher, Johannes, and Georg Schiemer. 2018. What Are Structural Properties? *Philosophia Mathematica*, 26/3, 295–323.
Leitgeb, Hannes, and James Ladyman. 2008. Criteria of Identity and Structuralist Ontology. *Philosophia Mathematica* 16(3), 388–396.
Lie, Sophus. 1888–93. *Theorie der Transformationsgruppen.* 3 vols. Leipzig: Teubner.
Linnebo, Øystein, and Richard Pettigrew. 2014. Two Types of Abstraction for Structuralism. *Philosophical Quarterly* 64(255), 267–283.
Lorenat, Jemma. 2015. Polemics in Public: Poncelet, Gergonne, Plücker, and the Duality Controversy. *Science in Context* 28(4), 545–585.
Marquis, Jean-Pierre. 2009. *From a Geometrical Point of View: A Study of the History and Philosophy of Category Theory.* New York: Springer.
McLarty, Colin. 2004. Exploring Categorical Structuralism. *Philosophia Mathematica* 12, 37–53.
Nagel, Ernest. 1939. The Formation of Modern Conceptions of Formal Logic in the Development of Geometry. *Osiris* 7, 142–223.
Plücker, Julius. 1828/1931. *Analytisch-geometrische Entwicklungen.* 2 vols. Essen: G. D. Baedeker.
Plücker, Julius. 1830. Über ein neues Coordinatensystem. *Journal für die reine und angewandte Mathematik* 5, 1–36.
Plücker, Julius. 1835. *System der analytischen Geometrie auf neue Betrachtungsweisen gegründet und insbesondere eine ausführliche Theorie der Curven dritter Ordnung enthaltend.* Berlin: Duncker und Humblot.

Plücker, Julius. 1846. *System der Geometrie des Raumes in neuer analytischer Behandlungsweise insbesondere die Theorie der Flächen zweiter Ordnung und Classe enthaltend*. Düsseldorf: Schaub'sche Buchhandlung.

Plump, Mechthild Ulrike. 2014. Julius Plücker: Leben und Werk eines analytischen Geometers im 19. Jahrhundert. Dissertation, Universität Wuppertal.

Reck, Erich. 2003. Dedekind's Structuralism: An Interpretation and Partial Defense. *Synthese* 137(3), 369–419.

Reck, Erich, and Price, Michael P. 2000. Structures and Structuralism in Contemporary Philosophy of Mathematics. *Synthese* 125(3), 341–383.

Resnik, Michael D. 1997. *Mathematics as a Science of Patterns*. New York: Oxford University Press.

Rowe, David E. 1989. The Early Geometrical Works of Sophus Lie and Felix Klein. In *The History of Modern Mathematics*, edited by David E. Rowe and John McCleary, vol. 1, pp. 209–273. Boston: Academic Press.

Rowe, David E. 1992. Klein, Lie, and the "Erlanger Programm." In *1830–1930: A Century of Geometry*, edited by, Luciano Boi, Domnique Flament, and Jean-Michel Salanskis, pp. 45–54. Berlin: Springer.

Schiemer, Georg. 2018. Cassirer and the Structural Turn in Modern Geometry. *Journal for the History of Analytical Philosophy* 6(3), 183–212.

Scholz, Erhard. 1980. *Geschichte des Mannigfaltigkeitsbegriffs von Riemann bis Poincaré*. Basel: Birkhäuser.

Shapiro, Stewart. 1997. *Philosophy of Mathematics: Structure and Ontology*. New York: Oxford University Press.

Shapiro, Stewart. 2008. Identity, Indiscernibility, and Ante Rem Structuralism: The Tale of i and –i. *Philosophia Mathematica* 3(16), 285–309.

Tobies, Renate. 1981. *Felix Klein*. Leipzig: Teubner.

Torretti, Roberto. 1978. *Philosophy of Geometry from Riemann to Poincaré*. Dordrecht: Reidel.

Wussing, Hans. [1984] 2007. *The Genesis of the Abstract Group Concept: A Contribution to the History of the Origin of Abstract Group Theory*. New York: Dover Publications.

6

The Ways of Hilbert's Axiomatics: Structural and Formal

Wilfried Sieg

It is a remarkable fact that Hilbert's programmatic papers from the 1920s still shape, almost exclusively, the standard contemporary perspective of his views concerning (the foundations of) mathematics; even his own, quite different work on the foundations of geometry and arithmetic[1] from the late 1890s is often understood from that vantage point. My essay pursues one main goal, namely, to contrast Hilbert's *formal axiomatic method* from the early 1920s with his *structural axiomatic approach* from the 1890s. Such a contrast illuminates the circuitous beginnings of the finitist consistency program and connects the complex emergence of structural axiomatics with transformations in mathematics and philosophy during the 19th century; the sheer complexity and methodological difficulties of the latter development are partially reflected in the well known, but not well understood correspondence between Frege and Hilbert. Taking seriously the goal of formalizing mathematics in an *effective* logical framework leads also to contemporary tasks, not just historical and systematic insights; those are briefly described as "one direction" for fascinating work.

1. Context

Hilbert gave lectures on the foundations of mathematics throughout his career. Notes for many of them have been preserved and are treasures of information; they allow us to reconstruct the path from Hilbert's logicist position, deeply influenced by Dedekind and presented in lectures starting around 1890, to the program of finitist proof theory in the early 1920s. The development toward *proof theory* begins, in some sense, in 1917, when Hilbert gave his talk "Axiomatisches Denken" in Zürich. This talk is rooted in the past and points to the future. As to the future, Hilbert suggested:

[1] Arithmetic is understood in this early work not as dealing with natural but rather with real numbers.

Wilfried Sieg, *The Ways of Hilbert's Axiomatics* In: *The Prehistory of Mathematical Structuralism*. Edited by: Erich H. Reck and Georg Schiemer, Oxford University Press (2020). © Oxford University Press.
DOI:10.1093/oso/9780190641221.003.0006

> We must—that is my conviction—take the concept of the specifically mathematical proof as an object of investigation, just as the astronomer has to consider the movement of his position, the physicist must study the theory of his apparatus, and the philosopher criticizes reason itself. (Hilbert 1918, 1115)

Hilbert recognized in the next sentence that "the execution of this program is at present, to be sure, still an unsolved problem." If one takes formalization of mathematical proofs as an important part of this program, then initial tentative steps were taken at the 1904 International Congress of Mathematicians in Heidelberg. Hilbert presented there an equational fragment of elementary number theory and used its formal structure as the basis for a syntactic consistency proof (by induction on derivations).

Four years earlier, Hilbert had articulated the need of a consistency proof for arithmetic in the Second Problem of his famous talk at the International Congress of Mathematicians in Paris; he wrote:

> I wish to designate the following as the most important among the numerous questions that can be asked with regard to the axioms: to prove that they are not contradictory, that is, that a finite number of logical steps based upon them can never lead to contradictory results. (Hilbert 1900b, 1104)

The axioms really concern analysis, i.e., the theory of complete ordered fields, and Hilbert points for their formulation to his paper *Über den Zahlbegriff*, which had been delivered at the meeting of the German Association of Mathematicians in September 1899. Its title indicates a part of the intellectual context, as Kronecker had published 12 years earlier a well-known paper with the same title (Kronecker 1887). In that paper, Kronecker sketched a way of introducing irrational numbers, without accepting the general notion. It is precisely to the *general concept* that Hilbert wanted to give a proper foundation—using the axiomatic method. The axiom system Hilbert formulated for the real numbers is not presented in the contemporary formal-logical style. Rather, it is given in an algebraic way and assumes that a system exists whose elements satisfy the axiomatic conditions; consistency proofs were to discharge that assumption. Because of this existence assumption, Hilbert and Bernays called this methodological approach *existential axiomatics* in the 1920s; I want to call it structural axiomatics and contrast it with formal axiomatics.

Section 2 of this chapter discusses structural axiomatics, whereas section 4 is devoted to the emergence and significance of formal axiomatics. The recognition of the dramatic difference between the two and the very character of the former is crucial for elucidating the different perspectives Frege and Hilbert expressed in their correspondence concerning Hilbert's *Grundlagen der Geometrie*; that topic is treated in the short interlude between sections 2 and 4. It is ironic that Frege saw a way of formulating Hilbert's view and the characteristic abstract element of modern mathematics, but insisted on a narrow *mis*understanding. What then is the methodological approach of structural axiomatics around 1900? How and for what purpose did Hilbert move, almost 20 years later, from it to formal axiomatics, using Frege's work as mediated by Whitehead and Russell's *Principia Mathematica* (1910 13)?

2. Structural Axiomatics

To begin with, Hilbert points out in *Über den Zahlbegriff* that the *axiomatic* way of proceeding is quite different from the *genetic method* used in arithmetic; it rather parallels the ways of geometry.

> Here [in geometry] one begins customarily by assuming the existence of all the elements, i.e., one postulates at the outset three systems of things (namely, the points, lines, and planes), and then—essentially after the model of Euclid—brings these elements into relationship with one another by means of certain axioms of linking, order, congruence, and continuity. [Hilbert should have included the axiom of parallels.] (Hilbert 1900a, 1092)

The geometric ways are taken over for the arithmetic of real numbers or rather, one might argue, are reintroduced into arithmetic by Hilbert; after all, they do have their origin in Dedekind's work on arithmetic and algebra. Hilbert frames and formulates the axioms for the real numbers in his (1900a) as follows: "We think a system of things, and we call them numbers and denote them by a, b, c, . . . We think these numbers to be in certain mutual relations, whose precise and complete description is obtained through the following axioms." Then the axioms for an ordered field are formulated and rounded out by the requirement of continuity via the Archimedean principle and the axiom of completeness.

This formulation is not only in the spirit of the geometric ways, but mimics Hilbert's contemporaneous and axiomatic presentation of *Grundlagen der Geometrie*, which is viewed even today as paradigmatically modern.

> We think three different systems of things: we call the things of the first system points and denote them by A, B, C, . . . ; we call the things of the second system lines and denote them by a, b, c, . . . ; we call the things of the third system planes and denote them by α, β, γ, . . . ; . . . We think the points, lines, planes in certain mutual relations . . . ; the precise and complete description of these relations is obtained by the axioms of geometry. (Hilbert 1899, 437)

Five groups of geometric axioms follow and, in the original Festschrift, the fifth group consists of just the Archimedean principle. In the French edition of 1900 and the second German edition of 1903, the completeness axiom is included. The latter axiom requires in both the geometric and the arithmetic case that the assumed structure is maximal, i.e., any extension satisfying the remaining axioms must already be contained in it. Hilbert's completeness formulations are frequently criticized as being metamathematical and, to boot, of a peculiar sort. However, they are just ordinary mathematical ones, if the abstract algebraic character of the axiom systems is kept in mind; they provide *structural definitions* of Euclidean space and the continuum, respectively. In the case of arithmetic we can proceed as follows: call a system *Acontinuous* when it satisfies the axioms of an ordered field and the Archimedean axiom, and call it *fully continuous* if and only if A is continuous and for any system B, if $A \subseteq B$ and B is continuous, then $B \subseteq A$. So Hilbert's axioms characterize fully continuous systems in analogy to the way in which Dedekind's conditions characterize simply infinite ones in (Dedekind 1888), or in which the axioms of group theory characterize groups.

Hilbert thought about axiom systems in this structural way already in his first lectures on the foundations of geometry. He had planned to give them in the summer term of 1893, but their presentation was shifted to the following summer term. Using the notions *System* and *Ding* so prominent in (Dedekind 1888), he formulated the central question as follows:

> What are the necessary and sufficient and mutually independent conditions a system of things has to satisfy, so that to each property of these things a geometric fact corresponds and conversely, thereby making it possible to completely describe and order all geometric facts by means of the above system of things? (Hilbert *1894, 72–73)

At a later point, Hilbert inserted the remark that this system of things provides a "complete and simple image of geometric reality." In the introduction to the notes for the 1898–99 lectures *Elemente der Euklidischen Geometrie*, this question is connected with Hertz's *Prinzipien der Mechanik*:

> Using an expression of Hertz (in the introduction to the *Prinzipien der Mechanik*) we can formulate our main question as follows: What are the necessary and sufficient and mutually independent conditions a system of things has to be subjected to, so that to each property of these things a geometric fact corresponds, and conversely, thereby having these things provide a complete "image" of geometric reality. (Hilbert *1898–99, 303)

One can see here the shape of a certain logical or set-theoretic structuralism in the foundations of mathematics and physics.[2] But what are the things whose system is implicitly postulated? As late as 1922 Hilbert articulated the *axiomatische Begründungsmethode* for analysis as follows:

> The continuum of real numbers is a system of things that are connected to each other by certain relations, so-called axioms.[3] In particular the definition of the real numbers by Dedekind cuts is replaced by two continuity axioms, namely, the Archimedean axiom and the so-called completeness axiom. In fact, Dedekind cuts can then serve to determine the individual real numbers, but they do not serve to define [the concept of] real number. On the contrary, conceptually a real number is just a thing of our system. . . . The standpoint just described is altogether logically completely impeccable, and it only remains thereby undecided, whether a system of the required kind can be thought, i.e., whether the axioms do not lead to a contradiction. (Hilbert 1922, 1118)

The remark "conceptually a real number is just a thing of our system" does not answer any question concerning the (nature of the) things making up the system, but it expresses a crucial element of structural axiomatics and is fully in line with Dedekind's views. In addition, the issue of consistency had been an explicit part of Dedekind's logicist program, and the further discussion of that issue will reveal details of Hilbert's position.

In the 19th century, logicians viewed the *consistency of a notion* from a semantic perspective as requiring a model. That is the way *we* put matters, whereas those earlier logicians, including Frege, saw themselves as facing the task of exhibiting a system that falls under the notion. Dedekind addressed the consistency problem for the notion of a simply infinite system exactly from such a traditional view. The methodological need for doing that is implicit in his (1872), but

[2] At this point one might also ask: What is the mathematical connection, in particular, between arithmetic and geometric structures? The informal comparison of the geometric line with the system of cuts of rational numbers in Dedekind's (1872) contains almost all the ingredients to establish these structures to be isomorphic; missing is the concept of mapping. That concept was available to Dedekind by 1879 and, with it, these considerations can be extended to show that arbitrary, fully continuous systems are isomorphic. The methodological remarks in (Dedekind 1888) about the arithmetic of natural numbers can now be extended to that of the real numbers.

[3] This is a peculiar formulation, even in the original German. As it happens, Hilbert formulated matters more precisely in his letter of September 22, 1900, addressed to Frege: "I am of the opinion that a concept can be logically determined only through its relations to other concepts. These relations, formulated in particular statements, I call axioms and thus I arrive at the view that the axioms . . . are the definitions of the concepts."

Here is the German text: "Meine Meinung ist eben die, dass ein Begriff nur durch seine Beziehungen zu anderen Begriffen logisch festgelegt werden kann. Diese Beziehungen, in bestimmten Aussagen formulirt [*sic!*], nenne ich Axiome und komme so dazu, dass Axiome . . . die Definitionen der Begriffe sind."

it is formulated most clearly in a letter to Keferstein dated February 27, 1890, a little more than a year after the publication of *Was sind und was sollen die Zahlen?*

> After the essential nature of the simply infinite system, whose abstract type is the number sequence N, had been recognized in my analysis (71, 73) the question arose: Does such a system exist at all in the realm of our thoughts? Without a logical proof of existence, it would always remain doubtful whether the notion of such a system might not perhaps contain internal contradictions. Hence the need for such a proof (articles 66 and 72 of my essay).

In article 66, Dedekind attempted to prove the existence of an infinite system within logic and, on the basis of that "proof," he provided in article 72 an example of a simply infinite system that was to guard against internal contradictions of the very notion.

Hilbert turned his attention to natural numbers around 1904 and used Dedekind's conditions for simply infinite systems, not as part of a structural definition, but as *formal* axioms. Until then he had taken for granted their proper foundation and focused on the notion of real numbers. Hilbert's retrospective remarks in (*1904) make this quite clear: the general concept of irrational number had created the "greatest difficulties," and Kronecker represented this point of view most sharply.[4] Those difficulties, Hilbert now claims, are overcome when the concept of natural number is secured, as the further steps toward real numbers can be taken without a problem. (It remains a puzzle why that was not as clear to Hilbert in 1899 as it had been to Dedekind in 1888; but see the discussion below.) This dramatic change of view raises the question, what did

[4] These issues are discussed in (Hilbert *1904, 164–167). The remark concerning Kronecker is found on pp. 165–166: "Die Untersuchungen in dieser Richtung [foundations for the real numbers] nahmen lange Zeit den breitesten Raum ein. Man kann den Standpunkt, von dem dieselben ausgingen, folgendermaßen charakterisieren: Die Gesetze der ganzen Zahlen, der Anzahlen, nimmt man vorweg, begründet sie nicht mehr; die Hauptschwiergkeit wird in jenen Erweiterungen des Zahlbegriffs (irrationale und weiterhin komplexe Zahlen) gesehen. Am schärfsten wurde dieser Standpunkt von Kronecker vertreten. Dieser stellte geradezu die Forderung auf: Wir müssen in der Mathematik jede Tatsache, so verwickelt sie auch sein möge, auf Beziehungen zwischen ganzen rationalen Zahlen zurückführen; die Gesetze dieser Zahlen andrerseits müssen wir ohne weiteres hinnehmen. Kronecker sah in den Definitionen der irrationalen Zahlen Schwierigkeiten und ging so weit, dieselben gar nicht anzuerkennen."

Here is the English translation: The investigations in this direction [concerning the foundations for real numbers] took the largest space for a long time. The standpoint from which they started can be characterized as follows: the laws for integers, the cardinal numbers, are taken for granted without any further justification; the main difficulty is seen in the extensions of the number concept (irrational and furthermore complex numbers). This standpoint was most strongly represented by Kronecker. He in fact required outright: in mathematics, we have to reduce every fact, however complicated it may be, to relations between whole rational numbers; the laws for these numbers, on the other hand, we have to accept without much ado. Kronecker saw difficulties in the definitions of irrational numbers and went so far as not to recognize them at all.

Hilbert see then as the "greatest difficulties" for the general concept of irrational numbers?

A somewhat vague, but nevertheless informative answer emerges from Hilbert's earlier discussion of a consistency proof for arithmetic; such a proof, Hilbert writes in *Über den Zahlbegriff*, should use "a suitable modification of familiar methods of reasoning." In the Paris lecture he suggested finding a *direct* proof and made "familiar methods of reasoning" more explicit:

> I am convinced that it must be possible to find a direct proof for the consistency of the arithmetical axioms [as proposed in *Über den Zahlbegriff* for the real numbers], by means of a careful study and suitable modification of the known methods of reasoning in the theory of irrational numbers. (Hilbert 1900b, 1104)

Hilbert believed at this point, it seems, that the genetic buildup of the real numbers could *somehow* be exploited to yield the blueprint for a semantic consistency proof in Dedekind's style. There are, however, difficulties with the genetic method that prevent it from easily providing a proper foundation for the general concept of irrational numbers. Hilbert's concerns are formulated most clearly in (Hilbert *1905, 10–11):

> It [the genetic method] defines things by generative processes, not by properties—what must really appear to be desirable. Even if there is no objection to defining fractions as systems of two integers, the definition of irrational numbers as a system of infinitely many numbers must appear to be dubious. Must this number sequence be subject to a law, and what is to be understood by a law? Is an irrational number being defined, if one determines a number sequence by throwing dice? These are the kinds of questions with which the genetic perspective has to be confronted.

Precisely this issue was to be overcome (or to be sidestepped) by the axiomatic method. In *Über den Zahlbegriff* Hilbert writes:

> Under the conception described above, the doubts that have been raised against the existence of the totality of real numbers (and against the existence of infinite sets generally) lose all justification; for by the set of real numbers we do not have to imagine, say, the totality of all possible laws according to which the elements of a fundamental sequence can proceed, but rather—as just described—a system of things whose mutual relations are given by the finite and closed system of axioms I–IV. (Hilbert 1900a, 1095)

In his Paris lecture he articulated that point and re-emphasized that "the continuum . . . is not the totality of all possible series in decimal fractions, or of all possible laws according to which the elements of a fundamental sequence may proceed." Rather, it is *any* system of things whose mutual relations are governed by the axioms; the completeness axiom, in particular, guarantees the continuity of the system without depending on any method of generating real numbers. The consistency proof is "the proof of the existence of the totality of real numbers." Hilbert expanded the second point by saying,

> In the case before us, where we are concerned with the axioms for real numbers in arithmetic, the proof of the consistency of the axioms is at the same time the proof of the mathematical existence of the totality [*Inbegriff*] of real numbers or of the continuum. Indeed, when the proof for the consistency of the axioms shall be fully accomplished, the doubts, which have been expressed occasionally as to the existence of the totality of real numbers, will become totally groundless. (Hilbert 1900b, 1105)

Could Hilbert think of addressing the consistency problem "by a careful study and suitable modification of the known methods of reasoning in the theory of irrational numbers," if he did not have in mind, ever so vaguely, the construction of a particular (Dedekindian) logical model?

Hilbert had known since 1897, through his correspondence with Cantor, about the difficulties in set theory and their impact on Dedekind's foundational work. Nevertheless, he did not move away from his programmatic position and the associated strategy for proving consistency until 1903 or 1904 at the latest. In the summer term of 1904, Hilbert lectured on *Zahlbegriff und Quadratur des Kreises*, and the notes written by Max Born reveal a significant change: Hilbert examines the paradoxes for the first time and sketches various foundational approaches. These discussions are taken up in his talk at the Heidelberg Congress in August of that year, where he presents a syntactic approach to the consistency problem. The goal is still to guarantee the existence of a suitable system, but the method of proof is inspired by one important aspect of the earlier investigations; he, in contrast to Dedekind, had formulated a quasi-syntactic notion of consistency already in his (1899) and (1900a); namely, no finite number of logical steps leads from the axioms to a contradiction. This notion is *quasi*-syntactic, as no deductive principles are explicitly provided.

Hilbert viewed the geometric axioms not only as characterizing a system of things that presents a "complete and simple image of geometric reality," but viewed them also in a very traditional way: the axioms must allow us to purely logically establish all geometric facts. Dedekind held such a view quite explicitly with respect to his "axioms" for natural numbers, i.e., the characteristic conditions for simply infinite systems; see his (1888, #73). Hilbert described this pivotal deductive role of axioms in the introduction to the Festschrift in a methodologically refined way:

> The present investigation is a new attempt at formulating for geometry a *simple* and *complete* system of mutually independent axioms; it is also an attempt at deriving from them the most important geometric propositions in such a way that the significance of the different groups of axioms and the import of the consequences of the individual axioms is brought to light as clearly as possible. (Hilbert 1899, 436)

The same perspective is expressed in the Paris lecture, where Hilbert states, first of all, that the totality of real numbers is "a system of things whose mutual relations are governed by the axioms set up and for which all propositions, and only those, are true that can be derived from the axioms by a finite number of logical inferences." Then, two fundamental problems have to be confronted for both geometry and arithmetic:

> The necessary task then arises of showing the consistency and the completeness of these axioms; i.e., it must be proved that the application of the given axioms can never lead to contradictions, and, further, that the system of axioms suffices to prove all geometric [and arithmetic] propositions. (Hilbert 1900a, 1092–1093)

It is not clear whether completeness of the axioms requires the proof of *all* true geometric (arithmetic) propositions or just of those that are part of the established corpora.

Independent of this issue is the question, which logical inferences are admitted in proofs? Frege criticized Dedekind on that point in the preface to his *Grundgesetze der Arithmetik*, claiming that the brevity of Dedekind's development of arithmetic in (Dedekind 1888) is only possible "because much of it is not really proved at all." He continues:

> Nowhere is there a statement of the logical or other laws on which he builds, and, even if there were, we could not possibly find out whether really no others were used—for to make that possible the proof must be not merely indicated but completely carried out.

Apart from demanding that the logical principles be made explicit, Frege hints at an additional aspect of such a systematic presentation that applies to Hilbert's *Grundlagen der Geometrie* as well. That aspect will be discussed in section 4, whereas the next section attempts to clarify, with the broader understanding of structural axiomatics we have gained, the main issue in the correspondence between Frege and Hilbert.

3. Interlude

My discussion is concerned exclusively with the six letters that were exchanged between Frege and Hilbert in the period from December 1899 to September 1900; they are all concerned with Hilbert's *Grundlagen der Geometrie* (and are found in Frege 1980). Frege wrote the opening letter to Hilbert on December 27, 1899; in it he seeks clarification on some important methodological questions pertaining to the *Grundlagen*. Frege reports that he had discussed parts of the work with his Jena colleagues Thomae and Gutzmer, and that they were not always clear about Hilbert's "real view" (*eigentliche Meinung*). As a start, Frege asks about Hilbert's use of "Erklärung" and "Definition"; they seem to be used for similar purposes, but by using both Hilbert presumably wants to indicate a difference—which is not clear to them. What makes matters even more difficult to understand, Frege points out, is the fact that *axioms* are taken to *define* relations under the heading *Erklärung*. Thus, it appears to Frege, Hilbert does not respect the sharp boundaries between axioms and definitions. Definitions are, after all, *Festsetzungen* ("determinations," "stipulations," or "agreements"), whereas axioms are true statements that are not be proved, as our knowledge of them arises from a source that is different from logic. That leads Frege to the observation that the truth of axioms guarantees that they do not contradict each other, and that no separate proof of consistency is required. Although that is of course a perspective different from Hilbert's, there seems to be some common ground when Frege remarks, in the context of independence proofs for the axioms, "You had to take a higher standpoint, from which Euclidian geometry appears as a special case of a more general [case]." (Frege 1980, 11)

In his response of December 29, 1899, Hilbert points out that, for example, the *Erklärung* for the concept "between" is indeed a proper definition, as its characteristic conditions (*Merkmale*) are given by the group of axioms II 1–II 5 that involve "between." If one wants to take "definition" in the exact traditional sense, he writes, then one would have to say:

> "Between" is a relation for the points of a line that satisfies the following characteristic conditions: II 1 . . . II 5. (Frege 1980, 11)

Later on, he emphasizes that he has absolutely no objection, if Frege wanted to simply call his axioms characteristic conditions (cf. footnote 3). Having discussed the striking and much-emphasized difference of their views concerning consistency and truth, Hilbert comes back to what he very strongly views as *the main issue* (*Hauptsache*) and asserts:

> The renaming of "axioms" as "characteristic conditions" is a pure formality and, in addition, a matter of taste—in any event, it is easily accomplished. (Frege 1980, 12).

This assertion holds sensibly for relations like "between," but not—as Hilbert then also claims—for the basic objects, e.g., points. The latter claim is in conflict with Hilbert's own view he describes next (Frege 1980, 13), namely that "any theory is only a framework [*Fachwerk*] or a schema of concepts together with the necessary relations between them." The basic elements (*Grundelemente*), Hilbert says, "can be thought in arbitrary ways."

Neither Hilbert nor Frege remembered that Dedekind presented in his (1888) under the heading *Erklärung* the definition of a simply infinite system: a system N is *simply infinite* if and only if there is an element 1 and a mapping Φ, such that the characteristic conditions (α)–(γ) hold for them.[5] This structural definition can be seen as providing a second-level concept in the sense in which Frege discusses it in his next letter of January 6, 1900 (Frege 1980, 17); Hilbert could have easily reformulated his *Erklärung* as a Dedekindian one: a triple of systems P, L, and E is a *Euclidian space* if and only if there are relations . . . , such that the characteristic conditions I–V (i.e., the geometric axioms in groups I through V) hold for them. Given such a common perspective, there would have been no reason for the fundamental disagreement Frege saw; indeed, there would have been a precise logical articulation of the abstract character of the emerging modern mathematics.[6]

4. Formal Axiomatics

Hilbert insisted that theorems in geometry or arithmetic must be established by a finite sequence of logical steps from the axioms; for the arithmetic of natural numbers Dedekind made exactly the same demand, considering as starting points of proofs the characteristic conditions for simply infinite systems. Since "axiom" can be taken for Hilbert as synonymous with "characteristic condition," Dedekind and Hilbert share this perspective on proof. Frege, starting with his

[5] These characteristic conditions "correspond" to the so-called Peano axioms and express the following: (α) – ϕ is a mapping from N to N; (β) – N is the chain of the system {1}; (γ) – 1 is not in the ϕ image of N; (δ) – ϕ is a similar (injective) mapping.

[6] There are important connections to 19th-century theories of concept formation, in particular to those formulated by H. Lotze in his *Logik* of 1843 as well as in the expanded editions of 1874 and 1880. There are good reasons to think that Dedekind was influenced by them already very early on in

1879 *Begriffsschrift*, precisely described the logical steps that can be taken in order to obtain "gapless" proofs and asserted later that in his logical system "inference is conducted like a calculation," but observed:

> I do not mean this in a narrow sense, as if it were subject to an algorithm the same as . . . ordinary addition or multiplication, but only in the sense that there is an algorithm at all, i.e., a totality of rules which governs the transition from one sentence or from two sentences to a new one in such a way that nothing happens except in conformity with these rules. (Frege 1984, 237)

In his 1902 review of Hilbert's *Grundlagen der Geometrie*, Poincaré radicalized the formal character of the axiomatic conditions and the algorithmic nature of logical rules, in a different context and for a different purpose; he writes:

> M. Hilbert has tried, so to speak, putting the axioms in such a form that they could be applied by someone who doesn't understand their meaning, because he has not ever seen either a point, or a line, or a plane. It must be possible, according to him [Hilbert], to reduce reasoning to purely mechanical rules.

Poincaré brings out this essential formal, mechanical aspect in a dramatic way and reinterprets the idea of strict formalization as machine executability.[7] Indeed, he suggests giving the axioms to a reasoning machine, like Jevons's logical piano, and observing whether all of geometry could be obtained. Such a mechanical formalization might seem "artificial and childish," Poincaré remarks, if it were not for the important question of completeness:

> Is the list of axioms complete, or have some of them escaped us, namely those we use unconsciously? . . . One has to find out whether geometry is a logical consequence of the explicitly stated axioms or, in other words, whether the axioms, when given to the reasoning machine, will make it possible to obtain the sequence of all theorems as output [of the machine].

his career; Dedekind's stay in Göttingen as a student and then Privatdozent (from 1850 to 1858) fell completely into the period Lotze was professor of philosophy there (from 1844 to 1880). The parallelism of Dedekind's general reflections on concepts in his (1854) and the expanding remarks on their significance in the preface to (Dedekind 1888) is rather striking, as are their view that arithmetic is a part of logic. However, a very distinctive notion of "abstraction" is centrally used by Lotze already in the 1843 *Logik* and allows a cohesive understanding of Dedekind's way of introducing "abstract" concepts. That has been worked out in a paper I wrote with Rebecca Morris. The paper was accepted for publication in 2015 and published as (Sieg and Morris 2018). (2018)

---

[7] How these considerations are woven into a broader philosophical and mathematical web is discussed in my paper *On Computability* (2009a), in particular on pp. 535–561.

The completeness problem is not formulated as a "mechanical" one in Hilbert's Festschrift, but the issue of what logical steps can be used in proofs is coming to the fore in Hilbert's lectures through references to logical calculi.

The syntactic approach to consistency proofs Hilbert suggested in his 1904 Heidelberg talk uses formal axioms and a logical calculus that is extremely restricted—it is purely equational! In the summer term 1905, Hilbert gave lectures under the title *Logische Prinzipien des mathematischen Denkens*; they are as special as those from 1904, but for a different reason: one finds in them a critical examination of logical principles and a realization that a broader logical calculus is needed that captures, in particular, universal statements and inferences.[8] In his subsequent lectures on the foundations of mathematics, Hilbert does not really progress beyond the reflections presented in his (*1905) until 1917: in the Zürich talk *Axiomatisches Denken* a new perspective emerges. In that essay, Hilbert remarks that the consistency of the axioms for the real numbers can be reduced, by employing set theoretic concepts, to that of integers. Hilbert continues:

> In only two cases is this method of reduction to another more special domain of knowledge clearly not available, namely, when it is a matter of the axioms for the *integers* themselves, and when it is a matter of the foundation of *set theory*; for here there is no other discipline besides logic to which it were possible to appeal.
>
> But since the examination of consistency is a task that cannot be avoided, it appears necessary to axiomatize logic itself and to prove that number theory and set theory are only parts of logic. (Hilbert 1918, 1113)

Hilbert remarks that Russell and Frege provided the basis for this approach.

The detailed study of *Principia Mathematica* began, however, already in 1913 and resulted in the remarkable 1917–18 lectures *Prinzipien der Mathematik*, the very first lectures on modern mathematical logic. All the tools for formally developing mathematics (number theory, but also analysis) were made available in these lectures and are in the background of the work of the Hilbert group during the 1920s. The material was published only 10 years later in (Hilbert and Ackermann 1928). As to the formalization issue, one finds this remark at the

[8] How important those lectures were can be seen from a letter Hilbert sent to his friend Hurwitz in late 1904 or early 1905, definitely after the Heidelberg talk: "It seems that various parties started again to investigate the foundations of arithmetic. It has been my view for a long time that exactly the most important and most interesting questions have not been settled by Dedekind and Cantor (and a fortiori not by Weierstrass and Kronecker). In order to be forced into the position to reflect on these matters systematically, I announced a seminar on the 'logical foundations of mathematical thought' for next semester." In Dugac (1976, 271).

very end of the lecture notes, after the beginnings of analysis had been developed and, in particular, the least upper bound principle had been established: "Thus it is clear that the introduction of the axiom of reducibility is the appropriate means to turn the ramified calculus into a system out of which the foundations for higher mathematics can be developed" (Hilbert *1917–18, 246).

The 1917–18 lectures gave a full and rigorous mathematical presentation of first- and higher-order logic, including a careful distinction between syntax and semantics.[9] There was, however, no immediate return to a syntactic approach to the consistency problem. Poincaré's incisive analysis of the "proof theoretic" approach in Hilbert (1905), but also Hilbert's own insight into its shortcomings, shifted his attention from the stand he had advocated in Heidelberg. Hilbert came back to it only in the summer semester of 1920. The notes from that term contain a consistency proof for the same fragment of arithmetic that had been investigated in 1904. Its formulation is informed by the investigations of the 1917–18 term: the language is more properly described; the combinatorial argument is sharper (albeit a bit different from that given in 1904), and it is further simplified in (Hilbert 1922). The details are important for (the development of) proof theory, but I emphasize here only the overarching strategic point of the modified argument; namely, Hilbert insists that Poincaré has been refuted.

> Poincaré's objection, claiming that the principle of complete induction can only be proved by complete induction, has been refuted by my theory. (Hilbert 1922, 167)

In the second part of this paper, the formal theory is expanded beyond the purely equational calculus. This expansion has one peculiarity, namely, that negation is applied only to identities. Hilbert gives as the reason for this severe restriction that the formal system is to be kept constructive. Thus, we can conclude that in (Hilbert 1922) the proper metamathematical direction of Hilbert's finitist program had not yet been taken.

The paper was based on talks Hilbert had given in the spring and summer of 1921 in Copenhagen and Hamburg. The first of three Copenhagen talks was devoted to the role of mathematics in physics and has been preserved as a manuscript in Hilbert's own hand. It is worth quoting its last paragraph in order to re-emphasize Hilbert's broad vision for mathematics.

[9] In Hilbert's lecture, a proof of the semantic completeness of the logical calculus for sentential logic is indicated; it is formulated and proved in Bernays' Göttingen *Habilitationsschrift* (Bernays 1918).

> We went rapidly through those chapters of theoretical physics that are currently most important. If we ask, what kind of mathematics do physicists consider, then we see that it is *analysis* that serves physicists in its complete *content* and *extension*. Indeed, it does so in two different ways: first it serves to *clarify* and *formulate* their ideas, and second—as an instrument of calculation—it serves to obtain quickly and reliably numerical results, which help to check the correctness of their ideas. Apart from this face seen by physicists, there is a completely different face that is directed toward philosophy; the features of that face deserve no less our interest. That topic will be discussed in my subsequent talks. (Hilbert *1921, 28–29)

In his "subsequent talks" Hilbert expounded his philosophical perspective, but argued also against the constructive stand of Brouwer and Weyl. In the 1922 essay he contrasts their constructivism with his own, claiming that Weyl has "failed to see the path to the fulfillment of these [constructive] tendencies" and that "only the path taken here in pursuit of axiomatics will do full justice to the constructive tendencies":

> The goal of securing a firm foundation for mathematics is also my goal. I should like to regain for mathematics the old reputation for incontestable truth, which it appears to have lost as the result of the paradoxes of set theory; but I believe that this can be done while fully preserving its accomplishments. The method I follow in pursuit of this goal is none other than the axiomatic method; its essence is as follows. (Hilbert 1922, 1119)

Having described the essential nature of the axiomatic method, he points to the task of recognizing the consistency of the arithmetical axioms including, at this point, axioms for number theory, analysis, and set theory. This task leads now to the investigation of formalisms, in which parts of mathematics can be carried out. The concepts of proof and provability are thus "relativized" to the underlying formal axiom system, but Hilbert emphasizes:

> This relativism is natural and necessary; it causes no harm, since the axiom system is constantly being extended, and the formal structure, in keeping with our constructive tendency, is becoming more and more complete. (Hilbert 1922, 1127)

Hilbert's version of constructivism comes in not only through the construction of ever more complete formalisms for the development of mathematics, but most importantly through their effective character; after all, it is the effectiveness of the basic concepts, in particular of the concept of (formal) proof, that makes it possible to investigate the formalisms from a restricted mathematical, "finitist" point of view.

The term *finite Mathematik* (finitist mathematics) appears for the first time in the 1921–22 notes.[10] Hilbert and Bernays give no philosophical explication; they rather develop finitist number theory, which they do not view as encompassing all of finitist mathematics. On the contrary, they envision a dramatic expansion in order to recognize why and to what extent "the application of transfinite inferences [in analysis and set theory] always leads to correct results." We have to expand, so they demand, the domain of objects that are being considered:

> I.e., we have to apply our intuitive considerations also to figures that are not number signs. Thus we have good reasons to distance ourselves from the earlier dominant principle according to which each theorem of pure mathematics is ultimately a statement concerning integers. This principle was viewed as expressing a fundamental methodological insight, but it has to be given up as a prejudice.
>
> We have to adhere firmly to one demand, namely, that the figures we take as objects must be completely surveyable and that only discrete determinations are to be considered for them. It is only under these conditions that our claims and considerations have the same reliability and evidence as in intuitive number theory. (Hilbert *1921–22, Part III, 4a–5a)

Hilbert and Bernays had thus arrived at a new standpoint that was to serve as the basis for consistency proofs, and formulated the goal of establishing the correctness of formally provable finitist statements.[11] The new approach involves induction and recursion principles for the broader class of "figures," that is, for effectively generated syntactic objects, like terms or formulas or

[10] What is the status of "finit" in "finite Mathematik" in historical regard? Was it introduced from a special philosophical perspective that emerged in the early 1920s? The way in which the concept is actually introduced in (*1921–22), very matter-of-factly, almost leads one to suspect that Hilbert and Bernays employ a familiar one. That suspicion is hardened by *aspects of the past* and an *attitude* that is pervasive until 1932: as to the attitude, finitism and intuitionism were considered as coextensional until Gödel and Gentzen proved in 1932 the consistency of classical arithmetic relative to its intuitionist version; as to aspects of the past, Hilbert himself remarked that Kronecker's conception of mathematics "essentially coincides with our finitist mode of thought."

The concrete background of the term "finitism" should be a topic of thorough historical analysis and definitely include Bernstein's paper (1918). I just state as a fact that in the lecture notes from the 1920s no detailed discussion of "finite Mathematik" is found. The most penetrating analysis is given in (Bernays 1930), still emphasizing the coextensionality of finitism and intuitionism. Indeed, Bernays interprets Brouwer's mathematical work as showing that considerable parts of analysis and set theory can be "given a finitist foundation." For a contemporary and informed discussion, see (Tait 1981) and (Tait 2002).

[11] The claim that consistency implies (mathematical) existence is no longer maintained; see in particular Bernays's later reflections in a note from between 1925 and 1928 that was published in Sieg (2002).

proofs. That is clearly articulated in the second half of the 1921–22 lectures and carried out with strikingly novel, genuinely proof-theoretic techniques. Hilbert and Bernays proved in these lectures the consistency of a quantifier-free fragment of formal elementary number theory, roughly what is now called primitive recursive arithmetic (PRA); the argument is sketched and the methodological approach is described in (Hilbert 1923)—a talk Hilbert gave in September 1922.

In the notes for other lectures from the early 1920s, one finds innovative meta-mathematical work, in particular, the introduction of the epsilon calculus and the associated substitution method, which tries to overcome in leaps and bounds the obstinate difficulties of giving finitist consistency proofs for strong formal theories, but in the end that work is unsuccessful. The reason for this failure was revealed already in 1931 for the theories that were of central interest, analysis and set theory: Gödel's second incompleteness theorem states for them that their consistency cannot be proved by means that are formalizable in those very theories. For the general formulation of the incompleteness theorems (as pertaining to *all* formal theories containing a modicum of number or set theory) Gödel needed an adequate notion of computability characterizing the "formality" of formal theories. In the 1964 postscriptum to his 1934 Princeton lectures, he argued that Turing's work provides such a notion of *mechanical procedure*, and that it is actually "required by the concept of formal system, whose essence it is that reasoning is completely replaced by mechanical operations on formulas" (Gödel 1964, 370). The second incompleteness theorem is usually taken in the way I formulated it earlier: finitist consistency proofs cannot be obtained for theories that are sufficiently strong; in other words, Hilbert's *finitist* program has been refuted for theories like analysis or set theory. The first incompleteness theorem is frequently taken to refute Hilbert's view that there is no *ignorabimus* in mathematics. However, that is not Gödel's view at all. In the 1964 postscriptum he explicitly states that the incompleteness theorems "do not establish any bounds for the power of human reason, but rather for the potentialities of pure formalism in mathematics" (370). For him, Hilbert's no-ignorabimus view is not connected to "pure formalism," as I'll point out in the next section.

5. One Direction

Gödel begins his (193?) by recalling Hilbert's famous words, "For any precisely formulated mathematical question a unique answer can be found." He takes these words to mean that for any mathematical proposition A there is a proof of either A or not-A, "where by 'proof' is meant something which starts from evident axioms and proceeds by evident inferences." He argues that the incompleteness theorems show that something is lost when one takes the step from this notion of proof to a formalized one: "It is not possible to formalize mathematical evidence even in the domain of number theory, but the conviction about which Hilbert speaks remains entirely untouched. Another way of putting the result is this: it is not possible to mechanize mathematical reasoning."

It is important to recognize early and deep roots of Hilbert's foundational thinking. His work in geometry and arithmetic around 1900 gave or indicated systematic developments, within the framework of structural axiomatics. A more formal presentation was sought already in Hilbert (*1905), but was viewed as extremely difficult. It is equally important to see that the study of *Principia Mathematica* raised the prospect of formalizing mathematics on the broad basis of type or set theory. In order to more closely reflect mathematical practice, Hilbert and Bernays even developed in (*1921–22) a new kind of logical calculus with axioms for all the logical connectives; these axioms were later basic for the introduction and elimination rules of Gentzen's natural deduction calculi.[12] But the more urgent proof theoretic issues surrounding the consistency problem shifted attention away from the formal representation of mathematical practice. With the advance of computer technology and the myriad problems that can be addressed mathematically, it is important, however, to construct formal frameworks in which mathematics can be formally developed not only "in principle," but actually and intelligibly.

To achieve that goal, it has been argued for a long time, computers have to take over routine parts of argumentation, so that human users can focus on the broader conceptual and strategic aspects of proof construction. In spite of much exciting contemporary work in (interactive) theorem proving, there is still no somewhat general theory of mathematical proof (as Hilbert had called for in 1917). I have taken the lack of a general theory as one central reason to formulate and implement strategies for *automated* proof search; the work I have been doing in this direction is described in (Sieg 2010). This is a first step not toward a general theory, but rather toward the more modest goal of finding intelligible proofs that reflect (and are inspired by) logical and mathematical understanding. Even this step already forces us, on the one hand, to make explicit the conceptual ingenuity underlying successful human proof construction; it asks us, on the other hand, to integrate it with proof-theoretic features of derivations (subformula properties of normal forms) for the sake of efficiency.

Coming back to the beginning of this essay, we clearly have to analyze concepts and articulate characteristic conditions for them, but we must also

12 This connection is sketched in (Sieg 2010, 197–198).

consider mathematical arguments as they present themselves "in experience," so to speak; that is how Dedekind in (Dedekind 1890) described his attitude toward the notion of natural numbers. That requires enriching suitable formal frames by *leading ideas* for particular parts of mathematics, thus, an effective conceptual organization that can be expressed through appropriate heuristics.[13] Saunders Mac Lane, one of the last logic students in Hilbert's Göttingen and a friend of Gentzen, wrote his thesis (Mac Lane 1934) with this general goal. He published an English summary (1935) that emphasizes the crucial programmatic features. In particular, it is pointed out that proofs are not "mere collections of atomic processes, but are rather complex combinations with a highly rational structure." When reflecting in 1979 on his early work, Mac Lane ended with the remark: "There remains the real question of the actual structure of mathematical proofs and their strategy. It is a topic long given up by mathematical logicians, but one which still—properly handled—might give us some real insight" (Mac Lane 1979, 66). It seems to me that we have the computational and logical tools to successfully tackle Mac Lane's "real question."

Appendix

The text that follows is a (small) part of the lectures Hilbert gave, with the assistance of Bernays, in the winter semester of 1917–18. As an example of the systematic presentation and penetrating analysis the axiomatic method affords, Hilbert discussed at first the axiom system for Euclidean geometry and then gave proofs of consistency and independence. It is the beginning of that section of the lectures that is presented here. The noteworthy fact is the emphasis on the *assumption* of a system of objects, etc., the core feature of structural axiomatics. There is no hint of a finitist proof-theoretic approach to the consistency problem, neither here nor later in these lectures when the system of arithmetic (for real numbers) is being discussed; at the very end, one rather finds the suggestion that the theory of types (with the axiom of reducibility) provides the appropriate means for developing the foundations of higher mathematics. This is an echo of the logicist leanings Hilbert had expressed in his Zürich lecture *Axiomatisches Denken* (Hilbert 1918).

[13] Three particular examples are discussed in my (2010): Gödel's incompleteness theorems, the Cantor-Bernstein theorem, and the Pythagorean theorem.

German Text (Hilbert 1917–18, 19–20)

Zu dem geometrischen Axiomensystem, dessen Aufstellung ich das letzte Mal beendet hatte, sei zunächst bemerkt, dass die Anordnung der Axiome im Einzelnen zwar eine gewisse Willkür aufweist, im grossen aber doch mit Notwendigkeit bestimmt ist. Bei Untersuchungen über mögliche Vereinfachungen dieses Axiomensystems hat man darauf zu achten, dass Kürzungen durch eine Reduktion der Annahmen nicht immer von Vorteil sind, insofern dadurch die Uebersicht leiden kann.

Wenden wir uns nun zur genaueren Diskussion des vorgelegten Systems der geometrischen Axiome, so ist zuerst die Frage der *Widerspruchslosigkeit* zu behandeln. Diese Frage ist darum die wichtigste, weil durch einen Widerspruch, zu dem die Konsequenzen aus den Axiomen führen würden, dem ganzen System seine Bedeutung genommen wäre. Das Axiomensystem ist ja so aufzufassen, dass über dem Ganzen die Annahme steht, es gebe drei Arten von Dingen, die wir als Punkte, Geraden und Ebenen bezeichnen und zwischen denen gewisse Beziehungen bestehen, welche durch die Sätze, die wir Axiome nennen, beschrieben werden. Diese Annahme wäre offenbar gegenstandslos, wenn man von den Axiomen durch richtige Schlussfolgerungen zu einem Satz und auch zu seinem Gegenteil gelangen könnte. Die Unmöglichkeit eines solchen Falles nennen wir die Widerspruchslosigkeit des Axiomensystems.

Den Beweis der Widerspruchslosigkeit für die Axiome der Geometrie werde ich führen durch Aufweisung eines Systems von Gegenständen, die miteinander in solcher Weise verknüpft sind, dass sich eine Zuordnung dieser Gegenstände und Verknüpfungen zu den in den geometrischen Axiomen vorkommenden Gegenständen und Beziehungen herstellen lässt, bei welcher sämtliche Axiome erfüllt sind. Die Gegenstände, auf die ich mich hierbei als auf etwas Gegebenes berufe, sind der Arithmetik entnommen, und das Beweisverfahren kommt also darauf hinaus, dass die Widerspruchslosigkeit der Geometrie auf die Widerspruchslosigkeit der Arithmetik zurückgeführt wird, indem gezeigt wird, dass ein Widerspruch, der sich bei den Folgerungen aus den geometrischen Axiomen ergäbe, auch innerhalb der Arithmetik einen Widerspruch zur Folge haben müsste.

Translation

As to the geometric axiom system whose exposition I completed last time, I would like to remark, first of all, that the particular ordering of the axioms shows in the small a certain arbitrariness, but in the large it is determined with necessity. When investigating possible simplifications of this axiom system one has to observe carefully that shortenings by reducing the [number of] assumptions is not always advantageous, as such a reduction may diminish the overall perspicuity.

When turning attention now to the more precise discussion of this system of geometric axioms, the question of consistency has to be addressed first. This question is the most important one, because the whole system would lose its significance if a contradiction could be inferred from the axioms. After all, the axiom system is to be understood as being completely covered by the assumption that there are three kinds of things we refer to as points, lines, and planes, and that certain relations obtain between them that are described by the statements we call axioms. This assumption obviously would be groundless if it were possible to obtain a statement and its negation from the axioms by correct inferences. The impossibility of such a case is called the consistency of the axiom system.

I will carry out the consistency proof for the axioms of geometry by exhibiting a system of objects that are connected to each other in a particular way; these objects and connections can be associated with the objects and relations that occur in the geometric axioms in such a way that all the axioms are satisfied. The objects to which I appeal as something given are taken from arithmetic, and the method of proof amounts to reducing the consistency of geometry to the consistency of arithmetic. We do this by showing that a contradiction that could be inferred from the geometric axioms must lead to a contradiction within arithmetic.

Acknowledgments

This chapter directly builds on my *Hilbert's Proof Theory* (2009b), but focuses on the dramatically different perspectives on the axiomatic method during the 1890s, culminating in the Paris address of 1900, and the early 1920s, when the finitist consistency program developed in a methodologically coherent way; that program was presented for the first time in Hilbert's talk in Leipzig in the fall of 1922. The lectures of Hilbert and Bernays from around 1920 were described in (Sieg 1999) and, finally, have been published in (Ewald and Sieg 2013). The chapter was originally published in *Perspectives on Science* 22 (2014), 133–157. It is being republished here with the permission of MIT Press; I made only minor modifications.

References

Translations in this chapter are my own, except when I am quoting directly an English source (sometimes with modifications). Unpublished lecture notes of Hilbert's are located in Göttingen in two different places, namely, the Staats- und

Universitätsbibliothek and the Mathematisches Institut. The reference year of these notes below is preceded by a "*"; their location is indicated by SUB xyz and MI, respectively. Many of them have been published or are being prepared for publication in *David Hilbert's Lectures on the Foundations of Mathematics and Physics, 1891–1933* (Heidelberg: Springer).

Bernays, Paul. 1918. *Beiträge zur axiomatischen Behandlung des Logik-Kalküls*. Habilitation. Göttingen: Georg-August-Universität.

Bernays, Paul. 1930. Die Philosophie der Mathematik und die Hilbertsche Beweistheorie. Reprinted in Bernays 1976, pp. 17–61.

Bernays, Paul. 1976. *Abhandlungen zur Philosophie der Mathematik*. Darmstadt: Wissenschaftliche Buchgesellschaft.

Bernstein, Felix. 1919. Die Mengenlehre Georg Cantors und der Finitismus. *Jahresbericht der DMV* 28, 63–78.

Dedekind, Richard. 1854. Über die Einführung neuer Funktionen in der Mathematik. Habilitationsvortrag. Printed in Dedekind 1932, pp. 428–438. Translated in Ewald 1996, pp. 754–762.

Dedekind, Richard. 1872. *Stetigkeit und irrationale Zahlen*. Braunschweig: Vieweg. Translated in Ewald 1996, pp. 765–779.

Dedekind, Richard. 1888. *Was sind und was sollen die Zahlen?* Braunschweig: Vieweg. Translated in Ewald 1996, pp. 787–833.

Dedekind, Richard. 1890. Letter to H. Keferstein. Translated in van Heijenoort 1967, pp. 98–103.

Dedekind, Richard. 1932. *Gesammelte mathematische Werke*. Vol. 3. Edited by R. Fricke, E. Noether, and Ö. Ore. Braunschweig: Vieweg.

Dugac, Pierre. 1976. *Richard Dedekind et les fondements des mathématiques*. Paris: VRIN.

Ewald, William B., ed. 1996. *From Kant to Hilbert: A Source Book in the Foundations of Mathematics*. 2 vols. Oxford: Oxford University Press.

Ewald, William B., and Wilfried Sieg, eds. 2013. *David Hilbert's Lectures on the Foundations of Mathematics and Physics, 1891–1933*. Vol. 3. Heidelberg: Springer.

Feferman, S., and Wilfried Sieg, eds. 2010. *Proofs, Categories and Computations: Essays in Honor of Grigori Mints*. London: College Publications.

Frege, Gottlob. 1893. *Grundgesetze der Arithmetik, begriffsschriftlich abgeleitet*. Jena: Pohle.

Frege, Gottlob. 1980. *Gottlob Freges Briefwechsel*. Edited by G. Gabriel, F. Kambartel, and C. Thiel. Hamburg: Meiner.

Frege, Gottlob. 1984. *Collected Papers on Mathematics, Logic, and Philosophy*. Edited by B. McGuinness. Oxford: Oxford University Press.

Gödel, Kurt. 1931. Über formal unentscheidbare Sätze der Principia Mathematica und verwandter Systeme I. In Gödel 1986, pp. 126–195.

Gödel, Kurt. 1934. On Undecidable Propositions of Formal Mathematical Systems. In Gödel 1986, pp. 346–369.

Gödel, Kurt. 193?. [Undecidable Diophantine propositions]. In Gödel 1995, 164–175.

Gödel, Kurt. 1964. Postscriptum (to 1934). In Gödel 1986, pp. 369–371.

Gödel, Kurt. 1986. *Collected Works*. Vol. 1. Oxford: Oxford University Press.

Gödel, Kurt. 1995. *Collected Works*. Vol. 3. Oxford: Oxford University Press.

Hallett, Michael, and Ulrich Majer, eds. 2004. *David Hilbert's Lectures on the Foundations of Mathematics and Physics, 1891–1933*. Vol. 1. Heidelberg: Springer.

Hertz, Heinrich. 1894. *Die Prinzipien der Mechanik*. Vol. 3 of *Gesammelte Werke*. Leipzig: Barth.

Hilbert, David. *1894. Die Grundlagen der Geometrie. SUB 541.

Hilbert, David. *1898–99. Grundlagen der Euklidischen Geometrie. Lecture notes taken by H. von Schaper, MI. In Hallett and Majer 2004, pp. 302–395.

Hilbert, David. 1899. Grundlagen der Geometrie. In *Festschrift zur Feier der Enthüllung des Gauss-Weber-Denkmals in Göttingen*, pp. 1–92. Leipzig: Teubner. Reprinted in Hallett and Majer 2004, pp. 436–525.

Hilbert, David. 1900a. Über den Zahlbegriff. *Jahresbericht der DMV* 8, 180–194. Reprinted in *Grundlagen der Geometrie*, pp. 256–262. 3rd ed. Leipzig, 1909. Translated in Ewald 1996, pp. 1089–1095.

Hilbert, David. 1900b. Mathematische Probleme. *Nachrichten der Königlichen Gesellschaft der Wissenschaften zu Göttingen*, 253–297. Translated in Ewald 1996, pp. 1096–1105. Reprinted with additions in *Archiv der Mathematik und Physik* 3, 1901.

Hilbert, David. 1900c. *Les principes fondamentaux de la géometrie*. Translation of Hilbert 1899 with some additions. Paris: Gauthier-Villars.

Hilbert, David. *1904. *Zahlbegriff und Quadratur des Kreises*. Lecture notes taken by M. Born, MI.

Hilbert, David. 1905. Über die Grundlagen der Logik und der Arithmetik. In *Verhandlungen des Dritten Internationalen Mathematiker-Kongresses*, pp. 174–185. Leipzig: Teubner. Translated in van Heijenoort 1967, pp. 129–138.

Hilbert, David. *1905. *Logische Prinzipien des mathematischesn Denkens*. Lecture notes taken by Hellinger, MI.

Hilbert, David. 1918. Axiomatisches Denken. *Mathematische Annalen* 78, 405–415. Translated in Ewald 1996, pp. 1105–1115.

Hilbert, David. *1917–18. *Prinzipien der Mathematik*. Lecture notes taken by P. Bernays, MI.

Hilbert, David. *1921–22. *Grundlagen der Mathematik*. Lecture notes taken by P. Bernays, MI.

Hilbert, David. *1921. Natur und mathematisches Erkennen. Talk given in Copenhagen on March 14, 1921. SUB 589.

Hilbert, David. 1922. Neubegründung der Mathematik. *Abhandlungen aus dem mathematischen Seminar der Hamburgischen Universität 1*, 157–177. Translated in Ewald 1996, pp. 1117–1134.

Hilbert, David. 1923. Die logischen Grundlagen der Mathematik. *Mathematische Annalen 88*, 151–165. Translated in Ewald 1996, pp. 1136–1148.

Hilbert, David, and Wilhelm Ackermann. 1928. *Grundzüge der theoretischen Logik*. Berlin: Springer.

Kronecker, Leopold. 1887. Über den Zahlbegriff. *Crelles Journal für die reine und angewandte Mathematik* 101, 337–355. Reprinted in Kronecker 1899, pp. 251–274. This is an expanded version of the essay with the same title that was published in *Philosophische Aufsätze, Eduard Zeller zu seinem fünfzigjährigen Doctor-Jubiläum gewidmet*, Leipzig, 1887; the shorter essay is translated in Ewald 1996, pp. 947–955.

Kronecker, Leopold. 1899. *Werke*. Vol. 3. Leipzig: Teubner.

Lotze, Hermann. 1843. *Logik*. Leipzig: Weidmann'sche Buchhandlung. There are two expanded editions of this book, with the same title, published in 1874 and 1880.

Mac Lane, Saunders. 1934. Abgekürzte Beweise im Logikkalkul. Dissertation. Göttingen: Georg-August-Universität.

Mac Lane, Saunders. 1935. A Logical Analysis of Mathematical Structure. *The Monist* 45, 118–130.

Mac Lane, Saunders. 1979. A Late Return to a Thesis in Logic. In *Selected Papers*, edited by I. Kaplansky, pp. 63–66. New York: Springer.

Poincaré, Henri. 1902. Review of Hilbert 1899. *Bulletin des sciences mathématiques* 26, 249–272.

Sieg, Wilfried. 1999. Hilbert's Programs: 1917–1922. Bulletin of Symbolic Logic 5, 1–44.

Sieg, Wilfried. 2002. Beyond Hilbert's reach? In *Reading Natural Philosophy: Essays in the History and Philosophy of Science and Mathematics*, edited by D. Malament, pp. 363–405 Peru, IL: Open Court.

Sieg, Wilfried. 2009a. On Computability. In *Philosophy of mathematics*, edited by A. Irvine, pp. 535–630. Amsterdam: Elsevier.

Sieg, Wilfried. 2009b. Hilbert's Proof Theory. In *Handbook of the History of Logic*, vol. 5, edited by D. M. Gabbay and J. Woods, pp. 321–384 Amsterdam: Elsevier.

Sieg, Wilfried. 2010. Searching for Proofs (and Uncovering Capacities of the Mathematical Mind). In Feferman and Sieg 2010, pp. 189–215.

Sieg, Wilfried, and Rebecca Morris. 2018. Dedekind's Structuralism: Creating Concepts and Deriving Theorems. In *Logic, Philosophy of Mathematics, and Their History: Essays in Honor of W. W. Tait*, edited by Erich Reck. London: College Publications, pp. 251-301.

Sieg, Wilfried, Richard Sommer, and Carolyn Talcott, eds. 2002. *Reflections on the Foundations of Mathematics: Essays in Honor of Solomon Feferman*. Natick, MA: Association for Symbolic Logic.

Tait, W. W. 1981. Finitism. *Journal of Philosophy* 78, 524–546.

Tait, W. W. 2002. Remarks on finitism. In Sieg, Sommer, and Talcott 2002, pp. 410–419.

van Heijenoort, Jean, ed. 1967. *From Frege to Gödel: A Sourcebook of Mathematical Logic, 1879–1931*. Cambridge, MA: Harvard University Press.

Whitehead, Alfred North, and Bertrand Russell. 1910–13. *Principia Mathematica*. 3 vols. Cambridge: Cambridge University Press.

7

Noether as Mathematical Structuralist

Audrey Yap

1. Introduction

Emmy Noether's student B. L van der Waerden wrote of her that the maxim by which she always let herself be guided was that "all relations between numbers, functions, and operations become clear, generalizable, and truly fruitful only when they are separated from their particular objects and reduced to general concepts." This chapter will show how Noether's emphasis on abstraction and generalization of frameworks and results contributed to the abstract conception of structure found in contemporary mathematics. Doing so will demonstrate her contribution to structuralist methodology, though she did not herself advocate many philosophical views that we now associate with articulations of structuralism, such as the idea that structures are the real objects of mathematical study. Instead, Noether can be seen as exemplifying what Reck and Price (2000) have called *methodological structuralism*, as opposed to *philosophical structuralism*. The former approach notes that many of the entities studied in mathematics, such as various different number systems and geometrical spaces, are studied primarily in terms of their structural features, and considers this to be the proper approach to mathematical practice. Further, it contends that it is of no real mathematical concern what the intrinsic nature of such mathematical entities might be above and beyond such structural features. What distinguishes this approach from philosophical structuralism is that the methodological structuralist is only purporting to make claims about how we ought to do mathematics, namely confining the scope of the view to mathematical *practice*. Philosophical structuralism goes beyond the claims about correct practice to ask what the further implications of a structuralist methodology might be:

> The way many contemporary philosophers of mathematics (as well as philosophers of language and metaphysicians) specify it further is this: How are we supposed to think about reference and truth along these lines, e.g., in the case of arithmetic? And what follows about the existence and the nature of the natural numbers, as well as of other mathematical objects, even if the answer doesn't matter mathematically? Put more briefly, what are the semantic

Audrey Yap, *Noether as Mathematical Structuralist* In: *The Prehistory of Mathematical Structuralism*. Edited by: Erich H. Reck and Georg Schiemer, Oxford University Press (2020). © Oxford University Press.
DOI:10.1093/oso/9780190641221.003.0007

> and metaphysical implications of a structuralist methodology? (Reck and Price 2000, 346–347)

We might answer these semantic and metaphysical questions in a variety of different ways, from "thin" views that reject the very question of the real nature of mathematical entities, perhaps in favor of a formalist or inferential characterization, to "thick" Platonist views that consider structures to be real, if nonphysical, entities. While a lot of focus in contemporary structuralism has to with these philosophical questions, they rest on a characterization of mathematical practice that is nevertheless underpinned by methodological structuralism. While one can articulate a methodological structuralist view without committing oneself to any particular version of philosophical structuralism, the converse would seem like a strange move. After all, endorsing philosophical structuralism without also believing it to be the correct, or at least an appropriate, way of doing mathematics would suggest that the correct answers to the philosophical questions rest on an ill-advised methodology. This, while perhaps logically consistent as a view, seems nonetheless to be self-undermining.

To return to Noether, then, the purpose of this chapter is to demonstrate how she contributes to this philosophical tradition by enabling the very mathematical developments that make it possible to be a methodological structuralist in the first place. I will do this by tracing her development as a mathematician and seeing the ways in which she came to exemplify a structuralist approach to mathematical practice and lay the technical groundwork for further work on mathematical structure. This biographical look at Noether will follow the periods into which Hermann Weyl divided her career and methodological styles when he delivered her obituary. First, in Noether's early work, she worked in an algorithmic, constructive style, having begun her career studying under Paul Gordan. But she truly grew into her own as an algebraist, having been encouraged to study abstract algebra by Ernst Fischer. In the second period Weyl identifies, Noether worked on invariant theory, some of which comprised her habilitation work, but then turned to the theory of ideals, which is arguably one of her most important mathematical contributions, and the most important for structuralism. This chapter will focus primarily on Noether's middle and later work rather than her work under Gordan, which she had a tendency to dismiss later on in life. Though in many ways, her contributions to ideal theory are generalizations of work that had already been done by others, most notably Dedekind, it is exactly her emphasis on generalization that embodies her pioneering approach to abstract algebra and contributed to the abstract conception of structure used in contemporary mathematics. For example, Noether's work on commutative rings was similar to Dedekind's *Theory of Algebraic Integers*, but proved the results for arbitrary integral domains and domains of general rings. And her work on non-commutative rings generalized work in representation theory. I also point out that it was not just in Noether's own work, but also in her influence on her students such as Mac Lane and van der Waerden, who went on to provide their own significant contributions to algebra, in which her structural approach to mathematics can be seen.

Several themes will emerge in outlining the development of Noether's methodological structuralism. One will be a commitment to abstraction and generalization—consistently finding ways to treat objects from a perspective that showcases the underlying concepts rather than relying on features of individual number systems. Another will be the use of axiomatic methods; indeed, the structural approach is often associated with the axiomatic approach in the historical literature, and in Noether's case in particular, we can see her use of axioms as exemplifying her commitment to working with structural definitions. Indeed, according to a well-known classification of axioms due to Feferman (1999), the type of axioms that Noether primarily uses are called *structural* axioms. These organize the practice of mathematics by providing the definitions of well-known and recurring types of structures. They can be contrasted with *foundational* axioms, which are taken to be universal throughout mathematics by providing definitions for fundamental concepts such as number and set. Finally, we can see what Koreuber (2015) has called "conceptual mathematics," an approach that has been described by Stein (1988) as follows: "*The role of a mathematical theory is to explore conceptual possibilities*—to open up the scientific *logos* in general, in the interest of science in general" (Stein 1988, 252). This point of view is often associated with Cantor's and Dedekind's advocating free creation in mathematics, but can be seen in Noether's methodology as well. We can see that she is not too preoccupied with the extent to which the concepts she studies are instantiated, preferring instead to focus on the relationships between them.

What follows, though, will be organized biographically rather than thematically, as we shall see how these tendencies emerge in Noether's thought as she develops as a mathematician. The next two sections will discuss the three epochs into which Weyl divided her work. The first will briefly discuss Noether's early work on invariant theory, starting with the formal and algorithmic approach influenced by Gordan, and moving on to her adoption of the Hilbert-style approach to invariants. The second section will consider her work in algebra and the development of the general theory of ideals as well as her contributions to non-commutative algebras. Throughout each of these periods we can see ways in which the themes of generality and axiomatization inform her approach. I will conclude the chapter by relating Noether's methodological structuralism to some contemporary philosophical structuralist views articulated by Schiemer (2014), Landry (2011), and Awodey (1996, 2004), and considering the extent to which they are compatible with each other.

2. Invariant Theory

Emmy Noether's doctoral dissertation was written under Paul Gordan at Erlangen, entitled "On Complete Systems of Invariants for Ternary Biquadratic Forms." Invariant theory is a branch of algebra whose early systematization can be attributed to Arthur Cayley, but which is now often associated with the work of Hilbert and Gordan. The development of Noether's structural approach to mathematics can be seen in her departure from Gordan-style work on invariants in favor of a Hilbert-style approach. As we will see, she did not start out as a methodological structuralist, having been trained in systems of complex symbolic calculations and equations by her supervisor.

Briefly, the study of *invariants* considers transformations of polynomial forms. An *invariant* of a polynomial form is an expression in its coefficients that changes only by a factor determined in a fixed manner by the transformation. This area of mathematical research arose from the work of Cayley, James Sylvester, and others, on the algebraic relationships that hold between the coefficients of higher-degree polynomial forms (Kosmann-Schwarzbach 2011, 29–30). To put this more precisely,[1] a polynomial form is a homogenous polynomial—one whose nonzero terms all have the same degree. This might be done by adding an extra variable. The *discriminant* of a polynomial is a fixed quantity determined by an equation on its coefficients. For example, the quadratic form is given by

$$F(x,y) = Ax^2 + Bxy + Cy^2$$

and its discriminant is given by $\Delta_F = B^2 - AC$. Now suppose that we transform our initial polynomial form by substituting the variables x, y with linear combinations of new variables x', y' and substitution coefficients $\alpha, \beta, \gamma, \delta$:

$$x = \alpha x' + \beta y'$$

$$y = \gamma x' + \delta y'.$$

This transformation defines a new form $F'(x', y')$ each of whose coefficients A', B', C' depends on the substitution coefficients as well as the initial coefficients A, B, C. In general, an *invariant* of $F(x, y)$ is an expression I_F in the coefficients of F such that any transformation of F into a form F', such as

$$F(x,y) = F(\alpha x' + \beta y', \gamma x' + \delta y') = F'(x', y'),$$

[1] This exposition is largely drawn from McLarty (2012).

is such that $I_{F'} = (\alpha\delta - \beta\gamma)^{n} I_F$. So the analogous expression $I_{F'}$ in the coefficients of F' is the product of I_F and some power of an expression in the substitution coefficients. As it happens, for the quadratic form, the discriminant $\Delta_F = B^2 - AC$ is an invariant—and in fact, all of its invariants are powers of the discriminant. This in some sense lets us think of the discriminant as providing a complete system of invariants for the quadratic form.

Gordan's best-known contribution to invariant theory was the solution of his eponymous problem: given any polynomial form in two variables of arbitrary degree, he was able to develop a method for calculating a finite complete system of invariants for that form. That is, he found a routine through which such a finite basis for the invariants of any binary polynomial form could be calculated. Its main drawback, however, was that actually carrying out these calculations for forms of higher degree proved to be relatively infeasible. A lot of Gordan-style mathematics involved applying symbolic transformation rules to complex equations; this frequently involved prohibitively long lists of formulas, and the development of routines that were impractical actually to carry out.

Noether did work on Gordan-style problems using these very methods for some time, but would abandon his algorithmic approach in favor of a new approach to invariant theory developed by David Hilbert. This is likely due to the influence of one of Gordan's successors, Ernst Fischer, who was a proponent of the Hilbert-style approach to invariants, and had a clear influence on Noether's development. This eventually led to her being invited to Göttingen by Hilbert and Felix Klein. The dramatic differences between Hilbert's and Gordan's respective approaches to invariant theory can be seen in Hilbert's own solution of the Gordan Problem, in which he provided a proof by contradiction of the existence of a finite basis for certain invariants. This means that he did not produce an actual finite basis, nor a procedure through which one could be determined. Instead, his proof by contradiction simply demonstrated that one must exist, whatever it may look like.

Upon reading the proof, Gordan is said to have remarked, "Das ist nicht Mathematik; das ist Theologie [This is not mathematics; this is theology]" (Kimberling 1981, 11), though it has been pointed out that the extent of Gordan's resistance to Hilbert's proof is often exaggerated (McLarty 2012). Certainly a non-constructive proof would have seemed illegitimate from the perspective of Gordan's algorithmic methodology, and he did not initially find Hilbert's proof to be clear. But our interest here lies in the fact that for his student Noether, this marked a turn toward such non-constructive approaches to invariant theory, and to mathematics more generally.

Noether's subsequent work in differential invariant theory, some of which constituted her 1919 habilitation work, proved to be extremely significant in theoretical physics— a connection she was able to develop further in Göttingen working with Hilbert, who had already discussed with Einstein the possibility of enlisting Noether's help on some open problems with general relativity. In particular, conservation laws such as the law of conservation of energy did not seem to work the same way within the framework of general relativity as they did in classical mechanics. Some connections between invariant theory and the conservation of quantities had already been made by mathematicians such as Joseph Lagrange, but, as Kosmann-Schwarzbach (2011) argues, it was with Noether that these connections were made in their full generality. Differential invariants are sought in the case of forms whose coefficients are functions; when they are not constant functions, their derivatives are found in the transformed expressions. In her approach to differential invariants, we can already see evidence of Noether's conceptual approach to mathematical problems:

> The second study, *Invariante Variationsprobleme*, which I have chosen to present for my habilitation thesis, deals with arbitrary, continuous groups, finite or infinite, in the sense of Lie, and derives the consequences of the invariance of a variational problem under such a group. These general results contain, as particular cases, the known theorems concerning first integrals in mechanics and, in addition, the conservation theorems and the identities among the field equations in relativity theory. (Noether 1919, quoted in Kosmann-Schwarzbach 2011, 49)

What this quotation illustrates is that the conservation laws in physics for which she is famous are special cases of more general theorems that she was able to prove about Lie groups. As Kosmann-Schwarzbach (2011) points out, the symbolic Gordan-style method of calculating these invariants could find solutions, but did not reveal any general connections. Instead, Noether's more conceptual view, in which the invariants in the conservation laws are seen as special cases of something more general, was the first full treatment of this problem. But beyond this important work that was crucial to modern physics, she did not continue this line of research for much longer, and turned instead to work in algebra and the theory of ideals, a domain that would further showcase her ability to think in terms of general concepts and the relationships between them, and continue the development of her methodological structuralism.

3. Rings and Ideals

The second period of Noether's mathematical work that I will explore covers her work in abstract algebra, especially her groundbreaking contributions to ideal theory in the 1920s. The important pieces here are her 1921 paper "Idealtheorie in Ringbereichen" and subsequent 1926 paper "Abstrakter Aufbau der Idealtheorie." Many of the foundations for ideal theory were laid well before then by Ernst Kummer, working on factorization problems in the cyclotomic integers, and further developed by Leopold Kronecker so that they could be extended to systems of complex numbers. However, a contrasting approach, explicitly rejecting Kummer's and Kronecker's more algorithmic methods, was taken by Richard Dedekind in several versions of his theory of algebraic integers, and it is the latter's work that is taken up and generalized by Noether, to become what we now think of as ideal theory proper. We will see through this history how Dedekind's structural approach was further refined and generalized by Noether.

In 1846, Kummer introduced ideal prime factors for the cyclotomic integers, which had turned out to be quite useful in the study of higher reciprocity laws.[2] Cyclotomic integers are integers of the form

$$a_0 + a_1\theta + \cdots + a_n\theta^n,$$

where the $a_i \in \mathbb{Z}$ and θ is a primitive p-th root of unity, a complex number $\neq 1$ such that $\theta^p = 0$. For such integers, Kummer discovered that unique factorization fails for $p = 23$, and published this result in 1844. This means that in rings of cyclotomic integers $\mathbb{Z}[\theta]$, where θ is a primitive p-th root of unity as previously described, Kummer was able to find distinct decompositions of some ring elements into irreducible factors. Kummer's development, then, of the notion of ideal prime factors was intended to restore some, albeit weakened, form of unique factorization to the rings he was studying.[3] But what he defined when he introduced them were not the ideal prime factors themselves, but rather the multiplicity by which they divided cyclotomic integers in the rings in question. The idea was that if we conjecture the existence of the divisors, we can provide rules for calculating divisibility by them. The methods for determining the calculations were also limited in their application, which sufficed for Kummer's purposes, since he was studying reciprocity laws rather than aiming to develop a theory of ideals in his own right (Edwards 1980, 1992). But for further applications, it was useful to develop a more general description of divisibility by these ideal factors, for which we turn to Kronecker. In Kummer's 1859 paper on reciprocity laws, in which the most general version of his own theory appeared, he wrote that

[2] At the time, Kummer just regarded these as a special kind of complex number, but now we have a geometrical interpretation of these kinds of integers which warrants the use of the term "cyclotomic integers," since the roots of cyclotomic polynomials lie on the unit circle in the complex plane.

[3] Though as it turned out, his work in this area was also applied to Fermat's last theorem. See Edwards (1977) for more details on Kummer's theory and its applications in that area.

> Kronecker would very soon (*nächstens*) publish a work "in which the theory of the most general complex numbers" [meaning, surely, the most general algebraic number field] "is completely developed with marvelous simplicity in its connection with the theory of decomposable forms of all degrees." (Edwards 1992, 7)

No such theory appeared until 1881, when Kronecker published his *Grundzüge einer arithmetischen Theorie der algebraischen Grössen*. In this work, Kronecker developed a theory of divisors, which did generalize Kummer's theory to some extent, applicable as it was to general algebraic number fields (though Kronecker, as a constructivist, would likely not have accepted several algebraic number fields that we do today). While Kronecker's theory, like Kummer's, was based on a divisibility test, the main difference between the two is that Kronecker's does not test for divisibility by an ideal prime factor, but for divisibility by the greatest common divisor, which may be ideal or prime (or both). Further, Kronecker does not make use of the notion of a prime because primality is relative to the particular field in question, while the idea of a greatest common divisor is not (Edwards 1980, 353). Then in Kronecker's version of divisor theory, he is able to determine, independent of the underlying field, whether or not the greatest common divisor of some numbers divides an algebraic integer, in a more general fashion than Kummer's theory can.

However, in the interim period between Kummer's announcement and the appearance of Kronecker's *Grundzüge*, Richard Dedekind went through several versions of his own theory of ideals, which would lay some important foundations for Noether's own work in the area. In contrast with Kummer and Kronecker, Dedekind's approach to ideal theory was to explicitly define the ideal divisors in terms of sets of numbers in the domain. So he does not focus, as Kummer and Kronecker do, on the multiplicity by which a given ideal divides a number. Rather, he focusses on the properties possessed by collections of numbers that are divisible by some given factor. In other words, for any algebraic integer a in our domain, we consider the collection of all multiples of a, denoted by $i(a)$. This is called the principal ideal (*Hauptideal*) generated by a. It is easy to see that these ideals satisfy certain closure properties. Namely,

(1) If b and c both belong to $i(a)$, then both $b + c$ and $b - c$ belong to $i(a)$.
(2) If b belongs to $i(a)$, then for any c in the domain, bc also belongs to $i(a)$.

But now, we realize that a did not have to be an algebraic integer in the first place. Even if it was one of Kummer's ideal prime factors, $i(a)$ would still satisfy (1) and (2). And indeed, these two conditions turn out to be both necessary and sufficient for characterizing the ideal numbers, as each algebraic integer can be identified with its unique principal ideal.

Now, Dedekind comments several times on his dissatisfaction with Kummer's theory and his reasons for developing his own theory in such a different way. In particular, he writes that

> the greatest circumspection is necessary to avoid being led to premature conclusions. In particular, the notion of *product* of arbitrary factors, actual or ideal, cannot be exactly defined without going into minute detail. Because of these difficulties, it has seemed desirable to replace the ideal number of Kummer, which is never defined in its own right, but only as a divisor of actual numbers ω in the domain $\mathfrak{o}$, by a *noun* for something which actually exists. (Dedekind 1877, 94)

So an improvement of Dedekind's theory over Kummer's is that the ideal divisors are now identified with things that actually exist and are defined in their own right. Dedekind also writes that it was this very consideration—that the mathematical objects should form the basis of the theory—that led him to develop his theory of ideals in his distinctive way. While Kummer (and Kronecker) have a divisibility test at the heart of their theory, at the heart of Dedekind's theory is the set-theoretic notion of an ideal. To obtain unique factorization, each ideal corresponds to a well-defined list of "prime ideals," each of which divides it with a particular multiplicity. The concepts of multiplication and division are also given set-theoretic interpretations.

An ideal A is a multiple of B, or is divisible by B, exactly when every number in A is also in B, or when A is a subset of B. Yet alongside that notion, we also have the definition of multiplication for ideals such that for ideals A and B, their product AB is defined to be the set of all numbers ab and their sums such that $a \in A$ and $b \in B$. Now, one way to see the central problem of the work is as the task of showing that divisibility in this sense coincides with multiplication in this sense. For Dedekind writes that we see immediately that AB is divisible by both A and B, but "establishing the complete connection between the notions of divisibility and multiplication of ideals succeeds only after we have vanquished the deep difficulties characteristic of the nature of the subject" (Dedekind 1877, 60). For certainly, the definition of multiplication suggests an alternate notion of divisibility (analogous to that in the integers) such that A is divisible by B exactly when there is another ideal R such that $A = BR$. And what Dedekind aimed at showing is that the two notions of divisibility coincide. The difference between Dedekind and Kummer's approaches to divisibility is an excellent illustration of the difference between the conceptual approach that we will see in Noether, and the algorithmic approach that Kronecker employed. As part of his criticism of Kummer, Dedekind wrote that

> Kummer did not define ideal numbers themselves, but only the divisibility of these numbers. If a number α has a certain property A, to the effect that α satisfies one or more congruences, he says that α is divisible by an ideal number corresponding to the property A. While this introduction of new numbers is entirely legitimate, it is nevertheless to be feared at first that the language which speaks of ideal numbers being determined by their products, presumably in analogy with the theory of rational numbers, may lead to hasty conclusions and incomplete proofs. And in fact this danger is not always completely avoided. On the other hand, a precise definition covering *all* the ideal numbers that may be introduced in a particular numerical domain $\mathfrak{o}$, and at the same time a general definition of their multiplication, seems all the more necessary since the ideal numbers do not actually exist in the numerical domain $\mathfrak{o}$. To satisfy these demands it will be necessary and sufficient to establish once and for all the common characteristic of the properties $A,B,C,\ldots$ that serve to introduce the ideal numbers, and to indicate, how one can derive, from properties A,B corresponding to particular ideal numbers, the property C corresponding to their product. (Dedekind 1877, 57)

Then the methodological issue that Dedekind has with the Kummer-style algorithmic approach is that it might lead to imprecise definitions or perhaps incoherent ones. Given that Dedekind is not a mathematical Platonist (though Kummer and Kronecker are no Platonists either), the importance of precise definitions in ensuring the proper, legitimate creation of mathematical objects is not to be underestimated. The underpinnings for Dedekind's structuralism are arguably based in the potential for precise logical definition (Reck 2003; Yap 2009), and this approach is continued by Noether in her own work (Yap 2017).[4]

Noether's paper "Idealtheorie in Ringbereichen" (Noether 1921) generalizes Dedekind's unique factorization results for the algebraic integers into the more abstract setting of arbitrary rings. Given the introduction of the ring axioms between Dedekind's work and Noether's, this was a natural extension of the conceptual approach that both favored. In Noether's case, the focus on finding the best definitions possible for the concepts was characteristic of her methodological structuralist approach. Now, since Noether's work greatly resembles and builds on Dedekind's, I will not go through many of the details here, though they are discussed in other places (Corry 2004; Yap 2017). She also defines ideals as

[4] Avigad (2006) and (Reck and Ferreirós, this volume) provide more in-depth treatments of Dedekind on ideal theory in particular, so we will return again to Noether.

sets, rather than focusing on ideal divisors, and defines concepts such as divisibility and decomposition in terms of set-theoretic concepts like inclusion and intersection.

The main difference between Noether's and Dedekind's contributions to ideal theory is in their generality, one of our central themes, though it is better described as an extension of Dedekind's methodological trajectory than as a change. In writing about Dedekind's own work on ideal theory, Avigad (2006) notes among the advantages of the axiomatic method that it allows for greater generality, and that it allows for a smoother transference of prior results. While other treatments of ideals, including Dedekind's, had relied on properties of algebraic integers that can be taken for granted, Noether was defining ideals in a more arbitrary setting. Rather than being able to rely on known facts about concrete mathematical entities, the decomposition theorems that Noether proved had to follow from general defining properties of sets of elements in a ring. The main thing in Noether (1921) that was taken for granted as a property of ideals was the ascending chain condition (a.c.c), which states that every chain of ideals ordered by inclusion has a maximal element. More precisely, if we have a chain of ideals $I_1 \subseteq I_2 \subseteq I_3 \subseteq \ldots$, then there is an index n after which all the ideals are equal, so $I_n = I_{n+1} = \ldots$. This condition was explicitly used to prove her decomposition results, but in 1921 was simply stated without proof. In contrast, Noether (1926) made it explicit that the a.c.c. was simply a condition on rings that Noether was interested in.

Her 1926 "Abstrakter Aufbau der Idealtheorie" made the axiomatic approach (another central theme) even more explicit and laid out the structural conditions that rings might satisfy at the outset. In this case, the a.c.c. was just one of the conditions that rings might or might not satisfy. Others included a multiplicative unit element and a lack of zero divisors. But in contrast to 1921, these were not assumed, but treated as contingent. This means that throughout the work, Noether considered rings that satisfied different conditions, so that we discover what follows from each one. Frequently over the course of the work, she will specify what type of ring $\mathfrak{R}$ is intended to be: whether it need only be a ring of some kind or whether it also needs to satisfy some other properties such as the a.c.c. These rings, then, are seen simply as instances of mathematical entities that satisfy various conditions. The use of the axiomatic method, then, is to facilitate the use of merely structural definitions of objects. It also allowed Noether to generalize, in that she could abstract away from different standardly assumed properties of mathematical entities, to consider more general cases of objects.

One other example of abstracting away from a standardly assumed property is commutativity. Though it is a standard property of algebraic integers, Noether in 1926 is careful to specify when a ring under discussion needs to be commutative, and when it only needs to satisfy the basic ring axioms or other conditions.

That approach was sufficiently fruitful that she eventually worked seriously on theories of non-commutative rings, as part of her continuing research trajectory toward studying entities of greater and greater generality. But even before her published work on the subject in the 1930s, we can see the structural approach at work, in which commutativity is only presupposed when it is required and theorems are proved with as much generality as possible. For instance, the theory of integers is initially introduced in terms of a commutative ring with no zero divisors and a multiplicative unit (Noether 1926, 29), while various isomorphism theorems presuppose nothing beyond the ring axioms (Noether 1926, 39).

The move toward the non-commutative setting, however, is importantly modern. In giving up commutativity for multiplication, we take a step away from the intended interpretation of ideals as ideal divisors for the algebraic integers, to consider what else rings as structures could be used to represent, allowing for a wider domain of application. The concept of module that she used in her work on ideals turned out to be a helpful device when it came to representation theory, a branch of algebra that uses vector spaces as representations of groups. Since the vector spaces used in representation theory can be seen as special cases of modules over rings, Noether was once again able to provide a more general structure to use as a mathematical tool (Noether 1929). The work on representation theory in hypercomplex numbers was also further extended into the domain of non-commutative algebras (Noether 1933).

Noether's move to a more general setting such as the theory of rings yielded the ability to make use of tools that describe very general relationships between structures, such as homomorphisms and isomorphisms. In 1926, she explicitly assumes only ring properties (and module properties, respectively), without any other axioms, in order to prove several isomorphism theorems, and theorems relating rings to their quotients. These results are then used for calculations with relatively prime ideals and subsequent decomposition results. Such theorems, as Noether herself notes, can be seen in Dedekind as well, but only as special cases of her own results (Noether 1926, 41). We will also see in the next section that this use of morphisms is further developed by some of Noether's students who went on to lay the foundations of category theory.

And although we find little in the way of autobiographical reflection on her approach to mathematics, Noether's colleagues and students provide a fairly uniform picture.[5] With respect to Noether's use of generalization as a way of developing mathematically fruitful connections, Weyl observes in his memorial address,

[5] Dedekind as well does not do much philosophical writing, and many of the philosophical positions we now attribute to him are extrapolated from his criticisms of other approaches and general methodological comments. So it seems fair to take a similar interpretive stance with respect to Noether.

> She possessed a most vivid imagination, with the aid of which she could visualize remote connections; she constantly strove toward unification. In this she sought out the essentials in the known facts, brought them into order by means of appropriate general concepts, espied the vantage point from which the whole could best be surveyed, cleansed the object under consideration of superfluous dross, and thereby won through to so simple and distinct a form that the venture into new territory could be undertaken with the greatest prospect of success. (Weyl 1981, 147)

This quotation could be taken to apply to both ideal theory and representation theory, as branches of algebra in which Noether was able to develop this more general vantage point. In the case of ideal theory, Noether was able to connect Fraenkel's definition of a ring to Dedekind's work on algebraic integers in order to give a more general treatment of the latter's factorization theorems. And in the case of representation theory, she connected work from Frobenius and Dickson in order to develop a general treatment of non-commutative algebras. Also, the full quotation that opened this chapter can be found in another of Noether's obituaries, in which her student van der Waerden writes,

> One could formulate the maxim by which Emmy Noether always let herself be guided as follows: *All relations between numbers, functions, and operations become clear, generalizable, and truly fruitful only when they are separated from their particular objects and reduced to general concepts.* For her this guiding principle was by no means a result of her experience with the importance of scientific methods, but an a priori fundamental principle of her thoughts. She could conceive and assimilate no theorem or proof before it had been abstracted and thus made clear in her mind. She could think only in concepts, not in formulas, and this is exactly where her strength lay. In this way she was forced by her own nature to discover those concepts that were suitable to serve as bases of mathematical theories. (van der Waerden 1981, 101)

Both Weyl and van der Waerden are consistent in their assessment of Noether as fundamentally committed to what I have called methodological structuralism; she was a mathematician who, at least in her mature work, preferred to think about the relationships between concepts rather than developing formulas or doing calculations. So while the generality of her thinking and fruitful use of axiomatics is apparent, situating Noether within a taxonomy of modern structuralist views is to some extent speculative, given the lack of her own philosophical writing. Nevertheless, we can consider which philosophical structuralisms are compatible with Noether's own methodological structuralism, and the extent to which her methodological views could support one philosophical picture over another.

4. Structuralism, Categories, and Invariants

There is no such thing as a single canonical philosophical structuralist view. They tend to at least have in common a view of mathematical objects as defined or determined by the structures to which they belong and some commitment to methodological structuralism. Sometimes, this means that mathematical objects are seen as "thin" or "incomplete" in the sense that they have no distinguishing properties other than those they possess in virtue of belonging to a particular mathematical structure. While mathematical objects may have other properties, such as the fact that the number one might have the property of being the number of moons of Earth, this is an accidental property that the number has, rather than one making it the thing that it is.

Within these relatively broad constraints, there are a range of positions, as well as a variety of different classifications of such views.[6] For our purposes, it will be most useful to compare Noether's mathematical methodology to two views that are closely connected to the areas of mathematics in which she worked: category-theoretic structuralism, as articulated by Awodey (1996, 2004) and Landry (2011), and invariant-based structuralism as outlined by Schiemer (2014). Noether's connection to category theory comes directly through her students and others who worked with her. For instance, Saunders Mac Lane, credited as one of the founders of category theory, studied with Noether in Göttingen, and is also an important figure in the history of structuralism (McLarty, this volume). Invariant-based structuralism builds on much of the work done by category-theoretic structuralists, but also accounts for issues raised for structuralism by, e.g., Carter (2008). Both are explicitly based on the idea of structure as it can be captured by various branches of abstract mathematics.

The reason for bringing in categories and invariants is the fact that the very concept of structure as it is used in mathematics can be hard to pin down as a single unified concept. Category theory has sometimes been discussed as a potential foundation for mathematics generally, but as Awodey describes it, it can also be used as a way to understand what we mean when we talk about mathematics as a field that deals essentially with structures. This might not be an easy task, because of the wide variety of mathematical structures and the number of different areas in mathematics that use them. The appeal of category theory as a kind of foundation for mathematics, then, is appealing because of its generality and flexibility in characterizing different kinds of mathematical structures. Landry (2011), for instance, gives a list of different categories that can be used to

[6] See Reck and Price (2000); Parsons (1990); Hellman (2005) for various classifications of different structuralist positions.

organize the mathematical structure involved in the concepts of group, set, and topological space, among others. All that we need to do is assign different kinds of entities to be objects and morphisms.

However, in offering categories as a means for the analysis of structure, the kind of foundation that category theory offers is not a foundation in the traditional sense—what Awodey calls "bottom up." Rather,

> The "categorical-structural" [approach] we advocate is based instead on the idea of specifying, for a given theorem or theory only the required or relevant degree of information or structure, the essential features of a given situation, for the purpose at hand, without assuming some ultimate knowledge, specification, or determination of the "objects" involved. (Awodey 2004, 56)

What this means is that the categorical foundations only need to be foundations insofar as they allow us to specify what is essential about the objects that we are interested in. So categories can serve as a foundation for mathematics because of the flexible way in which they permit the characterization of a diverse range of mathematical structures. This alternative approach also results in a different interpretation of the schematic nature of mathematical theories. I can illustrate this in terms of Noether's 1926 work, in which she is very careful to specify which properties of rings she is assuming in each section, for which definitions. For example, when Noether begins her introduction of prime and primary ideals, she simply says to let $\mathfrak{R}$ be a commutative ring, and is clear that no other assumptions are required. We can read this hypothetically, as stating that the definitions and theorems apply *if* an object satisfies the properties for being a commutative ring.

But this is unlike an eliminative structuralist view, or one in which we remove reference to individual mathematical objects by reinterpreting mathematical statements as being implicitly universally quantified. For in order for them to be interpreted in terms of universal quantification, there must be a preexisting domain over which we quantify. Rather, Awodey (2004) advocates for the indeterminacy in the objects being taken seriously, rather than taking a modal approach as does, for instance, Hellman (1989). Further, rather than the focus being on the relations between objects (as a focus on the relations presupposes the relata), morphisms in categories are a perfectly good autonomous concept on which to base the analysis of structure (Awodey 2004, 61). They are also a natural extension of the isomorphisms and homomorphisms on modules that Noether uses in 1926. In this more general situation, so long as the category of rings is sufficient to model the different types of rings, commutative, Dedekindian, etc., that Noether is interested in, it can form a perfectly good basis for her definitions. In that case, these various rings would simply be objects in the category ring, while ring homomorphisms are its morphisms.[7]

So even though Noether's work on ideals and rings preceded the development of categories, the top-down approach to category-theoretic foundations that Awodey and Landry advocate are a natural philosophical overlay atop Noether's mathematical structuralism. I have already noted Noether's deft use of axioms in her work on the theory of rings, and the extent to which it matured over time to place increased emphasis on the conceptual and structural of rings. For instance, the later work, such as her 1926 paper, focused on the connections between the properties of various rings and the theorems that could be proved about them, and this meshes nicely with many characterizations of mathematical practice to which category-theoretic structuralism claims to be faithful:

> The structural perspective on mathematics codified by categorical methods might be summarized in the slogan: The subject matter of pure mathematics is invariant form, not a universe of mathematical objects consisting of logical atoms. This trivialization points to what may ultimately be an insight into the nature of mathematics. The tension between mathematical form and substance can be recognized already in the dispute between Dedekind and Frege over the nature of the natural numbers, the former determining them structurally, and the latter insisting that they be logical objects. (Awodey 1996, 235)

The connection between Noether and Dedekind was famously emphasized by Noether herself, who was said to have remarked, "*Es steht alles schon bei Dedekind* (It is all already in Dedekind)" (quoted in Corry 2004, 250). While in this case she was talking about the decomposition results that she had proved, Awodey's characterization of his structuralist position suggests applying this remark to Noether's methodological views as well. After all, not only did Noether extend Dedekind's results to a more general setting, she also arguably extended his use of structural methods by treating the concept of mathematical structure with a greater degree of abstraction (Yap 2017). In fact, this very same move was arguably employed to extend Noether's work to more general settings by her student and category theorist Mac Lane (see McLarty, this volume), which makes category-theoretic structuralism a natural philosophical view to consider alongside Noether's methodological view. It is, after all, in category theory that many of the concepts that Noether worked with so fruitfully, such as morphisms, get treated in thoroughly general terms.

There are, however, some criticisms of structuralism that we might want to consider as well, which apply to both philosophical and methodological

[7] While I have argued in this chapter for the importance of morphisms to Noether's work, precursors for such ideas are also arguably in Dedekind (Reck and Ferreirós, this volume).

structuralisms. In particular, Carter (2008) considers the methodological claim that mathematics is the study of structure, arguing that this is inaccurate when considered as an overall view of mathematical practice. While she certainly agrees that mathematics deals with structures, it is unclear that there is any single sense of "structure" that will suffice and that can accurately be captured by a structuralist account. This ambiguity about the sense of "structure" can even be traced to the Noether school, at least according to some of its members. Mac Lane (1996) notes that the word "structure" was used in various informal ways by algebraists such as Noether and her students in the 1930s, and given the extent of its ambiguity, might not be able to form the basis of a philosophy of mathematics. Carter, following Mac Lane's discussion, gives examples of two distinct uses of structure that can be found in mathematical practice.

1. Structure over sets that is used to compute invariants of this set.
2. A case where "structure" is extracted in order to change relations between objects. (Carter 2008, 123)

In the first use of "structure," we want to obtain some information about a certain kind of mathematical object. In the case of Galois theory, we might want to determine whether a given polynomial is solvable by radicals. A permutation group based on invariance of the roots can be associated with this polynomial, which is called its Galois group. If the Galois group is solvable,[8] then the polynomial is also called solvable by radicals. So this is a case in which we obtain information about an object because of a certain structure that is associated with it, which we might say is a structure that the object or set has.

In the second use of "structure," we can consider cases in which we have to discover some information about certain structures in order to situate them among more general ones. For instance, we might have to determine how to treat some structures category-theoretically, and in order to do so, need to determine which category they should be subsumed under. In doing so, however, we in effect move objects from one structure to another, which has the following consequences:

> The fact that objects or "places" are moved between structures seems to go against the dictum that "places have no distinguishing features except those determined by the structure in which they have a place" which is taken as implying the claim that "places from different structures can not be identical." Firstly, we have seen that the properties of places or objects can be determined by different structures

[8] A solvable group is one that has a normal series whose normal factors are abelian.

> that they are part of. Secondly, the properties of an object in a given structure can be used to consider the object as part of another structure. (Carter 2008, 130)

To clarify, some features of mathematical practice seem at odds with some central structuralist claims about the identity of objects. If objects technically have different properties in different structures, then it is hard to make sense of moving objects between structures. So this suggests that there is something more to being a particular object than the structure to which it belongs. Carter, however, does not deny that structures are extremely important in mathematics, or even that they are central to mathematical practice, simply that structures cannot be all there is. And this does speak to some extent against Noether's tendency to generalize existing results to more abstract domains. After all, the category-theoretic way of modeling structure represents somewhat more of a difference between algebraic integers and abstract rings than might be warranted by Noether's remarks that it was all already in Dedekind.

I will now turn to an alternative characterization of structuralism based on abstract mathematics, namely Schiemer's structuralism based on invariants. The invariants that Schiemer considers are more general than isomorphisms; rather, they determine equivalence relations on objects that have a certain common property. The purpose of introducing them can be related to some of the issues that Carter raises with structuralist views, such as determining what counts as a structural property of mathematical objects in the first place, given that we might sometimes move an object to a different structure. Schiemer's solution to this is to give up on the idea of defining a structural property on its own, instead defining them relative to some invariant. This relates to Carter's first use of structure in mathematical practice, in which we might work with invariants to determine a property of an object or a set. But of course, different invariants can determine different property structures, where a pair $<S, P>$ (or simply set P) is a property structure of S iff

(i) there exists an invariant $f: S \to N$ and an equivalence relation $R \subseteq S \times S$ such that f determines R; and
(ii) P is the partition of S induced by R, i.e., $P = S/_R$. (Schiemer 2014, 84)

So this is what it means for a set P to be a property structure of another set S relative to an invariant f on S. This is, for instance, the idea behind the Galois group of a polynomial.

For Schiemer, this ultimately has the effect of defining structure in a higher-order set-theoretic fashion, in which structures are identified as classes of equivalence classes determined by some invariant on the objects. Now, whether or not this provides a solution to the problems with structuralism that Carter raises remains open. It does, however, provide a characterization of mathematical structure alternative to category-theoretic structuralism, but one that is nevertheless based in abstract mathematics and a formal definition of structure. One of the differences, however, is that this characterization of structure based in set theory relies on some background interpreted theory such as Zermelo-Fraenkel set theory, as opposed to a category-theoretic approach that, to borrow a turn of phrase from Landry, is structuralist "all the way down."

Schiemer's philosophical structuralism is not as obviously connected to Noether's methodological structuralism as its category-theoretic counterpart, as the latter has direct connections to her ongoing mathematical legacy, while her work on invariants was relatively early in her career. But it does mesh nicely with several aspects of Noether's view. For example, Noether's tendency toward taking a more general perspective on mathematical structures can be accommodated nicely, since these definitions lend themselves to comparisons between property structures, and one structure being more fine-grained than another. This is certainly one way to think of the relationship between ideals in the algebraic integers as Dedekind defined them and ideals on general rings as Noether defined them. If we want to talk about the sense in which the results that Noether proved are the same as Dedekind's, despite being in a different setting, we could consider them as being analogous results in a coarser-grained structure. Rings of algebraic integers are instances of rings that Noether considered, but in the latter setting, they are seen as a more general kind of mathematical object.

5. Conclusion

Ultimately, what version of philosophical structuralism, if any, to which Noether would have subscribed is speculative. While her methodological views are consistently described in structural and conceptual terms by her students, she did not articulate a considered philosophical position in her published work. However, of the various structural views in the literature, two good candidates that we can connect to Noether's work are characterizations of structure based on category theory and invariants. Both articulate the concept of structure in terms of areas of mathematics that Noether either contributed to (in the case of invariant theory) or directly influenced (in the case of category theory). So regardless of the exact field of mathematics that she might have considered to best articulate the concept of structure that she wanted to work with, a formal characterization of structure would likely have been appealing. Given her tendency to articulate concepts as precisely as possible, a metatheoretical articulation of the structure concept in terms of formal mathematics would be natural for her, philosophically.

Noether's methodological inclinations, which we can see borne out in her choices of areas to research, were to generalize given results to more abstract settings. This certainly influenced her students, many of whom went on to develop branches of abstract algebra such as category theory. So when we situate Noether within the history of structuralism as a view, not only can we see her as an excellent example of someone who used structural and axiomatic methods very successfully, we can also see her contributions to some of the mathematical theories underlying contemporary structuralist views, namely to methodological structuralism. In that case, even if it is somewhat open just what kind of structuralist Noether herself would have been, we at least know that she helped made it possible for others even to hold certain kinds of structuralist positions.

References

Avigad, J. 2006. Methodology and Metaphysics in the Development of Dedekind's Theory of Ideals. In *The Architecture of Modern Mathematics*, edited by J. Ferreirós and J. Gray. New York: Oxford University Press, pp. 159–186.

Awodey, S. 1996. Structure in Mathematics and Logic: A Categorical Perspective. *Philosophia Mathematica* 4(3), 209–237.

Awodey, S. 2004. An Answer to Hellman's Question: Does Category Theory Provide a Framework for Mathematical Structuralism. *Philosophia Mathematica* 3(12), 54–64.

Carter, J. 2008. Structuralism as a philosophy of mathematical practice. *Synthese* 163, 119–131.

Corry, L. 2004. *Modern Algebra and the Rise of Mathematical Structures*. Boston: Birkhauser Verlag.

Dedekind, R. 1877. *Theory of Algebraic Integers*. New York: Cambridge University Press.

Edwards, H. M. 1977. *Fermat's Last Theorem: A Genetic Introduction to Algebraic Number Theory*. New York: Springer.

Edwards, H. M. 1980. The Genesis of Ideal Theory. *Archive for the History of the Exact Sciences* 23, 321–378.

Edwards, H. M. 1992. Mathematical Ideas, Ideals, and Ideology. *Mathematical Intelligencer* 14(2), 6–19.

Feferman, S. 1999. Does Mathematics Need New Axioms? *American Mathematical Monthly* 106(2), 99–111.

Hellman, G. 1989. *Mathematics without Numbers: Towards a Modal-Structural Interpretation*. New York: Oxford University Press.

Hellman, G. 2005. Structuralism. In *The Oxford Handbook of Philosophy of Mathematics and Logic*, edited by S. Shapiro, pp. 536–562. New York: Oxford University Press.

Kimberling, C. 1981. Emmy Noether and Her Influence. In *Emmy Noether: A Tribute to Her Life and Work*, edited by J. W. Brewer and M. K. Smith, pp. 3–61. New York: Marcel Dekker.

Koreuber, M. 2015. *Emmy Noether, die Noether-Schule und die moderne Algebra*. Berlin: Springer.

Kosmann-Schwarzbach, Y. 2011. *The Noether Theorems: Invariance and Conservation Laws in the Twentieth Century*. New York: Springer.

Landry, E. 2011. How to Be a Structuralist All the Way Down. *Synthese* 179, 435–454.

Mac Lane, S. 1996. Structure in Mathematics. *Philosophia Mathematica* 4(3), 174–183.

McLarty, C. 2012. Hilbert on Theology and Its Discontents: The Origin Myth of Modern Mathematics. In *Circles Disturbed: The Interplay of Mathematics and Narrative*, edited by A. Doxiadis and B. Mazur, pp. 105–129. Princeton, NJ: Princeton University Press.

Noether, E. 1919. Die Endlichkeit des Systems der ganzzahligen Invarianten binärer Formen. *Göttingen Nachrichten*. 138–156.

Noether, E. 1921. Idealtheorie in Ringbereichen. *Mathematische Annalen* 83(1-2), 24–66.

Noether, E. 1926. Abstrakter Aufbau der Idealtheorie in algebraischen Zahl- und Funktionenkörpern. *Mathematische Annalen* 96(1), 26–61.

Noether, E. 1929. Hyperkomplexe Größen und Darstellungstheorie. *Mathematische Zeitschrift* 30, 641–692.

Noether, E. 1933. Nichtkommutative Algebra. *Mathematische Zeitschrift* 37, 514–541.

Parsons, C. 1990. The Structuralist View of Mathematical Objects. *Synthese* 84, 303–346.

Reck, E. 2003. Dedekind's Structuralism: An Interpretation and Partial Defense. *Synthese* 137, 369–419.

Reck, E., and Price, M. 2000. Structures and Structuralism in Contemporary Philosophy of Mathematics. *Synthese* 125, 341–383.

Schiemer, G. 2014. Invariants and Mathematical Structuralism. *Philosophia Mathematica* 22(1), 70–107.

Stein, H. 1988. *Logos*, Logic, and *Logistiké*: Some Philosophical Remarks on Nineteenth-Century Transformation of Mathematics. In *Minnesota Studies in the Philosophy of Science: History and Philosophy of Modern Mathematics*, vol. 9, edited by W. Aspray and P. Kitcher, pp. 238–259. Minneapolis: University of Minnesota Press.

van der Waerden, B. 1981. Obituary of Emmy Noether. In *Emmy Noether, 1882–1935*, edited by A. Dick, pp. 100–111. Boston: Birkhauser Verlag.

Weyl, H. 1981. Emmy Noether. In *Emmy Noether, 1882–1935*, edited by A. Dick, pp. 112–152. Boston: Birkhauser Verlag.

Yap, A. 2009. Logical Structuralism and Benacerraf's Problem. *Synthese* 171(1), 157–173.

Yap, A. 2017. The History of Algebra's Impact on the Philosophy of Mathematics. In *Innovations in the History of Analytical Philosophy*, edited by S. Lapointe and C. Pincock. New York: Palgrave Macmillan, pp. 333–358.

8
The Functional Role of Structures in Bourbaki

Gerhard Heinzmann and Jean Petitot

1. Introduction

From antiquity to the 19th century and even up to now, the following two theses are among the most debated in the philosophy of mathematics:

a) According to the Aristotelian tradition, mathematical objects such as numbers, quantities, and figures are entities belonging to different kinds.
b) Mathematical objects are extralinguistic entities that exist, independently of our representations, in an abstract world. They are conceived by analogy with the physical world and designated by singular terms of a mathematical language.

The Aristotelian thesis and that of ontological "Platonism" were countered by nominalism and early tendencies of algebraic formalization; but they became even more problematic when mathematicians, such as Niels Abel, thought of relations before their relata or when they, as Hermann Hankel (1867) pointed out, posited that mathematics is a pure theory of forms whose purpose is not that of treating quantities or combinations of numbers (see Bourbaki 1968, 317). In the 1930s, Bourbaki finally defended the view that mathematics does not deal with traditional mathematical objects at all, but that objectivity is solely based on the stipulation of structures and their development in a hierarchy.

In the history of 20th-century mathematical structuralism, the figure of Bourbaki is prominent; sometimes he is even identified with the philosophical doctrine of structuralism. However, the Bourbaki group consisted of pure mathematicians—among them the greatest of their generation—most of whom had a conflicted relationship to philosophy. This chapter proposes to explore this tension, following the current philosophical interest in scientific practice. The problem with properly assessing Bourbaki's importance is that he was at the same time the collective author of a monumental and long-lasting treatise (in a golden age of more than 30 years) and a pleiad of individual geniuses (including

Gerhard Heinzmann and Jean Petitot, *The Functional Role of Structures in Bourbaki* In: *The Prehistory of Mathematical Structuralism*. Edited by: Erich H. Reck and Georg Schiemer, Oxford University Press (2020). © Oxford University Press.
DOI:10.1093/oso/9780190641221.003.0008

five Fields medalists: Schwartz, Serre, Grothendieck, Connes, Yoccoz). The former had to faithfully conform to initial editorial choices, while the latter were at the cutting edge of innovation and creativity.[1] So it makes no sense to think that Bourbaki was not aware of mathematical advances, since his members were among the main agents in these advances.

Quite often interpreters focus only on Bourbaki's formal definition of structure, so as then to dismiss it. Our approach will be quite different. Our thesis is that the use of the concept of structure in Bourbaki is not so much logical and, in a philosophical sense, foundational as pragmatic and functional—"functional" not in the mathematical sense, but in a sense analogous to the relationship between structure and function in biology. We will illustrate the functional role of structures in Bourbaki's work, starting with Hilbert's axiomatics, which was developed to perfection by the Bourbaki group, and going up to category theory, thus to a higher level of structuralism, a path that Bourbaki initiated without yet actually engaging in it.[2]

2. Bourbaki, the *Éléments*, and the *Séminaires*

Nicolas Bourbaki was the pseudonym of a "collective mathematician,"[3] formed in 1934–35 by a group of young French mathematicians who graduated (with the exception of Mandelbrojt) from the École Normale Supérieure in Paris, who did research abroad, primarily in Germany (but also in Denmark, Italy, Hungary, Sweden, Switzerland, and the United States), and who taught mostly in Strasbourg, Nancy, and Clermont-Ferrand.[4] This group included Henri Cartan,

[1] Other mythical examples of such groups in French history include, in the middle of the 16th century, *La Pléiade*, which completely transformed the norms of poetic language (Du Bellay, Ronsard, Jodelle, Belleau, de Baïf, Peletier du Mans, de Tyard, etc.); and in the second part of the 18th century, the *Encyclopédie ou Dictionnaire raisonné des sciences, des arts et des métiers* (Diderot, d'Alembert, de Jaucourt, d'Holbach, Dumarsais, Quesnay, Turgot, etc., related to Montesquieu, Voltaire, Buffon, Mably, Condillac, Helvétius, etc.).

[2] For a philosophical discussion of the role of Bourbaki's concept of structure in the interpretation of category theory, see Krömer (2007).

[3] See Chevalley and Guedj (1985). The name of the French general Bourbaki, defeated in the French-German war of 1870–71, was part of the anti-militarist folklore at the École Normale Superieure long before the group chose it (Beaulieu 1989, 278ff).

[4] For a detailed bibliography and rich archive, see the site http://archives-bourbaki.ahp-numerique.fr of the Archives Henri-Poincaré, compiled by Liliane Beaulieu (supplemented by Ch. Eckes and G. Ricotier) and the site http://sites.mathdoc.fr/archives-bourbaki/feuilleter.php of the Association des Collaborateurs de Nicolas Bourbaki. An introduction to the corresponding history is Maurice Mashaal's *Bourbaki: A Secret Society of Mathematicians* (2006). Another interesting historical source is Amir Aczel's *The Artist and the Mathematician: The Story of Nicolas Bourbaki, the Genius Mathematician Who Never Existed* (2006). The two books were reviewed in 2007 by Michael Atiyah in the *Notices of the AMS*. At a theoretical level, a well-known reference is the 1992 essay by Leo Corry, "Nicolas Bourbaki and the Theory of Mathematical Structure."

Claude Chevalley, Jean Delsarte, Jean Dieudonné, Szolem Mandelbrojt, René de Possel, and André Weil, to which must be added Paul Dubreil and Jean Leray (who both, however, withdrew after some meetings), Jean Coulomb, and Charles Ehresmann (1935–36).[5] Their goal was "to define for 25 years the syllabus for the certificate in differential and integral calculus by writing, collectively, a treatise on analysis. Of course, this treatise will be as modern as possible" (Beaulieu 1989, 28). This was a revolt against the dominant French mathematics of the 1930s.

The initial group had no money nor any official administrative structure. Each draft of a chapter of their famous multi-form treatise, *Éléments de Mathématique* (the expression "Mathématique" in the singular emphasizes the unity of mathematics), was discussed largely in the group and had to be accepted unanimously by those present at the regular Bourbaki meetings.[6] Elected by consensus, after age 50 a Bourbakist had to leave the group, whose list of members was an (open) secret. Among the most prominent later Bourbakists were Hyman Bass, Armand Borel, Pierre Cartier, Alain Connes, Michel Demazure, Jacques Dixmier, Samuel Eilenberg, Roger Godement, André Gramain, Alexander Grothendieck, Jean-Louis Koszul, Serge Lang, Pierre Samuel, Laurent Schwartz, Jean-Pierre Serre, John Tate, Bernard Teissier, Jean-Louis Verdier, and Jean-Christophe Yoccoz, all also pursuing their own individual work.

The creation in 1962, by Grothendieck, of the group of algebraic geometry at the "Institut des Hautes Études Scientifiques," a "European Princeton Institute of Advanced Studies" (Bolondi 2009, 701) located in the Paris suburb Bures-sur-Yvette, was at the same time a continuation and an improvement of the Bourbaki perspective in mathematics in France and around the world. In a certain sense, the monumental *Éléments de Géométrie Algébrique* by Grothendieck and Dieudonné (1960–67) can be considered as a systematization of the same type as the *Éléments de Mathématique*: driven by the desire to optimize the framework of demonstration of great theorems and to attack major conjectures, especially the Weil conjectures. Indeed, its language was no longer that of classes of structures in a universe of set theory but that of full-fledged category theory.[7] But, as communicated by Jean-Pierre Ferrier, the project was a resurgence of the Bourbaki project, equally ambitious and original, greatly renewing mathematical

[5] Beaulieu (1989, 12–13). Beaulieu's dissertation is the most extensive description of the origin and the first 10 years of activity of the group. It is the sourcebook of all biographically oriented studies on Bourbaki (for works on Bourbaki see Beaulieu 2013). See also Weil (1992) for Weil's memories.

[6] The first publication of the *Éléments* was released in 1939 (Bourbaki 1939). An interesting document on Bourbaki's birth is the first issue of the *Journal de Bourbaki* handwritten by Jean Delsarte on November 15, 1935. Composed with a touch of humor, it refers to the creation of the group at the "congress" held at Besse-en-Chandesse in July and presents a first division of labor between Cartan, Delsarte, Dieudonné, Chevalley, Mandelbrojt, de Possel, and Weil (http://sites.mathdoc.fr/archives-bourbaki/PDF/deljb_001.pdf).

[7] On Grothendieck and the shift to category theory, see McLarty (2008), as well as the contribution, also by Colin McLarty, on Saunders Mac Lane and category theory in this volume.

thinking, as Bourbaki did in his time, and overcoming many difficulties raised by Bourbaki's initial choices.

In this essay we do not only report on the *Éléments* and its content. Bourbaki is a collective author, but, again, also a pleiad of unique individual masterminds who took up the most difficult mathematical challenges. This is confirmed by its encyclopedic *Séminaire*, which was unparalleled and continues until today. Started in 1948, it reached its 1,110th talk in June 2016. Almost all great mathematical results have been presented in it. As creative mathematicians, the members of Bourbaki were not only interested in the context of justification but also, and even more, in the context of discovery. Their conception of structures must be understood in this light. In particular, they were all working on very complex conceptual proofs of "big problems," and for them there existed a *complementarity* between general relevant structures and specific hard problems. One could say that this complementarity found its material expression in the complementarity of the *Éléments* and the *Séminaire*: the function of the *Éléments* was to offer to working mathematicians an extremely wide toolbox of axiomatized devices (structures), to be used as conceptual apparatuses in complex proofs, while the function of the *Séminaire* was to inform, in preview, about mathematical progress, thus being a preferred place to host creation.

Many controversial aspects of Bourbaki are well known, e.g., its overly formalist and algebraic setting or its lack of interest in logic. The first has been strongly criticized from the start by some great mathematicians who refused to be members of Bourbaki, while belonging to the same generation of the École Normale Supérieure as its founders. This is the case, e.g., for René Thom (1970), who accused Bourbaki of destroying geometric intuition, or for Roger Apéry, a constructivist mathematician inspired by the French constructivist school of Poincaré, Borel, Lebesgue, Fréchet, and Denjoy and opposed to Hilbertian formalism and axiomatics. The second aspect has been denounced, e.g., by Adrien Mathias in his 1992 paper "The Ignorance of Bourbaki," which analyzes the inadequate reflections of Bourbaki on foundational issues in set theory. For Matthias, Bourbaki's *Set Theory* "appeared to be the work of someone who had read *Grundzüge der Mathematik* by Hilbert and Ackermann,[8] and *Leçons sur les nombres transfinis* by Sierpinski, both published in 1928, but nothing since."[9] A lot of things have also been written about the folklore of Bourbaki, his legend, his dictatorial power, his dramatic impact on education with the introduction of "modern mathematics" in schools (see again Thom 1970). Our purpose

[8] It seems that Mathias means to refer to *Grundzüge der theoretischen Logik*.

[9] Mathias (1992, 5). Sometimes the ignorance seems to be intentional and polemical: thus Dieudonné says explicitly that his neglect of Gödel's result concerning a consistency proof for formal systems is not a consequence of ignorance, but of a "philosophical" position (see Heinzmann 2018).

here is quite different. We will try to explain the *functionality* of Bourbaki's structuralism.

3. Traditional Mathematical Objects versus Structures

Bourbaki inaugurated an axiomatic-structural point of view that could seemingly work without the need of metamathematics in Hilbert's sense. Indeed, given that metamathematics is "finitist" and contentual, it would be an exception to the slogan that mathematics is only about formal structures. The hypothetical-deductive foundations of Bourbaki were explicitly designed to be neutral with respect to philosophical foundations. However, it can engaged with along the lines of the philosophical interest in scientific practices that has been renewed recently: foundations as structural systematization.

The "working mathematician"[10] Henri Cartan, one of the founders of Bourbaki, wrote in 1943: "The mathematician does not need a metaphysical definition; he must only know the precise rules to which are subject the use he has in mind. . . . But who decides upon the rules?"[11] This may sound Wittgensteinian, but is not so in reality. According to Cartan, mathematical reasoning in a given area intuitively obeys certain rules at first; and if difficulties arise, the use is adapted, etc. Consequently, a mathematical reality is created through practice. What is the criterion for the practice and for the rules that result? In a historical notice on set theory, Bourbaki writes:

> [It was] recognized that the "nature" of mathematical objects is ultimately of secondary importance, and that it matters little, for example, whether a result is presented as a theorem of a "pure" geometry or as a theorem of algebra *via* analytical [Cartesian] geometry. In other words, the essence of mathematics . . . appeared as the study of *relations* between objects which do not of themselves intrude on our consciousness, but are known to us by means of *some* of their properties, namely those which serve as the axiom at the basis of their theory. (Bourbaki 1968, 316–317)

Bourbaki considered "the problem of the nature of beings" or of "mathematical objects" as deriving from a "naive point of view," "half-philosophical, half-mathematical" (Bourbaki 1948, 40). Indeed, it would be naive to presuppose that we can have a well-defined mathematical object at all, i.e., that it can be identified

[10] An expression used by Bourbaki (1949).
[11] Cartan (1943), transl. by Gerhard Heinzmann.

completely by specifying a property that characterizes it. It is only apparently well-defined according to the traditional theory of definition.[12]

Bourbaki henceforth abandoned the philosophical problem of object-individuation in favor of a premise that seems to have the same meaning today: the unity of mathematics (Houzel 2002, 3). The tool to achieve this unity was Hilbert's axiomatic method: it provides clarity and rigor in the register of reasoning (see Dieudonné 1939, 232b) by using a systematization of mathematical theories (Bourbaki 1948, 37). It allows one to obtain all kinds of axiom systems; not for all of classical mathematics, however, but only those domains that correspond to the hierarchy of structures classified as "simple," "complex," and "mixed." Indeed, to define a simple structure, we take a set "of elements whose nature is not specified," provide it with certain relationships, and formulate the axioms that satisfy them. And we define the structure as *algebraic* "if the relationships are the laws of composition," as *topological* "if the relations concern the intuitive concepts of neighborhood, limit and continuity," and as an *order structure* if the relations are of that type.

4. The Unity of Mathematics: Structures and Entangled Problems

Let us focus now on how structures were used by Bourbaki, in a process of clarification and unification, to further the discovery of new and unexpected results—as common to several systems of objects of very different origins, as indicative of deep and fruitful analogies between theories far removed from each other, and as a powerful heuristic for proofs. As two examples, Cartan's filters and Weil's uniform structures are among the greatest inventions of Bourbaki. The first illuminates the analogy between the convergence of sequences and that of functions, while the second illuminates the analogy between a metric and a family of pseudometrics. A third example produced directly a new result: the so-called *Banach-Alaoglu* compactness theorem (for the weak topology) of the dual unit ball of a normed space, which is also due, in the form that we know today, to

[12] H. Cartan gives the following example: "According to Lebesgue, the quantity

$$\lim_{m\to\infty}\left[\lim_{n\to\infty}\cos(m!\pi C)^{2n}\right]$$

is a well-defined number, when C is a well-defined real number, for example, Euler's constant. However, this quantity is equal to 0 if C is irrational, 1 if C is rational; and we are still today ignorant whether Euler's constant is rational or irrational. Thus, if C is Euler's constant, we obtain a well-defined number, but we do not know if it is equal to 0 or to 1" (Cartan 1943, 5; transl. Gerhard Heinzmann).

Bourbaki.[13] Everything in it is owed to the clarification by means of weak topologies, which revived a problem that could not be correctly formulated until then.

This functional aspect of structures, on which Bourbaki continually dwelt, is governed by the principle of the unity of mathematics, that is to say, by the very strong ability to translate pieces of one mathematical theory into another theory. Besides the deductive "vertical" dimension internal to every theory, taking into account the relevant structures can reveal a host of "horizontal" connections between different theories.[14] The resulting "horizontal" navigation between different theories involves (at least) two processes. On the one hand, there are analogies, intuitive at first, between structures of the same type in different areas, i.e., structures whose clarification and systematization often lead to new discoveries. On the other hand, there is the encounter of different structures within the same "crossroads" area, which allow for the unification of theories. We need both to tackle the complex proofs of intricate problems. (We will come back to this issue later.)

Let us clarify the importance of the unity of mathematics according Bourbaki further. In terms of category theory, many connections between theories correspond to the existence of functors and natural transformations of functors between categories (for example, between topological spaces and groups in algebraic topology); but many others are not simply functorial. In fact, conceptually complex proofs are very uneven, with rough and rugged multi-theoretical routes in a sort of "Himalayan chain" whose peaks seem inaccessible. They cannot be understood without the thesis of the unity of mathematics, because they are in some sense *holistic*. This holistic aspect of complex proofs has always been emphasized by Bourbaki. Thus, in his *Panorama des Mathématiques pures: le choix bourbachique* (1977, xii), Jean Dieudonné classifies theorems into six classes:

1. "Dead-born problems [les problèmes mort-nés]": particular problems for which a certain theoretical approach has failed.
2. "Problems without posterity [les problèmes sans postérité]": problems whose resolution did not generate any other problems.
3. "Problems bringing forth a method [les problèmes qui engendrent une méthode]": e.g., analytic number theory or finite group theory.

[13] Leon Alaoglu proved his generalization of Banach's 1932 theorem in 1940, but Jean Dieudonné claimed that it was already announced in Bourbaki in 1938. The point is controversial.

[14] Cf. Cavaillès' terminology of the "thematic" and "horizontal" construction of concepts (Cavaillès 1947, 27).

4. "Problems clustering around a general, fertile, and vibrant theory [les problèmes qui s'ordonnent autour d'une théorie générale, féconde et vivante]": e.g., Lie group theory or algebraic topology.
5. "Declining theories [les théories en voie d'étiolement]": e.g., the theory of invariants.
6. "Theories on the way to dilution [les théories en voie de délayage]": problems that try to modify the axioms of already known rich theories.

It is the third and fourth classes that deserve special attention in our context, since they manifest the enigmatic unity of mathematics. A typical example given by Dieudonné of difficult key results involving this unity, by intertwining several very heterogeneous theories, is that of *modular forms*:

> The theory of automorphic and modular forms has become an extraordinary crossroads where the most varied theories are reacting to each other: analytic geometry, algebraic geometry, homological algebra, non-commutative harmonic analysis, and number theory. (Dieudonné 1977, 87)

The notion of "crossroads" (*carrefour*) is crucial: "big problems" are problems where many structures of different type interact and became *entangled*. The systematization of structures in the *Éléments* can be thought of as a "disentanglement."

A spectacular confirmation of Dieudonné's claim has been the proof by Andrew Wiles and Richard Taylor, in 1993–95, of the Shimura-Taniyama-Weil conjecture (implying Fermat's Last Theorem via a theorem of Ribet). This proof uses modular forms in a central way, and it is the prototype of a complex proof whose deductive parts are widely *scattered* in the global unity of the mathematical universe.[15] Its holistic status has been emphasized by many specialists. For example, Israel Kleiner writes:

> Behold the simplicity of the question and the complexity of the answer! The problem belongs to number theory—a question about positive integers. But what area does the proof come from? It is unlikely one could give a satisfactory answer, for the proof brings together many important areas—a characteristic of recent mathematics. (2000, 33)

[15] For a summary of the proof, see Petitot (1993).

Similarly, Barry Mazur writes:

> The conjecture of Shimura-Taniyama-Weil is a profoundly unifying conjecture—its very statement hints that we may have to look to diverse mathematical fields for insights or tools that might leads to its resolution. (1991, 594)

To use the complementarity in physics between observed phenomena and measuring apparatuses as a metaphor, we could put it this way: For the Bourbakists, "big problems" and hard conjectures (the distribution of primes, linked to the zeroes of the zeta function and the Riemann hypothesis, etc.) were treated as key mathematical "given" phenomena that had to be looked at using appropriate formal "apparatuses"; and axiomatized structures are precisely such devices. Thus mathematics is *at the same time holistic and modular.*[16] Structures are modular, but key phenomena are holistic, since they have to be "observed" by using many completely different "apparatuses." The "scattered" character of complex proofs is due to this holistic/modular complementarity.

This complementarity illuminates some aspects of the axiomatic method that Bourbaki inherited from Hilbert: (i) the fact that axioms can be freely chosen and are prescriptive principles, as opposed to being descriptive of objects (in the physical metaphor, to treat structures as objects would be a confusion between objects and apparatuses); (ii) the fact that many genetically different mathematical objects can be analyzed using the same structures; (iii) the fact that, in order to avoid an irrelevant axiomatic game, relevant "interesting" structures must be discovered through a reflexive process from the practice.

5. René Thom and Bourbaki

It is interesting to return here to the evaluation of the *Éléments* by René Thom, a colleague of the Bourbakists first at the École Normale Supérieure (he was a PhD student of Henri Cartan, together with Jean-Pierre Serre) and, after 1963, at the Institut des Hautes Études Scientifiques. A good reference is Thom's 1970 paper, "Les mathématiques modernes: une erreur pédagogique et philosophique ?" (translated in 1971 for the *American Scientist*). Thom criticized the idea that axiomatization can be at the same time a tool for systematization *and* for discovery.

16 "Modular" not in the mathematical sense, but in a sense analogous to "modularity" in programming languages or in cognitive science ("modularity of mind").

> During the last few years many such opinions were being put forward about the importance of axiomatization as an instrument both of systematization and of discovery. Instrument of systematization for sure; but whether of discovery, that is a much more doubtful affair. (Bolondi 2009, 705)

And he based his critique of Bourbaki precisely on this point:

> It is characteristic that from the immense effort at systematization by Nicolas Bourbaki (which is not a formalization anyway, since Bourbaki uses a non-formalized meta-language) no new theorem of any importance has resulted. And if researchers in mathematics make reference to Bourbaki, they find food much more often in the exercises—where the author has repelled the concrete material—than in the deductive body of the text. (Thom 1971, 697–698)

Thus for Thom the *Éléments* offers a systematized toolbox of axiomatized structures whose real interest for mathematical practice lies outside of it, in "concrete problems." We agree; but we will see later that, in fact, Bourbaki himself was perfectly aware of this and thought that the purpose of the *Éléments* was to help in the resolution of concrete "big" problems.

In addition, Thom attributed to Bourbaki, and criticized, the idea that structures can be derived from set theory:

> The old Bourbakist hope, to see the mathematical structures emerge naturally from the hierarchy of sets, from their subsets and their combination, is no doubt a chimera. Reasonably, one can hardly escape the impression that important mathematical structures (algebraic structures, topological structures) appear as data fundamentally imposed by the external world, and that their irrational diversity finds its only justification in their reality. (Thom 1971, 699)

Here again, Bourbaki was in fact aware of this point and held, as we already pointed out, that relevant "interesting" structures must be discovered in a reflexive way from the practice and from the search for solutions to given "big" problems. Hence Thom's criticism is justified for a restricted formal conception of structures, but not for a more general approach emphasizing their functional role.

From this perspective, we will now comment further on (i) the restricted formal definition of structures in Bourbaki; and (ii) their functional role in a more general structuralist context. The latter ranged from Hilbert's axiomatic approach to Grothendieck's categorical approach, and it involved discovery and complex proofs.

6. Formal Definition of Structures: Set Theory and Category Theory

Initially, the key notion of structure in Bourbaki was supposed to be a noncontroversial concept; but the members of the group did not agree on the importance and priority it should be given. Especially the question of its definition was not conceived by everyone in the same manner. The options were to give either (1) a vague account of how to define a structure, formulated in the metatheory (Bourbaki had done so from the beginning), or (2) an explicit and general definition, to be referred to whenever a new structure is introduced. Liliane Beaulieu's PhD thesis bears witness to the hesitations of the first members of Bourbaki, in the 1930s, with respect to a formal definition of structure.[17] The definition was finally published 20 years later, [18] but was hardly respected or used in the released mathematical corpus, despite the principle to publish only what was unanimously accepted by the group.

Indeed, as pointed out by Leo Corry already (1992, 327), Bourbaki made a very revealing comment in this context. It can be found in the *Fascicule de résultats* (*Summary of Results*) of the treatise *Théorie des Ensembles*,[19] 3rd edition (1958), originally released as the first publication by Bourbaki (1939). In it, an informal definition of "structure" is used—well before the publication of chapter 4

[17] In the first plenary meeting in Besse, on July 1935, one can read in a resolution: "We warn the reader, once and for all, that the operations that will be applied to sets can be axiomatized and justified, provided that they are only carried out on sets we study in a mathematical theory" (Beaulieu 1989, 233). There is also a project, probably discussed during the 1936 plenary meeting of Chançay, entitled "Projet Laïus Scurrile" (the group used the term "scurrile" mostly for "what has to be done without enthusiasm, which leads to nothing, or what has a philosophical content." Thus, we find in the minutes of the Bourbaki meetings or in his writings the expressions "laius scurrile" (Beaulieu 1989, 228, note 37).) In the project description we can read: "The subject of a mathematical theory is a structure organizing a set of elements: the words 'structure', 'set', 'elements' are not likely definable, but constitute the basic concepts for all mathematicians. They take on clearer form once we have had the opportunity to define structures, as will be done from this chapter one. Thanks to a structure, one has the right to say that elements or parts of the set considered in a theory have some relationships between them or possess certain properties: the words 'part', 'relationship', 'property' are likely undefinable too, and are also basic notions. According to our principles, we should state the axioms that satisfy these notions: these axioms are those of set theory, and of any mathematical theory. Given the difficulties, until now not overcome, which stand in the way of the formulation of such axioms, we will assign temporarily to these words the meaning they have in ordinary language, and we will give in what follows general rules governing their use and how to switch from one to another. . . . We say that one has defined a structure on a fundamental set if properties of the (or relationships between the) elements of this set are given, or if one of those can be deduced by a combination of the above operations, and, eventually, by previously given auxiliary fundamental sets" (Beaulieu 1989, 561; transl. Gerhard Heinzmann).

[18] See chapter 4 of *Théorie des Ensembles* (Bourbaki 1957). The *Fascicule de résultats* of this volume had already been published in 1939, i.e., 18 years earlier!

[19] The *Summaries* are in principle attached to every volume of *Éléments de Mathématique*, and their goal is to give a "rough idea" of an entire book either for orientation before reading or for a hurried reader (Bourbaki 1939, vi).

of *Théorie des Ensembles* itself (1957). And in section 8 of the *Fascicule*, devoted to scales of sets and structures (*échelles d'ensembles et structures*), Bourbaki comments in a footnote:

> The reader may have observed that the indications given here are left rather vague; they are not intended to be other than heuristic, and indeed it seems scarcely possible to state general and precise definitions for structures outside of the framework of formal mathematics (see Chapter IV). (Bourbaki 1968, 384)[20]

In the *Fascicule de résultats*, four pages "summarize," or better anticipate, the 69 pages on the notion of structure in the fourth chapter of *Théorie des Ensembles*; and the footnote indicates that Bourbaki put chapter 4, entitled "Structures," in the "framework of formal mathematics," which is developed in chapter 1. This is even clearer in the introduction to chapter 4:

> The purpose of this chapter is to describe once and for all a certain number of formative constructions and proofs (cf. chapter I, §1, no. 3 and §2, no. 2)[21] which arise very frequently in mathematics. (Bourbaki 1968, 259)

Bourbaki will never resort to such formal "structures" in his other books. Indeed, as also noted by Corry, until the publication of chapter 4 in 1957 the only references are to the *Fascicule de résultats*, which gives simply an informal definition:

> Given for example, three *distinct* sets *E*, *F*, *G*, we may form other sets from them by taking their sets of subsets, or by forming the product of one of them by itself, or again by forming the product of two of them taken in a certain order. In this way we obtain *twelve* new sets. If we add these to the three original sets *E*, *F*, *G*, we may repeat the same operations on these fifteen sets, omitting those which give sets already obtained; and so on. In general, any one of the sets obtained by this procedure (according to an explicit scheme) is said to belong to the *scale of sets on E*, *F*, *G as base*. (Bourbaki 1968, 383)

[20] This footnote is the only change from previous editions with respect to section 8 ("Structures"); we therefore quote always the most accessible English edition of 1968.

[21] By "formative constructions and proofs," Bourbaki understands in chapter 1, entitled "Description of Formal Mathematics," the definition of a formula-calculus ("règles d'assemblages")—"terms are assemblings which represent objects, and relations are assemblings which represent assertions which can be made about these objects (20)— together with a formal description of derivations, defined as sequences of relations.

The discussion is rounded off in the following way:

> Thus being given a certain number of elements of sets in a scale, relations between . . .[22] elements of these sets, and mappings of subsets of these sets into others, all comes down in the final analysis to being given a *single element* of one of the sets in the scale.
>
> In general, consider a set *M* in a scale of sets whose base consists, for the sake of example, of three sets *E*, *F*, *G*. Let us give ourselves a certain number of explicitly stated properties of [an][23] element of *M*, and let *T* be the intersection of the subsets of *M* defined by these properties. An element *s* of *T* is said to define a *structure* of the *species T* on *E*, *F*, *G*. The structures of species *T* are therefore characterized by the schema of formation of *M* from *E*, *F*, *G*, and by the properties defining *T*, which are called the axioms of these structures. We give a specific name to all the structures of the same species. Every proposition which is a consequence of the proposition "$s \in T$" (i.e. of the axioms defining *T*) is said to belong to the *theory* of the structures of species *T*. (Bourbaki 1968, 383)

Bourbaki assumed not only to have written the previous chapters to meet these specifications, which remained an outline of the formal content of chapter 4, but was also working on filling them out. In addition, he needed to introduce structures with morphisms to talk about derived structures.

Nevertheless, Bourbaki did not wait until this chapter was written, because the expectations were clear. In particular, he had a clear idea of the three main types of structures, i.e., the "mother structures": algebraic structures, topological structures, and order structures. Thus in the introduction to the volume on *Algebra* it is noted:

> In conformity with the general definitions (*Théorie des Ensembles, IV, §1, no. 4* [entitled "Espèces de structures"], being given on a set one or several laws of composition or laws of action defines a structure on *E*; for the structure defined in this way we preserve precisely the name algebraic structures and it is the study of these which constitutes Algebra.
>
> There are several species (*Théorie des Ensembles, IV, §1, no. 4*) of algebraic structures, characterized, on the one hand, by the laws of composition or laws of action which define them and, on the other hand, by the axioms to which these laws are subjected. Of course, these actions have not been chosen arbitrarily, but are just the properties of most of the laws which occur in applications, such

[22] The available translation is "between generic elements," but "generic" is not in the original French text. We skip it because "generic" has a specific mathematical content that does not apply here.

[23] We skip again the word "generic" here.

> as associativity, commutativity, etc. Chapter I is essentially devoted to the exposition of these axioms and the general consequences which follow from them; also there is a more detailed study of the two most important species of algebraic structure: that of group (in which only *one* law of composition occurs), and that of a ring (with *two laws* of composition) of which a *field* structure is a special case. (Bourbaki 1974, xxii)

As we have seen, Bourbaki ranked the structures in a hierarchy at the base of which are the three "mother structures": algebraic structures are characterized by "laws of composition," as van der Waerden had already done[24]; order structures by an order relation; and the topological structures, again, by "an abstract mathematical formulation of the intuitive concepts of neighborhood, limit, and continuity, to which we are led by our idea of space" (Bourbaki 1948, 227). These basic structures are followed by "multiple structures" involving two or more mother structures (e.g., topological algebra), and at the top of the hierarchy are placed the "theories properly particular." The criteria of Bourbaki's hierarchy of structures for each kind of structures are simplicity, generality, and the number of axioms (229).

Actually, it is a contradiction to speak of a hierarchy within a particular structure. At most Bourbaki can compare the species of one kind of structure using the same scale, i.e., the same data for which the axioms set down properties. Thus groups are more general than commutative groups, which require an additional axiom while possessing the same scale. But groups and topological groups cannot be compared; the first are not more general than the latter: they are not defined on the same data (scales), and Bourbaki had to use what is now called the "forgetful" functor to reduce the scale of topological groups to the scale of mere groups. However, topological groups can be treated as mixed structures, i.e., as topological spaces provided in addition with a group structure whose operations are continuous. It is sufficient to consider the huge project of Lie groups here (where one uses the structure of a differentiable manifold in addition). But it is not clear whether, for Bourbaki, mixed structures were also full-fledged sui generis structures, which would be the case from a categorical perspective (the category of topological groups is a specific category). In any case, the categorical formalism necessary to compare species of structure was not yet fully available to Bourbaki. He began with structured sets and isomorphisms, so as then to add the most general relations between structured sets that amount to morphisms. This means in fact placing them in a category, but without using the term. The issue of the relationship between species of structures is not really addressed,

[24] From the beginning, for Bourbaki, *Modern Algebra* by B. L. van der Waerden (1930–31) was a model for the program in analysis, and then for mathematics as a whole (see Beaulieu 1989, 164).

which would mean considering *formally* functors, natural transformations, and categories of categories. In other words, in his pre-categorical framework Bourbaki introduced many categorical objects and constructions: morphisms, sub-objects, quotients, Cartesian products, projective and inductive limits, universal problems, and (implicitly) functorial objects, like the fundamental group $\pi_1(X, x)$ of a (pointed and arc-connected) topological space $(X, x \in X)$, but all in a universe of set theory and without the formal machinery of later category theory. For Bourbakists, categorical notions and operations became relevant and even inescapable in the 1950s (we only have to look at Cartan's seminar from 1948 onward); but for the *Treatise* category theory would have been too important an editorial transformation and, moreover, it was not really a foundational issue.

Why was the discrepancy between the formal definition of structure in chapter 4 of *Théorie des Ensembles* and the actual practice in applications never fixed by Bourbaki? And why was he not more interested in corresponding metamathematical questions (such as the question of consistency)? There is both a historical-mathematical and a systematic-philosophical explanation. The historical-mathematical explanation is that, even before being released, the chapter on structures had already been superseded, since it would have needed to consider categories.[25] Some members of Bourbaki did not agree with it, but Bourbaki could also not revise it for a silly material reason: Everything that had been printed so far would have to be thrown away.[26] Bourbaki confined itself, initially, to print just the *Fascicule de résultats* on the subject; and this is precisely because nothing else was needed for the main books of the *Élements*. Actually, the distance between the rest of the *Éléments* and its formal definition of structures was even greater. It also treated structures accurately defined but not in the formal sense of chapter 4. For example, in chapter 9 of his *General Topology* Bourbaki defines a normed space as a vector space "endowed with the structure defined by a given norm" (Bourbaki 1966, 170, 1st edition 1958), thus as a mixed structure (see the example of topological groups earlier). But, as communicated by Jean-Pierre Ferrier, there is no explicit reference to the formal definition of "structure" here; in fact, it is not explained in chapter 4 (1957) what the structure defined by a given norm is and what exactly "morphisms" between normed spaces could be.

[25] We emphasize: as already noted above, many categorical concepts are used more or less implicitly by Bourbaki. Categories were present between the writing of *Élément des Mathématiques* in 1939 and its publication in 1957. But the framework of *Éléments* is set theory and not category theory, because otherwise it would have meant a complete rewriting. Algebraic topology has been the main source of category-like reflections for Bourbakists, but, strangely enough, they postponed the writing of the volume *Algebraic Topology* (chapters 1–4) until 2016!

[26] A fuller historical account of this debate can be found in Krömer (2006).

But there is also a systematic-philosophical explanation. Namely, in some ways Bourbaki remained closer than his rhetoric suggests to the "geometric structuralism" of Poincaré than to that of Hilbert. According to both Hilbert and Poincaré, geometrical axioms and axiom schemas are not propositions, i.e., true or false, and there are no special ("ontologically" specific) objects that geometry should have to study. Rather, geometry is just a system of relations that can be applied to many kinds of objects. For Poincaré, the metric postulates in geometric systems are "apparent hypotheses" that are neither true nor false, i.e., they are conventions (see Poincaré 1898). For Hilbert, the axioms and axiom schemas in geometric systems are expressions that, again, are neither true nor false. But according to Hilbert, mathematical formalism requires a "finitist" metamathematics in order to demonstrate the consistency of formal mathematical systems. The failure of this program is well known (Gödel) and was known to Bourbaki (cf. note 9). In contrast, for Poincaré it is necessary to explain the hypotheses with respect to an informal standard that involves the unity of mathematics and preexists intuitively in our mind (at a first stage transformation groups, later the qualitative structure of topological spaces); and he takes a structuralist position without disengaging meaning and knowledge completely from ostension.[27] Poincaré's concept of structure is thus not the new Hilbertian one derived from his axiomatization of geometry, but constitutes a development of the traditional idea of geometrical invariances. For Bourbaki too, the mother structures have an informal background. And he also incorporates the metamathematical problems into mathematics, as it were, by adopting an empirical position and by sharing Poincaré's concern for the unity of mathematics. From a philosophical point of view, it is clearly the status of Hilbert's metamathematics (invalidated by Gödel) that makes it distinct from the shared position of Bourbaki and Poincaré.

From a practical point of view one can ask, finally, whether Bourbaki's "mother structures" are "natural" in the sense of common-sense habits. Here we agree with Piaget's analysis: "No subject, before he has learnt it, has the 'concepts' of what a group, lattice, topological homeomorphism etc. is: and in most cultural milieus, we do not come across such concepts before university or the upper classes of secondary school. Thus, it is not in the domain of reflective thought, considered from the subject's view-point, to ask whether these structures are 'natural' " (Beth and Piaget 1966, 167). In other words, to put such structures at the beginning of the mathematical edifice is not justified by socio-psychological practice, although elements of them can be used to describe parts of both mathematical practice and socio-psychological practice. Hence Bourbaki's mother structures are a sort of mix of normative standards and empirically confirmed tools.

[27] Compare here the contribution on Poincaré, by Janet Folina, in the present volume.

7. The Function of Structures: An Example from Weil

It must be emphasized that, maybe not for Bourbaki as a collective author but undoubtedly for Bourbaki as a pleiad of mathematical masterminds, structures and axiomatics were deeply linked with *analogies* and *intuitions*. This is remarkable since these two domains seem completely different, the first belonging to the formal world and the second, in this case, to creative imagination. However, the link is not so surprising if one takes into account that analogies are fundamental for discovering ways of solving "big problems." To explain this point further, let us consider one of the main examples of such problem-solving, namely the way in which André Weil—"primus inter pares" in Bourbaki—tackled the Riemann hypothesis. In his celebrated letter written in jail to his sister Simone (March 26, 1940), he described his procedure in natural language, thus leaving a rare and precious testimony of his way of thinking. Considering it will take us from Dedekind and Weber in the 19th to Alain Connes in the 21st century.[28]

7.1 The Initial Analogy by Dedekind-Weber

At the end of the 19th century, Richard Dedekind[29] and Heinrich Weber established a deep analogy between the theory of algebraic numbers and Riemann's theory of algebraic functions on algebraic curves over the field C of complex numbers (compact Riemann surfaces); see in particular their celebrated 1882 paper, "Theorie der algebraischen Funktionen einer Veränderlichen." One of their main ideas was to consider integers n as kinds of "polynomial functions" over the set P of primes, i.e., as "functions" globally defined and having a value and an order at every "point" p of P. The "value" is n modulo p, and the "order" is the power of p in the decomposition of n into prime factors. If the value at p is not 0, the order is 0, and if the value is 0, the order is at least 1. This is evident because, if we write n in base p, we get $n = p^{\text{order}(n)}(a_0 + a_1 p + \cdots + a_k p^k)$ with coefficients a_k between 0 and $p - 1$, $a_0 \neq 0$ For smooth functions on manifolds in the ordinary sense, the values and the orders at the points are local concepts. To find the equivalent of these concepts in the analogy, Dedekind and Weber had to define localization in a purely algebraic manner. This is the origin of the modern (crucial) concepts of *spectrum* and *scheme* in algebraic geometry.

28 For more details, see Petitot (2017).

29 Dedekind is one of the founders of axiomatic and structural methods in mathematics: cf. Sieg and Schlimm (2017), and also the contribution on Dedekind, by José Ferreirós and Erich Reck, to the present volume.

In his letter to Simone, Weil describes this analogy very well:

> [Dedekind] discovered that an analogous principle enabled one to establish, by purely algebraic means, the principal results, called "elementary," of the theory of algebraic functions of one variable, which were obtained by Riemann using transcendental [analytic] means. (Weil, [1940] 2005, 338).

He adds:

> At first glance, the analogy seems superficial. . . . [But] Hilbert went further in figuring out these matters. (228)

The simplest elements of the analogy can be summarized in table 1.

7.2 Hensel's *p*-adic Numbers

The analogy becomes deeper when we introduce a local/global dialectic. On **C**, we have analytic functions with Taylor expansions in the neighborhood of any point *z*. To extend this fact to arithmetic, it was necessary to find the equivalent of the Taylor expansion of a "function" in the neighborhood of a "point" *p*. For integers, the situation is very simple. In the same way as a polynomial is its own Taylor expansion at every point, an integer is its own "Taylor expansion" (its expansion in base *p*) at every prime *p*. But there are more functions than polynomials, which have different and infinite Taylor series at different points. To find an equivalent in the Dedekind-Weber analogy, one has to consider expansions in base *p* of *infinite* length, i.e., generalized numbers $n = p^{\text{order}(\tilde{n})}(a_0 + a_1 p + \ldots + a_k p^k + \ldots)$. Of course, such series are divergent (and therefore have no rigorous meaning) for the standard Archimedean metric on the integers. But they become defined and tractable if one introduces a new, quite strange, metric where the norm of p^k is $1/p^k$ and tends toward 0 when k goes to ∞.

This was the great achievement by Hensel with the invention of *p-adic numbers*. And exactly as **R** is the completion of **Q** for the natural Archimedean metric (via limits of equivalent Cauchy sequences), the *p*-adic numbers constitute a field $\mathbf{Q}_p$ of characteristic 0 that is the completion of **Q** for a specific ultrametric, non-Archimedean, *p*-adic metric. In Bourbaki's manifesto, "L'Architecture des mathématiques" (1948) Dieudonné emphasized (with a rather a posteriori conception of history) Hensel's unifying analogy:

> [In an] astounding way, topology invades a region which had been until then the domain *par excellence* of the discrete, of the discontinuous, *viz.* the set of whole numbers. (Bourbaki 1948, 228)

Table 1 The analogy between prime numbers and points on a Riemann surface

Primes	⇔	Points
Integers	⇔	Polynomials
Divisibility of integers	⇔	Divisibility of polynomials
Rational numbers (quotients of integers)	⇔	Rational functions (quotients of polynomials)
Algebraic numbers	⇔	Algebraic functions

7.3 Mixing Algebraic and Topological Structures

Of course, with any of its natural metrics Q is naturally embedded, as a topological subfield, in its corresponding completions Q_p and R (remember the analogy with polynomials that are their own Taylor expansion at every point). With its induced topology, it is by construction a dense subfield of all its completions; but it must be strongly emphasized that all these topologies on Q are completely heterogeneous: as a set endowed with an algebraic structure of a field, Q is everywhere the same. Yet as a topological space it is completely different for every metric, since the relations of neighborhood are completely different. We meet here a very good example of mixed structure: a single algebraic system compatible with an infinite number of different metric topologies. And we see how rich the "mixing" of structures of different types can be.

7.4 Places and Weil's "Birational" Approach

From this perspective, Q appears as what is called a *global field* with an infinite number of incommensurable completions, while Q_p and R are called *local fields*. In this context, R is often interpreted as Q_∞, that is, as the completion of Q for an "infinite" prime. This is of course just a manner of speaking. To conceptualize this remarkable geometrical intuition of "points" for finite and "infinite" primes in arithmetic, the specialists have coined the term "place" and speak of finite and infinite places.[30]

[30] The geometrical lexicon of Hensel's analogy can be rigorously justified by using the concept of *scheme* that we have already evoked: (i) finite primes p are the (closed) points of the spectrum Spec(Z) of the ring Z; (ii) the local rings $Z_{(p)}$ of rationals without any power of p in the denominator are the fibers of the structural sheaf O of Spec(Z); (iii) the finite prime fields F_p are the residue fields of the fibers of O; (iv) integers n are global sections of O; (v) Q is the field of rational functions on

The Dedekind-Weber analogy between arithmetic and geometry goes much further. The spectrum Spec(Z) of Z, i.e., the space whose (closed) points are the primes p, is an affine space and not a projective space. If one wants to extend to arithmetic the analogy with projective (birational) algebraic geometry of compact Riemann surfaces and transfer some of its results (the Riemann-Roch theorem, the Severi-Castelnuovo inequality, etc.), one has to work with *all* places at the same time. Indeed, in projective geometry the point ∞ is on a par with the other points. Weil emphasized this insight strongly from the start. Already in his 1938 paper, "Zur algebraischen Theorie des algebraischen Funktionen," he writes that he wants to reformulate Dedekind-Weber in a birationally invariant way. In his letter to Simone, he explains the problem as follows:

> In order to reestablish the analogy [lost by the singular role of ∞ in Dedekind-Weber], it is necessary to introduce, into the theory of algebraic numbers, something that corresponds to the point at infinity in the theory of functions. (Weil [1940] 2005, 339)

7.5. The Adelic Perspective

Unifying Archimedean and p-adic places is the origin of Weil's "adelic" approach. The problem is to consider families of local data indexed by all places together and to look at the possibility of gluing them into global entities. A first simple idea would be to take the elements of the infinite Cartesian product Π of all the completions Q_p and R. This would be a good example of a complex structure constructed as a product of simpler structures; but this idea turns out not to be so interesting. As Q_p and R are fields, Π is a ring; and as Q_p and R are normed fields, Π is a topological ring (it is another example of mixed structures); but its topology is rather pathological in the sense that it is not locally compact, where one says that a topological space is locally compact when every point has compact neighborhoods. We meet here a typical example of a Bourbakian reflection on what can be a relevant "good" structure: it is not the most formally general structure, but the most functionally general structure suitable for a particular purpose.

Spec(Z) (i.e., of global sections of the sheaf of fractions of O); and (vi) Spec(Z) plus the infinite place ∞ is like the "projectivization" of Spec(Z). In this context, Z_p and Q_p correspond to local restrictions of global sections around the "point" p of Spec(Z), analogous to what are called *germs* of sections in classical differential, analytic, or algebraic geometry. (For the Riemann surface C, they would correspond, respectively, to holomorphic functions on small disks around a point z and on small *punctured* disks around z.)

The lack of local compactness can be fixed using the concept of *adele*, a notion derived from the notion of *idele* introduced by Claude Chevalley in class field theory and coined by Weil (adele = additive idele, and the multiplicative group I_Q of ideles is recovered as the group $GL_1(A_Q)$). The core idea is to use the "restricted" product A_Q of the Q_p and R, where "restricted" means that almost all components, except a finite number, of an adele are p-adic *integers*. (Restricted products were already used by Chevalley for the ideles.) A_Q is a topological subring of Π, which has the fundamental advantage of being locally compact, because the ring Z_p of p-adic integers is compact in the locally compact field Q_p. Of course, the global field Q is naturally embedded diagonally in A_Q. (One associates to any rational r the adele a all of whose components are r; a is actually an adele since, for all p not dividing its denominator, r is a p-adic integer.) Due to the heterogeneity of the topologies induced on Q by its different completions, however, Q is naturally embedded in A_Q as a discrete subfield.

7.6. Locally Compact Structures

Now, why is being locally compact so important? The pragmatic reason is that the additive structure of A_Q is an abelian (i.e., commutative) locally compact topological group,[31] and such groups are naturally endowed with Haar measures (generalizing the Lebesgue measure on R), which allow integration and harmonic analysis. According to a theorem of Iwasawa,[32] this property belongs to the characterization of Q as a global field, the *arithmetic* of Q being correlated to the *analysis* of A_Q. As Alain Connes writes, referring to Weil (1967) and Tate (1950) in his "Essay on the Riemann Hypothesis" (2015, 5):

> It opens the door to a whole world which is that of automorphic forms and representations, starting . . . with Tate's thesis ["Fourier Analysis in Number Fields and Hecke's Zeta-Function," 1950] and Weil's book *Basic Number Theory*.

In chapter 9 of *Modern Algebra and the Rise of Mathematical Structures* (2004), Leo Corry discusses the fact that Weil's preference for a theory of integration à la Lebesgue on locally compact groups restrained the development of probability theory à la Kolmogorov. Indeed, the latter uses, e.g., for Brownian motion, measures, and integration theory on non-locally compact groups. In his

[31] Moreover, A_Q has the deep property of being "self-dual" for Pontryagin duality, i.e., it is isomorphic to the group of its characters.

[32] The fact that the topological ring A_K of adeles of a field K is locally compact, semi-simple (with trivial Jacobson ideal), K being cocompact in it, characterizes global fields.

autobiography, Laurent Schwartz testimonies that "Bourbaki stepped away from probability, rejected it, considered it to be unrigorous" (quoted in Corry 2004, 119). We see in this example how the selection of "good" relevant structures can depend heavily upon the "big problems" aimed at: the Riemann hypothesis is not Brownian motion.

7.7. The Rosetta Stone

His remarkable conceptual deepening of the Dedekind-Weber analogy enabled Weil to find a strategy for proving the Riemann hypothesis (RH) not for arithmetic, but for an analogous, more geometric world. Indeed, in characteristic 0 the only global fields are finite extensions K of Q (i.e., algebraic number fields). But there exist a lot of other global fields defined in characteristic p. They are the fields K of rational functions on algebraic curves over a *finite* field $F_q = Z/qZ$ with $q = p^n$, p prime. It is therefore natural, on the one hand, to try to transfer to these fields questions concerning algebraic number fields: Weil did it for RH. On the other hand, algebraic curves over a finite field must have something to do with algebraic curves over C, and it is also very natural to try to translate RH to their case. It is for this *intermediate* third world that Weil succeeded in proving RH. This was one of his greatest achievements. He overcame what he considered to be the main difficulty in the Dedekind-Weber analogy, namely: that the theory of Riemann surfaces is "too rich" and "too far from the theory of numbers," and that "one would be totally blocked if there were not a bridge between the two" (Weil [1940] 2005, 340). Hence his celebrated metaphor of the "Rosetta stone":

> My work consists in deciphering a trilingual text; of each of the three columns I have only disparate fragments; I have some ideas about each of the three languages: but I know as well there are great differences in meaning from one column to another, for which nothing has prepared me in advance. In the several years I have worked at it, I have found little pieces of the dictionary. (Weil [1940] 2005, 340)

7.8. The Riemann Hypothesis: From Hasse to Weil, Grothendieck, Deligne, and Connes

Before Weil, Emil Artin and Friedrich Karl Schmidt had already transferred the Riemann-Dirichlet-Dedekind zeta and L-functions from the arithmetic side to the side of algebraic curves over F_q. In this new context, Helmut Hasse proved RH for *elliptic* curves. Then Weil proved it for all algebraic curves over finite fields using mixed technical tools, such as divisors, the Riemann-Roch theorem for the curves and their squares, intersection theory, the Severi-Castelnuovo inequality coming from the classical geometric side (characteristic 0), and crucially, Frobenius maps coming from characteristic p (see Cartier 1993). It is well known that the attempts to generalize to higher dimensions Weil's proof of RH for curves over finite fields led him to his celebrated conjectures; and to find a strategy for proving them has been at the origin of the monumental program of Grothendieck (schemes, sites, topoi, etale cohomology, etc.), culminating in 1973 with Deligne's proof.

But the original Riemann hypothesis remained, and still remains, unsolved. A few years ago, Alain Connes proposed a new strategy, consisting in constructing a new geometric framework for arithmetics in which Weil's proof in the intermediary case of curves over finite fields could be transferred by analogy. His fundamental discovery is that a way forward could be to work in a new "new world," namely the strange world of "tropical algebraic geometry in characteristic 1." In his 2015 essay he explains that the strategy is

> to find a geometric framework for the Riemann zeta function itself, in which the Hasse-Weil formula, the geometric interpretation of the explicit formulas, the Frobenius correspondences, the divisors, principal divisors, Riemann-Roch problem on the curve and the square of the curve all make sense. (Connes 2015, 8)

8. Conclusion: Structures and Mathematical Discovery

From Weil to Grothendieck and Deligne, and from Grothendieck to Connes, we see how crucial and permanent the long-term functional role of structural analogies as a method of discovery is. As Weil strongly stressed from the outset in his letter to Simone:

> *If one follows it in all of its consequences*, the theory alone permits us to reestablish the analogy at many points where it once seemed defective: it even permits us to discover in the number field simple and elementary facts which however were not yet seen. (Weil [1940] 2005, 339)

Thus, a structural clarification of an analogy yields more understanding and allows to go further.

Indeed, structures enable us to imagine strategies for solving hard problems. It is amusing to see how Weil used a lot of military metaphors—"find an opening for an attack (please excuse the metaphor)," "open a breach which would permit one to enter this fort (please excuse the straining of the metaphor)," "it is necessary to inspect the available artillery and the means of tunneling under the fort (please excuse, etc.)"—when explaining to his sister that finding a proof is actually a strategy. He added:

> It is hard for you to appreciate that modern mathematics has become so extensive and so complex that it is essential, if mathematics is to stay as a whole and not become a pile of little bits of research, to provide a unification, which absorbs in some simple and general theories all the common substrata of the diverse branches of the science, suppressing what is not so useful and necessary, and leaving intact what is truly the specific detail of each big problem. This is the good one can achieve with axiomatics (and this is no small achievement). This is what Bourbaki is up to. It will not have escaped you (to take up the military metaphor again) that there is within all of this great problems of strategy. (Weil [1940] 2005, 341)

This illustrates that Bourbaki's structures concern much more than mere "simple and general" abstractions. They have a functional role, a strategic and creative function, namely "leaving intact what is truly the specific detail of each big problem."

This pragmatic functionality of structures is really the key point for our purposes. Bourbaki was a group of creative mathematicians, not of philosophers. The true philosophical meaning of their structuralist approach is rooted deeply in their practice and must be extracted from there. To evaluate it, it is not sufficient to criticize their more or less clever or educated philosophical claims. The fundamental relation between, on the one hand, their holistic and "organic" conception of the unity of mathematics and, on the other hand, their thesis that some analogies and crossroads can be creative and lead to essential discoveries is a leitmotiv for Bourbaki since the 1948 manifesto, "L'Architecture des mathématiques." The continued insistence on the "immensity" of mathematics and on its "organic" unity; the claim that "to integrate the whole of mathematics into a coherent whole" (222) is not a philosophical question, as it was for Plato, Descartes, Leibniz, or "logistics"; the constant criticism of the reduction of mathematics to a tower of Babel juxtaposing separated "corners"—these are not vanities of philosophically ignorant mathematicians. They have a very precise technical function: to construct complex proofs navigating in this holistic, conceptually coherent world.

Hence: "The 'structures' are tools for the mathematician"; "each structure carries with it its own language"; and to discover a structure in a concrete problem "illuminates with a new light the mathematical landscape" (Bourbaki

1948, 227) (compare again the example of the locally compact adelic ring). Leo Corry has formulated this key point well:

> In the *L'Architecture des mathématiques* manifesto, Dieudonné also echoed Hilbert's belief in the unity of mathematics, based both on its unified methodology and in the discovery of striking analogies between apparently far-removed mathematical disciplines. (Corry 2004, 304)

And indeed, Dieudonné claimed:

> Where the superficial observer sees only two, or several, quite distinct theories, lending one another "unexpected support" through the intervention of mathematical genius, the axiomatic method teaches us to look for the deep-lying reasons for such a discovery. (Bourbaki 1948, 230)

Structures are guides for intuition and allow to overcome "the natural difficulty of the mind to admit, in dealing with a concrete problem, that a form of intuition, which is not suggested directly by the given elements, . . . can turn out to be equally fruitful" (Bourbaki 1948, 230). Thus for Bourbaki "more than ever does intuition dominate in the genesis of discovery" (228). And intuition is guided by structures.

After his 1948 manifesto, Bourbaki deepened this vision considerably. The structural systematization made by the *Éléments* allowed clarification of many difficulties, opened up good prospects, and led to fruitful angles of attack, which helped to solve difficult and entangled problems. In the combination of, on the one hand, systematizing and clarifying formal operations in the context of justification and, on the other hand, implementing proof strategies in the context of discovery rests, in our opinion, Bourbaki's main contribution. Thus the philosophical scope of Bourbaki's concept of structure goes far beyond its formal presentation in *Théorie des Ensembles*. Its coherence has to be found not in foundational issues, but in the extraordinary corpus of technical results the Bourbakists produced and inspired. To understand Bourbaki's "philosophy," one has to take seriously, and discuss philosophically, the statements, reflections, and testimonials concerning how they thought about the operational practice involving structures for the creative imagination in pure mathematics. Very few philosophers have addressed these issues.[33]

[33] A remarkable exception was Albert Lautman; cf. Heinzmann (2019) and Petitot (1987). Compare also Zalamea (2012) and Chevalley (1987).

Acknowledgments

Some parts of this chapter are the result of collective work discussed in seminars, symposia, and a forum of the ANR/DFG project *Mathematics: Objectivity by Representation*. We are especially grateful to Jean-Pierre Ferrier, Alain Genestier, Erich Reck, Georg Schiemer, and David Stump for their many precious suggestions and numerous linguistic corrections, which enabled us to improve the text.

References

Aczel, Amir D. 2006. *The Artist and the Mathematician: The Story of Nicolas Bourbaki, the Genius Mathematician Who Never Existed*. New York: Thunder's Mouth Press.

Atiyah, Michael. 2007. Bourbaki, a Secret Society of Mathematicians and the Artist and the Mathematician. *Notices of the AMS* 54(9), 1150–1152.

Beaulieu, Liliane. 1989. Bourbaki: Une histoire du groupe de mathématiciens français et de ses travaux (1934–1944). Thesis, University of Montréal.

Beaulieu, Liliane. 2013. *Bibliographie Bourbaki Bibliography. Travaux portant sur Bourbaki, Works about Bourbaki*. https://poincare.univ-lorraine.fr/sites/poincare.univ-lorraine.fr/files/users/documents/fichier_page/b-biblio-b-iii2013.pdf.

Beth, Evert W., and Jean Piaget. 1966. *Mathematical Epistemology and Psychology*. Translated by W. Mays from *Épistémologie mathématique et psychologie: Essai sur les relations entre la logique formelle et la pensée réelle*. Paris: Presses Universitaires de France, 1961.

Bolondi, Giorgio. 2009. La France du XXe siècle: le phénomène Bourbaki. In *La Mathématique*, vol. 1, *Les Lieux et les Temps*, edited by C. Bartocci and P. Odifredi. Paris: CNRS Éditions, pp. 689-715.

Bourbaki, Nicolas. 1939. *Théorie des ensembles (Fascicule de résultats)*. Paris: Hermann. English translation in Bourbaki 1968.

Bourbaki, Nicolas. 1948. L'architecture des mathématiques. In *Les grands courants de la pensée mathématique*, edited by F. Le Lionnais, pp. 35–47. Paris: Blanchard. English translation The Architecture of Mathematics, *American Mathematical Monthly* 57(4) (1950), 221–232.

Bourbaki, Nicolas. 1949. Foundations of Mathematics for the Working Mathematician. *Journal of Symbolic Logic* 14(1), 1–8.

Bourbaki, Nicolas. 1957. *Éléments de mathématique*. Vol. 1, *Théorie des ensembles, 4. Structures*. Paris: Hermann.

Bourbaki, Nicolas. 1966. *Elements of Mathematics: General Topology, Part 2*, Paris: Hermann; Palo Alto, CA: Addison-Wesley.

Bourbaki, Nicolas. 1968. *Elements of Mathematics: Theory of Sets*. Paris: Hermann.

Bourbaki, Nicolas. 1974. *Algebra I, Chapters 1–3*. Paris: Hermann; Palo Alto, CA: Addison-Wesley.

Cartan, Henri. 1943. Sur le fondement logique des mathématiques. *Revue Scientifique* 1, 1–11.

Cartier, Pierre. 1993. Des nombres premiers à la géométrie algébrique (une brève histoire de la fonction zeta). *Cahiers du Séminaire d'Histoire des mathématiques*, 2nd series, 3, 51–77.

Cavaillès, Jean. 1947. *Sur la logique et la théorie de la science*. Paris: Presses Universitaires de France. 3rd ed. Reprinted Paris: Vrin, 1976. English translation by T. S. Kisiel in *Phenomenology and the Natural Sciences*, edited by J. Kockelmans and T. Kisiel, pp. 357–409. Evanston, IL: Northwestern University Press, 1970.

Chevalley, Catherine. 1987. Albert Lautman et le souci logique. *Revue d'Histoire des Sciences* 40(1), 49–77.

Chevalley, Claude, and Denis Guedj. 1985. Nicolas Bourbaki, Collective Mathematician. An Interview with Claude Chevalley. Translated by Jeremy Gray. *Mathematical Intelligencer* 7(2), 18–22.

Connes, Alain. 2015. An Essay on the Riemann Hypothesis. In *Open Problems in Mathematics*, edited by John Nash and Michael Rassias, pp. 225-257.

Corry, Leo. 1992. Nicolas Bourbaki and the Concept of Mathematical Structure. *Synthese* 92(3), 315–348.

Corry, Leo. 2004. *Modern Algebra and the Rise of Mathematical Structures*. 2nd ed. Basel: Birkhäuser.

Dedekind, Richard, and Heinrich Weber. 1882. Theorie der algebraischen Funktionen einer Veränderlichen. *J. Reine Angew. Math.* 92, 181–290.

Dieudonné, Jean. 1939. Les méthodes axiomatiques modernes et les fondements des mathématiques. *Revue scientifique* 77, 224–232.

Dieudonné, Jean. 1977. *Panorama des mathématiques pure: le choix bourbachique*. Paris: Gauthier-Villars.

Grothendieck, Alexandre, and Jean Dieudonné. 1960–67. Éléments de Géométrie Algébrique. *Publications Mathématiques de l'IHÉS* 4 (1960), 5–228; 8 (1961), 5–222; 11 (1961), 5–167; 17 (1963), 5–91; 20 (1964), 5–259; 24 (1965), 5–231; 28 (1965), 5–255; 32 (1967), 5–361.

Hankel, Hermann. 1867. *Vorlesungen über die complexen Zahlen und ihre Functionen*, Leipzig: Voss.

Heinzmann, Gerhard. 2018. Correspondance Dieudonné-Cavaillès (1939). *Revue de Métaphysique et de Morale*, forthcoming.

Heinzmann, Gerhard. 2019. Early French Structuralism Reconsidered: Poincaré and Lautman. The Structuralist Roots of Mathematical Understanding. Forthcoming.

Hilbert, David, and Wilhelm Ackermann. 1928. *Grundzüge der theoretischen Logik*. Berlin: Springer.

Houzel, Christian. 2002. Bourbaki und danach. *Mathematische Semesterberichte* 49, 1–10.

Kleiner, Israel. 2000. From Fermat to Wiles: Fermat's Last Theorem Becomes a Theorem. *Elemente der Mathematik* 55, 19–37.

Krömer, Ralf. 2006. La "machine de Grothendieck," se fonde-t-elle seulement sur des vocables métamathématiques? Bourbaki et les catégories au cours des années cinquante. *Revue d'Histoire des Mathématiques* 12, 119–162.

Krömer, Ralf. 2007. *Tool and Object: A History and Philosophy of Category Theory*. Basel: Birkhäuser.

Mashaal, Maurice. 2006. *Bourbaki: A Secret Society of Mathematicians*. Translated by Anna Pierrehumber. Providence, RI: American Mathematical Society. Translation of Bourbaki: une société secrète de mathématiciens, *Pour la Science*, May 2000.

Mathias, Adrien R. D. 1992. The Ignorance of Bourbaki. *Mathematical Intelligencer* 14(3), 4–13.

Mazur, Barry. 1991. Number Theory as Gadfly. *American Mathematical Monthly* 98(7), 593–610.

McLarty, Colin. 2008. "There Is No Ontology Here": Visual and Structural Geometry in Arithmetic. In *The Philosophy of Mathematical Practice*, edited by P. Mancosu, pp. 370–406. New York: Oxford University Press.

Petitot, Jean. 1987. Refaire le "Timée"; Introduction à la philosophie mathématique d'Albert Lautman. *Revue d'Histoire des Sciences* 15(1), 79–115. Full version at hal-01130394.

Petitot, Jean. 1993. The Unity of Mathematics as a Method of Discovery: Wiles' Example. http://jeanpetitot.com/ArticlesPDF/Petitot_TSW_Wiles.pdf.

Petitot, Jean. 2017. Axiomatics as a Strategy for Complex Proofs: The Case of the Riemann Hypothesis. In *Axiomatic Thinking*, Académie Internationale de Philosophie des Sciences, Lisbon, October 11–14. Cf. http://jeanpetitot.com/ArticlesPDF/Petitot_AIPS_2017.pdf.

Poincaré, Henri. 1898. On the Foundations of Geometry. *The Monist* 9, 1–43.

Razafindrakoto, Ando. 2013. What is . . . Bourbaki? http://math.sun.ac.za/wp-content/uploads/2013/03/Bourbaki.pdf.

Sieg, Wilfried, and Dirk Schlimm. 2017. Dedekind's Abstract Concepts: Models and Mappings. *Philosophia Mathematica* 25(3), 292–317.

Tate, John. 1950. *Fourier Analysis in Number Fields and Hecke's Zeta-Function*. thesis Princeton University. published n *Conference Proceedings: Algebraic Number Theory*, edited by John W.S. Cassels and Albrecht Fröhlich. Washington, DC: Thompson Book Company, 1967.

Thom, René. 1970. Les mathématiques modernes: une erreur pédagogique et philosophique? In *L'Age de la Science*, vol. 3, part 3, pp. 225–242. Paris: Dunod.

Thom, René. 1971. "Modern" Mathematics: An Educational and Philosophical Error? *American Scientist* 59, 695–699.

Van der Waerden, Bartel Leendert. 1930–31. *Moderne Algebra: Unter Benutzung der Vorlesungen von E. Artin und E. Noether*. 2 vols. Berlin: Springer.

Weil, André. 1938. Zur algebraischen Theorie der algebraischen Funktionen. *Crelle Journal* 179, 129–138.

Weil, André. [1940] 2005: Letter to Simone Weil, March 26, 1940. In *Collected Papers*, vol. 1, 244–255. English translation by Martin Krieger, *Notices of the AMS*, 52(3), , 334–341.

Weil, André. 1967. *Basic Number Theory*. New York: Springer.

Weil, André. 1992. *The Apprenticeship of a Mathematician*. Basel: Birkhäuser. Originally published as *Souvenirs d'apprentissage*. Basel: Birkhäuser, 1991.

Zalamea, Fernando. 2012. *Synthetic Philosophy of Contemporary Mathematics*. New York: Sequence Press.

9

Saunders Mac Lane: From *Principia Mathematica* through Göttingen to the Working Theory of Structures

Colin McLarty

1. Mac Lane Overall

Saunders Mac Lane (1909–2005) attended David Hilbert's weekly lectures on philosophy in Göttingen in 1931. He utterly believed Hilbert's declaration that mathematics will know no limits: *Wir müssen wissen; wir werden wissen*—We must know, we will know.[1] Mac Lane had a room in Hermann Weyl's house and worked with Weyl revising Weyl's book *Philosophy of Mathematics and Natural Science* (1927). At the same time he absorbed a structural method from Emmy Noether. Mac Lane always linked mathematics with philosophy, but he was disappointed in his own Göttingen doctoral dissertation (1934) trying to streamline the logic of *Principia Mathematica* into a practical working method for mathematics. He had wanted to do that since he was undergraduate at Yale.[2] Now he saw it could go nowhere. He lost interest in philosophic arguments for or against philosophic ideas about mathematics.

Mac Lane learned a new standard for philosophy of mathematics from Hilbert and Weyl: Which ideas advance mathematics? Which help us solve longstanding problems? Which help us create productive new concepts, and prove new theorems? In other words: which ideas work? The Göttingers taught him that a philosophy of *form*, or *structure*, is key to the productivity of modern mathematics.

He urged this direction for logic research in a talk to the American Mathematical Society in 1933 published in the *Monist* (Mac Lane 1935). He continued promoting logic and writing reviews for the *Journal of Symbolic Logic*. He always tried to move logic research closer to other mathematics. The

[1] Mac Lane (1995a, 1995b).

[2] Philosophy instructor F. S. C. Northrop, a Whitehead student like Quine, turned Mac Lane toward *Principia Mathematica*. See Mac Lane (1996b, 6) and Mac Lane (1997a, 151).

Colin McLarty, *Saunders Mac Lane* In: *The Prehistory of Mathematical Structuralism*. Edited by: Erich H. Reck and Georg Schiemer, Oxford University Press (2020). © Oxford University Press.
DOI:10.1093/oso/9780190641221.003.0009

last single-author book he completed in his lifetime (1986) aimed to recruit philosophers to looking at mathematics this way: Which ideas work?

Look at his relation with Quine. He and Quine were both decisively influenced as undergraduates by *Principia Mathematica*, at elite liberal arts schools, he at Yale and Quine at Oberlin. Both did doctoral dissertations based on that and spent years studying in Europe. Both were founding members of the Association for Symbolic Logic. They often spoke as faculty colleagues at Harvard from 1938 to 1947 but in decisively different departments. Mac Lane felt "the impressive weight of *PM* had continued to distort Quine's views on the philosophy of mathematics" (1997a, 152); and he rejected Quine's "undue concern with logic, as such" (Mac Lane 1986, 443).

He published on several topics in his early career including logic but focused on technical problems in algebra aimed at number theory. His solution to one of these was a strange family of groups. Samuel Eilenberg knew these same groups solve a problem in topology. When Eilenberg (who, by the way, liked philosophy a great deal less than Mac Lane did) learned of Mac Lane's result, the two of them agreed this could not be a coincidence. They set out to find the connection. They spent the next 15 years calculating a slew of specific relations between topology and group theory and building these relations into the new subject of *group cohomology*.[3] The work stood out immediately, and during that time Mac Lane became president of the Mathematical Association of America and chair of Mathematics at the University of Chicago.

Eilenberg and Mac Lane also believed the following:

> In a metamathematical sense our theory provides general concepts applicable to all branches of mathematics, and so contributes to the current trend towards uniform treatment of different mathematical disciplines. (Eilenberg and Mac Lane 1945, 236)

The concepts were category, functor, and natural isomorphism. They expected this to be the only paper ever needed on these ideas (Mac Lane 1996a, 3).

Within a few years these concepts were standard in topology, abstract algebra, and functional analysis such as (Grothendieck 1952). By 1960 they were central to cutting-edge algebraic geometry. In differential geometry, they were the right tool for Adams (1962) to show exactly how many different vector fields there can be on spheres of any finite dimension. Soon categories, functors, and natural transformations (including natural isomorphisms, but not only isomorphisms)

[3] Mac Lane (1988) is a gentle introduction to group cohomology and Washington (1997) is a more current precis. An earlier innovator on this was another Noether student, Heinz Hopf, but Mac Lane could not contact him during the war years.

became textbook material. They became the standard mathematical framework for structural mathematics.

Structuralists in philosophy of mathematics talk more often about Bourbaki's theory of structures. Indeed Bourbaki (1949, 1950) promoted their view as a philosophy, while Eilenberg and Mac Lane did not.[4] But Bourbaki's theory of structures (1958, chap. 7), which they created as a conscious alternative to categories and functors, never worked for them or anyone else. Several members of Bourbaki became major innovators in category theory. This, together with Daniel Kan (1958) defining *adjoint functors*, secured category theory as a theory in its own right. Systems biologist Rosen was the first person to use the term "category theory" in print (1958, 340).

Mac Lane (1948) had pioneered the idea that categorical tools are also useful in defining some very simple structures. Yet he was surprised in 1963 to meet Eilenberg's graduate student William Lawvere, who was describing such basic things as the natural numbers and function sets categorically. Lawvere had even axiomatized set theory in categorical form. Mac Lane found this absurd and said you need sets to define categories in the first place—until he read Lawvere's paper. As a member of the National Academy of Science, Mac Lane sent it to the *Proceedings*, where it became Lawvere (1964). Lawvere's ideas on many aspects of category theory launched a new phase in Mac Lane's career and brought him back to looking more at philosophy and logic than he had since the 1930s. Mac Lane's last doctoral student was philosopher Steve Awodey in 1997.

2. Structuralist Philosophy of Mathematicians, 1933

Philosophy for mathematicians in 1930s Göttingen meant phenomenology. And this was not only in Göttingen. When Carnap (1932, 222) lists four ways to describe word meanings, the first is his own, which he claims is correct, the next two use what he calls the language of logic and epistemology, and he calls the fourth one "philosophy (phenomenology)." Mac Lane will have known Carnap's paper, as he thought of going to study logic with Carnap in Vienna (Mac Lane 1979, 64). In fact Eilenberg and Mac Lane later took the word "functor" from Carnap's logic (Mac Lane 1971, 30). All of these people meant roughly Husserl's phenomenology. Husserl was widely respected by mathematicians since he had studied mathematics with Weierstrass and Kronecker and had written a doctoral dissertation in mathematics.

[4] Mac Lane gave a hint of his philosophy by titling his classic textbook *Categories for the Working Mathematicians* (1971) in response to "Foundations of Mathematics for the Working Mathematician" (Bourbaki 1949).

Hilbert brought Husserl onto the Göttingen faculty against resistance from other philosophers there (Peckhaus 1990, 56f). When Husserl left Göttingen for Freiburg, another Hilbert protégé and phenomenologist, Moritz Geiger, took his place. Mac Lane studied Geiger (1930) as part of his degree requirements.[5]

Geiger admired Husserl's phenomenological method while rejecting Husserl's idealism (Spiegelberg 1994, 200). The method aims to understand many attitudes toward being, without taking one or another of them as correct. For Geiger, the *naturalistic attitude* recognizes physical objects and considers anything else merely psychical/subjective. The *immediate attitude*, more widely used in daily life, recognizes psychical objects like feelings, and social objects like poems, and more including mathematical objects. Mathematics for Geiger belongs to the immediate attitude since its objects are neither physical nor subjective. They exist as *forms* (*Gebilde*), which may or may not be forms of physical objects. Years later Mac Lane's *Mathematics: Form and Function* (1986) would say, in the title among other places, mathematics studies *forms*, which may be applied in physical sciences but need not.

Geiger applied his philosophy in a *Systematic Axiomatics for Euclidean Geometry* (1924), aiming to go beyond Hilbert's axioms by drawing out their real connections as ideas. Compare Mac Lane (1986) sketching several proofs for a given theorem, then singling one out as "the reason" for it.[6]

Weyl explains the role of forms by quoting an influential textbook by Hermann Hankel on complex numbers and quaternions:

> [This universal arithmetic] is a pure intellectual mathematics, freed from all intuition, a pure theory of forms [*Formenlehre*] dealing with neither quanta nor their images the numbers, but intellectual objects which may correspond to actual objects or their relations but need not.

Weyl approves Husserl saying: "Without this viewpoint . . . one cannot speak of understanding the mathematical method."[7]

Hankel says this to help students learn. Weyl approves it because it helps mathematicians discover and prove theorems—where Weyl's favorite example is Hilbert. Of course Husserl's paradigms were his teachers Kronecker and Weierstrass. This is what works in modern mathematics.

Philosophers today might feel this account privileges abstract mathematics from Göttingen over more computational Berlin mathematics. But in fact both Husserl and Hankel were Berlin mathematicians trained by Kronecker and

[5] Much more on Geiger, Weyl, and Mac Lane is in McLarty (2007a).
[6] For example pp. 145, 189, 427, 455.
[7] Hankel (1867, 10) and Husserl (1922, 250) quoted by Weyl (1927, 23).

Weierstrass. Hankel's book is all about calculating with complex numbers and quaternions. And Weyl famously supported concrete calculational mathematics over abstract axioms. Conversely, the archetypal Göttingen algebraist Emmy Noether saw her algebra as advancing calculation. In the middle of her work on abstract ideal theory she supervised a doctoral dissertation devising algorithms to apply her theory in the case of polynomial rings (Hermann 1926). It is still cited for that today (Cox et al. 2007). All of these mathematicians believed correct focus on form facilitates computation. They only disagreed over how abstract a correct focus would be!

Mac Lane heard all this from Weyl himself. I do not know whether Mac Lane noticed Hans Hahn's admiring yet barbed review of Weyl's book, or Hahn's conclusion:

> Most of this eloquent exposition concerns that which, according to Wittgenstein's teaching, cannot be said at all, or to express it in a less radical way: what can only be said in a beautiful style and not in dry formulas. (Hahn 1928, 54)

I do know Mac Lane had no inkling that he would soon create a mathematical theory of form and preservation of form, specifically of *homomorphisms* and *isomorphisms*, that is expressible in quite dry formulas and would go on to organize huge amounts of mathematical research and writing.

3. Method, Methodology, and Who is a Philosopher

> All the Second Philosopher's impulses are methodological, just the thing to generate good science.... She doesn't speak the language of science "like a native"; she *is* a native. (Maddy 2007, 98, 308)

Maddy's character the Second Philosopher is a native science speaker. Yet she is also a philosopher because she articulates scientific methods and brings her methodological impulse to "traditional metaphysical questions about what there is" and how we know it (Maddy 2007, 410). In just these ways the Second Philosopher matches Hilbert, Weyl, and Mac Lane. But Mac Lane's philosophy was also shaped by Emmy Noether, a mathematician who herself was no philosopher.

Her best-known comment on her own method was to say no one including her talks about it:

> My methods are working and conceptual methods, and so they penetrate everywhere anonymously. (Letter to Helmut Hasse, November 12, 1931, quoted in Lemmermeyer and Roquette 2006, 8 and 131)

She showed a method. We may say she gives a methodophany rather than methodology, by analogy to theophany/theology.

4. Noether on Structures

4.1. On Not Understanding Noether

> I heard from Noether about the use of factor sets, but did not then understand them. Much later I did.
>
> —Mac Lane (1998b, 870)

There are two different ways to not understand factor sets: You might not see how to use them. Or you might feel there must be more than you yet see. Mac Lane certainly did understand them in that first sense. He used them well in the paper (Mac Lane and Schilling 1941) that got him into the collaboration with Eilenberg. What he means in this quotation is that he felt he had not seen deeply enough what they really are. He achieved that understanding, to his satisfaction, years later by reformulating factor sets in categorical terms with Eilenberg (Mac Lane 1988, 33).

To put the matter in correct historical order we must say Eilenberg and Mac Lane (1942a) spoke of *natural isomorphisms*. Their term *functor* first saw print a few months later, in a paper further explaining natural isomorphisms (1942b). Their first printed use of *category* is in (1945), giving the general definition of functors. For more relating Mac Lane to Noether see Koreuber (2015); Krömer (2007); Mac Lane (1981, 1997b); McLarty (2006, 2007a).

4.2. From Equations to Structures

Noether brought stunningly swift insights to a perspective going back to Gauss and Dedekind, and even to Galois: it is often productive to replace solutions to equations by maps between structures. Clearly motivated algebra replaces long, incomprehensible calculations. It makes theorems of arithmetic easier to find and prove in the first place and makes the proofs easier for students to learn.

For a simple illustration consider these two groups: The group of integers modulo 12, written $\mathbb{Z}/(12)$, is often popularized as "clock face arithmetic." On a 12-hour clock, five hours past nine o'clock is two o'clock, as 2 is the remainder of 14 by 12. The members of $\mathbb{Z}/(12)$ are the integers from 0 to 11 (with 12 taken as equal to 0), and $5 + 9 = 2$ in $\mathbb{Z}/(12)$. The group of integers modulo 3, written $\mathbb{Z}/(3)$ consists of $\{0,1,2\}$ with addition defined by taking remainders on division by 3:

$$1+0=1 \qquad 1+1=2 \qquad 1+2=0 \qquad 2+2=1.$$

Two facts about mappings between $\mathbb{Z}/(3)$ and $\mathbb{Z}/(12)$ both express the fact that 3 divides 12:

Theorem 1. *There is an injective group homomorphism* $i\colon \mathbb{Z}/(3) \to \mathbb{Z}/(12)$. *Here injective means* $i(x) = i(y)$ implies $x = y$.

Proof. Group homomorphisms preserve 0 and +, so define i by

$$i(0) = 0 \quad i(1) = 4 \qquad i(2) = i(1) + i(1) = 8 \qquad \text{in } \mathbb{Z}/(12).$$

Since $1 + 2 = 0$ in $\mathbb{Z}/(3)$ we must check that $i(1) + i(2) = 0$ in $\mathbb{Z}/(12)$. Indeed:

$$i(1) + i(2) = 4 + 8 = 0 \qquad \text{in } \mathbb{Z}/(12).$$

Theorem 2. *There is an onto group homomorphism* $h\colon \mathbb{Z}/(12) \to \mathbb{Z}/(3)$. *Here onto means every* y *in* $\mathbb{Z}/(3)$ *is* $h(x)$ *for some* $x \in \mathbb{Z}/(12)$.

Proof. Define $h\colon \mathbb{Z}/(12) \to \mathbb{Z}/(3)$ by $h(0) = h(3) = h(6) = h(9) = 0$. Preserving + means we must then say $h(3x + 1) = 1$ in $\mathbb{Z}/(3)$ for every $x \in \mathbb{Z}/(12)$. And $h(3x + 2) = 2$. In words, this works because counting up by 3s leads to 0 modulo 12, since 12 is divisible by 3.

Then $3 \cdot 4 = 12$ becomes a group isomorphism $\mathbb{Z}/(3) \times \mathbb{Z}/(4) \approx \mathbb{Z}/(12)$. Of course the practical payoff is when isomorphisms of richer groups reveal deeper arithmetic (Dedekind 1996).

Noether radically sharpened, articulated, and generalized Dedekind's insight in her *homomorphism and isomorphism theorems*, using what she called her "set-theoretic" conception (McLarty 2006, esp. 217–220). This was not the long-familiar idea that groups are sets of elements. To the contrary, she would focus as little as possible on the elements $0, x, y, z \ldots$ and operations $x + y$ or $x - y$ of a group G. She would focus as directly as possible on homomorphisms between G and other groups, and especially isomorphisms. One of her best students wrote:

> Noether's principle: base all of algebra so far as possible on consideration of isomorphisms. (Krull 1935, 4)

Mac Lane bought Krull's book and left marginal notes that seem to date from many different years.

Mac Lane saw Noether at the peak of her career. She had moved beyond her early 1920s work on axioms in commutative algebra to more intricate applications in group representation theory. Much of Mac Lane's work in the 1930s was close to themes in her plenary address at the International Congress of Mathematicians in Zurich (Noether 1932).

4.3. Making the Theorems Yet More Structural

Mac Lane (1948) used categories to make Noether's homomorphism and isomorphism theorems even more structural by removing elements from the very definitions of injective and onto homomorphisms.[8] When the following definitions are applied to groups they are equivalent to saying $\mathbf{0}$ is a one-element group while $g: G \to H$ is one-to-one and $h: H \to G$ is onto. And they are more directly useful in proving theorems than the element-based definitions are:

(1) A *zero group* is any group $\mathbf{0}$ such that every group G has exactly one homomorphism $G \to \mathbf{0}$ and exactly one homomorphism $\mathbf{0} \to G$.

(2) A *zero homomorphism* $0: H \to K$ is any homomorphism that factors through a zero group.

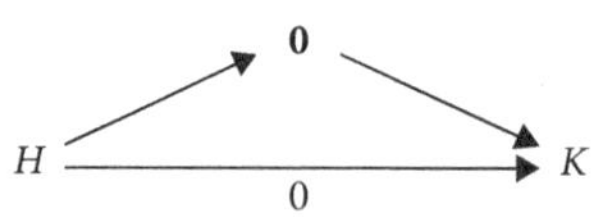

(3) Homomorphism $g: G \to H$ is *monic* if, whenever a composite gf is zero then already f is zero.

[8] Bypassing elements in the definitions and theorems was especially handy in work with the new idea of *sheaves*. Elements of sheaves are much more complicated than group elements while the patterns of homomorphisms between sheaves are very similar to those between groups

If 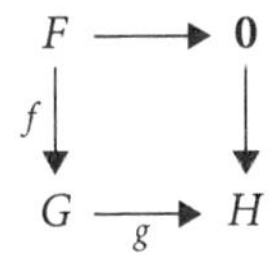 then 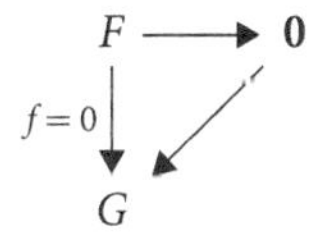

(4) Homomorphism $h: H \to G$ is *epic* if, whenever a composite *fh* is zero then already f is zero.

If 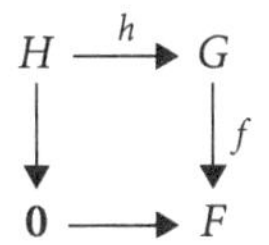 then 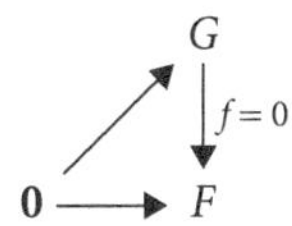

Notice the definition of epics is just that of monics with the arrows reversed. So monics are called *dual* to epics. Turning the arrows around in the definition of zero group just gives the same definition, so zero groups are self-dual. These ideas were much expanded over time, notably by Grothendieck in his theories of *abelian categories* and *derived categories*, and new aspects of that are still being developed today (Gelfand and Manin 2003).

5. Natural Isomorphisms

The phrase "second integral simplicial homology group of the torus" tells a topologist how to construct a unique group (up to isomorphism) but is no explicit description of the result. Explicitly, that group is (up to isomorphism) just the integers $\mathbb{Z}$ with addition. Often a mathematician has a construction like that and wants an explicit description.

Often it helps to find another construction of the same thing. But that one will rarely give the exact same thing. More often its result is *naturally isomorphic* to the first. "Natural isomorphism" was a common expression in mid-20th-century algebra and topology. Eilenberg and Mac Lane leaned hard on the idea and so had to say exactly what they meant by it. In principle they only had to be precise about their specific uses but in fact they came to see they had captured very much of the whole preexisting informal idea. They frequently put "natural" in quote marks to emphasize that they give "a clear mathematical meaning" to a colloquial idea (Eilenberg and Mac Lane 1942b, 538).

To simplify, Eilenberg and Mac Lane were in this situation: they had constructions C, C' that each apply to an arbitrary topological space S to yield groups $C(S)$ and $C'(S)$. These results were not exactly the same but were always isomorphic, $C(S) \approx C'(S)$. And much more than that was true.

First, the constructions did not apply only to spaces, but also applied to maps. Each map $f : S \to T$ of topological spaces induced a specific group homomorphism from $C(S)$ to $C(T)$, call this $C(f) : C(S) \to C(T)$. They dubbed such constructions *functors* from the *category* of topological spaces and maps, to the category of groups and group homomorphisms. Full definitions of category and functor are too easily available in print and online for us to linger on them here.

Second, there not only existed isomorphisms $C(S) \approx C'(S)$. Each space S had a specifiable isomorphism $i_S : C(S) \xrightarrow{\sim} C'(S)$ compatible with all the maps. For any map $f : S \to T$, isomorphism i_S followed by homomorphism $C'(f)$ is the same as homomorphism $C(f)$ followed by isomorphism i_T.

$$\begin{array}{ccccc} S & \quad & C(S) & \xrightarrow[i_S]{\sim} & C'(S) \\ f\downarrow & & C(f)\downarrow & & \downarrow C'(f) \\ T & & C(T) & \xrightarrow[i_T]{\sim} & C'(T) \end{array}$$

Again, full details are widely published and available online.

These concepts did not solve Eilenberg and Mac Lane's problems by themselves. Years of massive calculations remained. Each single one of these calculations had to summarize how some infinite family of interrelated groups and group homomorphisms all contribute to solving one problem about one topological space. Each such family would be organized into one infinite *diagram* of arrows between points—where each point represents one group and each arrow one group homomorphism. Then natural transformations between entire diagrams would yield the actual answer to the problem.

The new concepts organized the calculations. They showed how to shortcut some and bypass many others, and so they made the project feasible. These concepts have been working ever more widely across mathematics ever since.

6. Basic Constructions and Foundations

Because categories were invented for otherwise infeasible calculations on infinite diagrams, simple ideas like the Cartesian product $A \times B$ of two groups were not addressed in 1945. Simple ideas did not need category theory. But then Mac Lane (1948) saw how $A \times B$ and the injective and onto homomorphisms as described earlier could profitably be put in categorical terms. He began to see categories and functors as a way to organize advanced mathematics as a whole.

6.1. Bourbaki

The Bourbaki group in France had set out before the war to do just that, organize the whole of university mathematics. To this end they sketched a theory of *structures* in (Bourbaki 1939) and around 1950 they turned to creating it in full. The group considered what they could get from category theory for several years but finally produced their own theory of structured sets and structure preserving functions (Bourbaki 1958, chap. 7).

Neither they nor anyone ever used that theory. Corry (1992) documents at length that Bourbaki never used it in their series *Elements of Mathematics*, let alone for research, and how they argued over this. Several leading members of Bourbaki took up categories in their own work. Member Alexander Grothendieck created roughly half the topics of today's category theory: abelian categories, derived categories, and topos theory.[9]

Bourbaki's theory was extremely complicated and few people have ever read it. But the real problem was that the theory is "decidedly narrow in the shoulders" (Grothendieck 1987, 62–78). Even if mathematics is founded on set theory, so every object is by definition a set, the maps between structures need not be structure-preserving functions. Already in 1950 important examples of maps that are not simply functions included partial functions, equivalence classes of partial functions, functions that go "the wrong way," combinations of these, and other constructs that are not even like functions.[10]

The theory would need impossibly many extensions to capture the maps used today. And further extensions would soon be needed. There is no limit to what might serve as mappings. Category theory does not try to say what maps can be. The category axioms merely say that maps must include identity maps, and must compose associatively.

Mac Lane admired Bourbaki's project but found their theory of structures "a cumbersome piece of pedantry" (Mac Lane 1996c, 181).

[9] McLarty (2016) illustrates the mathematics. For history and conceptual discussion see McLarty (2007b).

[10] McLarty (2007a, 80–81) gives historically relevant examples.

6.2. Lawvere

Mac Lane met Lawvere as a graduate student with a program to unify all mathematics from the simplest to the most advanced in categorical terms. Mac Lane, like Eilenberg, thought it was absurd to axiomatize sets as a category. Then he read how Lawvere did it. He came to find this and many other of Lawvere's innovations extremely valuable.[11]

Mac Lane admired Lawvere's set theory precisely because it was *not* a novel conception of sets. Rather Lawvere gave expression, better than earlier set theories had done, to what mathematicians already know and use about sets. Leinster (2014) is a recent explanation of this.

As the paradigm case, earlier set-theoretic treatments of the natural numbers were clever, but merely technical. They did not focus on what we really want to know and use about arithmetic. This is exactly what Benacerraf (1965) complained about in the paper that launched current structuralism in philosophy of mathematics. Lawvere's definition of natural numbers, to the contrary, was almost verbatim Theorem 126 of Dedekind (1888) on inductive definition of functions from the natural numbers, though Lawvere did not know that at the time.

Definition 1. *A natural number object is a set* $\mathbb{N}$, *a* function $s\colon \mathbb{N} \to \mathbb{N}$, *and an element* $0 \in \mathbb{N}$, *such that for any set S, and function* $f: S \to S$, and element $x \in S$ there is a unique function $u: \mathbb{N} \to S$ with $u(0) = x$ *and* $us = fu$.

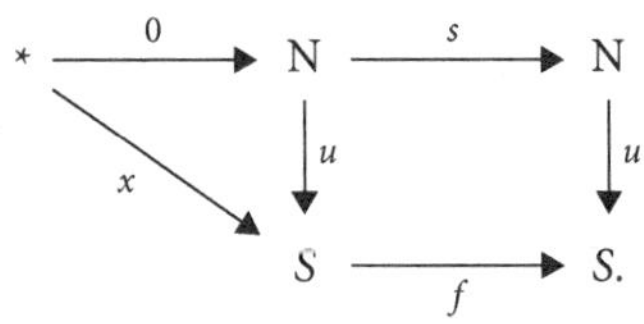

So u is a sequence in S with $u(0) = x$ and $u(s0) = f(x)$ and $u(ss0) = f(f(x))$ and so on. Every mathematician knows and uses this way of defining sequences in a set S. Few ever hear of the von Neumann or Zermelo natural numbers in ZFC.

Dedekind (1888) knew this fact was the key to his Theorem 132, which in modern terms proves Dedekind's definition of simply infinite systems is isomorphism invariant. Lawvere proves his natural number objects are isomorphism invariant the same way Dedekind did: all natural number objects are isomorphic

[11] Examples we will not discuss include functorial algebraic theories, Cartesian closedness, and comma categories. Some of these appear in Mac Lane (1971), and see McLarty (1990).

and anything isomorphic to a natural number object is one. Mac Lane liked the way this set theory gets directly to the mathematical point of the various constructions. See his enthusiastic exposition in (1986, chap. 11).

Mac Lane always valued logical foundations, not as starting points or rational justifications for mathematics, but as "proposals for the organization of mathematics" (Mac Lane 1986, 406). After 1964 he consistently urged Lawvere's "Elementary Theory of the Category of Sets" (ETCS) for this role (Lawvere 1964, 1965).[12]

His outlook led him to assimilate ETCS to Lawvere's other foundational axiom system, published in "Category of Categories as a Foundation for Mathematics" (CCAF) (1966). There is a great difference, as all the objects in ETCS are sets. They form a category, axiomatized entirely in categorical terms. But they are sets. Categories within ETCS (like everything in ETCS) are defined in terms of sets, and ETCS posits no category of sets as an entity any more than the universe of all sets is an entity in ZFC. On the other hand CCAF axiomatizes categories directly, not defining them via sets, and does posit a category of sets as an entity—though no actual category of all categories.[13]

From Mac Lane's point of view, though, they are alike since each provides

> an effective foundation by category theory.... The categorical foundation takes functors and their composition as the basic notions and it works very effectively. (Mac Lane 2000, 527)

6.3. Set-Theoretic Foundations of Category Theory

Eilenberg and Mac Lane (1945) already cared enough about logical foundations to note that the category of all groups or the category of all sets are illegitimate objects in set theory. However:

> The difficulties and antinomies here involved are exactly those of ordinary intuitive *Mengenlehre* [set theory]; no essentially new paradoxes are apparently involved. Any rigorous foundation capable of supporting the ordinary theory of classes would equally well support our theory. Hence we have chosen to adopt the intuitive standpoint, leaving the reader free to insert whatever type of logical foundation (or absence thereof) he may prefer. (Eilenberg and Mac Lane 1945, 246)

[12] See Mac Lane (1986, chap. 11; 1998a, Appendix; 1992; 2000) and Mac Lane and Moerdijk (1992, VI.10).

[13] See Lawvere (1963, 1966) and McLarty (1991).

They sketch ways to talk around the problem, and formal approaches via type theory and Gödel-Bernays set theory.

As the mathematics developed, though, the issue mattered more. Grothendieck adopted a systematic approach to these logical issues in using *universes* in algebraic geometry. These are sets so large they roughly speaking look like the set of all sets. Standard set theories, whether ZFC or ETCS, do not prove universes exist.

Mac Lane both did research on this and supported other research. See the papers by him and by Georg Kreisel and Solomon Feferman in (Mac Lane 1969). His preferred technical fix was an axiom positing one universe (Mac Lane 1998a, 21–22).

7. Sameness of Form

7.1. What Is an Isomorphism?

Philosophers should know that before about 1950 "There was great confusion: the very meaning of the word 'isomorphism' varied from one theory to another" (Weil 1991, 120). The word *isomorphism* often used to mean any homomorphism, and there was no general term for structurally identical things. It was hardly obvious that one notion of "sameness of structure" could work for all the different kinds of structures.

Today model theory gives a uniform notion of sameness of structure for models of any given first-order theory, and indeed it is called isomorphism of models. Few mathematicians learn this definition because it is nowhere near general enough to cover most of the structures used in practice. Bourbaki (1958) had a much more general notion that was still not general enough.

The current general definition of *isomorphism* turns out to be as simple as the idea of a morphism that does nothing plus the idea of two morphisms undoing each other. It came from Eilenberg and Mac Lane. And it is easy to picture in diagrams.

First, in any category, each object A has an identity morphism $1_A : A \to A$ defined by this property: composing it with any other morphism to or from A just leaves that other morphism.

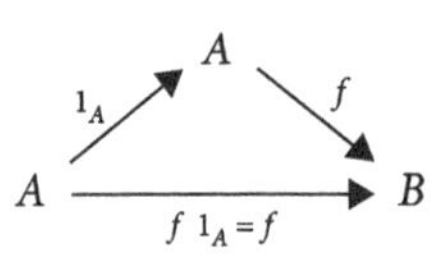

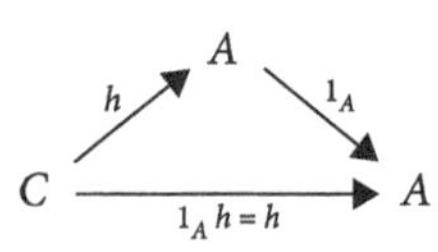

Think 1_A does nothing.

Then, a morphism $f : A \to B$ is an *isomorphism* if some morphism $g : B \to A$ has composite *gf* equal to $1_A : A \to A$ and composite *fg* equal to $1_B : B \to B$.

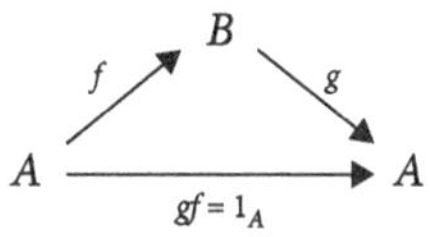

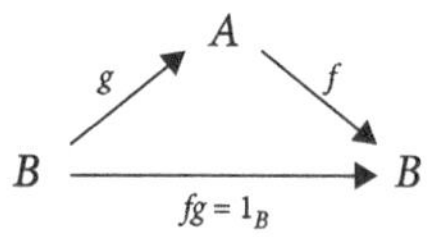

Think *f* and *g* undo each other.

Of course the definition of one or another specific kind of morphism may be somewhat complicated—for example, smooth maps as morphisms between manifolds in differential geometry are somewhat complicated. But that is not the business of category theory. Category theory applies to whatever morphisms you choose to supply, so long as they satisfy the few Eilenberg–Mac Lane axioms. Defining isomorphism as a general term did, in fact, become the business of category theory. The resulting simple abstract definition unifies all the many specific traditional versions that came before. For one thing, it agrees with the model theorists' notion of isomorphism, when elementary embeddings of models are taken as the morphisms. But notice this definition is relative to a category. And this is important in practice. Consider three claims, all well known in 1870 as they are today:

(1) Every elliptic curve is a torus.[14]
(2) Every torus is isomorphic to every other.
(3) Elliptic curves are not all isomorphic to each other.

The appearance of contradiction comes from confusing isomorphisms in two different categories. Correct statements are more explicit:

(2′) Every torus is topologically isomorphic to every other (i.e., isomorphic in the category of topological spaces).
(3′) Elliptic curves are not all analytically isomorphic to each other (i.e., isomorphic in the category of complex manifolds).

[14] Elliptic curves are not ellipses. They are surfaces. They are called "curves" because they are algebraically one-dimensional over the complex numbers (McKean and Moll 1999).

Karl Weierstrass (1863) worked this example out in his beautiful classification of analytically different elliptic curves. His classification rests on the fact that all these curves are topologically equivalent, as clarified by Bernhard Riemann (1851).

Riemann and Weierstrass got these facts straight in an ad hoc way without category theory. But ad hoc approaches became ever more burdensome as they proliferated. The explosion of structural mathematics produced category theory as the easy, uniform way to keep all such facts straight.

The bare categorical notions of identity morphism and composition of morphisms turned out to give an account of "sameness of form" that works all across mathematics. The philosophic relevance is highlighted by our next topic.

7.2. Nonidentity Automorphisms

Kouri (2015) takes a position in the philosophic structuralist debate over *automorphisms*. An automorphism of a structure S is any isomorphism of S to itself. Many structures S in mathematics have nonidentity automorphisms. In other words they have isomorphisms $S \xrightarrow{\sim} S$ to themselves different from the identity $1_S\colon S \xrightarrow{\sim} S$. Do these somehow challenge structuralism?

As a central example, I believe all structuralist philosophers up to now have agreed *complex conjugation* is an automorphism of the complex numbers C. Write complex numbers as $a+bi$ where a,b are real numbers and the complex unit i is defined by $i^2=-1$. Conjugation takes any $a+bi$ to $a-bi$. In other words it leaves every real number a,b fixed, and turns i into $-i$. Of course also $(-i)^2=-1$. An automorphism should leave all structural properties unchanged, and yet conjugation takes i to $-i$ and vice versa. Does this show that, even though $i \neq -i$ the two are structurally identical so that structuralists cannot tell which one is which? *Should* structuralists (or anyone else) be able to tell which one is which?[15]

Mathematicians face questions close to these. They are not philosophical quibbles. For this very reason, though, mathematicians have rigorous answers

[15] Kouri (2015) emphasizes as I do that "automorphism of the complex numbers" is ambiguous. She contrasts what she calls "the complex field" and "the complex algebra," which she argues should be considered different structures because they admit different automorphisms. I believe mathematicians more often discuss this contrast as one structure C occurring in two categories: the category of fields and the smaller category of real algebras. But this contrast is rarely mentioned in any terms, so it is hard to document the usage. (Complex conjugation is an automorphism in both of these contexts.) On the other hand the contrast between C as real algebra and C as complex manifold comes up often, and the standard explicit usage says one field C occurs in two categories. Results on C proved in one category are applied in the other. McKean and Moll (1999) work these contexts together like a symphony, leading to results in number theory.

that work in daily practice. These answers are systematically unlike the ones discussed by structuralist philosophers up to now. The very claim that complex conjugation is an automorphism of ℂ is an oversimplification. In practice complex conjugation is an automorphism of the complex numbers, and is not, depending on context.

Algebra textbooks say conjugation is an automorphism of ℂ. Complex analysis texts deny it. Conversely, analysis texts say for each complex number z_0 there is an automorphism of ℂ taking each $z \in \mathbb{C}$ to $z + z_0$. Algebra books deny that.

The algebraists and analysts do not disagree. They are often the same people. Algebraic and analytic facts on ℂ are both used in both algebra and analysis. Rather, algebra looks at ℂ in the category of *real algebras* and *algebra homomorphisms*. Complex conjugation is a morphism in that category, and is its own inverse. Adding a constant z_0 is not an algebra homomorphism unless $z_0 = 0$. Analysts look at C in the category of *complex manifolds* and *holomorphic maps*. Complex conjugation is not a morphism in that category but adding any fixed $z_0 \in \mathbb{C}$ is, with subtracting z_0 as its inverse.

The definition of holomorphic maps makes i and $-i$ geometrically distinct because i lies on the imaginary axis counterclockwise around 0 from 1 on the real axis, while $-i$ is clockwise around 0 from 1. This is a standard picture, as, e.g., in Mazur (2003, 190).[16] Complex conjugation flips the plane over, turning clockwise into counterclockwise, so it is not holomorphic. It is not a morphism in the category of complex manifolds at all, and a fortiori not an automorphism.

On the other hand, i and $-i$ have all the same real-algebraic relations, since complex conjugation is an automorphism in the category of real algebras. That category suits the algebraists' purposes, and algebraists never have any reason to tell which is i and which is $-i$ per se. But when more than one pair of conjugates is in question there are algebraic reasons, and means, for linking the choices between pairs. These more advanced problems are as algebraically intricate as they are productive for concrete number theory.[17]

To sum up, mathematicians track the difference between i and $-i$ using the usual tools of structural mathematics: categories, functors, and the associated apparatus. For substantial geometric and number-theoretic reasons they place

[16] Take $a + bi$ as a point $<a, b>$ in the real coordinate plane. The standard convention we all met in high school places $<0, 1>$ on the vertical axis counterclockwise around the origin from $<1, 0>$ on the horizontal. Formally, analysts specify an inclusion of complex manifolds into the category of oriented real manifolds, using the fact that holomorphic maps preserve orientation. This is textbook material as in. e.g.. Miranda (1995, 5–6).

[17] E.g., define ω, $\bar{\omega}$ as the roots of $X^2 + X + 1$ so ω, $\bar{\omega}$ are algebraically indistinguishable, just as $i, -i$ are. Yet $\omega + i$ and $\omega - i$ differ, as one provably has absolute value > 2.8, the other absolute value < 1.2. It is just not provable which is which. Given a choice of ω, the usual convention chooses i to make the absolute value of $\omega - i$ smaller than that of $\omega + i$. See Lang (2005, 465ff.) for the algebraic theory of absolute values.

$\mathbb{C}$ into several categories, some of which admit conjugation as an automorphism while some do not. And mathematicians in fact make different distinctions between i and $-i$ in these different contexts. Mac Lane derived his ontology of structures from that kind of mathematics.

8. Structural Ontology

> On the philosophical side, the structuralist ontology is often presented as a response to the "multiple reductions" problem raised in Benacerraf (1965). On the hermeneutic side, the structuralist ontology is said to be faithful to the discourse and practice of mathematics (Gasser 2015, 1).

Mac Lane was on a third side, the mathematical side. He was not faithful to the discourse or practice of mathematics. He changed both. To be clear: category theory has in fact been a central part of changes to both over the past 75 years now. And he did not respond to any form of the multiple reduction problem.[18] He first responded to technical questions in number theory and topology and later to the unanticipated reach of those same methods across the rest of mathematics.

Gasser argues (1) philosophical structuralist accounts so far fail to explain why only structural properties are *essential* in mathematics, while (2) mathematical objects do have some nonstructural properties, as, for example, 4 is the number of Galilean moons of Jupiter:

> A more subtle distinction between essential and nonessential properties of mathematical objects is necessary to spell out the structuralist view: it won't do to claim mathematical objects only have structural properties, or that these are the only properties they could coherently be said to possess. (2016, 6)

These issues are beside the point of Mac Lane's structural ontology. Like Maddy's Second Philosopher, Mac Lane does not start with philosophic terms and try to apply them to mathematics. He starts with mathematics and tries to answer traditional philosophic questions. His mature philosophy drew on his whole career, so summarizing it will draw on everything already presented.

Gasser very aptly says philosophers put the key claim of structuralism this way: "Mathematicians only care about things 'up to isomorphism'" (2015, 5). Mac Lane could say more or less these same words. But philosophers take their

[18] That is, unless you count it as a response when Mac Lane (1986, 407) endorsed Weyl's aphorism that set theory "contains far too much sand." That is, set theory loads mathematics with unnecessary bulk, though he felt ETCS does this less than ZFC.

notion of isomorphism from model theory, or possibly Bourbaki, neither of which is widely used in mathematics. Eilenberg and Mac Lane (1942a) began with the working notions from group theory and topology, and over several years pared those down to the categorical definition in section 9.7, which is now the explicit standard in most of mathematics. Unlike common philosophic notions of isomorphism, the mathematical one does not let you take a structure (say, the complex numbers) and talk about isomorphisms to or from it, without specifying a category.

Further, Mac Lane knew far too many mathematicians to dream of encapsulating what they "care about." Different people care about very different things. Mac Lane's ontology aims at the specifics of mathematical research and teaching. During World War II, and after it, he was often charged to write government reports on what is and what should be the direction of mathematics, both for funding purposes and in pedagogy (Steingart 2013). While his reports inevitably reflect his and other people's motives, they focus on specific achievements in mathematics and mathematical projects, not on felt motivations. So does his ontology.

Through his career he saw mathematics turn ever more to explicitly structural methods and eventually to category theory. He saw how over time more and more mathematics research and publication and teaching were organized around homomorphisms and isomorphisms. Through the 1950s the notions of homomorphism in widespread use got more and more general, far outside Bourbaki's structure theory. By the 1980s the research and textbook norm for organizing this was—certainly not advanced category theory—but the plain language of categories and functors. While research and textbooks rarely get down to the level of logical foundations, Mac Lane had known since the 1960s that rigorous logical foundations can be given in categorical terms and these terms bring logical foundations closer than ever before to what mathematicians normally do. The "trend towards uniform treatment of different mathematical disciplines" went deeper than he or Eilenberg had dreamed in (Eilenberg and Mac Lane 1945, 236).

Mac Lane got his ontology from the specific mathematics of his time. By the 1980s that meant the objects of mathematics are *structures* in the sense that all their properties are isomorphism invariant, and isomorphism means categorical isomorphism.[19] The ontology of current mathematics is categories, functors, and the objects and arrows of categories.

[19] Categorical foundations easily treat ZF sets as mathematical objects in this way, although ZF sets have many properties not invariant under bijections, i.e., under isomorphism in the category of sets. The suitable context was already worked out by ZF set theorists representing set membership in terms of well-founded, extensional ordered sets. ZF set theorists use these orders precisely to relate set membership to other order structures isomorphic to these in the category of ordered sets (Kunen 1983, 108–109 and passim). Categorical set theorists interpret ZF sets by these well-founded extensional orders, whose properties are isomorphism invariant in the category of ordered sets (Mac Lane and Moerdijk 1992, 331ff.).

Without asking what is essential to mathematical objects, Mac Lane observes the properties used in current mathematics are isomorphism invariant. That prominently includes applied mathematics like counting moons of Jupiter. Being the number of Galilean moons of Jupiter may well be nonstructural in some philosophic sense. But the statement "4 is the number of Galilean moons of Jupiter" is plainly invariant under isomorphisms of the natural numbers. If the 4 in one version of the natural numbers works in counting those moons, then the 4 in any isomorphic version works as well. As noted in section 9.6.2 this is precisely the point of Benacerraf (1965).

Philosophic training in Göttingen prepared Mac Lane to hold that, since mathematicians consistently work with structures in this sense, these structures are the ontology of mathematics. That same philosophic training taught him:

> A thorough description or analysis of the form and function of Mathematics should provide insights not only into the Philosophy of Mathematics but also some guidance in the effective pursuit of Mathematical research. (Mac Lane 1986, 449)

References

Adams, J. F. 1962. Vector Fields on Spheres. *Bulletin of the American Mathematical Society* 68, 39–41.

Benacerraf, P. 1965. What Numbers Could Not Be. *Philosophical Review* 74, 47–73.

Bourbaki, N. 1939. *Théorie des Ensembles, Fascicules de résultats*. Paris: Hermann.

Bourbaki, N. 1949. Foundations of Mathematics for the Working Mathematician. *Journal of Symbolic Logic* 14, 1–8.

Bourbaki, N. 1950. The Architecture of Mathematics. *American Mathematical Monthly* 57, 221–232. Reprinted in *From Kant to Hilbert: A Source Book in the Foundations of Mathematics*, edited by W. Ewald, pp. 1265–1276. New York: Oxford University Press, 1996.

Bourbaki, N. 1958. *Théorie des Ensembles*. 3rd ed. Paris: Hermann.

Carnap, R. 1932. Überwindung der Metaphysik durch logische Analyse der Sprache. *Erkenntnis* 2, 219–241.

Corry, L. 1992. Nicolas Bourbaki and the Concept of Mathematical Structure. *Synthese* 92(3), 315–348.

Cox, D. A., J. Little, and D. O'Shea. 2007. *Ideals, Varieties, and Algorithms: An Introduction to Computational Algebraic Geometry and Commutative Algebra*. New York: Springer.

Dedekind, R. 1888. *Was sind und was sollen die Zahlen?* Braunschweig: Vieweg. Translated in *Essays on the Theory of Numbers*. New York: Dover Reprints, 1963.

Dedekind, R. 1996. *Theory of Algebraic Integers*. Edited and translated by J. Stillwell. New York: Cambridge University Press. Originally published in French in 1877.

Eilenberg, S., and S. Mac Lane. 1942a. Group Extensions and Homology. *Annals of Mathematics* 43, 757–831.

Eilenberg, S., and S. Mac Lane. 1942b. Natural Isomorphisms in Group Theory. *Proceedings of the National Academy of Sciences* 28, 537–543.

Eilenberg, S., and S. Mac Lane. 1945. General Theory of Natural Equivalences. *Transactions of the American Mathematical Society* 58, 231–294.

Gasser, M. 2015. Structuralism and Its Ontology. *Ergo* 2(1), 1–26.

Geiger, M. 1924. *Systematische Axiomatik der euklidischen Geometrie*. Augsburg: B. Filser.

Geiger, M. 1930. *Die Wirklichkeit der Wissenschaften und die Metaphysik*. Bonn: F. Cohen.

Gelfand, S., and Y. Manin. 2003. *Methods of Homological Algebra*. 2nd ed. New York: Springer.

Grothendieck, A. 1952. Résumé des résultats essentiels dans la théorie des produits tensoriels topologiques et des espaces nucléaires. *Annales de l'Institut Fourier Grenoble* 4, 73–112.

Grothendieck, A. 1985–87. *Récoltes et Semailles*. Université des Sciences et Techniques du Languedoc, Montpellier. Published in several successive volumes.

Hahn, H. 1928. H. Weyl: *Philosophie der Mathematik und Naturwissenschaft. Monatsheft für Mathematik und Physik* 35, 51–58.

Hankel, H. 1867. *Vorlesungen über die complexen Zahlen und ihre Funktionen*. Leipzig ; L. Voss.

Hermann, G. 1926. Die Frage der endlich vielen Schritte in der Theorie der Polynomideale. *Mathematische Annalen* 95, 736–788. Translated by Michael Abramson as The Question of Finitely Many Steps in Polynomial Ideal Theory in ACM SIGSAM *Bulletin* 32(3) (1998), 8–30.

Husserl, E. 1922. *Logische Untersuchungen*. 2nd ed. Halle: Max Niemeyer.

Jacobson, N., ed. 1983. *E. Noether: Gesammelte Abhandlungen*. Berlin: Springer.

Kan, D. 1958. Adjoint Functors. *Transactions of the American Mathematical Society* 87, 294–329.

Koreuber, M. 2015. *Emmy Noether, die Noether-Schule, und die Moderne Algebra*. Mathematik im Kontext. Berlin: Springer.

Kouri, T. 2015. A Reply to Heathcote's "On the Exhaustion of Mathematical Entities by Structures." *Axiomathes* 25, 345–357.

Krömer, R. 2007. *Tool and Object: A History and Philosophy of Category Theory*. Basel: Birkhäuser.

Krull, W. 1935. *Idealtheorie*. Berlin: Springer.

Kunen, K. 1983. *Set Theory: An Introduction to Independence Proofs*. Amsterdam: NorthHolland.

Lang, S. 2005. *Algebra*. Reading, MA: Addison-Wesley.

Lawvere, F. W. 1963. Functorial Semantics of Algebraic Theories. PhD thesis, Columbia University. Published with author commentary in *Reprints in Theory and Applications of Categories*, No. 5 (2004), 1–121, available online at http://138.73.27.39/tac/reprints/articles/5/tr5abs.html.

Lawvere, F. W. 1964. An Elementary Theory of the Category of Sets. *Proceedings of the National Academy of Sciences of the U.S.A.* 52, 1506–1511.

Lawvere, F. W. 1965. An Elementary Theory of the Category of Sets. Lecture notes of the Department of Mathematics, University of Chicago. Reprint with commentary by the author and Colin McLarty in *Reprints in Theory and Applications of Categories*, No. 11 (2005), 1–35, available online at http://138.73.27.39/tac/reprints/articles/11/tr11abs.html.

Lawvere, F. W. 1966. The Category of Categories as a Foundation for Mathematics. In *Proceedings of the Conference on Categorical Algebra, La Jolla, 1965*, edited by S. Eilenberg et al., pp. 1–21. New York: Springer.

Leinster, T. 2014. Rethinking Set Theory. *American Mathematical Monthly* 121(5), 403–415.

Lemmermeyer, F., and P. Roquette. 2006. *Helmut Hasse und Emmy Noether*. Göttingen: Universitätsverlag Göttingen.

Mac Lane, S. 1934. Abgekürzte Beweise in Logikkalkul, PhD thesis, Georg August Universität zu Göttingen. Reprinted in *Selected Papers*, edited by I. Kaplansky, pp. 1–62. New York: Springer-Verlag, 1979.

Mac Lane, S. 1935. A Logical Analysis of Mathematical Structures. *Monist* 45, 118–130.

Mac Lane, S. 1948. Groups, Categories and Duality. *Proceedings of the National Academy of Sciences of the U.S.A.* 34, 263–267.

Mac Lane, S, ed. 1969. *Reports of the Midwest Category Seminar III*. Number 106 in Lecture notes in mathematics. New York: Springer.

Mac Lane, S. 1971. *Categories for the Working Mathematician*. Springer.

Mac Lane, S. 1979. A Late Return to a Thesis in Logic. In *Selected Papers*, edited by I. Kaplansky, pp. 63–66. New York: Springer.

Mac Lane, S. 1981. Mathematics at the University of Göttingen, 1931–1933. In *Emmy Noether: A Tribute to Her Life and Work*, edited by J. Brewer and M. Smith, pp. 65–78. New York: Marcel Dekker.

Mac Lane, S. 1986. *Mathematics: Form and Function*. New York: Springer.

Mac Lane, S. 1988. Group Extensions for 45 Years. *Mathematical Intelligencer* 10(2), 29–35.

Mac Lane, S. 1992. Is Mathias an Ontologist? In *Set Theory of the Continuum*, edited by H. Judah, W. Just, and H. Woodin, pp. 119–122. New York: Springer.

Mac Lane, S. 1995a. Mathematics at Göttingen under the Nazis. *Notices of the American Mathematical Society* 42, 1134–1138.

Mac Lane, S. 1995b. A Matter of Temperament. Letter, with reply by Freeman J. Dyson. *New York Review of Books* 42(15), 56.

Mac Lane, S. 1996a. The Development and Prospects for Category Theory. *Applied Categorical Structures* 4, 129–136.

Mac Lane, S. 1996b. Stephen Cole Kleene—a Reminiscence. *Annals of Pure and Applied Logic* 81, 3–7.

Mac Lane, S. 1996c. Structure in Mathematics. *Philosophia Mathematica* 4, 174–183.

Mac Lane, S. 1997a. Despite Physicists, Proof Is Essential in Mathematics. *Synthese* 111, 147–154.

Mac Lane, S. 1997b. Van der Waerden's Modern Algebra. *Notices of the American Mathematical Society* 44, 321–322.

Mac Lane, S. 1998a. *Categories for the Working Mathematician*. 2nd ed. New York: Springer.

Mac Lane, S. 1998b. Mina Spiegel Rees (1902–1997). *Notices of the American Mathematical Society* 45, 869–872.

Mac Lane, S. 2000. Contrary Statements about Mathematics. *Bulletin of the London Mathematical Society* 32, 527.

Mac Lane, S., and I. Moerdijk. 1992. *Sheaves in Geometry and Logic*. New York: Springer.

Mac Lane, S., and O. Schilling. 1941. Normal Algebraic Number Fields. *Transactions of the American Mathematical Society* 50, 295–384.

Maddy, P. 2007. *Second Philosophy*. New York: Oxford University Press.

Mazur, B. 2003. *Imagining Numbers (particularly the square root of minus fifteen).* New York: Farrar Straus and Giroux.

McKean, H., and V. Moll. 1999. *Elliptic Curves: Function Theory, Geometry, Arithmetic.* New York: Cambridge University Press.

McLarty, C. 1990. The Uses and Abuses of the History of Topos Theory. *British Journal for the Philosophy of Science* 41, 351–375.

McLarty, C. 1991. Axiomatizing a Category of Categories. *Journal of Symbolic Logic* 56, 1243–1260.

McLarty, C. 2006. Emmy Noether's "Set Theoretic" Topology: From Dedekind to the Rise of Functors. In *The Architecture of Modern Mathematics: Essays in History and Philosophy*, edited by J. Gray and J. Ferreirós, pp. 211–235. New York: Oxford University Press.

McLarty, C. 2007a. The Last Mathematician from Hilbert's Göttingen: Saunders Mac Lane as a Philosopher of Mathematics. *British Journal for the Philosophy of Science* 58, 77–112.

McLarty, C. 2007b. The Rising Sea: Grothendieck on Simplicity and Generality I. In *Episodes in the History of Modern Algebra (1800–1950)*, edited by J. Gray and K. Parshall, pp. 301–326. American Mathematical Society.

McLarty, C. 2016. How Grothendieck Simplified Algebraic Geometry. *Notices of the American Mathematical Society* 63(3), 256–65.

Miranda, R. 1995. *Algebraic Curves and Riemann Surfaces.* Providence, RI: American Mathematical Society.

Noether, E. 1932. Hyperkomplexe Systeme in ihren Beziehungen zur kommutativen Algebra und Zahlentheorie. In *Verhandlungen des Internationalen Mathematiker-Kongresses Zürich*, vol.1, pp. 189–194. Zurich: Orell Füssli. In Jacobson 1983, pp. 636–641.

Peckhaus, V. 1990. *Hilbertprogramm und Kritische Philosophie, Das Göttinger Modell interdisziplinärer Zusammenarbeit zwischen Mathematik und Philosophie.* Göttingen: Vandenhoeck & Ruprecht.

Riemann, B. 1851. Grundlagen für eine allgemeine Theorie der Functionen einer veränderlichen complexen Grösse. Inauguraldissertation, Universität Göttingen. Reprinted in *Bernhard Riemann's Gesammelte mathematische Werke*, edited by R. Dedekind and H. Weber, pp. 3–45. Leipzig: B.G. Teubner, 1876.

Rosen, R. 1958. The Representation of Biological Systems from the Stand-point of the Theory of Categories. *Bulletin of Mathematical Biophysics* 20, 317–342.

Spiegelberg, E. 1994. *The Phenomenological Movement: A Historical Introduction.* 3rd ed. Dordrecht: Kluwer Academic.

Steingart, A. 2013. Conditional Inequalities: American Pure and Applied Mathematics, 1940–1975. PhD thesis, Massachusetts Institute of Technology.

Washington, L. 1997. Galois Cohomology. In *Modular Forms and Fermat's Last Theorem*, edited by G. Cornell, J. Silverman, and S. Stevens, pp. 101–120. New York: Springer.

Weierstrass, K. 1863. Vorlesungen über die Theorie der elliptischen Funktionen. in vol. 5, *Gesammelte Werke*. Berlin: Mayer und Müller, 1894–1927.

Weil, A. 1991. *Souvenirs d'apprentissage*. Basel: Birkhäuser.

Weyl, H. 1927. *Philosophie der Mathematik und Naturwissenschaft.* Munich: R. Oldenbourg.

PART II

LOGICAL AND PHILOSOPHICAL REFLECTIONS

10
Logic of Relations and Diagrammatic Reasoning: Structuralist Elements in the Work of Charles Sanders Peirce

Jessica Carter

1. Introduction

This chapter presents aspects of the work of Charles Sanders Peirce, illustrating how he adhered to a number of the pre-structuralist themes characterized in the introduction to this volume. I shall present aspects of his contributions to mathematics as well as his philosophy of mathematics in order to show that relations occupied an essential role. When writing about results in mathematics he often states that they are based on his "logic of relatives," and he refers to the reasoning of mathematics as "diagrammatic reasoning." Besides pointing to structural themes in Peirce's work, much of this exposition will be devoted to explaining what is meant by these two phrases.

In a recent article Christopher Hookway (2010) places Peirce as an *ante rem* structuralist.[1] In support of this claim Hookway refers to some of Peirce's writings on numbers (also to be treated here). In addition he spends some time analyzing what Peirce means by the phrase "the form of a relation." These considerations involve an in-depth knowledge of Peirce's categories and their metaphysical implications. In contrast I will focus on methodological aspects, in particular Peirce's writings on *reasoning* in mathematics, stressing that mathematics consists of the *activity* of drawing necessary inferences. This leads to a position that resembles methodological structuralism as it is characterized in Reck and Price (2000). Furthermore I find that Peirce's position is similar in spirit to the contemporary categorical structuralist views, in particular, as formulated by Steve Awodey (2004). Still I resist characterizing Peirce as a structuralist since I do not find that this label captures the richness of his views as presented here.

[1] But see Pietarinen (2010) arguing that Peirce's continuum cannot be a structure.

Jessica Carter, *Logic of Relations and Diagrammatic Reasoning* In: *The Prehistory of Mathematical Structuralism*. Edited by: Erich H. Reck and Georg Schiemer, Oxford University Press (2020). © Oxford University Press.
DOI:10.1093/oso/9780190641221.003.0010

To mention one point, besides claiming that mathematics is the science of necessary reasoning, Peirce has something to say about how this necessity is achieved.

The chapter consists of two main parts. The first documents Peirce's extensive knowledge of, and contribution to, the mathematics of his time. Areas include arithmetic, set theory, algebra, geometry (including non-Euclidean and topology)—and logic. Examples, together with indications of what drove his engagement with them, will be given from his work in geometry, arithmetic, set theory, and algebra. In relation to the pre-structuralist themes it can be mentioned that he presented different axiomatizations of the natural numbers. Furthermore his insistence on the inappropriateness of the characterization of mathematics as "the science of quantity" will be addressed. Finally we shall see that he draws a clear distinction between pure and applied mathematics—both in arithmetic and in geometry. I further note that Peirce's use of formal methods and his view of mathematics as an autonomous body places him as an early modernist according to the characterization given by Jeremy Gray (2008).

The second part is concerned with Peirce's philosophy of mathematics. It addresses Peirce's description of mathematical reasoning as diagrammatic reasoning. A diagram to Peirce is an iconic sign that represents rational relations. In order to explain what is contained in "diagrammatic reasoning" the chapter therefore includes a few relevant parts of Peirce's semiotics. In addition one example of a proof will be given in order to explain how mathematical, that is, necessary, reasoning proceeds by constructing and observing diagrams.

2. Mathematics and the Logic of Relations

A few biographic details are relevant.[2] Charles Sanders Peirce was born in Cambridge, Massachusetts, in 1839, as the son of Benjamin Peirce, a distinguished professor of mathematics at Harvard and a leading social figure. Benjamin Peirce taught his children mathematics and Charles certainly was very talented—as he was talented in so many fields. (Peirce had one sister and three brothers, of whom the oldest, James Peirce, became professor in mathematics at Harvard.) C. S. Peirce studied chemistry at Harvard and (in 1859), obtained a job at the US Coast and Geodetic Survey, and later, in 1879, was appointed a lecturer in logic at the Department of Mathematics at Johns Hopkins University. In 1884 his contract with Johns Hopkins was not renewed, and in 1891, due to disagreements, he was also forced to resign from his post at the US Coast and Geodetic Survey.

[2] This biographic information is based mainly on the introduction of Peirce (1976, vol. 1) and Gray (2008). I also recommend Brent (1998).

Peirce is often referred to as having a somewhat asocial behavior, something he admits and blames on his upbringing by his father focusing mainly on his formal training: "In this as in other respects I think he underrated the importance of the powers of dealing with individual men to those of dealing with ideas and with objects governed by exactly comprehensible ideas, with the result that I am today so destitute of tact and discretion that I cannot trust myself to transact the simplest matter of business that is not tied down to rigid forms" (NE IV, v).[3] Another peculiarity to mention is his habit of adopting his own terminology, e.g., calling relations "relatives" and writing "semeiotic" for "semiotics."

Two further things regarding Peirce's early years are worth mentioning here. First is Kant's influence on his thinking. Much of Peirce's thought is developed in reaction to the ideas of Kant; it is certainly the case that many of the ideas dealt with in this chapter are presented by Peirce with reference to Kant. Peirce writes (commenting on a text from 1867 introducing his categories) about his early influence by Kant, stating that he by 1860 "had been my revered master for three or four years" (CP 1.563). Second is Peirce's passion for logic. According to Peirce this passion was aroused by reading Whateley's *Logic*: "It must have been in the year 1851, when I must have been 12 years old, that I remember picking up Whateley's Logic in my elder brother's room and asking him what logic was. I see myself, after he told me, stretched on his carpet and poring over the book for the greater part of a week for I read it through. . . . From that day to this logic has been my passion although my training was chiefly in mathematics, physics and chemistry" (NE IV, vi).

There are two distinct periods in Peirce's contributions to logic (see Dipert 2004). The first is algebraic, using algebraic tools in order to formulate a calculus of the logic of relations with inspiration from (among others) Boole and de Morgan. A seminal paper in this period is his "Description of a Notation for the Logic of Relatives, Resulting from an Amplification of the Conceptions of Boole's Calculus of Logic" published in 1870 (reprinted in CP 3.45–148). The second and later period is characterized as "diagrammatic." In this period Peirce develops his existential graphs (see Roberts 1973 or Shin 2002).

An important part of Peirce's characterization of mathematics is his statement that mathematics is the science of necessary reasoning concerning hypothetical states of things. He attributes this claim to his father, writing: "It was Benjamin Peirce, whose son I boast myself, that in 1870 first defined mathematics as 'the

[3] Citations of Peirce follow traditional standards. (NE I, 3) refers to the collections *New Elements* edited by Carolyn Eisele (Peirce 1976) volume I, page 3. Similarly (CP 4.229) refers to the *Collected Papers of Peirce* edited by Hartshorne and Weiss (1931–1967) volume 4, paragraph 229. (EP 2, 7) refers to *Essential Peirce*, volume 2, page 7. I sometimes include a reference to the year the paper was written/published. This is available from R. Robin's catalog; see http://www.iupui.edu/~peirce/robin/robin_fm/toc_frm.htm.

science which draws necessary conclusions.' This was a hard saying at the time; but today, students of the philosophy of mathematics generally acknowledge its substantial correctness" (CP 4.229). The reference to 1870 is to *Linear Associative Algebra*, which opens with the statement C. S. Peirce quotes (B. Peirce [1870] 1881, 97). Peirce states at various places that the necessity of mathematical conclusions is obtainable precisely due to the hypothetical nature of mathematical statements, characterizing mathematics as the science "which frames and studies the consequences of hypotheses without concerning itself about whether there is anything in nature analogous to its hypotheses or not" (NE IV, 228). We shall return to these claims about mathematics throughout the chapter.[4]

2.1. Geometry

A good place to learn about the extent of his knowledge of geometry is his (unpublished) book *New Elements of Geometry Based on Benjamin Peirce's Works and Teaching*, which fills most of the second volume of the *New Elements of Mathematics* (Peirce 1976). As the title indicates, the book is an extension of his father's *Elementary Treatise on Geometry* (published in 1837), but it contains much more—apparently so much more that the publisher in the end refused to publish the book. When Peirce was forced to retire from his position in the US Coast and Geodetic Survey in 1891 he turned to textbook writing as a possible source of income. Ginn, the publisher of the American Book Company, made enquiries regarding an update of his father's book in 1894 (NE II, xiv). The introduction of NE II makes clear that Peirce worked for long on (versions of) the book while corresponding with the publisher, who did not see the need for publishing all the topics and sections Peirce wanted to include.[5] From the introduction it is possible to gain insight into Peirce's motivation for extending it as he wished to do. Given the developments of geometry during the 19th century, he found a substantial revision necessary. He lists a number of ways that geometry had "metamorphosed" since 1835: Given the acceptance of non-Euclidean geometries, Peirce claims, "geometry has two parts; the one deals with the *facts* about real space, the investigation of which is a physical, or perhaps metaphysical, problem, at any rate, outside of the purview of the mathematician, who

[4] Although Peirce claims that mathematics consists of the drawing of necessary conclusions, in some places he considers including the process of forming the hypotheses from which to reason as part of mathematics. See, for example, CP 4.238, where he praises the ingenuity of Riemann for developing the idea of a Riemann surface.

[5] See the correspondence between publisher, C. S. Peirce, and his brother, James (Jem) Peirce, professor of mathematics at Harvard (NE II, xiv–xxvii), also providing information about the different versions of the book.

accepts the generally admitted propositions about space, without question, as his *hypotheses*, that is, as the ideal truth whose consequences are deduced in the second, or mathematical part, of geometry" (NE II, 4). I return to this claim later. It is evident that Peirce was well informed about the various versions of non-Euclidean geometry formulated by Bolyai and Lobachewsky and even worked on both elliptic and hyperbolic geometry himself, claiming that space was hyperbolic (NE III, 710). To mention another thing, Peirce reviewed Halsted's[6] translation of Lobachevsky's geometry in *The Nation* (54, February 11, 1892), calling it an excellent translation. The next topic Peirce mentions among the areas that had not previously been included in his father's book is the new branch of geometry of Listing, named topology, which "deals with only a portion of the hypotheses accepted in other parts of geometry; and for that reason, as well as because its relative simplicity, it should be studied before the others."[7] The subsequent topic is what is today denoted as projective geometry. He then mentions "metrical geometry," which, he writes, was revolutionized after 1837 based on the contributions of Gauss's students Lobachewsky, Riemann, and Bolyai (building on the works of Lambert and Saccheri).[8] Finally, Peirce mentions the work of Cantor and others who "have succeeded in analyzing the conceptions of infinity and continuity, so as to render our reasonings concerning them far more exact than they had previously been" (NE II, 5).

Throughout his writings one finds explicit statements separating pure geometry, which traces the consequences of hypotheses, from "applied geometry," which makes enquiries about the properties of real space and so is a branch of physics.[9] At other places the distinction is implicit, as in the paper "Synthetical Propositions À Priori" (NE IV, 82–85). The aim of this paper is to show—opposing Kant—that mathematical propositions are not synthetic. He remarks that it is *possible* that the propositions of geometry could be regarded as statements concerning physical space, but consistent with his general claims

[6] Georg Bruce Halsted was a student of Sylvester's from John Hopkins University and became professor at the University of Texas in 1884. According to Eisele "Halsted was spearheading in his publications on the new geometry the effort to bring to mathematicians in America the awareness of the revolution in mathematical thought" (NE II, ix).

[7] Peirce also made contributions to topology (see Havenel 2010 for an account of this). Furthermore, Havenel notes that topology is "*par excellence* the mathematical doctrine that is incompatible with the widespread idea that mathematics is the science of nothing else than quantities, geometrical quantities, and numerical quantities, for the topological properties do not involve measurement" (Havenel 2010, 286).

[8] He makes references to Cayley (in 1854) and Klein (in 1873). In 1854 Cayley published a paper on finite groups, showing which multiplication tables are possible for a given number of elements of the group. Later, as Peirce indicates, Klein used the concept of a group and definitions of a metric (due to Cayley) to propose that the different geometries could be defined in terms of the invariance of properties of figures under a group of transformations, what is known as the Erlangen program.

[9] For explicit statements about the distinction between pure and applied geometry see NE IV, 359 and NE III, 703–709.

of mathematics, he concludes that to the mathematician they are simply held to be *hypotheses*: "nothing but ignorance of the logic of relatives has made another option possible" (NE IV, 82). He implicitly refers to the introduction of Riemann's *Über die Hypothesen, welche der Geometrie zu Grunde liegen* ([1854] 1892), calling it "Riemann's greatest memoir." According to Peirce, Riemann writes that geometrical propositions are matters of fact, and as such not necessary, but only empirically certain; they are hypotheses. Referring to the last part of this statement, Peirce comments: "This I substantially agree with. Considered as *pure* mathematics, they define an ideal space, with which the real space approximately agrees" (84–85).[10]

2.2. Foundations of Arithmetic

When Peirce writes about arithmetic, he distinguishes between different versions. The first is arithmetic as used in counting and calculations, which he denotes "vulgar" arithmetic (see NE I, xxxv and CP 1.291) or practical arithmetic. The other is pure arithmetic, concerning the abstract dealings with the properties of (the operations on) numbers. What is in particular worth mentioning in this context is that he bases both on axioms, and that proofs use what he denotes the "logic of relatives." (Note that the use of "axioms" here is my terminology. Peirce refers to them as "primary propositions" or "definitions." In general he is wary of using the label "axiom," which at the time referred to propositions held to be indubitably true.)[11] Note that one may find statements claiming that even practical arithmetic is based on (ideal) hypotheses: "2 and 3 is 5 is true of an idea only, and of real things so far as that idea is applicable to them. It is nothing but a form, and asserts no relation between outward experiences" (NE IV, xv).

Peirce's axiomatizations of (practical) arithmetic intend to prove that arithmetical propositions are logical consequences of a "few primary propositions," that is, countering Kant's view that arithmetical propositions are synthetic. In "The Logic of Quantity" from 1893 (CP 4.85–93) Peirce addresses Kant's

[10] See the article by J. Ferreirós (2006) for an interpretation of how Riemann understood "hypothesis" (and foundations). According to Ferreirós, Riemann used the word "hypothesis" instead of "axiom" precisely to emphasize that they are not evident.

[11] "The science which, next after logic, may be expected to throw the most light upon philosophy is mathematics. It is a historical fact, I believe, that it was the mathematicians Thales, Pythagoras, and Plato who created metaphysics, and that metaphysics has always been the ape of mathematics. Seeing how the propositions of geometry flowed demonstratively from a few postulates, men got the notion that the same must be true in philosophy. But of late mathematicians have fully agreed that the axioms of geometry (as they are wrongly called) are not by any means evidently true. Euclid, be it observed, never pretended they were evident; he does not reckon them among his κοιναὶ ἔννοιαι or things everybody knows, but among the αἰτήματα, postulates, or things the author must beg you to admit, because he is unable to prove them" (CP 1.130).

characterization of analytic judgments, finding that Kant's thought "is seriously inaccurate" (even calling it "Monstrous"). The distinction between analytic and synthetic judgments depends on whether a predicate is *involved* in the subject, or "whether a given thing is consistent with a hypothesis." Peirce accuses Kant of, due to insufficient knowledge of logic, confusing a question of logic with psychology when he writes that being *involved* in the conception of the subject is the same as being *thought* in it (CP 4.86). According to Peirce the question is easily resolved if one is familiar with the logic of relatives. Its solution does not depend on "a simple mental stare or strain of mental vision. It is by manipulating on paper, or in the fancy, formulae or other diagrams—experimenting on them, *experiencing* the thing" (CP 4.86). That is, whether a judgment is analytic or not can be determined by the use of logic and is an objective fact, not something depending on our thoughts. Note here also Peirce's formulation "experimenting on a diagram" and "experiencing the thing," which are central to his characterization of mathematical and diagrammatic reasoning, as I shall explain in the second part of the chapter. Concerning the status of arithmetic, he continues: "the whole of the theory of numbers belongs to logic; or rather it would do so, were it not, as pure mathematics, *prelogical*, that is, even more abstract than logic" (CP 4.90). Peirce holds that the different sciences can be ordered according to the generality of the objects they concern. In this philosophical system Peirce places mathematics at the top level, being the science that draws necessary conclusions, and logic as part of philosophy just below. Logic, according to Peirce, studies the drawing of necessary conclusions done in mathematics in order to formulate "laws of the stable establishment of beliefs" (CP 3.429). One may therefore note that Peirce, in contrast to, for example, G. Frege, who also took an interest in the foundations of arithmetic, did not claim that arithmetic is reducible to logic.[12] Another interesting point is Peirce's remark that there is no one unique way to found arithmetic on the logic of relations: There "are even more ways in which arithmetic may be conceived to connect itself with and spring out of logic" (CP 4.93). To document this claim Peirce refers to some of the texts presented in what follows.

Another motivation for providing axioms for arithmetic is to counter the empiricism of J. S. Mill, referred to in the introduction to the first paper presented in

[12] It should be noted, though, that Peirce in his 1881 paper introducing the numbers writes things that could be construed as approaching a logicist position. He first writes that the aim of the paper is to show that the truths of arithmetic are consequences of a few primary propositions. He states about these propositions (calling them definitions), "the question of their logical origin . . . would require a separate discussion" (CP 3.252). For a more precise formulation of the interrelation between logic and mathematics in terms of how the different subjects, i.e., mathematics, philosophy, logic, etc., relate in Peirce's system see Stjernfelt (2007, 11–12). Peirce writes, for example, the following about the relations between the subjects ("sciences") in his system: "The general rule is that the broader science [e.g., mathematics] furnishes the narrower with principles by which to interpret its observations while the narrower science furnishes the broader science with instances and suggestions" (NE IV, 227).

section 10.2.2.1, a position quite influential at the time. (Peirce implicitly refers to Mill as "a renowned English logician.") In what follows three examples of characterizations of arithmetic are given, two of pure arithmetic and one of the "counting numbers."

2.2.1. Basing Pure Arithmetic on the Logic of Relations

The first example comes from Peirce's article "On the Logic of Number," published in the *American Journal of Mathematics* 4 (1881).[13] (The paper is reproduced in CP 3.252–288.) The system presented here has been called the first successful axiom system for the natural numbers (see Mannoury 1909 and Shields 1997, 43).[14] In the introduction Peirce writes that the aim of the paper is "to show that [the elementary propositions concerning number] are strictly syllogistic consequences from a few primary propositions" (CP 3.252). Peirce remarks that the inferences drawn are not exactly like syllogistic consequences but that they are of the same nature.

The numbers, or as Peirce refers to them, "a system of quantities," is introduced as a collection together with a particular relation defined on it. The natural numbers are defined as a totally ordered (discrete) set with a minimum element, fulfilling the axiom of induction. In Peirce's terminology they are a *semi-infinite, discrete and simple system of quantity.* In contemporary terms a simple system of quantity is a totally ordered set (i.e., the relation defined on the set is transitive, reflexive, and anti-symmetric and fulfills trichotomy). Furthermore, "discrete" and "semi-infinite" mean the set has a minimum element (called a semi-limited system of quantity) and fulfills the axiom of induction. In what follows we shall see how Peirce defines these notions. His use of notation (or perhaps lack thereof) may be a bit confusing to a modern reader. He uses the expression "one thing is said to be r of another," meaning that one thing is *related* to another. In contemporary technical terms we would instead write that ArB or $(A,B)\in r$ for A being "one thing" and B "another." Peirce also uses the formulation that "the latter be r'd by the former." Listing properties that hold for the relation "less than

[13] Peirce sketches a characterization of the natural numbers even earlier than 1881 in the paper "Upon the Logic of Mathematics," dated 1867 and published in the *Proceedings of the American Academy of Arts and Science*, vol. 7 (CP 3.20-3.44). It is based on his modifications of "the logical calculus of Boole." He defines, e.g., "logical identity" and "addition," corresponding at first to operations on classes. That is, addition corresponds to taking the union of two classes. Identity between two classes states they consist of the same elements. Toward the end of the article Peirce notes that if one considers a kind of abstraction on classes—"numerical rank"—identity will play the role of equality and by considering the operation disjoint union one obtains the rules of arithmetic, for example, that $a+b=b+a$.

[14] Shields presents Peirce's axioms formulated in a modern way and compares his axiomatization to Dedekind's and Peano's. One thing Shields points out as worthy of attention is that Peirce chose a transitive relation as the basic relation when defining the numbers instead of the successor relation. Peirce later formulates systems based on an equivalent of the successor relation.

or equal to" and using "r" and "q" to stand for this relation, he states the fundamental properties of a system of quantities as follows: "In a system in which r is transitive, let the q's of anything include that thing itself, and also every r of it which is not r'd by it. Then q q may be called a fundamental relative of quantity" (CP 3.253). That is, Peirce defines a system of quantity to be a collection Q on which there is defined a relation, q, which fulfills that q is transitive and reflexive, and for any A in the collection, AqB holds for all B s for which BqA is not the case. (If one thinks of the relation $\leq$ on the numbers, the last property states that for any two numbers A and B, if not $B \leq A$ then $A \leq B$.) Peirce continues to list the properties of q, stating that "it is transitive; second, that everything in the system is q of itself, and, third, that nothing is both q of and q'd by anything except itself." The last is anti-symmetry. A relation fulfilling these three properties defined on a set is usually called a partial order on that set. He defines a simple system (of quantity) to be one in which it is the case that for any two elements, A and B , it is the case that either ArB or BrA(i.e., trichotomy).[15] Simple systems can be discrete, which means "every quantity greater than another is next greater than some quantity (that is greater than without being greater than something greater than)" (CP 3.256). For simple and discrete systems of quantity he introduces semi-limited systems, i.e., systems that have a limit, often an absolute minimum element (which he calls "one"). Finally Peirce considers this class of quantities (that is, a simple, discrete system with a minimum element), noticing that "an infinite system may be defined as one in which from the fact that a certain proposition, if true of any number, is true of the next greater, it may inferred that that proposition if true of any number is true of every greater" (3.258). That is, Peirce notes that what is today called the induction axiom characterizes the natural numbers.[16] Elsewhere Peirce denotes this principle by Fermatian inference.

In the next paragraph Peirce continues to study "ordinary number," which can be defined as a semi-infinite (that is semi-limited and infinite), discrete, and simple system of quantity, defining addition and multiplication (using the notion of predecessor), and he shows how one may then prove a number of fundamental propositions of arithmetic by induction, e.g., associativity, commutativity of addition, and the distributive law.[17] I present one of Peirce's proofs that addition is

[15] In modern terminology a set on which there is defined a partial order fulfilling trichotomy is denoted a totally ordered set.

[16] Note the unfortunate choice of terminology calling a system of quantities for which the induction axiom holds an "infinite" system. The reason behind this might be that Peirce contrasts infinite systems with finite systems at the end of the paper.

[17] In a later paper from 1901–1904 (NE IV, 2–3) Peirce has introduced more notations, for example "G" denoting the successor function, but essentially maintains the same characterization of numbers. In this paper he shows that the associative law holds for numbers, where numbers are defined as an ordered system on which induction holds. Two numbers are defined to be equal in terms of the relation "greater than," where A = B means that A is at least as great as B and B is as least as great as A. Furthermore he notes that "as least as great as" is transitive and reflexive and if $N \geq M$, then

commutative; that is, using$x$ x and y X to denote natural numbers, $x+y=y+x$ (see CP 3.267). Addition is defined by the following two rules (here adding "s" to denote successor): $1+y=s(y)$ and $x+y=s(x'+y)$, where x' denotes the predecessor of x. The proposition is proved using induction twice, and it employs the associative rule, $x+(y+z)=(x+y)+z$, that Peirce proves first. In the first step it is shown that the statement holds for $y=1$, namely that $x+1=1+x$. I omit the details from this part. For the general proposition, one may now note that $x+y=y+x$ has been proven for $y=1$. In order to conclude the statement by induction, it thus remains to show that if the statement holds for $y=n$, then it holds for $y=1+n$. We suppose that $x+n=n+x$ and consider $x+(1+n)$. Calculating on—or manipulating—this expression, we obtain the following (which Peirce would refer to as a diagram):

$$x+(1+n)=$$

$$(x+1)+n=$$

$$(1+x)+n=$$

$$1+(x+n)=$$

$$1+(n+x)=$$

$$(1+n)+x.$$

Here we have used associativity of addition, the result that $x+1=1+x$, , and the induction hypothesis. The diagram displays that if $x+n=n+x$, , then $x+(n+1)=(n+1)+x$ holds. Combining this with the fact that $x+1=1+x$ and using the principle of induction, the result follows.

Note that the (natural) numbers are defined as a *relational system*, that is, as a collection on which is defined a certain order relation. Peirce formulates properties of relations, e.g., transitive and "quantitative" relations, in his language of

$GN \geq GM$. Finally he formulates the axiom of induction: "whatever is true of zero and which if true of any number N, is also true of GN the ordinal number next greater than N, is true of all numbers" (NE IV 2). The addition of numbers is defined as follows: (i) 0 + 0 = 0, (ii) GM + N = G(M + N), (iii) M + GN = G(M + N). By successive use of these definitions and induction, he is able to prove the stated proposition.

the logic of relatives in many of his papers. To mention a couple of examples, he expresses formally a one-one relation in "On the Logic of Number" (1881) and properties of transitive relations in "The Logic of Quantity" (1893).

In a collection of papers called *Recreations in Reasoning* dated around 1897 Peirce defines the numbers by something very close to the Dedekind-Peano axioms. More precisely he adopts notation for what plays the role of a successor and states basic properties of this relation. These properties together with what Peirce derives as a consequence of them constitute what is now known as the Dedekind-Peano axioms. Another thing to mention about this system is that Peirce *derives* the principle of induction from this system, and thus calls it the *Fundamental Theorem of Pure Arithmetic* (CP 4.165). Peirce continues to define the relation "greater than" on this system and deduces some properties so that this system is comparable to the first-mentioned example of defining the natural numbers via an order relation.

2.2.2. Practical Arithmetic or Demonstration That Arithmetical Propositions Are Analytic

The fundamental theorem in practical arithmetic, serving as the foundation for counting, is denoted "The Fundamental Theorem of Arithmetic."[18] It states that "if the count of a lot of things stops by the exhaustion of those things, every count of them will stop at the same number" (NE IV, 82). Peirce contrasts this to the Fundamental Theorem for Pure Arithmetic. In the text "Synthetical Propositions à Priori" (NE IV, 82–85) he demonstrates that "5 + 7 = 12" follows from the fundamental proposition of arithmetic. Therefore "5 + 7 = 12" is not "synthetical" (but corollarial, since it follows directly from the definitions). From the fundamental proposition Peirce deduces the principle of associativity (A + B) + C = A + (B + C) and then that the equality 5 + 7 = 12 follows. The article continues to prove the fundamental proposition using the language of the "logic of relatives" (demonstrating, according to Peirce, that it is not synthetic). The main steps are listed here. First a finite collection is defined. A collection, A, is finite, if whenever there is (here using modern notation) a one-to-one function $\lambda : A \to A$, then it is necessarily onto.[19] (That is, there is no one-to-one correspondence

[18] Peirce explicitly writes that the proposition 5 + 7 = 12 is analytic (NE IV, 84). He here explains that the proposition is analytic since it follows by necessity from the definitions. In contrast are propositions that he calls "theorematic," to which we return in the second part of the chapter. See Levy (1997) for a discussion concerning the relation between the synthetic and analytic distinction and corollarial vs. theorematic proofs. See also Otte (1997) on the analytic-synthetic distinction in Peirce's philosophy.

[19] In Peirce's terminology it is expressed as "Suppose a lot of things, say the As, is such that whatever class of ordered pairs λ may signify, the following conclusion shall hold. Namely, if every A is a λ of an A, and if no A is λ'd by more than one A, then every A is λ'd by an A. If that necessarily follows, I term the collection of As a *finite* class" (NE IV, 83).

between A and a proper subset of A.) In the next step it is shown that if a collection is counted, it is finite. To count a collection, according to Peirce, means to establish a one-to-one correspondence with the objects in the collection (taken in some order) and an initial segment of the natural number sequence. Finally, in the last step, it is demonstrated that if one assumes that a count of a sequence results in two different numbers, and the relation "next followed in the counting by" is employed, then there will be no least number in the sequence. This is a contradiction.[20]

2.3. Foundations: Set Theory

It is clear that underlying Peirce's conception of the various systems of numbers is a form of naive set theory. The same holds for his work in logic. Following Boole, Peirce's algebraic logic deals with classes. As an example one may point to his early paper "On an Improvement of Boole's Calculus of Logic" (1867), where Peirce uses letters to refer to classes of things or occurrences. It is then possible to define operations corresponding to addition and multiplication (and their inverses) on these classes (addition corresponding to, at first, union, and later to disjoint union). Such classes then form the basis for formulating the laws of arithmetic, as noted in note 13.

In later papers Peirce develops what can be denoted as versions of transfinite set theory along the lines of Cantor (and Dedekind)[21]—although he disagrees with Cantor on certain points. He often mentions the theory of multitudes, which is how he refers to cardinal numbers, and is well aware of Cantor's work on set theory, but developed some ideas independently. In particular Peirce claims credit for two results. One is the proof of the theorem that "there is no largest multitude," which in contemporary terms is that the cardinality of a set is strictly less than the cardinality of its power set. I return to this result later. The second is the definition of a finite collection as a collection for which the syllogism of transposed quantity holds.[22] In connection with his work on logic Peirce realized that the validity of inference rules depends on the size of the collection they are applied to (see "On the Logic of Number" from 1881 or the letter to Cantor in NE

[20] In the paper "On the Logic of number" (1881), referred to in section 10.2.2.1, a similar, although more complicated, proof is made.

[21] Peirce notes that his approach is closer to Cantor's since they both start with cardinal numbers, whereas Dedekind is concerned with ordinals.

[22] In his later years Peirce expresses his frustration (e.g., in CP 4.331) that he has not received more credit for his original ideas. For one thing, he accuses Dedekind of not giving him credit for his definition of a finite collection. Peirce writes in 1905 that he sent his 1881 paper, where he defines a finite collection, to Dedekind. There is no evidence, however, that Peirce's definition served as inspiration for Dedekind since he formulated his definition of an infinite set as early as 1872 (see Ferreirós 2007, 109).

III(2), 772). The syllogism of transposed quantity is the following—using one of Peirce's own examples:

Every Texan kills a Texan.
Nobody is killed but by one person.

Every Texan is killed by a Texan.

This syllogism is only valid when applied to finite collections (of Texans), so a finite collection may be defined as a collection for which this syllogism is valid. If one translates the premises and conclusion to expressions using functions, it states the same as the definition given earlier. The first premise is that there is a function, $k: \text{Texans} \rightarrow \text{Texans}$, the second that this is one-to-one. The conclusion is that the function is onto.

A further thing to note is that Peirce, like others at the time, struggled to find a proper definition of a collection.[23] Such a characterization could serve as a hypothesis from which the properties of sets would follow, similar to what he had accomplished for the numbers. One definition offered is the following: "We may say that a collection is an object distinguished from everything which is not a collection by the circumstance that its existence, if it did exist, would consist in the existence of certain other individual objects, called its members, in the existence of these, and not in that of any others; and which is distinguished from every other collection by some individual being member of the one and not a member of the other; and furthermore every fact concerning a collection will consist in a fact concerning whatever members it may have" (NE IV, 9).

The paper "Multitude and Number," dated 1897, presents in some detail Peirce's contribution to the theory of multitudes (see CP 4.170–226). These notes start out by defining a relation "being a constituent unit of" that can be regarded as a membership relation. Via this relation he defines a collection, as "anything which is *u*'d by whatever has a certain quality or general description and by nothing else" (CP 4.171). Having defined collections, he defines the notion of multitude to "denote that character of a collection by virtue of which it is greater than some . . . others, provided the collection is discrete" (CP 4.175). A collection is discrete if its constitutive units are or may be distinct as opposed to a continuous collection. Equality of collections is defined in terms of one-to-one relations: That the "collection of M's and the collection of N's are equal is to say: There is a one-to-one relation, c, such that every M is c to an N; and there is a one-to-one

[23] See Dipert (1997) for a discussion of Peirce's philosophical conception of sets. Noting the difficulty of providing a characterization of a set, Dipert furthermore presents Peirce's subtle criticism of Dedekind's definition of an infinite collection. For this criticism see (CP 3.564).

relation, d, such that every N is d to an M" (CP 4.177). Before dealing with the different types of multitudes, Peirce addresses a question of which kinds of relations are meaningful on collections, mentioning in particular what we would denote as trichotomy. Peirce's classification of multitudes can be compared to Cantor's treatment of cardinal numbers (Peirce also refers to his papers, e.g., in CP 4.196). But Peirce disagrees with his names, calling them enumerable (finite), denumerable (countable), primipostnumeral (first uncountable), secundopostnumeral, etc. When dealing with the countable collections he shows standard propositions, e.g., that the product of two denumerable multitudes is a denumerable multitude. Furthermore he uses Cantor's notation for cardinal numbers, i.e., the alef, ℵ. Whenever moving on to the next multitude, Peirce writes that the problem is to determine the smallest multitude exceeding the previous (CP 4.200). For example, the section on the "primipostnumeral" begins: "Let us now enquire, what is the smallest multitude which exceeds the denumerable multitude?" Interestingly he finds that a way to obtain a primipostnumeral collection is by taking the collection of subsets of a denumerable set, and so he implicitly accepts the Continuum Hypothesis. He shows that this has the same multitude as, e.g., the collection of quantities between zero and one. He also argues that the size of this is $2^{\aleph}$ and that in general larger multitudes can be obtained by taking further powers.

Taking the collection of subsets as a larger collection corresponds to Peirce's theorem, namely that there is no largest multitude. Peirce seems to be particularly fond of this theorem as he presents many different proofs of it. The proofs are often used to illustrate various points: In the "Prolegomena for an Apology to Pragmaticism" the proof serves as an example of "diagrammatic reasoning." In other places it is given as an example of "theorematic" reasoning, something Peirce contrasts with "corollarial" reasoning. I return to these notions in the last part of the chapter.

In addition to studying multitudes, Peirce engages himself with a characterization of the continuum that he in the paper just treated argues cannot be a multitude. The reason is the stated property, that there is no greatest multitude. For one thing Peirce finds that it is possible "in the world of non-contradictory ideas" to consider the aggregate of all postnumeral multitudes and that this aggregate cannot be a multitude. It must instead be a continuous collection. There are both mathematical and philosophical angles to Peirce's thoughts on the continuum. Here I will make a few remarks pertaining to the mathematical ones.[24] First, one

[24] Scholars have explained how "continuity" is fundamental to Peirce's mature philosophy; see Hookway (1985), Stjernfelt (2007) and Zalamea (2010). Moore (2015) evaluates Peirce's description from a mathematical point of view. Dauben (1982) presents in some detail Peirce's conception of the continuum from the point of view of set theory.

may mention that Peirce's conception of the continuum has little to do with the project of rigorization of analysis, which led, e.g., Weierstrass and Dedekind to formulate their versions of the mathematical continuum, although he is aware of these developments. He is critical of the replacement of infinitesimals with the "cumbrous" method of limits pointing to the odd formulations mathematicians made, such as defining a limit "as a point that can 'never' be reached," stating that "This is a violation not merely of formal rhetoric but of formal grammar" (CP 4.118). Furthermore he objects to the characterization of the continuous line as composed of points and mentions topology and projective geometry as areas where continuous quantity in this sense does not enter at all (see CP 4.218–225, 3.526).

Peirce's and Cantor's motivations for engaging in set theory are thus quite different, and both had motives and sources of inspiration besides the mathematical. It is usually said that Cantor's initial inspiration came from analysis, where he worked on the conditions for unique representation of functions by trigonometric series. Peirce, on the other hand, was first influenced by his work in logic and later his interest in mathematics in general. He also had a philosophical motive, and doubly so, since mathematics (and logic) served as a foundation for his philosophical system.

2.4. Algebra

In this section I address another theme from what can be denoted Peirce's use of the axiomatic method. More importantly, the examples from Peirce's writings on algebra illustrate his emphasis on the inadequateness of the claim that mathematics is the "science of quantity." He writes things like "To this day, one will find metaphysicians repeating the phrase that mathematics is the science of quantity,—a phrase which is a reminiscence of a long past age when the three words 'mathematics,' 'science,' and 'quantity' bore entirely different meanings from those now remembered. No mathematicians competent to discuss the fundamentals of their subject any longer suppose it to be limited to quantity. They know very well that it is not so" (NE IV, 228–229). Furthermore, I will note his characterization of algebra as a system of symbols functioning as a calculus, i.e., a language to reason in. Part of Peirce's knowledge of algebra stemmed from his father, Benjamin Peirce, including his monograph *Linear Associative Algebra*, first published in 1870, the same year as Peirce's remarkable paper on the logic of relatives (that is, his "Description of a Notation for the Logic of Relatives, Resulting from an Amplification of the Conceptions of Boole's Calculus of Logic," CP 3.45–149). Peirce remarks that he and his father discussed the contents of both with each other, writing: "There was no collaboration, but there were frequent conversations on the allied subjects, especially about the algebra" (NE III, 526). Inspiration for this work clearly comes from the British algebraists and the emerging way of designing algebras by detaching symbols of their traditional meaning (denoting numbers), and simply focusing on the rules of combinations. This work had a boost from Hamilton's introduction of the quaternions, where it turned out that multiplication is not commutative. In a sense B. Peirce's *Linear Associative Algebra* can be seen as a generalization of the work of Hamilton, dealing in general with systems—or algebras—of expressions formed as linear combinations of a given number of elements. In C. S. Peirce's writings there are numerous examples from linear associative algebra, but the examples to be considered here concern the imaginary numbers and permutation groups.

In a section of the paper "The Logic of Quantity" Peirce discusses the imaginary number i. This (long) paper starts out with the criticism of Kant's claim that mathematical propositions are synthetic, as referred to in section 10.2.2.[25] Peirce starts by praising Cauchy for giving the first "correct logic of imaginaries," but regrets that the rule-of-thumbists "do not understand it to this day" (CP 4.132). They object that there cannot be a quantity that is neither positive nor negative and that the square of a quantity is always positive. Despite this Peirce explains how it is possible to introduce a quantity whose square is negative. The mathematician "would reason indirectly: that is the mathematician's recipe for everything" (CP 4.132). The algebraist simply states that he needs a quantity whose square root is -1, noting: "there is no such thing in the universe: clearly then, I must import it from abroad" (CP 4.132). Peirce's explanation displays his use of the axiomatic method. He lists the fundamental properties of numbers[26] (quantities), stating that "If there is one of those laws which requires a quantity to be either positive or negative, find out which it is and delete it. If you have a system of laws which is self-consistent, it will not be less so when one is wiped out." Peirce deduces that the property "(16) $x > 0$ or $x < 0$ or $x = 0$" is required in order to prove that the square of all (nonzero) numbers is positive. The conclusion is that if this property is deleted, one may introduce the hypothesis that there is a quantity, i, defined as the square root of -1. The symbols so introduced have no other meaning than given by the hypotheses,[27] i.e., the meaning of i is that $i^2 + 1 = 0$

[25] "The Logic of Quantity" is dated 1893. It was supposed to be included in Peirce's book *The Grand Logic*. It is a long paper starting out with criticisms of the positions of Kant and Mill on mathematics. The ensuing sections deal with the logic of quantity, that is, expressing properties of quantitative relations in his language of relations and deriving their consequences. Toward the end are sections treating the imaginary quantities, quaternions, and a section of measurement and infinitesimals.

[26] The listed properties of quantities include, for example, the commutative and associative properties of addition and multiplication and properties of the relation "less than."

[27] In CP 4.314 a similar statement is made, i.e., that symbols have no meaning other than that we give them. The example in this case concerns developing an algebra of three elements.

: "the meaning of a sign is the sign it has to be translated into" (CP 4.132). In this way the system of symbols of algebra becomes a calculus; "that is to say, it is a language to *reason in*" (CP 4.133). He continues: "To say that algebra means anything else than just its own forms is to mistake an *application* of algebra with the *meaning* of it" (CP 4.133).

In order to define a complex number, reference to numbers (and so quantities) is required. But a complex number goes beyond quantities since relations must be introduced that do not fulfill the properties of relations defining quantities, e.g., transitivity and the like: "[It] is readily seen that what is called an imaginary quantity or a complex quantity is not purely quantity" (NE IV, 229). To show that there are examples from mathematics that have nothing to do with quantity whatsoever, Peirce presents the notion of a group: "By a 'group,' mathematicians mean the system of all the relations that result from compounding certain relations which are fully defined in respect to how they are compounded" (NE IV, 229). As an example of a group where these relations have nothing to do with quantity he presents what is essentially a group of permutations, where elements are permutations on the four letters A, B, C, and D. The group is presented as containing relations that, composed by themselves four times, give the identity. That is, if l represents such a relation, it fulfills that $l^4 = \text{Id}$. One such relation is $D:A+C:B+A:C+B:D$, meaning that that D maps to A, A to C, C to B, and finally B to D. Today this could be written in cyclic notation as $(DACB)$. He notes there are more such relations, 24 in total, that they have converses (i.e., inverses), and refers to the product of such relations—which he notes has a logical meaning having nothing to do with quantity (similarly to the use of "+" above in the presentation of the permutation). The totality of these 24 relations thus forms a group. Furthermore he talks about smaller sub-collections of the 24 elements that will also form a group (NE IV, 227–234, 1905–6).

2.5. Conclusions: Peirce, Pre-structuralist Themes, and Relations

Summing up on Peirce's adherence to a number of pre-structuralist views, I have noted Peirce's distinctions between physical geometry and mathematical geometry on the one hand and practical and pure arithmetic on the other. Regarding the first distinction, he remarks that the hypotheses in pure geometry are studied irrespective of whether they apply to the real world or not. I showed that it is possible to find similar comments referring even to practical arithmetic. In the writings on algebra I also noted that Peirce several times explicitly rejects the characterization of mathematics as the science of quantity, producing examples that have nothing to do with quantity.

Furthermore I have shown that Peirce uses formal methods in arithmetic to determine which hypotheses are sufficient in order to derive the properties in question. One aim was to argue that the statements of arithmetic are logical consequences of certain definitions, or hypotheses. His use of the "axiomatic method" in arithmetic can be likened to the process described by Hilbert ([1918] 1996) where, given a collection of propositions, a certain collection of axioms can be identified so that the given propositions can be derived from them—what Hilbert calls "deepening of foundations." This seems to fit well with Peirce's procedure. His method thus has two interrelated aims. Focusing on reasoning and inference rules, the point is on the one hand to formulate "a few primary propositions" of the numbers so that properties of them follow by necessity. On the other hand, focusing on the propositions, the aim is to determine the postulates sufficient for deriving the propositions of arithmetic. Peirce's discussions of the meaning of "postulates" and "hypotheses" reflect these concerns: "For what is a postulate? It is the formulation of a material fact which we are not entitled to assume as a premiss, but the truth of which is requisite to the validity of an inference" (CP 6.41). A further similarity to Hilbert's method is Peirce's claim that there are multiple ways of organizing the propositions of arithmetic (cf. CP 4.93). One could take as basic the propositions defining the numbers via the successor function or the definition of numbers as a certain ordered collection.

In the case of the imaginary quantity, I indicated how Peirce traces out the consequences of a body of fundamental properties of the numbers, in order to determine which of these contradicts a desired property (i.e., that the square of a quantity is negative). In this case, he mentions the property of a collection of axioms of "being internally consistent." It does not seem, however, that he is concerned with further metamathematical considerations such as consistency in general, independence, and completeness. He appears to be quite confident in the mathematical method, writing in numerous places "in mathematics there are no mistakes and no (deep) disagreement" (CP 3.426).

Peirce's use of formal methods as well as his distinction between pure and applied versions of mathematics places him as an early modernist, characterized by J. Gray (2008) as "an autonomous body of ideas, having little or no outward reference, placing considerable emphasis on formal aspects of the work and maintaining a complicated—indeed anxious—rather than a naïve relationship with the day-to-day world, which is the de facto view of a coherent group of people, such as a professional or discipline based group that has a high sense of what it tries to achieve" (1).

After the many of examples of the mathematics of Peirce we may better understand what is meant when stating that a result or theory is based on the logic of relations. The first thing to note is that Peirce finds that relations of various sorts play a key role in the definition of mathematical objects. Having seen the examples presented here, we must concur. We have seen that, for example, an order relation is used to define the numbers and a bijective correspondence is used to define multitudes as well as "a count." The properties of these can be formulated in his language of the logic of relations. Second, to Peirce the main activity of mathematics is reasoning, that is, the practice of drawing necessary conclusions. Logic, according to Peirce, includes the study of (the methods of) such inferences. Peirce notes that he together with other logicians like de Morgan (NE IV, 1) early realized that the previous versions of logic came up short when trying to capture the structure of the statements of mathematics.[28] To formulate definitions as well as statements in mathematics thus requires reference to relations, so reasoning in mathematics must take into account how one draws inferences from statements involving relations.

3. Philosophy: Diagrammatic Reasoning

I now turn to focus on how Peirce proposes the necessity of reasoning is achieved, namely through diagrammatic reasoning. The description given here draws mainly on Peirce's 1906 paper "Prolegomena for an Apology to Pragmatism" (PAP), published in *The Monist* (reprinted in CP 4.530–582), and a draft of this (NE IV, 313–330).[29] But others of Peirce's writings will also be referred to. My presentation focuses on how diagrammatic reasoning applies to mathematics. It thus complements the contributions of Stjernfelt (2007) on diagrammatic reasoning in general and Shin's (2002) account of his existential graphs. I also refer to Marietti (2010) for a more detailed account than I am able to give here.

There are two key points to bear in mind when addressing "diagrammatic reasoning." The first is that Peirce thinks of a diagram as a certain type of sign. An important property of this sign, the diagram, is that it is *observable*. Peirce explains that the necessity of the conclusion of a proposition is established because it can be perceived in the diagram. The second key point is that his definition of a "diagram" applies to objects that one would not normally count as diagrams. I mention three possible sources of inspiration for Peirce's view of reasoning as linked to observing diagrams: First, Peirce's work on logic contributed to this view. I will return to this point at the end of this section. Second, the reasoning based on

[28] In Peirce's early papers on logic (see, e.g., volume 3 of CP) there are sections on the Aristotelian syllogisms. But these are not used when he turns to his algebra of logic. One may also find comments as to the shortcomings of the syllogisms; see CP 4.426 in relation to Euclid's *Elements*.

[29] The last part of PAP consists of a presentation of the existential graphs. The paper also includes an explanation of which types of signs these graphs are. The iconic existential graphs were supposedly meant to pave the way for a proof of his pragmaticism: "For by means of this, I shall be able almost immediately to deduce some important truths of logic, little understood hitherto, and closely connected with the truth of pragmaticism" (CP 4.534). See also EP 2, xxvvii–xxix and Shin (2002).

diagrams in Euclid's *Elements*—a source Peirce is familiar with and often cites from—proceeds in a way that is compatible with the description of diagrammatic reasoning. Third, Peirce explicitly mentions Kant in connection with the characterization of mathematical reasoning. According to Kant reasoning in mathematics proceeds by constructions, or the drawing of diagrams, formed in intuition. Peirce remarks that this view is partially correct, since it focuses on the method of mathematics rather than stating what mathematics is about, and he agrees that mathematics deals with constructions—but not in intuition (CP 3.556, 1898). Peirce claims the necessity of mathematical reasoning is due to the procedure of constructing "a diagram, or visual array of characters or lines. Such a construction is formed according to a precept furnished by the hypothesis. Being formed, the construction is submitted to the scrutiny of observation, and new relations are discovered among its parts, not stated in the precept by which it was formed, and are found, by a little mental experimentation, to be such that they will always be present in such a construction" (CP 3.560). That is, although he agrees with Kant that reasoning is done by constructions, as I have noted, he disagrees with Kant that this construction invokes intuition and depends on "thought"—although a diagram might be considered in one's imagination. As noted, it is essential for Peirce that the relations discovered are observed.[30]

When Peirce refers to a "diagram" he does not only understand it in its common sense, that is, as a figure mainly composed of points, lines, and circles, since he also describes it as a "visual array of characters or lines." To Peirce "diagram" refers to a sign that represents (intelligible) relations: "a Diagram is an Icon of a set of rationally related objects . . . the Diagram not only represents the related correlates, but also, and much more definitely represents the relations between them" (NE IV, 316–317, 1906). Mentioning an "icon," he refers to his semiotics. The next section therefore extracts a few points from his theory of signs. This introduction will be followed by an example of a proof together with a further elaboration of how to understand his characterization of necessary reasoning as diagrammatic reasoning.

3.1. Signs: Tokens and Types; Icons, Indices, and Symbols

Early on Peirce attached importance to signs, conceiving of them as the vehicles of thought. His theory of signs is interrelated with his categories (at first developed as a response to Kant's 12 categories, see, for example, CP 1.545–567 from 1867). According to Peirce there are only three types of categories. The categories consist

[30] That relations are seen to hold because they are observed brings mathematics on a par with natural science. See Marietti (2010) for an elaboration of this point.

of feeling, reaction, and law—or as he also called them, possibility, existence, and habit.[31] One way Peirce arrives at these categories is in terms of his logic of relations. Any given relation applies to a fixed number of relata, and so a relation may be monadic, dyadic, or triadic, and so on. Peirce claimed that he could prove that higher-order relations are reducible to relations taking only one, two, or three relata.[32] The monadic relations (predicates) correspond to the first category (feeling or quality), dyadic to the second (reaction or existence), and irreducible triadic relations to the third (law or habit). Later Peirce referred to the categories more abstractly in his phaneroscopy as firstness, secondness, and thirdness.

A sign, according to Peirce, is an irreducible triadic relation (corresponding to the three categories): it relates the sign, the object that is represented by the sign, and the interpretant of the sign. The last is important, in that Peirce holds that a sign is not a sign unless it is interpreted as such: "a sign (stretching that word to its widest limits), as *anything which, being determined by an object, determines an interpretation to determination, through it, by the same object*)" (PAP CP 4.531). In Peirce's early classification of signs, each of these three, that is, the sign, the relation between the sign and object, and the interpretant, is considered in terms of the previously mentioned three categories: possibility, existence, and law.[33] I only mention two of these here. Peirce's first division concerns the nature of the sign itself. This division includes the well-known notions of a *token* and a *type*: "A common mode of estimating the amount of matter in a MS. or printed book is to count the number of words. There will ordinarily be about twenty *the*'s on a page, and of course they count as twenty words. In another sense of the word 'word,' however, there is but one word 'the' in the English language; and it is impossible that this word should lie visibly on a page or be heard in any voice, for the reason that it is not a Single thing or Single event. It does not exist; it only determines things that do exist. Such a definitely significant Form, I propose to term a *Type*. A Single event which happens once and whose identity is limited to that one happening or a Single object or thing which is in some single place at any one instant

[31] The paper "What Is a Sign" (Peirce 1894, EP 2, 4–10) explains the three categories in terms of possible ways experience can be had: The first, most immediate, is *feeling*, e.g., thinking about the color red. Second is *reaction*, as when we are startled by a loud noise and try to figure out its origin. The second category thus requires "two things acting on each other" (EP 2, 5). Third is thought, or *reasoning*, formulating a law based on our immediate experiences and actions. This is described as "going through a process by which a phenomenon is found to be governed by a general rule" (EP 2, 5). Note also that the third category mediates between the other two. See also Hoopes (1991).

[32] See Misak (2004, 21), Burch (1997), and the paper "Detached Ideas Continued and the Dispute between Nominalists and Realists" (NE IV, 338–339).

[33] Around 1903 (see *Syllabus* 1903, published in EP 2, 289–299) Peirce presents his classification of signs into 10 different classes. Later, after introducing a more elaborate theory of interpretants and a distinction between the immediate and the dynamic object, he is able to produce 66 classes of signs. Peirce refers to both of these additions in PAP. In addition to PAP, see Hoopes (1991) and Short (2007) for an elaboration of the development of Peirce's semeiotics. Bellucci and Pietarinen (n.d.) give an account in relation to logic and Carter (2014) in relation to use in mathematics.

of time, such event or thing being significant only as occurring just when and where it does, such as this or that word on a single line of a single page of a single copy of a book, I will venture to call a *Token*" (CP 4.537). The sign corresponding to the first category is named a *quality* or a *tone*. A diagram is to be taken as a *type*, but a type can only be shown through a replica of it, that is, a token.

The second division is his division of signs into *icons, indices*, and *symbols*. They appear as answers to the question: In what capacity does the sign represent the object? The sign may represent because of similarities (likeness) between the object and the sign, in which case the sign is an icon: "Anything whatever, be it quality, existent individual, or law, is an icon of anything, insofar as it is like that thing and used as a sign of it" (EP 2, 291). Simple examples of icons used in mathematics are geometric objects, such as drawn triangles and circles. Icons do not only represent by visual resemblance; an important, and a characterizing, property of the icon is that it reveals new facts about the object that it represents. As such they are essential to mathematics: "The reasoning of mathematicians will be found to turn chiefly upon the use of likenesses, which are the very hinges of the gates of their science. The utility of likenesses to mathematicians consists in their suggesting, in a very precise way, new aspects of supposed states of things" (Peirce 1894, 6). As will be shown below, icons may represent relations. Note also that most icons used in mathematics involve conventional (symbolic) elements.[34] If I wish to prove something about an odd number, I could represent it iconically as "$2 \cdot k + 1$," for some number k, using the symbols "$\cdot$" and "+". Subsequently I will represent the statement that "a number divides another number" by the icon "$p \cdot k = a$."

The index represents its object because of some existent (causal) relation between the two. Peirce mentions as an example a weathercock, which, as a result of the wind blowing, tells us about the direction of the wind, so that the weathercock becomes an index of the direction of the wind. The type of index just mentioned represents due to some causal relation between the sign and the object. A pure index represents because of some purposeful association of it with what it represents, as one does in mathematics. Peirce mentions the geometers assigning of letters to geometric figures, naming places on such figures, so that one may reason about these places, points, lines, etc., via these letters.[35] This is obviously done in mathematics in general, as will be noted in the examples to follow.

[34] Peirce (CP 3.363) refers to the shading in Venn diagrams as a symbolic, or conventional element. See Carter (2018) for further examples of iconic representations in mathematics.

[35] In a paper published in 1885 Peirce characterizes an index as follows: "the sign [index] signifies its object solely by virtue of being really connected with it. Of this nature are all natural signs and physical symptoms. I call such a sign an *index*. . . . The index asserts nothing; it only says 'There!' It takes hold of our eyes, as it were, and forcibly directs them to a particular object, and there it stops. Demonstrative and relative pronouns are nearly pure indices, because they denote things without describing them; so are the letters on a geometric diagram, and the subscript numbers which in algebra distinguish one value from another without saying what those values are" (CP 3.361).

Finally, the sign could represent by virtue of a law, or a habit, stating that the particular sign refers to a certain kind of object. These are symbols. Examples of symbols are words; in mathematics we use symbols like "+," "π," etc.

3.2. Diagrammatic Reasoning

I now return to Peirce's description of the process of reasoning in mathematics. Reasoning consists of three steps: "following the precepts," (1) one constructs a diagram representing the conditions of a proposition and (2) one "experiments" on it until (3) one is able to read off the conclusion from the resulting diagram. This description seems to fit well (part of) the proof procedure in Euclid's *Elements*. Take, for example, proposition I.32, where it is proved that the sum of angles in a triangle is equal to two right angles. In order to prove this, a triangle ABC is drawn. In the next step, "experimenting on it," one extends the base line, say AB, and, from the starting point of this extended line, B, one draws a line parallel to AC. Reasoning in this diagram, one comes to see that the conclusion holds. What is remarkable is that Peirce finds that the above characterization also holds for mathematics in general, where the notion of "diagram" extends according to the preceding usage: "for even in algebra, the great purpose which the symbolism subserves is to bring a skeleton representation of the relations concerned in the problem before the mind's eye in a schematic shape, which can be studied much as a geometric figure is studied" (CP 3.556). (See also NE IV, 158.) The example of diagrammatic reasoning given by Peirce in PAP is the proof of the above-mentioned theorem that there is no largest multitude.[36] I present instead a (simpler) algebraic proof, proving that "if an integer divides two other integers, then this integer divides any linear combination of the two."[37] Introducing indices, a, b and p standing for the numbers and the symbol "|" to denote "divides,"[38] the proposition can be expressed as

For p, a and b being integers, if $p \mid a$ and $p \mid b$ then $p \mid sa + tb$ for any integers s and t.

[36] Peirce has a number of different formulations of this theorem in PAP, for example, "the single members of no collection or plural, are as many as are the collections it includes, each reckoned as an single object" (CP 4.532).

[37] Note that this example is not taken from Peirce. It is introduced by the author in order to explain "diagrammatic reasoning."

[38] $n|m$ means that there exists a number k such that $kn = m$. Using this notation it is for example the case that $2|8$, $-2|8$, and $3|-39$.

In order to prove this theorem we follow the three steps given previously. First we have to "form a diagram according to a precept of the hypothesis." That is, considering the antecedent of the proposition and translating the definition(s) used, we write down, in this particular case, the relations stated to hold between the numbers p,a and p,b respectively (cf. "calculating with a system of algebraic symbols"). The diagram thus obtained is that there exist numbers k and l such that

$$kp = a \text{ and } lp = b.$$

In the second step this diagram is experimented on; the signs are manipulated by using relevant (and valid) algebraic formulas:

$$\text{If } kp = a \text{ and } lp = b \text{ then } skp = sa \text{ and } tlp = tb.$$

Combining (adding) the last two we see that

$$sa + tb = skp + tlp = (sk + tl)p.$$

Noting that $sk + tl$ must be an integer since s, k, t, and l are all integers, one is able to observe that p divides the linear combination of a and b. It is thus possible to read off the conclusion of the proposition in the final line—corresponding to the third step.

Combining the preceding and leaving out the explanatory text so that it is in fact a "visual array of characters" makes it easier to appreciate why Peirce insists on calling it a diagram. p | a and p | b is represented as

$\mathrm{kp} = \mathrm{a}$ and $\mathrm{lp} = \mathrm{b}$.
$kp = a$ and $lp = b$ implies that $skp = sa$ and $tlp = tb$.
$sa + tb = skp + tlp = (sk + tl)p$.

Observation of the last line tells us that p divides the linear combination, which is the conclusion.

A further, and most important, point is that by going through this diagram[39] one should be able to see that the conclusion follows by necessity from the stated condition. Relations referred to thus subsist on two different levels, as indicated by the following explanation: "a Diagram is an Icon of a set of rationally related objects . . . the Diagram not only represents the related correlates,

[39] In fact Peirce urges the reader to construct a diagram herself while following the instructions of the proof.

but also, *and much more definitely represents the relations between them*" (NE IV, 316–317, 1906, my emphasis). In the first stage of constructing the diagram relations referred to are relations that hold between numbers, the main relation used being the relation of "a number dividing another." At the second level are what can be denoted *logical* relations. Recall that a major interest for Peirce when studying mathematics was to extract the principles of drawing necessary conclusions. The stated purpose of the "Prolegomena" is precisely to argue that all necessary reasoning is diagrammatic reasoning, assuming that mathematical reasoning is necessary reasoning. What is achieved by the process of diagrammatic reasoning is that one comes to *see* the necessary relation that holds between the hypothesis and the conclusion of the proposition, that is, what I here refer to as a logical relation. In support of this view, in a passage telling us how to do proofs in mathematics (again referring to this as an activity) by constructing a diagram, making alterations to it, and comparing these two diagrams, Peirce writes that finally "the book . . . will make it quite plain and evident to you that the relation *always will* hold exactly" (NE IV, 200). This last use of "relation" refers to the logical relation in question.[40] Recall also the proof given in section 2.2 that addition is commutative. I remarked that the signs produced constituted a diagram. The purpose of that diagram was to allow us to see (or deduce) that if $x+n=n+x$ then $x+(1+n)=(1+n)+x$ follows by necessity.

In the different versions of PAP, Peirce analyses which type of sign is involved in diagrammatic reasoning in order to address a number of issues, such as how the necessity of reasoning, and generality of the conclusions, are obtainable.[41] In these papers Peirce mentions his extended theory of interpretants.[42] According to Peirce the drawn diagram is a sort of hybrid sign. He stresses that a diagram is an icon, but of a special kind. A diagram *shows* that a consequence follows "and more marvellous yet, that it *would* follow under all varieties of circumstances accompanying the premisses" (NE IV, 318). Peirce explains that this is achieved since diagrams are *schemas*. Being drawn and so capable of being perceived, they are tokens. But they are at the same time representations of symbolic statements (actually the interpretant of a symbol) and so general: the diagrams

[40] As further support of this claim, the paragraphs CP 4.227–240 link Peirce's characterization of mathematics as the science that draws necessary conclusions with a description of diagrammatic reasoning.

[41] See Stjernefelt (2007, chap. 4) for a more elaborate explanation of these issues.

[42] The extension made by Peirce includes different interpretants, in PAP named the *immediate*, *dynamic*, and *final* interpretant. The immediate interpretant is how it "is revealed in the right understanding"—the meaning of the sign; the dynamic interpretant is the actual effect the sign has on some interpretant. The final interpretant is "the manner in which the Sign tends to represent itself to be related to its Object" (CP 4.536). Another addition is that all of these can partake in either firstness, "emotional," secondness, "energetic," or thirdness, "logical" or "thought" (CP 4.536).

are representations of (symbolic) statements like "The sum of the angles of a triangle is equal to two right angles" or "If a number divides two numbers, then it will divide any linear combination of those two numbers." In Peirce's words: "the Iconic Diagram and its Initial Symbolic Interpretant taken together constitute what we shall not too much wrench Kant's term in calling it a *Schema*, which is on the one side an object capable of being observed while on the other side it is General" (NE IV, 318).

Referring to "experimenting on a diagram" brings us to Peirce's distinction between corollarial and theorematic reasoning. In corollarial reasoning, the consequences of the hypotheses can be read off directly from the constructed diagram. Furthermore the proof only makes use of the definitions of the concepts presented in the proposition, whereas this is not the case for theorematic reasoning. Corollarial reasoning "consists merely in carefully taking account of the definitions of the terms occurring in the thesis to be proved. It is plain enough that this theorematic proof we have considered differs from a corollarial proof from a methodeutic point of view, in as much as it requires the invention of an idea not at all forced upon us by the terms of the thesis" (NE IV, 8). The theorematic proof referred to is a proof of his theorem that the cardinality of a collection is less than the cardinality of its power set. Another example of a theorematic proof is the proof of Euclid I.32, since additional lines have to be drawn.[43] The deductions of the properties of numbers are corollarial proofs (as well as the example mentioned in note 18).

I finally note that Peirce also worked with diagrams (closer to the ordinary meaning of diagram) in relation to logic. In several places Peirce notes the schematic shape of the presentation of arguments (as in the syllogism of the transposed quantity). As early as 1885 Peirce refers to syllogisms as "diagrams," stating that their purpose is to make it possible to observe the relations among the parts (CP 3.363). The reason this has not been noticed before, Peirce assumes, is that the constructions of logic are so simple that they are overlooked: "Why do the logicians like to state a syllogism by writing the major premiss on one line and the minor below it, with letters substituted for the subject and predicates . . . he has such a diagram or a construction in his mind's eye" (CP 3.560, 1898). Later, in Peirce's so-called diagrammatic period in logic,[44] the representations of logical propositions and inferences *were* diagrams, that is, figures composed of lines. See Figure 1 for an example of such a diagram (that is also an example of an existential graph).[45]

[43] Various interpretations have been proposed regarding the distinction between theorematic and corollarial reasoning; see Hintikka (1980) for a logical interpretation and Levy (1997) for a specific interpretation concerning the theorem that there is no largest multitude.

[44] See Dipert (2004).

[45] In this period Peirce studied the well-known diagrams of Euler and Venn making it possible to visualize the validity of arguments and used these as inspiration for developing his own systems

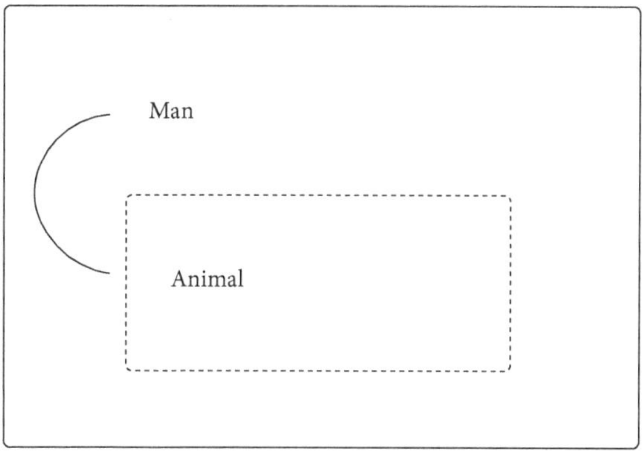

Figure 1 This diagram represents the statement "Any man would be an animal" or that nothing is both a man and not an animal. A box around p means "not p." A line joining p and q means p is related to q, in the sense that "some p is q."

In conclusion note the following. When writing down proofs in mathematics something *like* a diagram is formed. They are not exactly like the diagrams used in the *Elements*, but (relations between) concepts are represented in a schematic form that allows us to do things to them so that new relations become visible. Furthermore when Peirce managed to formulate his systems of logic, as in Figure 1, the representations *are* composed of letters and lines. In PAP Peirce comments that if all steps of a proof were to be spelled out, they would be reproducible by his graphs (NE IV, 319). The conclusion is that reasoning in mathematics involves (representations of) relations and that in the existential graphs these are displayed using diagrams so mathematical reasoning (which is necessary reasoning) is diagrammatic reasoning.

4. Structuralist Elements

In the introduction I proposed that Peirce could be interpreted as a methodological structuralist. Reck and Price (2000) characterize such a position by two principles. The first states that mathematicians "study the *structural features* of"

of logic, his existential graphs (see, e.g., CP 3.456–498, CP 4.347–371, and Bellucci and Pietarinen, n.d.). Besides Euler and Venn diagrams, other visual tools used in mathematics, chemistry, and their combination served as inspiration for these systems. In the mid-19th century "diagrammatic" notation was being developed and used both in chemistry and in graph theory. It was even proposed by Sylvester (a colleague of Peirce at Johns Hopkins) and Clifford to combine work in chemistry and the algebra of graphs around 1877 (see Biggs et al. 1976). Peirce was aware of the developments in both areas as well as the proposed link.

the entities assumed in their everyday practices, such as the various number systems, algebraic structures, various spaces, etc. Second, "it is (or should be) of *no* real concern in mathematics what the *intrinsic nature* of these entities is, beyond their structural features" (Reck and Price 2000, 45). Besides the emphasis on structure and structural features, this description resonates well with Peirce's emphasis that a mathematician only cares about deriving the consequences of her hypotheses. I have furthermore shown that the hypotheses, or definitions, formed by Peirce often characterize objects (e.g., the number systems) as relational systems. But I have also stressed, in particularly referring to the numbers, that Peirce found that there are different ways to define them, that is, there are multiple ways to logically organize the theory of numbers.

I have noted that Peirce did not seem to be interested in the foundations of mathematics, being convinced of the rigorousness of the reasoning of mathematicians and placing mathematics at the top of his philosophical system. These three elements, i.e., an "anti-foundationalist" view of mathematics,[46] the methodological structuralism, and the (relativism of) logical structure can also be found in the contemporary categorical structuralist view of Steve Awodey (2004). One component of Awodey's position is to "avoid the whole business of 'foundations'" (Awodey 2004, 55). Categorical structuralism rejects the idea of having a foundational system consisting of enough objects of some type, e.g.. sets, from which all mathematical objects may be built, and a collection of "laws, inference rules, and axioms to warrant all of the usual inferences and arguments made in mathematics about these things" (Awodey 2004, 56). In contrast structuralists advocate the "idea of specifying, for a given theorem or theory only the required or relevant degree of information or structure . . . for the purpose at hand, without assuming some ultimate knowledge, specification, or determination of the 'objects' involved. The laws, rules, and axioms involved in a particular piece of reasoning, or a field of mathematics, may vary from one to the next, or even from one mathematician or epoch to another" (Awodey 2004, 56). Awodey illustrates this top-town, or schematic, approach by the following example. Say one wishes to prove that if $x^2 = -1$ then $x^5 = x$.. The result follows in a field and a consequence is that $i^5 = -1$. A proof can also be found based on the axioms of a ring. Assuming even less, it can be proved in a semi-ring with identity that $x^2 + x + 1 = x$ implies $x^5 = x$. From a foundational (bottom up) point of view, one has to presuppose that the construction of the complex numbers as well as rings and semi-rings have been made in order to state these propositions. From a structuralist perspective the propositions are schematic statements about any structure (ring or semi-ring) fulfilling the appropriate conditions. There are also differences between Awodey's categorical structuralism and Peirce's position.

46 I borrow this term from Pietarinen (2010).

To mention one, the basic entity of categorical structuralism is the morphism, whereas Peirce still refers to relations and their relata. Another is Peirce's study of mathematics in order to extract, for logic, its method of drawing valid inferences. It thus appears that he believes in the objectivity and reality of such inference rules. He would presumably not, as does Awodey, accept the arbitrariness of inference rules.

5. Conclusion

In this chapter I have documented Peirce's impressive knowledge of and contributions to the mathematics of his time. Examples of his contributions to geometry, set theory, and the foundations of arithmetic and his discussions on algebra have been given. These examples also served to illustrate a number of pre-structuralist themes, such as Peirce's distinction between pure and applied mathematics, e.g., his claim that applied geometry does not belong to mathematics. In addition I mentioned his objection to the characterization of mathematics as the science of quantity.

In a number of papers Peirce characterizes mathematics as the science that draws necessary conclusions from stated hypotheses. In the case of arithmetic we saw that he was able to deduce the properties of numbers from a system of axioms or, as he referred to them, "a few primary propositions." A key element of Peirce's position was to acknowledge the role of relations in mathematics both as used in the definition of mathematical objects (such as the numbers) and when formulating mathematical statements in general. We saw, e.g., that he defines the natural numbers as a relational system, and I noted that he formulates the properties of relations in his language of the logic of relatives. I have also presented Peirce's notion of "diagrammatic reasoning," that is, his explanation of how the necessity of reasoning is achieved by constructing, experimenting on, and observing diagrams. In this connection I proposed that these diagrams allow us to see the necessary relation, that is, a logical relation, holding between the antecedent and conclusion of a proposition.

In the final section I identified two structuralist positions that have some common elements with Peirce's views as presented here. Peirce defined a system of quantity as a relational system, that is, as collections on which is defined a specific order relation. His motive was to show that the properties of numbers follow by necessity from this characterization. That is, in structural terms, one may say that they are structural properties. In this way Peirce may be construed as a methodological structuralist. Furthermore I find that his anti-foundationalism, the claim that there are multiple ways to organize a mathematical theory and his insistence that mathematics concerns hypotheses, led to a view that is similar in spirit to categorical structuralism.

References

Awodey, Steve. 2004. An Answer to Hellman's Question: "Does Category Theory Provide a Framework for Mathematical Structuralism?" *Philosophia Mathematica* 12 (3): p. 54–64.

Bellucci, Francesco and Ahti-Veikko Pietarinen. n.d. Charles Sanders Peirce: Logic. *The Internet Encyclopedia of Philosophy*. Accessed February 25, 2016. http://www.iep.utm.edu/peir-log/.

Biggs, Norman L., E. Keith Lloyd, and Robin J. Wilson. 1976. *Graph Theory, 1736–1936*. Oxford: Oxford University Press.

Brent, Joseph. 1998. *Charles Sanders Peirce: A Life*. Bloomington: Indiana University Press.

Burch, Robert. 1997. Peirce's Reduction Thesis. In *Studies in the Logic of C.S. Peirce*, edited by Nathan Houser, Don D. Roberts, and James van Evra, pp. 234–251. Bloomington: Indiana University Press.

Carter, Jessica. 2014. Mathematics Dealing with "Hypothetical States of Things." *Philosophia Mathematica* 22(2), 209–230.

Carter, Jessica. 2018. Graph-Algebras—Faithful Representations and Mediating Objects in Mathematics. *Endeavour* online, August 31, 2018.

Cayley, Arthur. 1854. On the Theory of Groups, as Depending on the Symbolic Equation $\theta^n = 1$. *Philosophical Magazine* 7, 40–47. In Cayley's collection of mathematical papers, vol. 2. Cambridge: Cambridge University Press, 1889.

Dauben, Joseph. 1982. Peirce's Place in Mathematics. *Historia Mathematica* 9, 311–325.

Dipert, Randall D. 1997. Peirce's Philosophical Conception of Sets. In *Studies in the Logic of C.S. Peirce*, edited by Nathan Houser, Don D. Roberts, and James van Evra, pp. 53–76. Bloomington: Indiana University Press.

Dipert, Randall D. 2004. Peirce's Deductive Logic: Its Development, Influence, and Philosophical Significance. In *The Cambridge Companion to Peirce*, edited by Cheryl Misak, pp. 287–324. New York: Cambridge University Press.

Ferreirós, José. 2006. Riemann's Habilitationsvortrag at the Crossroads of Mathematics, Physics, and Philosophy. In *The Architecture of Mathematics*, edited by Jeremy Gray and José Ferreirós, pp. 67–96. Oxford: Oxford University Press.

Ferreirós, José. 2007. *Labyrinth of Thought: A History of Set Theory and Its Role in Modern Mathematics*. 2nd ed. Berlin: Birkhäuser Verlag.

Gray, Jeremy. 2008: *Plato's Ghost. The Modernist Transformation of Mathematics*. Princeton, NJ: Princeton University Press.

Havenel, Jérôme. 2010. Peirce's Topological Concepts. In *New Essays on Peirce's Mathematical Philosophy*, edited by Matthew E. Moore, pp. 283–322. Chicago: Open Court.

Hilbert, David. [1918] 1996. "Axiomatic Thought." In *From Kant to Hilbert: A Source Book in the Foundations of Mathematics*, edited by William B. Ewald, pp. 1105–1115. Oxford: Oxford University Press.

Hintikka, Jaakko. 1980. C. S. Peirce's "First Real Discovery" and Its Contemporary Relevance. *The Monist* 65, 182–188.

Hookway, Christopher. 1985. *Peirce*. London: Routledge and Kegan Paul.

Hookway, Christopher. 2010. "The Form of a Relation": Peirce and Mathematical Structuralism. In *New Essays on Peirce's Mathematical Philosophy*, edited by Matthew E. Moore, 19–40. Chicago: Open Court.

Hoopes, James. 1991. *Peirce on Signs*. Chapel Hill: University of North Carolina Press.

Levy, Stephen H. 1997. Peirce's Theorematic/Corollarial Distinction and the Interconnections between Mathematics and Logic. In *Studies in the Logic of C.S. Peirce*, edited by Nathan Houser, Don D. Roberts, and James van Evra, pp. 85–110. Bloomington: Indiana University Press.

Mannoury, Gerrit. 1909. *Methodologisches und Philosophisches zur Elementar-Mathematik*. Haarlem: P. Visser.

Marietti, Susanna. 2010. Observing Signs. In *New Essays on Peirce's Mathematical Philosophy*, edited by Matthew E. Moore, pp. 147–167. Chicago: Open Court.

Misak, Cheryl. 2004. Charles Sanders Peirce (1839–1914). In *The Cambridge Companion to Peirce*, edited by Cheryl Misak, pp. 1–26. New York: Cambridge University Press.

Moore, Matthew E., ed. 2010. *New Essays on Peirce's Mathematical Philosophy*. Chicago: Open Court.

Moore, Matthew E. 2015. Peirce's Topical Theory of Continuity. *Synthese* 192, 1055–1071.

Otte, Michael. 1997. Analysis and Synthesis in Mathematics from the Perspective of Charles S. Peirce's philosophy. In *Analysis and Synthesis in Mathematics: History and Philosophy*, edited by Michael Otte and Marco Panza, pp. 327–362. Dordrecht: Kluwer.

Peirce, Benjamin. [1870] 1881. Linear Associative Algebra. *American Journal of Mathematics* 4(1), 97–229.

Peirce, Charles Sanders. 1894. What Is a Sign. In *The Essential Peirce: Selected Philosophical Writings*, vol. 2 (1893–1913), edited by the Pierce Edition Project, pp. 4–10. Bloomington: Indiana University Press.

Peirce, Charles Sanders. 1931–1967. *Collected Papers of Charles Sanders Peirce*. 4 vols. First printing 1931–1933, third printing 1965–1967. Edited by Charles Hartshorne and Paul Weiss. Cambridge, MA: Belknap Press of Harvard University Press.

Peirce, Charles Sanders.1976. *The New Elements of Mathematics*. 4 vols. Edited by Carolyn Eisele. The Hague: Mouton.

Peirce, Charles Sanders. 1992. *The Essential Peirce: Selected Philosophical Writings*. Vol. 1 (1867–1893). Edited by Nathan Houser and Christian Kloesel. Bloomington: Indiana University Press.

Peirce, Charles Sanders. 1998. *The Essential Peirce: Selected Philosophical Writings*. Vol. 2 (1893–1913). Edited by the Pierce Edition Project. Bloomington: Indiana University Press.

Pietarinen, Ahti-Veikko. 2010. Pragmaticism as an Anti-foundational Philosophy of Mathematics. In *Philosophical Perspectives on Mathematical Practice*, edited by Bart Van Kerkhove, Jonas De Vuyst, and Jean Paul Van Bendegem, pp. 155–184. Milton Keynes, UK: College Publications.

Reck, Erich H., and Michael Price. 2000. Structures and Structuralism in Contemporary Philosophy of Mathematics. *Synthese* 125, 341–383.

Riemann, Bernhard. [1854] 1892. Über die Hypothesen, welche der Geometrie zu Grunde liegen. In *Bernhard Riemann's Gesammelte Mathematische Werke und*

Wissenschaftlicher Nachlass, edited by Heinrich Weber, pp. 272–287. Leipzig: B.G. Teubner.

Roberts, Don. 1973. *The Existential Graphs of Charles S. Peirce*. The Hague: Mouton.

Shields, Paul. 1997. Peirce's Axiomatization of Arithmetic. In *Studies in the Logic of C.S. Peirce*, edited by Nathan Houser, Don D. Roberts, and James van Evra, pp. 43–52. Bloomington: Indiana University Press.

Shin, Sun-Joo. 2002. *The Iconic Logic of Peirce's Graphs*. Cambridge, MA: MIT Press.

Short, Thomas L. 2007. *Peirce's Theory of Signs*. New York: Cambridge University Press.

Stjernfelt, Frederik. 2007. *Diagrammatology: An Investigation on the Borderlines of Phenomenology, Ontology and Semiotics*. Dordrecht: Springer.

Zalamea, Fernando. 2010. A Category-Theoretic Reading of Peirce's System: Pragmaticism, Continuity, and Existential Graphs. In *New Essays on Peirce's Mathematical Philosophy*, edited by Matthew E. Moore, pp. 203–233. Chicago: Open Court.

11
Poincaré and the Prehistory of Mathematical Structuralism

Janet Folina

1. Introduction and Historical Background

"Structuralism" denotes a family of views united by a common conception of the subject matter of mathematics. According to this approach, mathematics is not "about" mathematical objects, such as the number 2; nor is it even about specific mathematical systems, such as the Dedekind-Peano natural numbers. Rather, mathematics is about something more abstract: mathematical structure.

As a fully-fledged philosophy of mathematics, structuralism is young. Its birth is associated with Benacerraf's famous "What Numbers Could Not Be," and its current form has been shaped by subsequent work by Hellman, Resnik, Shapiro, and others.[1] Well before any of this recent philosophical work, however, a more general structuralist conception emerged from mathematical practice, that is, from mathematicians reflecting on their methodology and subject matter. We can call these earlier views "methodological" structuralism (Reck and Price 2000). Some questions about methodological structuralism include the following. How far into past mathematics does it go? How does philosophical structuralism arise from it? Is there really a sharp distinction between methodological and philosophical structuralism?

For this chapter I take for granted that one can distinguish methodological structuralism from further philosophical views about mathematical structures. The former is a simpler, working view about the general subject matter and methodology of mathematics, independent of any specific metaphysical and/or epistemological views about structures. For example, it is a commonplace that the development of non-Euclidean systems made geometry more abstract: the subject matter ascended to a more general perspective, to accommodate multiple geometric systems. Basic methodological structuralism solves any concern about this change by viewing geometry as the study of (possible) geometric structure

[1] See Benacerraf (1965), Hellman (1989), Resnik (1997), and Shapiro (1997) for a start.

Janet Folina, *Poincaré and the Prehistory of Mathematical Structuralism* In: *The Prehistory of Mathematical Structuralism*. Edited by: Erich H. Reck and Georg Schiemer, Oxford University Press (2020). © Oxford University Press.
DOI:10.1093/oso/9780190641221.003.0011

rather than as the attempt to provide a (true) theory of space. Another example concerns the way symbolical algebra developed out of initiatives for teaching calculus in Britain in the 19th century. Detaching the calculus symbols from any particular interpretation is a move toward abstraction that, again, fits well with a structuralist view of the subject matter. In both of these cases, however, neither the shifting subject matter nor its mathematical treatment depended on or derived from philosophical structuralism, such as specific views about the metaphysical nature of structures.

That said, it seems clear that methodological structuralism has been in the air for quite some time. Indeed, some of the success of key historical mathematical figures can be correlated with this new way of thinking about their subject matter. For example, Hankel writes in 1867 that mathematics is

> purely intellectual, a pure theory of forms, which has for its object not the combination of quantities or their images, the numbers, but things of thought to which there could correspond effective objects or relations, even though such a correspondence is not necessary. (Kline 1972, 1031)

Earlier, Gauss articulates a similar view:

> One quantity in itself cannot be the object of a mathematical investigation; mathematics considers quantities only in their relation to one another. . . . Now, mathematics really teaches general truths concerning the relations of quantities. (Gauss 1829, paragraphs 2, 3)

And even earlier, in the mid-18th century, both D'Alembert and Maclaurin express similar ideas in defending the calculus.[2] Both emphasize the *method* of calculus to justify its subject matter rather than vice versa (as, for example, Berkeley [1734] appears to have demanded). And the method highlights relations—supported by the clear conception of, and evidence for, those relations—over the existence and nature of specific types of objects. The term "methodological" structuralism is thus apt.

One might object that emphasizing mathematical relations is not structuralism, since relations are too specific. For example, though the "greater than" relation between numbers is different from the "older than" relation between

[2] For example, Maclaurin writes: "The mathematical sciences treat of the relations of quantities to each other, and of . . . every thing of this nature that is susceptible of a regular determination. We enquire into the relations of things, rather than their inward essences, in these sciences. . . . It is not necessary that the objects of the speculative parts should be actually described, or exist without the mind; but it is essential, that their relations should be clearly conceived, and evidently deduced" (1742, Ewald, 116; 51 in original).

possible physical objects, such differences are beside the point for a structuralist. Structuralism abstracts not only from the referents of singular terms (objects) but also from the meanings of relational terms and properties. So an emphasis on mathematical relations does not add up to structuralism.

However, both emphasizing relations over objects, and focusing on method over content, reference, and meaning, are important steps toward a more robust structuralism. The idea that a theory can be grounded in *what it does* rather than *what it is about* is significant, for it naturally leads to the view that mathematics need not have any particular (object level) subject matter. Finally, some (also beyond Hankel)[3] explicitly connect the relational nature of mathematics to the view that it is about "form" rather than content.[4]

I cannot argue for these general claims, nor am I attempting to answer a historical question with any precision. Whether a cause or a response, methodological structuralism emerges from a close connection to mathematical practice; and it goes back at least to early defenses of the calculus. So it was clearly in the air before Poincaré.[5]

Like many others, Poincaré expresses this basic structuralist view:

> Mathematicians study not objects, but relations between objects; the replacement of these objects by others is therefore indifferent to them, provided the relations do not change. The matter is for them unimportant, the form alone interests them. ([1902] 1952, chap. 2, 44)

His point concerns the subject matter of mathematics—that which mathematicians "study." The emphasis on form as well as relations justifies classifying him as at least a basic methodological structuralist.

As we'll see, however, Poincaré professes further philosophical views about the *nature* of mathematical structures. These are clearest in his remarks about groups and the group concept. I will argue that his conception of mathematical intuition can be understood similarly. That is, Poincaré's views about the nature and knowledge of mathematical structure are philosophically significant, extending beyond methodological structuralism.

3 Compare, e.g., the article on Grassmann by Paola Cantù in the present volume.

4 For example, Gauss's emphasis on relations between *quantities* might seem to cast him as a mere pre-structuralist. But Gauss also writes in defense of complex numbers: "The mathematician abstracts totally from the nature of the objects and the content of their relations; he is concerned solely with the counting and comparison of the relations among themselves" (1831, paragraph 22). Gauss thus emphasizes abstraction generally—from relations as well as objects. Mathematics is about neither individual objects nor particular relations; it makes comparisons between relations, yielding insights into relation-forms. The view is thus genuinely structuralist. For more on Gauss, see the chapter by Ferreirós and Reck in the present volume.

5 Others whose work seems in the structuralist spirit around this time include Serret (1819–1885), De Morgan (1806–1871), and Galois (1811–1832).

I will proceed as follows. In section 1, I focus on Poincaré after briefly sketching some basics on structuralism in the philosophical literature. I aim to show that Poincaré endorses structuralist ideas, arguing further that in some ways his view best aligns with a "strong" version of structuralism, the "structure first" view. In section 2 I address the question of how extensive this strong view is and whether it is consistent with some of Poincaré's other philosophical commitments. Thus, before concluding, I consider his semi-Kantian views about mathematical intuition, his structural realist views in natural science, and whether they can all be consistently combined with the strong, "structure first" view about some aspects of mathematics.

2. Structuralism: A Basic Taxonomy and Poincaré's Place in It

2.1. Some Basic Types

Structuralism is the view that mathematics is about abstract structures rather than specific mathematical objects or even specific systems of objects. For example, consider the natural number 3 as defined by Zermelo. Contrast this with the set of Zermelo natural numbers, and also with the natural number structure. The latter can be thought of as the form of all systems of natural numbers, regardless of how the particular systems and objects are defined or construed. The natural number structure is what the Zermelo numbers, the von Neumann numbers, and the Frege numbers have in common. Structuralism is the view that mathematics is about this sort of thing.

As noted, this view about the subject matter of mathematics does not entail any specific metaphysical views about the *nature* of that subject matter, nor about how we *know* mathematical structures. Thus, the basic structuralism/nonstructuralism distinction is different from realism/anti-realism disputes about the (independent) existence of mathematical objects. A structuralist may or may not think mathematical objects exist independently of mathematicians; she also may or may not think structures so exist. One can also be a realist, or Platonist, about structures, believing that they exist independent of the minds/constructions of human mathematicians. Alternatively one can be a Platonist about mathematical objects and systems, but not abstract structures. One can of course also be anti-realist about all abstract objects. So basic structuralism is simply a view about the subject matter of mathematics, remaining neutral about the nature of structures.[6]

[6] Independent or dependent, and if the latter, dependent *on what*.

Philosophical structuralism, on the other hand, aims to articulate and defend some of these further properties. It is important to note, however, that the differences between structuralist philosophies are generally not explained in terms of the familiar issue of the dependence or independence of mathematical objects/reality on *mathematicians* (and their constructions, proofs, etc.). Instead, these further views are typically characterized in terms of the relationship between mathematical structures on the one hand and the mathematical *objects/systems* that instantiate them on the other. That is, people don't play any immediate role in a typical basic structuralist taxonomy.

For example, Shapiro explains three main structuralist views as follows:

> Any of the usual array of philosophical views on universals can be adapted to structures. One can be a Platonic *ante rem* realist, holding that each structure exists and has its properties independent of any systems that have that structure. On this view, structures exist objectively, and are ontologically prior to any systems that have them (or at least ontologically independent of such systems). Or one can be an Aristotelian *in re* realist, holding that structures exist, but insisting that they are ontologically posterior to the systems that instantiate them. Destroy all the natural number systems and, alas, you have destroyed the natural number structure itself. A third option is to deny that structures exist at all. Talk of structures is just a convenient shorthand for talk of systems that have a certain similarity. (Shapiro, n.d., Part 1)

Like ordinary Platonism, Platonic *ante rem* structuralism espouses a kind of realism about structures in that the structures are independent of both objects and systems of objects for their existence. So *ante rem* structuralism, one might say, simply adds another kind of universal to the old Platonic universe: mathematical structure.

A common analogy to explain this view involves the distinction between places or offices and the objects that can occupy those places. For example, reference to the US president might be to a particular person, as in "The president is tired." But it may also refer to the position independent of who occupies it, as in "The president heads the executive branch of the government."

With this distinction in mind, *ante rem* structuralists generally view mathematical assertions as more like the latter than the former—as assertions about offices rather than occupants of those offices. Furthermore, the truth-value of such assertions is held to be indifferent to whether or not the places in the structure *have* occupants. So, for example, arithmetic studies the natural number structure, which exists independently of any individual natural numbers as well as any particular systems (definitions) of the numbers. For the Platonic *ante rem* structuralist, abstract structures are what concern mathematicians.

Although Shapiro considers *in re* structuralism another (though weaker) form of realism about structures, it also bears some similarity to ordinary constructivism in the philosophy of mathematics. Constructivists think mathematical objects exist, but only dependently—on the constructions carried out by mathematicians. Similarly, on the *in re* view, structures exist, but only dependently—on the existence of systems of mathematical objects instantiating those structures. (So systems rather than human constructions.) Whereas the *ante rem* view asserts the ontological *priority* and *independence* of structures from objects and systems, the *in re* view asserts the ontological *posteriority* of structures and their *dependence* on mathematical objects/systems. For example, the *in re* view is that "2 + 3 = 5" is a truth about the natural number structure because it is true of *any occupants* of the "offices" 2, 3, and 5.

The third main option, eliminativism (e.g., fictionalism and modal structuralism), is a form of anti-realism, or nominalism, about structures. On this view, structures don't actually exist. Whether or not mathematical objects or systems exist independently of mathematicians, talk of mathematical structure is simply a convenient way to speak.

The point for us is that these possible views about structure are differentiated with respect to underlying mathematical objects/systems, rather than mathematicians and their activities. Issues of dependence or independence thus do not correspond to the ordinary Platonism-constructivism debates in the philosophy of mathematics. Even the eliminativist option is expressed as an anti-realism or constructivism about structure only; it appears that one could be fictionalist about structure and realist about particular mathematical systems or objects. With this in mind I will argue that in this taxonomy, Poincaré's views about mathematical structure most closely match Shapiro's *ante rem* category. With the *ante rem* structuralist, Poincaré advocates the priority and independence of some structures to their systems—despite the fact that he is a constructivist, not a Platonist, about mathematics. This "structure first" view is why I consider him as holding a position one might call "constructivist *ante rem* structuralism." (I will return to the apparent incongruity of this position, in section 2.)[7]

[7] Because of this interpretation, my argument will involve pointing out that the taxonomy referenced here is incomplete. (This is not a criticism; Shapiro did not claim to provide a complete taxonomy.) As noted, since the issues of priority and independence are articulated relative to other mathematical objects, they don't engage in the ordinary discourse regarding realism versus constructivism. In particular, the priority of structures over systems seems detachable from metaphysical realism; that is, one need not be a Platonist to endorse the *ante rem* relationship between structures and systems.

2.2. Poincaré as a Structuralist

An important mathematician during a significant time (1854–1912), Poincaré reflected on the increasing abstraction of mathematics and its impact on both the subject matter and our knowledge of it. Structure is a central concept guiding his understanding of these changes. The two main purposes of this section are (i) to clarify the nature of his structuralist views, and (ii) to show that they were not casual, or tangential to the rest of his philosophy. His views about structure are entwined with several themes in his philosophy of mathematics. We will begin with some general structuralist sympathies, which emerge from his reflections on mathematical understanding. We will then work toward more specific, and stronger, philosophical views about the nature of mathematical structures, which appear in his thoughts about geometry and group theory. Like Shapiro, Poincaré contrasts two main philosophical views about the group structure in terms of whether or not it should be considered as prior to, and independent of, its relevant mathematical systems. Also with Shapiro, and somewhat surprisingly, Poincaré explicitly endorses the priority view about the group structure.

2.2.1. Remarks on Mathematical Understanding

Poincaré famously comments on mathematical understanding and insight, referring to phenomena such as "seeing the whole" and the view "from afar." Some of these are vague, negative remarks against the role of logic in mathematics (sometimes against logicism more specifically), while others seem more positive, as genuine attempts to articulate the nature of mathematical understanding. That logical reasoning alone does not constitute understanding seems obvious. The hard task is to say what more is needed.

Starting with the negative, Poincaré complains that "the logician cuts up, so to speak, each demonstration into a very great number of elementary operations." But as we all know, following individual inference steps does not amount to understanding even a straightforward proof: "we shall not yet possess the entire reality; that I know not what, which makes the unity of the demonstration" (1900, V, 1017). His point is that understanding a proof is something over and above understanding the individual inferences. Obviously, making individual proof-inferences will not suffice for creating new mathematics; here Poincaré asserts that the same holds even for understanding an existing proof.[8]

[8] This view may call to mind Wittgenstein's remarks about proofs needing to be surveyable (see Wittgenstein 1989). However, Poincaré does not propose surveyability as a requirement for *proofs*; he connects it only to *understanding* proofs. (Of course, interpreting Wittgenstein on this and similar issues is not simple.)

He provides a famous analogy to make this point about the holistic nature of mathematical understanding: "A naturalist who never had studied the elephant except in the microscope, would he think he knew the animal adequately?" ([1908] 1982, Book 2, chap. 2, sec. 6, 436). In addition to making a part-whole contrast, Poincaré is alluding to a "big picture" or the "forest for the trees" idea. Analyzing elephant cells does not provide understanding of the animal as a whole, an understanding that can be gained only by observing the living animal. Similarly, it is not that Poincaré saw no value in attending to local logical inferences; rather, his point is that this type of focus does not *suffice* for—or constitute the whole of—understanding a proof, or a mathematical fact. "This view of the aggregate is necessary for the inventor; it is equally necessary for whoever wishes really to comprehend the inventor. Can logic give it to us? No" (1900, V, 1018).

What can give a view of the whole, if not logic? That is much harder to articulate. In these and similar passages Poincaré sets up a contrast between *rigor* and *understanding* in mathematics. "Rigor" here means focusing on the "parts"—the formal, symbolic definitions, explicit deductive inferences, etc. "Understanding," in contrast, is presented as something that involves the "whole," something that transcends rigor concerning the parts. This includes grasping the unity of a proof (1900, V), the historical origins of precise definitions (1900, IV; [1908] 1982, Book 2, chap. 2), the point of a mathematical question (1900, IV), and the ability to invent (1900, V). The claim is that to understand and create new mathematics, one needs this perspective of the whole.

At times Poincaré mentions intuition in this context. "We need a faculty which makes us see the end from afar, and intuition is this faculty" (1900, V, 1018). The appeal to intuition here may seem psychologistic, and certainly it is distinct from his semi-Kantian appeal to mathematical intuition (which will be addressed later). Though Poincaré's remarks are vague, both the metaphors and the reference to intuition point to the idea of transcending individual results and local logical inferences. How is this related to structuralism?

Consider, for example, the metaphor of "seeing from afar"; plausibly this includes the ability to connect distinct results and even different areas of mathematics. To do so requires a more abstract, higher-level, perspective—a perspective that seems generally structural. At the very least, structure is something that different areas of mathematics *can* have in common. For example, as we will later see, Poincaré cites the group structure as what is common to various mathematical systems. And in a chapter on the relation between mathematics and physics he writes rather poetically:

> What has taught us to know the true, profound analogies, those that the eyes do not see but reason divines?

> It is the mathematical spirit, which disdains matter to cling to pure form. This it is which has taught us to give the same name to things differing only in material, to call by the same name, for instance, the multiplication of quaternions and that of whole numbers. . . . He sees best who stands highest. ([1905] 1958, chap. 5, II, 77–78)

The view from afar, or above, is where one can "see" structural, relational similarities between systems "differing only in material."[9]

Structure also underpins the perception of beauty according to Poincaré, which, in turn, supports understanding. When we perceive the beauty of a piece of mathematics, he thinks, we understand it better, and vice versa. Further, he argues that mathematical beauty involves the more primitive properties of order and unity. Good ideas, the impression of elegance, the use of analogy, and the importance of generality all depend on perceptions of order and unity, in his view. Successful creative work, he argues, is guided (consciously or unconsciously) by "the feeling of mathematical beauty, of the harmony of numbers and forms, of geometric elegance" ([1908] 1982, Book 1, chap. 3, 391). For him, these "feelings" are all grounded in unity and order, which is both aesthetically pleasing and useful in "guiding" the mind to fruitful results. This view—that perceptions of order and unity facilitate understanding—fits well with basic structuralism, since structure is an organizing tool. Poincaré's conception of mathematical understanding is thus harmonious with the view that mathematics is (at least largely) about abstract structure.[10]

2.2.2. The Subject Matter of Mathematics

In addition to the epistemic view that the perception of structure facilitates mathematical understanding, Poincaré also expresses structuralist views about the subject matter of mathematics. In fact these claims seem parallel. Just as the general "overview" perspective is essential for mathematical understanding, so a "bigger picture" perspective is critical for the subject matter, since the significant results are general and have broad scope. In addition, as we saw earlier, Poincaré makes the standard structuralist point that mathematics is "about" relations rather than objects. His emphasis on both general truths over particular truths, and relations over particular objects, reflects a structuralist vision of the subject matter and methodology of mathematics.

[9] One may perceive here an early expression of something like the category-theoretic perspective. For more on the path toward category theory, see the chapters on Noether (by Audrey Yap), Bourbaki (by Gerhard Heinzmann and Jean Petitot), and Mac Lane (by Colin McLarty) in this volume.

[10] I attempt to more fully address this connection between mathematical structure and understanding in Folina 2018.

Regarding the importance of generality, he remarks: "So a chess player, for example, does not create a science in winning a game. There is no science apart from the general" (1894, II, 975). Further, he aligns generality in mathematics with infinity, claiming that without the idea of mathematical infinity "there would be no [mathematical] science, because there would be nothing general" (1894, V, 979).

About relations, he famously says of Dedekind cuts:[11]

> Mathematicians study not objects, but relations between objects; the replacement of these objects by others is therefore indifferent to them, provided the relations do not change. The matter is for them unimportant, the form alone interests them.
>
> Without recalling this, it would scarcely be comprehensible that Dedekind should designate by the name incommensurable number a mere symbol, that is to say, something very different from the ordinary idea of quantity, which should be measurable and almost tangible. ([1902] 1952, chap. 2, 44–45)

The first paragraph seems a canonical statement of methodological structuralism. However, what about the negative tone of the second paragraph?

The context is relevant. In this section of the chapter Poincaré not only explicates, he also criticizes, work he considers reductionistic.[12] For example, he is unhappy about attempts to define the continuum "without using any material other than the whole number" (44). He objects as follows after explaining Dedekind cuts.

> But to be content with this would be to forget too far the origin of these symbols; it remains to explain how we have been led to attribute to them a sort of concrete existence, and, besides, does not the difficulty begin even for the fractional numbers themselves? Should we have the notion of these numbers if we had not previously known a matter that we conceive as infinitely divisible, that is to say, a continuum? (45–46)

Poincaré is thus not just *citing* Dedekind cuts as an example of the fact that mathematics is about abstract structure; he is *criticizing* Dedekind cuts as a theory of the real numbers. He seems to regard it as too abstract and too formal, or

[11] For Dedekind, including a further discussion of this remark by Poincaré about his use of cuts, see again the chapter by Ferreirós and Reck in this volume.

[12] He cites the "Berlin school" here, and "Kronecker in particular," for these sins, but then goes on to discuss Dedekind in some detail (who is usually not considered as having belonged to the Berlin school). His point seems to have been against reductionist programs generally without differentiating between the various motives and origins of specific projects.

symbolic.[13] At least, it is insufficient if a theory should provide understanding. Since structuralism is associated with the increasing abstraction and formalization of the content of mathematics, this critique of Dedekind (and similar projects) may appear to weaken the structuralist interpretation of Poincaré. Let me flesh out this concern before attempting to assuage it.

In contrast with a formal/symbolic theory of the real numbers, as one might regard Dedekind cuts, Poincaré insists that understanding the mathematical continuum comes from our relating it to experience. He makes the following familiar argument: we experience a physical continuum, but this leads to contradictions owing to our limited senses. That is, we can experience three lengths or weights as A = B and B = C. But we can also tell that A is longer or heavier than C; so A > C; but now this is inconsistent. We solve this by citing the limited nature of our sense perceptions; so we suppose that even though A seemed the same weight as B and B seemed the same weight as C, at least one of these measurements was not quite right. That is, we *conceive* the things measured in terms of quantities that are further divisible—beyond our capacities to sensibly distinguish such differences. In addition, once we interpolate between two given quantities, "we feel that this operation can be continued beyond all limit" ([1902] 1952, chap. 2, 48). We thus suppose the operations are indefinitely iterable, which leads to the conception of everywhere dense sets like the rationals. Irrationals are then postulated owing to theoretical gaps in the rationals. Poincaré states the point thus: "the mind has the faculty of creating symbols. . . . Its power is limited only by the necessity of avoiding all contradiction; but the mind only makes use of this faculty if experience furnishes it a stimulus thereto" (49). Returning to our issue, the problem with Dedekind cuts is that the theory can be presented completely in its abstract, formal guise; and this would make the real number system seem independent of both geometry and experience. But analysis emerges from the union of geometry with the needs of physics and arithmetic, which the subject matter should reflect.[14]

In fact, Poincaré was generally suspicious of mathematics that is detached from history and experience, and when applied to this case, this may seem contrary to structuralism or the structuralist enterprise.[15] For example, regarding nowhere differentiable continuous functions he says: "Instead of seeking to reconcile analysis with intuition, we have been content to sacrifice one of the two, and as analysis must remain impeccable, we have decided against intuition"

[13] Poincaré *was* anti-logicist, but his concern here seems more general. He is questioning our ability to understand formal, symbolic definitions without supplementary information, whether or not they are part of a logicist program.

[14] His critique implies that he thought the connections to experience must go beyond merely motivating or teaching the theory.

[15] Of course in another sense, attention to the "larger" view is an expression of structuralism—as just argued.

([1902] 1952, chap. 2, 52). The remark is clearly a complaint or a regret; it is not merely a *report* of mathematical progress by increasing formalization. I think it is fair to say that Poincaré was a bit ambivalent about some of the changes in mathematics associated with the development of the structuralist viewpoint.

We can now return to our concern, which is that Poincaré's canonical statement of methodological structuralism is accompanied by critical remarks about the central example of Dedekind cuts.[16,17] More evidence and more particulars regarding the nature of his structuralist commitments will clarify and improve our case that Poincaré genuinely embraced structuralism. Geometry provides this support.

2.2.3. Geometric Conventionalism

Poincaré's conception of the subject matter of geometry provides some key, added evidence of his structuralism. Now, Shapiro connects both axiomatics and the idea of implicit definition with structuralism, singling out Hilbert's 1899 *Grundlagen der Geometrie* as "the culmination of a trend toward structuralism within mathematics" (sec. 1).[18] Why is the axiomatic method associated with structuralism? Because along with the idea of implicit definition, the axiomatic method *lifts* the subject matter of mathematics up to a higher, more abstract, level. For example, rather than thinking of geometry as having a single, object-level subject matter, which the axioms are *about*, the "axiomatic view" is that axioms are about *whatever* systems fulfill the criteria they jointly stipulate. So the axiomatic method changes our perspective from *single* subject matter to a *multiplicity*, or set, of possible interpretations. Indeed, the ascendance, or abstraction, of subject matter was important for 19th-century developments in both geometry and algebra.[19]

Poincaré clearly conceives the subject matter of geometry similarly to Hilbert—in this "elevated" way. He is famous for claiming that geometric axioms are implicit definitions, or conventions:

[16] To elaborate a bit on Poincaré's mixed feelings in these passages: he recognized that the more formal, abstract "structuralist" perspective enables advances in mathematics, so it is crucial. But he also recognized that abstraction makes even ordinary mathematics harder to understand. Thus, more formal, symbolic methods are desired at times; but these must be supplemented to facilitate "understanding."

[17] One might respond here by pointing out that these criticisms of Dedekind are strictly *epistemic*; they do not undermine the general structuralist view regarding the *subject matter* of mathematics. Nevertheless, more evidence will solidify my interpretation. Additionally, the epistemology of mathematics should cohere with its subject matter.

[18] I assume that "structuralism within mathematics" is (essentially) what I (after Erich Reck and others) have been calling "methodological structuralism."

[19] This is of course historically complex and interesting, though I cannot here address it further.

> The axioms of geometry, therefore, are neither synthetic *a priori* judgments nor experimental facts.
>
> They are conventions; our choice among all possible conventions is guided by experimental facts; but it remains free and is limited only by the necessity of avoiding all contradiction. . . .
>
> In other words, the axioms of geometry (I do not speak of those of arithmetic) are merely disguised definitions.
>
> Then what are we to think of that question: Is the Euclidean geometry true?
>
> It has no meaning.
>
> As well ask whether the metric system is true and the old measures false. ([1902] 1952, chap. 3, 65)

That geometry is based more on choice than truth expresses the fact that like Hilbert, Poincaré viewed geometric axioms as "disguised" or implicit definitions. The comparison to measurement systems—for which the main criteria for acceptability are consistency and convenience rather than truth—shows that Poincaré sees geometry as at least partly detached from a truth-determining subject matter.

It may be worth elaborating a bit on the differences between geometry and arithmetic, as Poincaré saw it. To him, arithmetic has an intuitively grounded subject matter. In contrast, intuition does not anchor geometry in a similar way.[20] We have no direct intuition of points: "What is a point of space? Everybody thinks he knows, but that is an illusion" ([1902] 1952, chap. 5, 89–90). Nor does intuition decide what is straight: "I grant, indeed, that I have the intuitive idea of the side of the Euclidean triangle, but I have equally the intuitive idea of the side of the non-Euclidean triangle. Why should I have the right to apply the name of straight to the first of these ideas and not to the second?" ([1905] 1958, chap. 3, I, 37–38). This is why conventional choices, or implicit definitions, are necessary. We decide which axiom system is most convenient, and this decision determines what lines will be considered straight when using that system; that is, the axiom system is the implicit definition of "straight line." In this way, geometry is no longer seen as *reflecting* a single definite subject matter (though it was once so regarded).

What we might call "mathematical geometry" thus occupies a more abstract, structural, position than ordinary Euclidean geometry. In contrast, ordinary geometry—working within a particular geometric system—is in a sense closer

[20] This is not to say intuition does not anchor geometry at all. Indeed geometry is supported both by the intuitive continuum and by the intuition of indefinite iteration. (He even refers to "geometric intuition" in later work (e.g., [1913] 1963, 26–27and 42–44). The difference is that, unlike arithmetic, intuition does not yield any particular geometric system as *true*.

to applied mathematics, since it lies at a lower level of abstraction. In any case, like Hilbert, Poincaré's conception of geometric axioms as implicit definitions provides further evidence of his structuralist perspective, at least regarding geometry.

Before moving on, I'll note that in addition to his view of axioms as implicit definitions, Poincaré also calls certain key concepts "implicit axioms." Here he draws attention to the existence of concepts or principles that unite different systems. While changing an axiom creates a different axiom system, Poincaré's point is that despite the differences, there are often important connections, or relations, between the systems ([1902] 1952, chap. 3, 60–62). For example, rigid body motion is presupposed by several geometric systems; but its possibility is neither self-evident nor analytically true. So Poincaré considers rigid body motion to be an "implicit axiom," in that it acts as a unifying principle for the geometries of constant curvature. As we shall see, the group concept plays a similar role. My point is that it is not only the *differences* between systems—e.g., the proliferation of geometries—that can be linked to the structuralist perspective. Emphasizing the *links* between the different systems—the unifying concepts and principles—also expresses structuralism. Indeed, in my view, Poincaré's emphasis on unifying concepts provides an even stronger connection to structuralism than his view of axiom systems as implicit definitions. Let us now turn to a key example—that of the unifying concept of group.

2.2.4. The Group Concept

In addition to conceiving the geometric axioms as implicit definitions, Poincaré further emphasizes geometric *form*, and the group concept is central here. An interesting twist is that the group concept is a priori for Poincaré—not as an intuition, or form of sensibility, but as a "form of our understanding" ([1902] 1952, chap. 4, 79). It underpins our ability to conceive geometry from the more abstract perspective, which, in turn, helps us make sense of multiple possible geometric systems.

> What we call geometry is nothing but the study of formal properties of a certain continuous group; so that we may say, space is a group. The notion of this continuous group exists in our mind prior to all experience; but the assertion is no less true of the notion of many other continuous groups; for example, that which corresponds to the geometry of Lobatchevski. (1898, Conclusions, 1010)

A variety of continuous groups are a priori possible, and are studied in mathematics. The choice for a theory of physical space (Euclidean or non-Euclidean,

3-dimensional or 4-dimensional) is conventional, depending on experience, science, and other factors.[21] Because the mathematics behind geometry is common to a variety of options, it—*mathematical* geometry—lies at a more abstract level than work within particular geometric systems. The group concept, and the idea that the various geometries are simply different continuous groups, facilitates the "ascendance" to this more abstract mathematical perspective. So Poincaré's appeal to the group concept, and its role in articulating the abstract perspective of mathematical geometry, provides even more evidence of his methodological structuralism.

Crucially, however, Poincaré's view about groups goes further. It not only furnishes a clear statement of basic structuralism; it also includes properly philosophical views about the metaphysical nature of groups as well as our knowledge of them. Thus, his view here clearly advances beyond methodological structuralism to a more philosophical position about (at least some) mathematical structure.

One addition is an epistemological claim, noted previously. The apriority of the group concept, and the view that this a priori status provides mathematics with a unifying ideal, transcends basic methodological structuralism (which mainly concerns the general subject matter of mathematics). For Poincaré, the group concept provides a perspective from which to consider, compare, and unify the different geometries (as well as other structures). His views about the group concept thus address the epistemology of geometry, in its new, more abstract, guise.

But he also makes an ontological claim. That is, in addition to the apriority, and unifying role, of the group concept, Poincaré adds that the group structure is *prior to* the systems falling under it. The following remark, in particular, shows him asserting a view similar to Shapiro's "*ante rem*" structuralist (in the "structure first" sense):

> We must distinguish in a group the form and the matter. For Helmholtz and Lie the matter of the group existed previously to the form. . . . The number of dimensions is therefore prior to the group. For me, on the contrary, the form exists before the matter. The different ways in which a cube can be superposed upon itself, and the different ways in which the roots of a certain equation may be interchanged, constitute two isomorphic groups. They differ in matter only. The mathematician should regard this difference as superficial, and he should no more distinguish between these two groups than he should between a cube

[21] There is a large literature on this; for example see the anthology de Paz and DiSalle (2014).

> of glass and a cube of metal. In this view the group exists prior to the number of dimensions. (1898, Form and Matter, 1009–1010)

That the important mathematical properties of geometry concern form rather than matter is simply methodological structuralism. But that the "form exists before the matter" is a much stronger view, one that coincides with the "structure first" view of *ante rem* structuralism. For form to exist *before* matter—for it to be prior— it must also be *independent* of matter or specific systems.

For Poincaré, the group structure, or form, is relatively independent—independent of the mathematical objects or systems exemplifying it. Now, Poincaré was not a realist about mathematical existence, so his view is not that of (Shapiro's) Platonic *ante rem* structuralism. Yet his view perfectly matches the independence and priority aspects of the *ante rem* view. Because the group concept is a priori, the group structure is epistemically prior to any particular group. And because, as he asserts, the form of a group exists prior to any specific group, the group structure is prior in (some sense of) existence as well. Detached from Platonism, the category of *ante rem* structuralism simply indicates the *relative* independence and priority of structures to their systems and objects. As we just saw, this is precisely what Poincaré asserts about the group structure.[22]

3. Structuralism and Other Issues in Poincaré's Philosophy

Supposing that Poincaré's view of groups matches that of *ante rem* structuralism, how does this fit with his other philosophical commitments? Are there any other structures that come "first," or are groups unique on this matter? Was Poincaré consistently anti-realist in his philosophy? If so, how does his mathematical constructivism, or anti-realism, combine with this view about groups; that is, can one be a constructivist and still think that any mathematical structures are prior to and independent of their instances?

I will now take up these last three questions, each in its own section. Starting with the nature and extent of Poincaré's constructivism, I will present his semi-Kantian conception of mathematical intuition as structuralist. That is, I'll argue that intuition for Poincaré regards mathematical structures, and moreover, that the intuitive structures come "first" in a way similar to that of the group structure. Intuition on my reading thus adds to the stock of abstract structures that come "first." The second question is whether or not Poincaré is consistently anti-realist. Here I note that his philosophy of natural science is generally considered

[22] Again, I will come back to this issue subsequently.

a form of structural realism. I argue, however, that his views about mathematical structures and those about natural structures are independent of one another. So his realism about some scientific structures is consistent with his general constructivist, or anti-realist, position on mathematical existence.[23] Last, I address the apparent tension between his anti-realism about mathematics, on the one hand, and the view that mathematical structures can come "first," on the other; that is, I try to make sense of how one can be a "constructivist *ante rem* structuralist."

3.1. Intuition and Structuralism

It may be easy to understand how group theory fits with structuralism, but intuition may seem more puzzling. As for Kant, Poincaré sees intuition as necessary for both the subject matter of mathematics and our knowledge of it. Yet, as argued earlier, he also endorses a structuralist view of the subject matter of mathematics. If intuition governs the content and our knowledge of mathematics, and mathematics is about structure, then intuition must provide insight into structure. Intuition for Poincaré delivers an epistemology of mathematics that complements the new structuralist conception of its subject matter.

Poincaré is clearly a "constructivist" given his repeated claims to support a semi-Kantian conception of mathematics, including mathematical intuition. But his view is distinctive. For Kant, intuition in mathematics is spatiotemporality, the a priori form of all experience; there is, for Kant, no specifically *mathematical* intuition. Furthermore, the role of intuition in mathematics is quite complicated, having to do with the necessity of input from space and time to instantiate the mathematical concepts via a process he calls "construction of concepts."

Like Kant, Poincaré defends two a priori intuitions in relation to mathematics; but instead of space and time, he cites the intuition of the continuum and the intuition of indefinite iteration. These are more abstract and closer to mathematical intuitions. In some ways they are like "stripped down" versions of space and time: spatiotemporality minus most of the sensorial aspects we associate with it. To put it another way, intuition for Poincaré is more cognitive and less connected to ordinary sense experience than space and time are for Kant. Also in contrast with Kant, there is no explicit reliance on "construction of concepts" in Poincaré.[24]

[23] More than consistent, I actually find the two views mutually supporting despite the appearance of a contrast between them.

[24] For Kant constructing concepts is distinctive of mathematical methodology, and means something quite specific having to do with considering/exhibiting arbitrary instances of mathematical

3.1.1. Arithmetic and Intuition

In arguments against programs like logicism, Poincaré emphasizes both the centrality of the principle of induction and its basis in intuition. As mentioned, unlike Kant, Poincaré appeals to indefinite iteration rather than time. But as in Kant, intuition's role in grounding mathematical knowledge makes that knowledge synthetic a priori. So the dependence of induction on intuition makes our knowledge of induction, as well as any knowledge it yields, synthetic a priori for Poincaré.

This is clear in his early writings on the nature of mathematical reasoning, where he argues that the power of mathematical reasoning—its ability to transcend the merely tautological—springs from the principle of induction. And since induction is grounded in intuition, it's really intuition that gives mathematical reasoning this power.

> Why then does this judgement force itself upon us with an irresistible evidence? It is because it is only the affirmation of the power of the mind which knows itself capable of conceiving the indefinite repetition of the same act when once this act is possible. The mind has a direct intuition of this power, and experience can only give occasion for using it and thereby becoming conscious of it. (1894, VI, 979–980)

The intuition of indefinite iteration is a mental capacity that allows us to conceive of certain sets, or certain sequences, by conceiving of the way they can be produced by us: iteratively, step by step. Poincaré's skepticism about transfinite cardinals, as well as any philosophy that accepts actual infinity, is related to this idea that sets must be conceived (if not strictly constructed) by envisioning our producing, or "running through," their elements in a stepwise fashion.

Though essential for conceiving discrete infinite sets, Poincaré's main explanation of the importance of indefinite iteration focuses on how we understand induction rather than how we construct the mathematical sets that induction might target. Further, his explanation of why intuition is required for induction articulates the intuition as insight into a certain type of mental process. It is the focus on *type* in these and surrounding arguments that supports the connection to structuralism. Poincaré claims the same intuition is the basis for inferences about very different sorts of objects. What unifies these, what the different systems of objects have in common, is their structure, or order type, not their

concepts in space and/or time. Mathematical "constructions" are central to Poincaré's view, but their role is less specific; sometimes "construction" simply means defining and perhaps comprehending. (See later in this chapter for one such use regarding the continuum.)

content.[25] Thus the main intuitive basis for arithmetic is that which provides cognition of structure. Let me explain this in a bit more detail.

For Poincaré, the paradigm case of knowledge that requires the intuition of iteration is inductive knowledge. "The essential characteristic of reasoning by recurrence [induction] is that it contains, condensed, so to speak, in a single formula, an infinity of syllogisms" (1894, V, 978). The puzzle is this: if induction implicitly contains an infinity of inferences, how can we recognize it as yielding truth? The answer for Poincaré is intuition. We can arrange some initial inferences as the following: property P is true of 0, and if true of n then true of n + 1, so property P is true of 1. Since P is true of 1, and if true of n then true of n + 1, the property P is also true of 2. And so on. Poincaré notes that when the inferences are ordered this way, they are "arranged in 'cascade' "; because of this arrangement we can see that they will continue to be valid indefinitely, or infinitely. And intuition, the mind's ability to conceive "the indefinite repetition of the same act when once this act is possible" (1894, VI, 979), is what provides the necessary insight into the infinite chain of modus ponens steps constituting the "cascade."

Not only do we see induction as thus leading to truth, we recognize it as necessary. Unlike empirical induction, "Mathematical induction, that is, demonstration by recurrence, on the contrary, imposes itself necessarily because it is only an affirmation of a property of the mind itself" (1894, VI, 980). Induction affirms the property of the mind associated with the intuition of iteration. Thus, the paradigm case of intuitive knowledge is not (acquaintance) knowledge *of* a series of objects, such as the natural numbers, but (propositional) knowledge *that* a type of inference preserves truth. The intuitive basis for arithmetic is thus more distant from sensibility than an account focused on the potential construction of finite objects (such as sequences of strokes for cardinal numbers). In short, it is more epistemic, more abstract, and less ontological.

Moreover, Poincaré defends his semi-Kantianism about arithmetic against challenges from logicism and formalism/axiomatics by arguing that different types of induction are really the same principle, all depending on the same intuition.[26] The kernel of the argument is expressed at least by 1894: "If we look closely, at every step we meet again this mode of reasoning, either in the simple form we have just given it, or under a form more or less modified" (1894, IV, 978). He also claims at this time that induction cannot be demonstrated in a non-circular way (1894, VI, 979), though he doesn't provide an argument for this until a later series of circularity objections to logicism (Poincaré [1905] 1996, [1906a] 1996, and [1906b] 1996).

25 Their order type being that of a simply infinite system.

26 That is, induction, or really iteration, acts as another unifying principle for Poincaré.

I cannot go into the details of his circularity arguments here.[27] Though his point at first regards the different ways we reason inductively about numbers, he later adds that *metatheoretic* uses of induction depend on the same underlying intuition of iteration.[28] If metatheoretic uses of induction and induction on numbers depend on the *same* intuition, intuition is formal, and independent of any particular content.

For example, in discussing any proof that the arithmetic axioms are consistent, he writes, "recourse must be had to procedures where in general it is necessary to invoke just this principle of complete induction which is precisely the thing to be proved" ([1905] 1996, IV, 1027). So a consistency proof *about* arithmetic uses "precisely" the same principle as when reasoning inductively *in* arithmetic. Since the two uses obviously will concern different *objects*, any precise sameness must regard form or type. His associating intuition so closely with a form of reasoning applicable to various domains, rather than (just) a method for constructing objects to constitute a domain, gives this intuition a more abstract feel. That is, similarly to how the group concept facilitates the ascendance to a more abstract way of thinking about geometry, intuition is appealed to here to explain a more abstract way of thinking about inductive reasoning and the domains to which it applies. In short, what grounds and guides induction is intuition of structure.

Indeed, Poincaré explains why this form of reasoning can be used in such different contexts by invoking structuralist terms—by reference to the common underlying structure as an "ordinal type."

> Thus one envisages a series of reasonings succeeding one another, and one applies to this succession, regarded as an ordinal type, a principle that is true for certain ordinal types, called finite ordinal numbers, and which is true for these types precisely because these types are by definition those for which it is true. ([1906a] 1996, XXIII, 1043)

Though the remark is a little cryptic, I take it as supporting a strong connection between intuition and structure. Whether it is induction about numbers, or about sequences of inferences, the reasoning depends on the fact that intuition enables both the understanding of induction and the cognition of the simply infinite systems to which induction can be applied. Without this intuition, this insight, we (finite thinking beings) would not be able to do mathematics about discrete infinite systems, and we would not be able to see that the principle of induction is true.

[27] But see Folina (2006) and Goldfarb (1985) for opposing views on these arguments.
[28] This is how he justifies the earlier circularity claim.

On my interpretation, then, Poincaré thought of the natural number *structure* as—like the basic group structure—more fundamental than any particular system of natural numbers. His stress on the equivalence of the various uses of induction ([1906a] 1996, 1050) supports this interpretation, as does his description of intuition as insight into an abstract, or type of, mental capacity. Intuition for Poincaré is a *format*—for producing, understanding, and reasoning about various systems of objects. Since it is what enables us to produce infinite sets (insofar as we can—in our conceiving them) it is prior to those sets. Thus, like the group concept, the natural number structure, as given by a priori arithmetic intuition, is another structure that comes "first."

3.1.2. Geometry, Analysis, and Intuition

Despite his famous conventionalism about geometry (more specifically, the choice of a geometry for physics), Poincaré also endorses "geometric intuition" ([1913] 1963, 43–44). But "geometric" is a bit of a misnomer here. Geometric intuition does not yield a particular set of geometric truths, or knowledge that a particular geometric structure is true. So it is not an intuition of geometry. Rather, it provides more general cognitive access to physical and mathematical continua, via the "intuitive notion of the continuum."

> I shall conclude that there is in all of us an intuitive notion of the continuum of any number of dimensions whatever because we possess the capacity to construct a physical and mathematical continuum; and that this capacity exists in us before any experience, because, without it, experience properly speaking would be impossible and would be reduced to brute sensations. . . . And yet this capacity could be used in different ways; it could enable us to construct a space of four just as well as a space of three dimensions. ([1913] 1963, 44)

Like indefinite iteration, the intuition that lies behind the more "spatial" areas of mathematics, such as geometry, analysis, and topology, is a mental capacity—a capacity for constructing various types of abstract spaces.[29]

Poincaré associates this intuition most closely with "analysis situs," or topology, "the true domain of geometric intuition" ([1913] 1963, 42). But it also supports other areas of mathematics that are "spatial," including geometry and analysis. For example, in assessing Hilbert's axiomatic approach to geometry, Poincaré argues that the axioms of order are genuine intuitive truths ([1913] 1963, 43). They are central to topology, and they also play a fundamental role in our cognition of ordinary (metric) geometry.

[29] Though in rather Kantian style he also asserts that it is a form of experience—necessary for experience as we know it.

Because the intuitive continuum functions as a template for cognizing and defining various mathematical and physical continua, it is structural in a similar way to iteration. For example, it too *precedes* any particular instantiation, since the intuition is what "enable[s] us to construct" the continuous space we wish to consider. Thus, the intuitive continuum joins arithmetic intuition and the group concept to provide a third example of a mathematical structure that comes "first," i.e., before its instances.

To conclude this section, I aimed to show that Poincaré's conception of mathematical intuition harmonizes with the structuralist interpretation. Indeed, it strengthens it by adding new mathematical structures that come "first": the simply infinite structure and the (n-dimensional) continuum. Like the group concept, intuition precedes its uses. Recall that the group concept is a priori and the group structure is prior to its instances. Similarly, mathematical intuition is a priori; it enables synthetic a priori mathematical knowledge about infinite domains by providing insight into infinite structures, which are prior to their instances. The structures supplied by intuition are prior for Poincaré because we need the intuitive structures in order to "construct" (or conceive) the mathematical domains for which they provide the template. Thus, Poincaré's conception of intuition strengthens and adds to the "structure first," *ante rem*, interpretation.

3.2. Structural Realism and Mathematical Structuralism

Before turning to the question of *how* Poincaré (or anyone) could combine a semi-Kantian anti-realist view of mathematics with a "structure first" view, we will first briefly consider the fact that he is also commonly associated with realism—structural realism in the philosophy of (natural) science. This is roughly the view that although science does not generally provide us with absolute truths about objects in nature, it can yield knowledge of structures in nature. Structural realists acknowledge the so-called pessimistic meta-induction—scientific theories come and go—and they concede from this that it is naive to think that any one scientific theory provides eternal knowledge or insight into the essences of things. But this does not mean that science provides no knowledge at all. Instead, they maintain, there is evidence that science yields knowledge of the structure of reality. The success of science and the persistence of form through theory change are two supporting arguments commonly deployed by structural realists.

Poincaré has been cited as one of the first to articulate structural realism.[30] As I see it, the structuralist perspective guided his work in both mathematics

[30] See the classic piece by Worrall (1989); but also see Brading and Crull (2017) for a more modest, middle position on the "realism" in Poincaré's structural realism.

and physics, providing a cornerstone for his overall scientific epistemology. Structural realism about science and structuralism about mathematics are thus in a sense two sides of one epistemic coin.

One can find at least four arguments from Poincaré supporting structural realism. There are two familiar "negative" arguments against naive realism: (i) the privacy of acquaintance knowledge and the resulting weakness of direct realism,[31] and (ii) the acknowledgment of scientific change—the so-called pessimistic meta-induction ([1905] 1958, chap. 6). Poincaré also provides two common "positive" arguments aiming to rescue scientific knowledge from a more skeptical viewpoint, to which the two negative arguments might seem to lead. These are claims about (iii) the success of science ([1902] 1952, Introduction, 28) and (iv) the persistence of form through theory change ([1902] 1952, chap. 10, 153). His "rescue" leads to a type of structural realism.

Poincaré agrees that science is neither a direct reflection of reality nor is it simply cumulative: revisions and revolutions are part of science. Yet he resists skepticism. That is, despite the fact that scientific theories often change (the basis for the pessimistic meta-induction) Poincaré was optimistic about scientific knowledge. For example, though he emphasizes scientific conventions, it is a mistake to think this is the view that science is *just*, or *mainly*, based on decisions. That is, overemphasizing the freedom of conventions makes science seem arbitrary.

> If this were so [if science were arbitrary], science would be powerless. Now every day we see it work under our very eyes. That could not be if it taught us nothing of reality. Still, the things themselves are not what it can reach, as the naïve dogmatists think, but only the relations between things. Outside of these relations there is no knowable reality. ([1902] 1952, Introduction, 28)

There may be other realms of truth, or other ways of knowing reality; certainly the reality we can know is limited. Direct acquaintance knowledge is not objective for it is not even intersubjective; and things in themselves cannot be known at all.[32] But from the success of science Poincaré concludes that we do have objective knowledge—that of general, relational facts.

[31] "The sensations of others will be for us a world eternally closed. We have no means of verifying that the sensation I call red is the same as that which my neighbor calls red. . . . In compensation, what we shall be able to ascertain is that, for him as for me, the cherry and the red poppy produce the same sensation. . . . The relations between the sensations can alone have an objective value" ([1905] 1958, chap. 11, 136).

[32] Along these lines, just as logical positivism can be seen as an adjustment of Kant's vision, one can see Poincaré's structural realism similarly.

In addition, there is evidence that science uncovers the structure of reality in particular: this is the persistence of equations, or equation-forms, across theory change.

> Not only do we discover new phenomena, but in those we thought we knew, unforeseen aspects reveal themselves. . . . Nevertheless the frames are not broken. . . . Our equations become, it is true, more and more complicated, in order to embrace the complexity of nature; but nothing is changed in the relations which permit the deducing of these equations one from the other. In a word, the form of these equations has persisted. ([1902] 1952, chap. 10, 153)

Putting some of this together, we can fill in our picture a bit. Essences of things, or things in themselves, are not knowable. This resonates with his broadly Kantian epistemic vision, and it explains why the "images" of things shift with changes in scientific theory. But skepticism on these grounds is "superficial" according to Poincaré ([1902] 1952, chap. 10, p. 140). Instead, he believes that science reveals relations that actually exist in nature: "equations express relations, and if the equations remain true it is because these relations preserve their reality" ([1902] 1952, 140). Together, these views express a fairly straightforward version of scientific structural realism.[33]

Now, structural realism about science is neither necessary nor sufficient for structuralism about mathematics. Yes, mathematics is the "language" of science; but that scientific theory reveals the structures, or relations, of *nature* is simply different from the view that the subject matter of *mathematics* is abstract structure. Of course, the two views are not completely independent, or merely consistent. Though Poincaré was realist about (some) scientific structure and anti-realist about mathematics, the emphasis on structure in both views makes them harmonious and mutually supportive. In particular, they share a compelling view about perspective. The perspective of object-level content is "superficial" both in

[33] Putting "Kantian" in the same paragraph with "realism" may jar some readers. However, I do think Poincaré held versions of both views. With Kant we cannot know the things in themselves; also with Kant, mathematics provides a synthetic a priori foundation for scientific knowledge. Unlike Kant, Poincaré expresses confidence that the persistent structures and relations revealed by scientific inquiry reflect the way things are in nature, rather than just the way we are constituted to experience and/or conceptualize nature. There is a hint of Darwinism in this view, reminiscent also of Hume's "pre-established harmony" between nature and ideas (Hume, *Enquiry* Part V, last two paragraphs). For Poincaré, general, simple laws are most interesting and most beautiful—perhaps because we are constituted to appreciate them; but they are also necessary for science. If there were no general laws in reality, if there were 60 million chemical elements or only individuals but no biological species, "In such a world there would be no science; perhaps thought and even life would be impossible, since evolution could not there develop the preservational instincts" ([1905] 1958, Preface, 5; see also chap. 10, sec. 3, 115–122, for the view that there are prescientific "crude" facts, which science merely "translates" rather than "creates"). Though his arguments focus mostly on epistemological issues, his conclusions clearly endorse realism about at least some structural facts.

natural science (e.g., [1902] 1952, 140) and in mathematics (e.g., 1898, Form and Matter, p. 1009). However, this epistemic point—that the higher-level, structural perspective is crucial in both natural science and mathematics—is indifferent to the metaphysical question of whether or not the relevant structures exist independently of the scientists and mathematicians.

3.3 Constructivism versus "Structure First"?

In section 1, I argued that within Shapiro's basic taxonomy, the category of *ante rem* structuralism best fits Poincaré's further philosophical assertions about the group structure. And in section 3.1, I argued that we can extend the *ante rem* interpretation to the structures given by mathematical intuition. Indeed, as I reconstruct Poincaré's vision, the main a priori elements of mathematics—the group concept, the intuition of iteration, and the intuitive continuum—are each associated with what appear to be the fundamental mathematical structures. The group structure, the simply infinite structure, and continua are singled out as known a priori and as existing prior to the mathematical systems and constructions that instantiate them, which the a priori structures make possible.

A question about the consistency of my interpretation can now be addressed more clearly. The priority and independence of form over matter aligns Poincaré with the *ante rem* "structure first" view. However, in Shapiro's taxonomy, *ante rem* structuralism appears only as a form of realism,[34] and Poincaré was anti-realist about mathematics. (Despite his structural realism about natural science, Poincaré's appeal to mathematical intuition, his claims to defend Kant, and his views on mathematical existence all show this.)[35] Is this consistent? Is the structures-first, *ante rem* view consistent with the semi-Kantian constructivist view of mathematics that Poincaré defends? How can structures exist prior to, and/or independent of, mathematical objects and systems for an anti-realist?

One way to understand Poincaré is that the priority of structures to their objects and systems is merely epistemic and not ontological.[36] After all, as a constructivist, the ontology of mathematics will be constrained by its epistemology. If so, if the priority of structures is just epistemic, then one may object that his conception of structure better fits the eliminativist view than the *ante rem* view, since it is the main anti-realist alternative in the basic structuralist taxonomy. What about this alternative?

[34] While this is not asserted, the only type of *ante rem* structuralism he discusses is "Platonic."

[35] For example, he argues repeatedly against the existence of actual infinities because infinity just means there is "no reason for stopping" the generation of elements in a set.

[36] This appears to be Heinzmann's inclination (2014).

On the one hand, the emphasis on epistemology is right. The structures that are "prior" in Poincaré's philosophy are grounded in a priori intuitions and concepts, which, of course, lie more in the category of epistemology than ontology. On the other hand, eliminativism is the view that structures don't exist at all, that talk of "structure" is a mere manner of speech. And this does not fit Poincaré's views about mathematical structures. For him, structure is the core of the subject matter of mathematics; structure is the most important form of mathematical existence; indeed, as we saw above, focusing on the matter, or specific mathematical systems, rather than the form is "superficial" to him. This directly opposes the eliminativist view, according to which objects and systems may exist but structures do not.

Admittedly, Poincaré's rhetoric can be confusing. Intuitions and concepts do seem to concern epistemology; for example, the intuitions of iteration and continuity are "faculties" that enable the construction of simply infinite systems and physical and mathematical continua. Yet, for Poincaré intuition can also have a realist "feel."

> It is the intuition of pure number, that of pure logical forms, which illuminates and directs those we have called analysts. This it is which enables them not only to demonstrate, but also to invent. By it they *perceive at a glance* the general plan of a logical edifice. (1900, VI, 1020, my emphasis)

Here intuition is presented as like a telescope; it "illuminates"; it enables us to "perceive at a glance" things we couldn't otherwise perceive—pure logical forms, or structures. It is still epistemic, but it is articulated in terms of the ontology to which it provides access, rather than the activities by which we construct that ontology. And this may encourage a picture of the ontology as independent, or "out there." That is, a telescope does not create the objects we see with its aid;[37] thus, this way of describing intuition may encourage a similar view of mathematical structure—a more realist view of structure than would be consistent with constructivism. There are similar remarks about geometric intuition: because mathematics includes spaces of more than three dimensions, "there is surely *an intuition about the* continua of more than three dimensions" ([1913] 1963, 42–43, my emphasis). Finally, Poincaré explicitly says that the group structure "exists" prior to its instances. Recall that he says Helmholtz thinks the form of a group is posterior to, or dependent on, the matter, whereas in contrast, for him "the form *exists* before the matter" (my emphasis). At points like this, the structures of interest seem to preexist in a more ontological, rather than a merely epistemic, way.

[37] *Pace* a more strident form of scientific nominalism than would suit most, including Poincaré.

Is such talk just sloppy? Perhaps. To interpret Poincaré as consistent we must start with his mathematical anti-realism. That much is clear. Given this, *no* mathematical objects exist in a mind-independent sense, not even the fundamental structures. The "existence" of the group and intuitive structures is therefore something like existence in the minds of mathematicians. The priority in each case—a priori intuition and the a priori group concept—can be understood as a kind of mental template that enables specific instantiations, or constructions, in mathematical practice. So the priority of structures seems consistent with constructivism as long as "priority" and "existence" can be interpreted as reliant on minds (or finite thinking beings).

Furthermore, though they are generally associated with each other, *ante rem* structuralism does not require metaphysical realism. Note that Shapiro adds "Platonic" to the label "*ante rem* structuralism." This implies that the *ante rem* aspect—priority—does not entail Platonism, or realism; otherwise "Platonic" would be redundant. And this, in turn, implies that there can be a non-realist version of *ante rem* structuralism too.

That is, the term "*ante rem*" just indicates the priority in existence/reality of a general to its particulars. This is consistent with constructivism. Though for constructivists no mathematical entities exist absolutely independent of the minds and activities of mathematicians, some things can exist prior to others. Some templates can be required for some constructions, and some constructions can be required for others. An anti-realist version of the priority of structure would precisely fit Poincaré's conception of the a priori elements in mathematics—the elements that I have highlighted in my interpretation of his structuralism.

Of course, as mentioned, the taxonomy with which we began, and which is fairly common in the literature, is incomplete. Poincaré is not a Platonic *ante rem* structuralist because he is not a Platonist. He is not an *in re* structuralist because for him structures are not posterior to and dependent on systems; rather they are prior to and independent of the systems instantiating them. And he is not an eliminativist about structure because he believes that they—as well as many other mathematical objects—do exist (though we may have to do mathematics to make them exist). Eliminativism cannot be the only option for a structuralist who rejects realism about mathematical existence. Constructivist versions of both the *ante rem* and the *in re* views therefore seem coherent options.

4. Concluding Thoughts

Like many others at the time, Poincaré endorsed the basic methodological structuralist view. What's unusual for his time is that he expressed further philosophical views about the metaphysics and epistemology of mathematical structures. I have attempted to articulate these views and to situate them with respect to some of his other main commitments. In particular, I have argued that Poincaré's views about structure should be understood in a "strong" *ante rem* way, and that they are, nevertheless, consistent with his general constructivist approach to mathematics.

As a constructivist, Poincaré endorses a close connection between the epistemology and the ontology of mathematics. So his philosophy does not permit the same type of independence that one sees in traditional Platonism—that is, independence of mathematical reality from the minds and activities of mathematicians. But this is not the type of independence that characterizes *ante rem* structuralism. *Ante rem* structuralism only requires that structures be independent of, and prior to, their instances, which is exactly what Poincaré asserts.

For Poincaré, the form exists before the matter in that—to speak crudely—we need the form in order to cognize the matter. Though no mathematical entities or structures exist absolutely independent of the work and minds of mathematicians, the fundamental structures—the group structure, the natural number structure, and the continuum—have *more* independence than the domains they yield. In other words, they are prior to their instantiating systems.[38]

The fundamental mathematical structures, given by a priori concepts and intuitions, are thus, for Poincaré, a kind of cognitive blueprint necessary for conceiving and instantiating mathematical systems. Systems are the result of definitions; but both the definitions and our understanding of what they yield are guided by our "blueprints." The intuition of indefinite iteration guides our cognition of simply infinite systems as well as our inductive inferences about them. The intuitive continuum "enables" us to define physical and mathematical continua and to work with them in mathematics and science. The a priori group concept is a "form" that exists prior to the matter of any group, providing a unifying mathematical ideal and a foundation for geometry and group theory. In each of these cases the form "exists" prior to the domains, or systems.[39] Additionally, structures are more significant than matter or particular systems, which Poincaré dismisses as "superficial." This is the sense in which structures "preexist" for Poincaré: in the mind as an a priori intuition of structure, or as an a priori concept, rather than as independently existing Platonic reality.[40] Provided we agree that *ante rem* structuralism does not require Platonism, we can appreciate Poincaré's "constructivist *ante rem* structuralism"—a view that fits Shapiro's *ante rem* category, minus the "Platonism" typically attached to it.

[38] And possibly other, less fundamental, structures.

[39] Indeed, epistemologically speaking, each of these forms (the fundamental structures) are a priori, not just relatively prior.

[40] Whose mind? Interestingly not just humans for Poincaré; he implies that at least some a priori elements of mathematics (concepts, intuitions) are common to all finite beings who can conceive of infinity or space ([1902] 1952, p. 39; [1908] 1982, 427–428).

Acknowledgments

Multiple layers of thanks go to Erich Reck and Georg Schiemer: for inviting me to participate in this volume, for their detailed and extremely helpful comments on an early draft (and more wonderful comments on a later draft), and for their patience. Thanks also to María de Paz, who also generously provided helpful feedback on an early draft.

References

Benacerraf, P. 1965. What Numbers Could Not Be. *Philosophical Review* 74, 47–73.

Berkeley, G. 1734. The Analyst; or A Discourse Addressed to an Infidel Mathematician. In Ewald 1996, pp. 60–92.

Brading, K., and E. Crull. 2017. Epistemic Structural Realism and Poincaré's Philosophy of Science. *HOPOS* 7(1), 108–129.

De Morgan, A. 1835. Review of Peacock's *A Treatise on Algebra*. *Quarterly Journal of Education* 9, 91–110 (part 1) and 293–311 (part 2).

De Paz, M., and R. DiSalle, eds. 2014. *Poincaré, Philosopher of Science*. Dordrecht: Springer.

Ehrlich, P., ed. 1994. *Real Numbers, Generalizations of the Reals, and Theories of Continua*. Dordrecht: Springer.

Ewald, W., ed. 1996. *From Kant to Hilbert: A Source Book in the Foundations of Mathematics*. 2 vols. Oxford: Clarendon Press.

Folina, J. 2006. Poincaré's Circularity Arguments for Mathematical Intuition. In Friedman and Nordmann 2006, pp. 275–293.

Folina, J. 2014. Poincaré and the Invention of Convention. in De Paz and DiSalle 2014, pp. 25–45.

Folina, J. 2018. Towards a Better Understanding of Mathematical Understanding. In *Truth, Existence and Understanding*, edited by M. Piazza and G. Pulcini, pp. 121–146. New York: Springer.

Friedman, M., and A. Nordmann. 2006. *The Kantian Legacy in Nineteenth Century Science*. Cambridge, MA: MIT Press.

Gauss, C. 1829. Letter to Bessel. In Ewald 1996, p. 301.

Gauss, C. 1831. Notice on the Theory of Biquadratic Residues. In Ewald 1996, pp. 307–313.

Goldfarb, W. 1985. Poincaré against the Logicists. In *History and Philosophy of Modern Mathematics*, edited by W. Aspray and P. Kitcher, pp. 61–81. Minneapolis: University of Minnesota Press.

Heinzmann, G. 2014. Does the French Connection (Poincaré, Lautman) Provide Some Insights Facing the Thesis That Meta-mathematics Is an Exception to the Slogan That Mathematics Concerns Structures? In de Paz and DiSalle 2014, pp. 113–124.

Hellman, G. 1989. *Mathematics without Numbers*. New York: Oxford University Press.

Hume, D. [1748] 1988. *An Enquiry Concerning Human Understanding*. Editorial arrangement by Antony Flew. Chicago: Open Court.

Kline, M. 1972. *Mathematical Thought from Ancient to Modern Times*. New York: Oxford University Press.

Maclaurin, C. [1742] 1996. Excerpt from *A Treatise of Fluxions*. In Ewald 1996, vol. 1, pp. 95–122.

Poincaré, J. H. 1894. On the Nature of Mathematical Reasoning. Halsted, transl., in Ewald 1996, pp. 972–981.

Poincaré, J. H. 1898. On the Foundations of Geometry, T. J. McCormack transl. (reprint of original translation in the *Monist*), in Ewald 1996, pp. 982–1011.

Poincaré, J. H. 1900. Intuition and Logic in Mathematics, Halsted, transl., in Ewald 1996, pp. 1012–1020.

Poincaré, J. H. [1902] 1952. *Science and Hypothesis*, G. Halsted transl., references to reprint in *The Foundations of Science*, University Press of America, Washington, D.C., 1982.

Poincaré, J. H. [1902] 1994. Review of Hilbert's *Foundations of Geometry*, E. Huntington transl., in Ehrlich 1994.

Poincaré, J. H. [1905] 1958. *The Value of Science*, George Halsted transl., references to Dover edition, New York (which includes Poincaré's preface to the translation).

Poincaré, J. H. [1905] 1996. Mathematics and Logic I. Translated by G. Halsted and W. Ewald. In Ewald 1996.

Poincaré, J. H. [1906a] 1996, Mathematics and Logic II. Translated by G. Halsted and W. Ewald. In Ewald 1996.

Poincaré, J. H. [1906b] 1996. Mathematics and Logic III. Translated by G. Halsted and W. Ewald. In Ewald 1996.

Poincaré, J. H. [1908] 1982. *Science and Method*. Translated by G. Halsted. In *The Foundations of Science*. Washington, D.C.: University Press of America.

Poincaré, J. H. [1913] 1963. *Mathematics and Science: Last Essays*. John Bolduc transl., Dover, New York.

Reck, E. 2003. Dedekind's Structuralism: An Interpretation and Partial Defense, *Synthese* 137, 369–419.

Reck, E., and M. Price. 2000. Structures and Structuralism in Contemporary Philosophy, *Synthese* 125, 341–383.

Resnik, M. 1997. *Mathematics as a Science of Patterns*. Clarendon Press.

Shapiro, S. 1997. *Philosophy of Mathematics: Structure and Ontology*. Oxford University Press, 1997.

Shapiro, S. n.d. Mathematical Structuralism. *Internet Encyclopedia of Philosophy*. https://www.iep.utm.edu/m-struct/.

Wittgenstein, L. 1983. *Remarks on the Foundations of Mathematics*. MIT Press.

Worrall, J. 1989. Structural Realism: The Best of Both Worlds? *Dialectica* 3(1–2) 99–124.

12

"If Numbers Are to Be Anything At All, They Must Be Intrinsically Something": Bertrand Russell and Mathematical Structuralism

Jeremy Heis

Russell's philosophy of mathematics is often opposed to structuralism for a number of reasons. First, Russell is a paradigm logicist (indeed, perhaps the most thoroughgoing and systematic defender of logicism ever), and structuralism is often defended as an alternative to logicism. Second, Russell's famous definition of cardinal numbers as classes of equinumerous classes has the very feature that structuralists deny is necessary: it goes beyond the "structural" features of numbers and attributes to them an intrinsic character (namely, as classes). Third, Russell forcefully defended his logicist definition of real numbers over Dedekind's, by accusing him of engaging in "theft over honest toil"—postulating the existence of objects that fulfill a certain structural description, without first proving that there are such objects (Russell 1919, 71).[1] In the century since Russell first wrote these words, this accusation has been a standard objection to at least some versions of structuralism, and overcoming this objection has been a source of ongoing work for many of structuralism's contemporary adherents.

Nevertheless, this chapter will show that Russell's relationship to structuralism is not entirely negative. Russell defended—and in some cases even introduced into philosophy—many ideas that were essential for the full articulation and defense of structuralism. (Indeed, some of Russell's ideas were explicitly appropriated in Ernst Cassirer's philosophical defense of structuralism.) Of course, Russell was a critic of mathematical structuralism—the most thoroughgoing and trenchant critic of structuralism in the early twentieth century. As this

[1] Russell is here discussing the definition of real numbers in terms of Dedekind cuts. He argues that Dedekind himself simply laid down an axiom that *postulates* that any segment of the series of rationals has a bound; he advocates instead for *constructing* the reals as sets of segments of the series of rationals.

Jeremy Heis, *"If Numbers Are to Be Anything At All, They Must Be Intrinsically Something"* In: *The Prehistory of Mathematical Structuralism.* Edited by: Erich H. Reck and Georg Schiemer, Oxford University Press (2020). © Oxford University Press.
DOI:10.1093/oso/9780190641221.003.0012

chapter will show, Russell's criticisms of structuralism are manifold and subtle, going well beyond the well-known ideas I mentioned in the opening paragraph.

This chapter has two parts. In the first part ("Russell's Positive Contribution to Structuralism," section 1), I identify three theses of Russell's philosophy of mathematics that could be—and indeed have—been employed as key parts of structuralism. In the second part ("Russell's Criticism of Dedekind's Structuralism," section 2) I show how Russell, between the years 1898 and 1901, returned again and again to the structuralist idea in Dedekind's philosophy of arithmetic, and developed four series of criticisms of this structuralism.

Two clarifications before we begin. First, the main topic of this chapter is Russell's relation to "non-eliminative" versions of structuralism, such as the version in Dedekind's *Was sind und was sollen die Zahlen?* (Dedekind [1888] 1963).[2] As other philosophers have made clear (Reck and Price 2000), the core idea of structuralism, that mathematics is about positions in structures, can be developed in multiple, incompatible ways. For non-eliminative structuralism, mathematical objects are just positions in structures: that is, all of the essential properties of, say, a particular natural number are irreducible relational properties between it and the other natural numbers. On this view, the positions in the structure are distinct from any of the systems of objects that have that structure. For example, the number 4 in the natural number series is an object in its own right, distinct from any particular things that have the fourth position in some system (e.g., the fourth planet in the solar system, or the fourth child of J. S. Bach). This clarification is necessary, since (as I will argue later) some of Russell's philosophy of mathematics is quite close to certain eliminative versions of structuralism. Second, beyond the quip about theft and honest toil in *Introduction to Mathematical Philosophy*, there is little substantial discussion of recognizably structuralist ideas in Russell's writings in the philosophy of mathematics after his 1903 *Principles of Mathematics* (*POM*). What's more, throughout *POM*, and in Russell's various papers and drafts that he wrote while composing *POM*, Russell returns to Dedekind's version of non-eliminative structuralism repeatedly. For this reason, my focus in this chapter will be on *POM* and Russell's papers in the years immediately preceding its publication.

1. Russell's Positive Contribution to Structuralism

In this section, I identify three theses of Russell's philosophy of mathematics that could be—and indeed, as I will show, have—been employed as key parts of a fully articulated structuralism. First, *the logic of relations makes it possible to conceive*

[2] On Dedekind as a non-eliminative structuralist, see Reck (2003).

structures abstractly, without any reference to space, time, or empirical properties. Second, *Russell is one of the first philosophers (if not the first) to explicitly separate pure from applied mathematics in such a way that all of the rival metric geometries become parts of pure mathematics.* Third, *Russell introduced the concept of a "relational type" and distinguished the various areas of pure mathematics according to the specific relational type that they study*—an approach that provides a concrete way of cashing out the idea that the various branches of pure mathematics concern distinct "structures." I take each of these theses in turn.

1.1. The Logic of Relations and Abstract Structures

The core idea of structuralism is that all the essential properties of mathematical objects are their relational properties to other mathematical objects within the structure. This core idea is incompatible with the view that spatial, temporal, intuitive, or empirical properties are essential properties of mathematical objects. Consider spatial properties (by which I mean properties of an object in relation to "physical" space, the space occupied by concrete bodies). Spatial properties involve essential relations to things in space, since it is the fact that physical space is occupied by concrete bodies that distinguishes it from, say, color space or abstract mathematical "spaces." A similar point holds for temporal, intuitive, and empirical properties: temporal properties involve relations to events in the physical world, intuitive properties involve relations to our sensibility, and empirical properties involve relations to empirical (and so non-mathematical) objects. Thus, structuralism requires that the concept of a structure does not depend conceptually on spatial, temporal, intuitive, or empirical concepts. In short, the objects of mathematics are *abstract* structures (or positions in abstract structures).

But is it possible to conceive structures abstractly, without any reference to space, time, intuitive, or empirical properties? Consider our paradigm structuralist theory, Dedekind's philosophy of arithmetic. Dedekind defines the natural number numbers by first defining a simply infinite system, or in Russell's language, a "progression." A progression is a structure with a distinguished element, 0, and a successor map that takes each position in the structure to the "next" position. But is this notion spatial, temporal, intuitive, or empirical? Certainly, the word "next" suggests such an origin. More generally, in chapter 31 of *POM*, Russell considers the following constellation of ideas, which he attributes to Leibniz and Meinong: progressions are a kind of *series; series* presupposes *order*, which in turn presupposes *distance*; distances are magnitudes, but *magnitude* is an empirical notion. This is a natural line of reasoning. After all, if, say, A, B, and C are ordered in such a way that B is between A and C, what else could this mean than that the distance from A to B is less than the distance from A to C? So, our objector concludes, the concept of a progression ultimately has an empirical origin.

Russell's reply to this objection depends on his definition of order, and ultimately on his new logic of relations. In chapter 24, he isolates six distinct ways of generating a series. For example, elements may be ordered into a series using the notion of *distance*, or the notion of *between*, or the notion of *separation*. In chapter 25, he argues in detail that these methods for generating a series can be reduced to one single method:

> The minimum ordinal proposition, which can always be made wherever there is an order at all, is of the form "y is between x and z"; and this proposition means: "There is some asymmetrical, transitive relation which holds between x and y and between y and z." (§207)

(In the case of the natural numbers, this asymmetrical, transitive relation is $n < m$, and "m is between n and o" means "$n < m$ and $m < o$"). And so the objection is defeated, since the notion of order depends ultimately on the concept of an asymmetrical transitive relation—not on the notion of distance or magnitude. Russell concludes further that the concept of an asymmetrical transitive relation, being a *logical* notion, does not depend conceptually on any spatial, temporal, intuitive, or empirical concepts. And this is just what the defender of mathematical structuralism needed.[3]

Russell's analysis of the notion of series depends, then, on the concepts that he had developed in the logic of relations. Russell developed (independently of Frege) an original version of modern polyadic higher-order quantificational logic in the fall of 1900, and published his first version of it as "The Logic of Relations" (Russell 1901c). This paper (see also *POM* §§27–30; chap. 9) distinguishes kinds of relations—as say, transitive or intransitive, symmetrical, asymmetrical, or anti-symmetrical—in the now standard way, in many cases introducing the terms that we use today. Russell made the logic of relations independent of the theory of classes, thus avoiding the artificiality that beset the logic of relations done in the Boolean tradition by DeMorgan, Schröder, and Peirce. Unlike Frege, who thematized the function/argument analysis when arguing for the originality of his polyadic quantificational logic, Russell repeatedly pointed to the relational character of his logic to explain its originality and significance. And, most importantly for our purposes, he loudly proclaimed the centrality of

[3] I have spoken of the *conceptual independence* of the *concept* of an asymmetrical transitive relation. Russell would of course also held that certain abstract relations are *ontologically independent* of anything empirical, spatial, temporal, or intuitive.

the logic of relations for understanding mathematics: "the logic of relations has a more immediate bearing on mathematics than that of classes or propositions, and any theoretically correct and adequate expression of mathematical truths is only possible by its means" (*POM*, §27).

Of course, Russell himself was not a non-eliminative structuralist (see section 2). But a philosopher could draw on Russell's ideas to defend and elaborate structuralism. Not only *could* Russell's theory of relations be used to shore up structuralism, but in fact it *was* so used. Ernst Cassirer was, arguably, the first philosopher to give an explicit articulation and defense of a thoroughgoing non-eliminative mathematical structuralism (see Cassirer 1907, which is a very positive review of Russell's *POM*, and Cassirer [1910] 1923, chaps. 2 and 3). Though Cassirer finds the structuralist point of view paradigmatically in Dedekind's philosophy of arithmetic (Cassirer [1910] 1923, 39),[4] he self-consciously draws on ideas from Russell in this articulation and defense. In Cassirer 1907 (§II), Cassirer endorses Russell's idea that the reals, and more generally, continuity, can be defined entirely in terms of order; and that order, being definable using concepts from the logic of relations, does not presupposes space, distance, or magnitude. "One recognizes in this connection," Cassirer writes, "the value and necessity of the new foundation on which Russell is seeking to place logic. Mathematics in his treatment is nothing other than a special application of the general logic of relations" (Cassirer 1907, 7). Indeed, Cassirer claims, Russell's point of view is confirmed in Dedekind's structuralist philosophy of arithmetic (Cassirer 1907, 7).

1.2. Russell on Pure and Applied Geometry

According to structuralism, the objects of pure math are abstract structures. Concrete structures, then, are the concern of *applied* mathematics only (Parsons 2008, §14). Now, "physical" space, the space occupied by concrete bodies, is itself a concrete structure. And so, a thoroughgoing non-eliminative mathematical structuralism will have to identify some other subject matter for geometry besides physical space. The standard way for structuralists to address this issue is by distinguishing pure from applied geometry: only applied geometry is concerned with physical space; pure geometry concerns some family of abstract structures.

The pure/applied geometry distinction has played an important role in the emergence of mathematical structuralism through a more specific historical route. By the 1860s, mathematicians had proven that there are other consistent theories of metrical geometry besides classical Euclidean geometry. In the early

[4] On Cassirer's structuralism, see Erich Reck's chapter in this volume. On Cassirer's reception of Dedekind, see also Yap (2017).

1870s, Klein discovered deep interrelations between these non-Euclidean geometries and projective geometry and group theory.[5] In the 1880s, Poincaré used non-Euclidean geometry to prove some very important results in complex analysis. These results convinced mathematicians by the end of the 19th century that the non-Euclidean geometries were just as much a part of pure mathematics as classical Euclidean geometry. What, philosophically, could justify this attitude? How could mathematicians accept, as equally legitimate, contradictory theories of space? (In what follows, I'll call this "the puzzle of non-Euclidean geometry.") The structuralist has a ready answer: only applied geometry is concerned with physical space, and so whether it turns out to be Euclidean or not is a question for physics, not pure mathematics; pure geometry, on the other hand, concerns certain kinds of abstract structures, some of which are Euclidean and some of which are not.

Structuralism's ability to justify the mathematicians' attitude toward the rival metric geometries was a chief argument in its favor.[6] Once again, this argument was presented very clearly by Cassirer ([1910] 1923, chap. 3, sec. 4; [1921] 1923, 432), thereby extending the non-eliminative structuralism he found in Dedekind's philosophy of arithmetic to pure geometry (Schiemer 2018; Heis 2011). Structuralists such as Cassirer solve the philosophical puzzle posed by non-Euclidean geometry, then, in four steps: first, distinguish pure from applied geometry; second, argue that the question of the metric of physical space is a question for the latter only; third, conclude that therefore the subject matter of pure geometry is something other than physical space; and, fourth, propose abstract structures as the subject matter of pure geometry. The first three steps have now become standard in the philosophy of mathematics, even among those philosophers who do not take the final distinctively structuralist step. But it is essential to recognize that very few, if any, philosophers or mathematicians prior to Russell took these three steps.

In fact, the first philosopher to clearly take these first three steps, and thereby justify the equal legitimacy of the rival geometries as pure mathematical theories independent of physical space, was arguably Russell himself.[7] He first articulated the idea in Russell (1902), which was written around December 1898:

[5] On Klein, see Georg Schiemer's chapter in this volume.

[6] This historical point is presented in detail in Shapiro (1997), chap. 5, "How We Got Here", especially sections 2 and 3. Shapiro, unfortunately, does not mention Cassirer, who in fact presents this argument for structuralism very clearly.

[7] Russell was, as far as I know, the first *philosopher* to take these three steps. There were *mathematicians* before Russell who distinguished pure from applied geometry, and denied that physical space is the subject of pure geometry. These include Grassmann, Pieri, and Whitehead (Grassmann [1844] 1894, 23–24; Pieri 1898; Whitehead 1898, vii, 370).

> We have seen that there are a number of possible Geometries, each of which may be developed deductively with no appeal to actual facts. But no one of them, *per se*, throws any light on the nature of our space. Thus geometrical reasoning is assimilated to the reasoning of pure mathematics, while the investigation of actual space, on the contrary, is found to resemble all other empirical investigations as to what exists. There is thus a complete divorce between Geometry and the study of actual space. . . . It points out a whole series of possibilities, each of which contains a whole system of connected propositions; but it throws no more light upon the nature of our space than arithmetic throws upon the population of Great Britain. (Russell 1902, 503)

One year later (in Russell 1901a, written in December 1900 or January 1901), this solution to the puzzle of non-Euclidean geometry motivated[8] a new way of characterizing the distinction between pure and applied mathematics:

> Pure mathematics consists entirely of assertions to the effect that, if such and such a proposition is true of anything, then such and such another proposition is true of that thing. It is essential not to discuss whether the first proposition is really true, and not to mention what the anything is, of which it is supposed to be true. Both these points would belong to applied mathematics. (Russell 1901a, 366)

On this view, the sentences of pure mathematics are all "formal implications," sentences of the form *for all x*, $\varphi(x) \supset \psi(x)$.[9] Thus, a sentence of Euclidean geometry, understood as a branch of pure mathematics,[10] would be *for all* $x_1, \ldots, x_n$, if the axioms of Euclidean geometry are true of $x_1, \ldots, x_n$, *then such and such is also true of* $x_1, \ldots, x_n$. Russell characterizes the antecedent of these generalized conditionals as definitions: in the case of Euclidean geometry, "φ(x)" would be the definition of a Euclidean space, and so a sentence of pure Euclidean geometry is equivalent to the sentence "ψ is true of every Euclidean space." In parallel passages in the following years,[11] Russell clarifies that "φ" and "ψ" contain only

[8] Russell cites the puzzle about non-Euclidean geometry as the decisive argument for his definition of pure mathematics in the introduction to the 1937 second edition of *POM* (vii) and earlier in a January 1902 letter to Couturat (Russell 2002, 220).

[9] Russell allows that the quantifiers in formal implications be higher order. On formal implications, see *POM*, §§40–45.

[10] I speak here of Euclidean geometry, understood as a branch of pure mathematics. However, there are passages in *POM* where Russell asserts that metric geometry is an empirical science and so "does not belong to pure mathematics" (*POM*, §411; cf. Gandon 2012, 72). These passages have led Gandon to conclude that there was no fundamental break in Russell's philosophy of geometry between Russell 1897 and *POM*, as I am claiming (2012, 53). Unfortunately, space considerations preclude the extended discussion that Gandon's claims merit.

[11] Draft of Part I of *Principles of Mathematics* (Russell 1901b, 185, 187), written in May 1901; Part I of *POM* (§1), which Russell composed in May 1902.

logical constants. A sentence of applied mathematics, then, results from a sentence of pure mathematics when the universal quantifier is instantiated by a constant that is not a logical constant (or analyzable into logical constants); when the antecedent of the conditional is asserted outright for some nonlogical constant; or when some new primitive, nonlogical vocabulary is added.

Once again, not only *could* Russell's use of the pure/applied mathematics distinction to solve the puzzle about non-Euclidean geometry be used to motivate a structuralist theory of pure geometry, in fact it *was* used in precisely this way. Cassirer, in the section of Cassirer ([1910] 1923) on non-Euclidean geometry, draws the pure/applied geometry distinction and solves the puzzle about non-Euclidean geometry in precisely Russell's way. The axioms of the various metric geometries, Cassirer says, simply pick out different "pure logico-mathematical forms" ([1910] 1923, 109). He criticizes other possible solutions to the puzzle, such as empiricist solutions or Poincaré's conventionalist solution. And of course we know that Cassirer had studied Russell's *POM* very closely just a few years earlier (Cassirer 1907). Furthermore, Carnap's *Der Raum*, which articulates a structuralist philosophy of pure geometry, explicitly points to Russell's distinction between pure and applied geometry for inspiration, and draws on Russell's characterization of pure geometry for his theory of "formal space."[12]

1.3. Relational Types

For a structuralist, it is not enough to characterize the sentences of mathematics as conditionals of the form "if axioms, then theorems": for a structuralist, the axioms characterize abstract structures. But what are abstract structures? How can we pick out the distinctly *structural* properties of a system of entities? Russell's logic of classes and relations provides a ready language for characterizing these structural properties. Moreover, the structuralist holds that the various areas of pure mathematics are distinguished from one another by the kind of structure they study: number theory studies the structure of progressions, analysis studies the structure of the continuum, etc. But how do we individuate structures? Once again, Russell's logic of relations and classes provides a means.

Russell picks out "structural" properties and distinguishes structures through his notion of a "relational type," which he defines in the following way:

[12] See Schiemer's chapter on Carnap for details. Carnap, like Cassirer (see section 12.1.1), also points to Russell's logic of relations to show that formal space, inasmuch as it is a "pure theory of relations," is "free of non-logical (intuitive or experiential) components" (Carnap 1922, 8).

> Now a *type* of relation is to mean, in this discussion, a class of relations characterized by the above formal identity of the deductions possible in regard to the various members of the class; and hence, a type of relations, as will appear more fully hereafter, if not already evident, is always a class definable in terms of logical constants. We may therefore define a type of relations as a class of relations defined by some property definable in terms of logical constants alone. (*POM*, §8; cf. §412)

In fact, Russell argues that the "true subject matter" of mathematics is relational types (§27), and he engages in a detailed program of analyzing the various branches of existing mathematics as each concerned with a different relational type.

An example will make Russell's analysis of mathematics vivid. In chapter 46 of *POM*, Russell gives an axiomatization of "descriptive geometry":

1. There is a class of relations *K*, whose field is defined to be the class *point*.
2. There is at least one point.

If *R* be any term of *K* we have

3. *R* is an aliorelative (i.e., for all *x*, ~*Rxx*).
4. R^{-1} is a term of *K*.
5. $R^2 = R$ (i.e., for all *x*, *y*, *z*, if *Rxy* and *Ryz*, then *Rxz*).
6. The points in the domain or range of R^{-1} are also in the domain or range of *R*.
7. Between any two points there is one and only one relation of the class *K*.
8. If *a*, *b* be points in the domain or range of *R*, then either *aRb* or *bRa*.

Descriptive geometry, intuitively, is the geometry of directed line segments. "*Rxy*" means "*y* comes after *x* on the directed line segment *R*"; every relation *R* represents a directed line segment, R^{-1} is the same line segment directed in the opposite way. But note that this axiomatization does not make mention of lines or directions: it simply picks out various classes *K* of relations that have the specified logical properties. The only nonlogical word is "point," which is actually just a shorthand for "object in the domain or range of some relation *R* in some class *K* of relations satisfying the axioms." Any two classes of relations *K* and *K′* that each satisfy the axioms share a relational type, and descriptive geometry is the theory of this relational type. Russell summarizes his procedure in this way:

> We saw that the above method enabled us to content ourselves with one indefinable, namely the class of relations *K*. But we may go further, and dispense

> altogether with indefinables. The axioms concerning the class *K* were all capable of statement in terms of the logic of relations. Hence we can define a class *C* of classes of relations, such that every member of *C* is a class of relations satisfying our axioms. The axioms then become parts of a definition, and we have neither indefinables nor axioms. If *K* be any member of the class *C*, and *k* be the field of *K*, then *k* is a descriptive space, and every term of *k* is a descriptive point.... This affords a good instance of the emphasis which mathematics lays upon relations. To the mathematician, it is wholly irrelevant what his entities are, so long as they have relations of a specified type. It is plain, for example, that an instant is a very different thing from a point; but to the mathematician as such there is no relevant distinction between the instants of time and the points on a line. (§378)

This procedure is not exactly what a structuralist would adopt. For her, once the relational type of descriptive spaces has been identified, she would pick out (perhaps by an act of "Dedekind" abstraction) the *structure* exemplified by all descriptive spaces. This structure for the non-eliminative structuralist is an individual (as are positions in this structure), and is distinct from any concretum that has this structure. Russell does not seem to make this move: *POM* suggests two alternatives, neither of which would be palatable to the non-eliminative structuralist. On one alternative—which is suggested by his definition of pure mathematics—a sentence of descriptive geometry is just a universally quantified conditional: for all *K*, if *K* is a collection of relations that satisfies the axioms of a descriptive space, then $\psi(K)$. No individual is mentioned here and there is no object *the relational type of descriptive spaces*; instead we have the higher-order propositional function *x is a collection of relations that satisfies the axioms of a descriptive space*. In fact, this alternative is really a kind of eliminative structuralism. More precisely, it is a kind of modal eliminative structuralism, where the modal operator means "it is a logical truth that . . ." The modal character derives from Russell's insistence that the relational types be characterized using purely logical vocabulary, and that the sentences of pure mathematics be logical.[13]

The second alternative interpretation of relational types is suggested by his definition of a relational type at §8 and by §378, quoted earlier. On this alternative, a sentence of descriptive geometry expresses a relation between two

[13] For a reading of early Russell as an eliminative structuralist: Reck and Price (2000, 354–361). For a contemporary defense of modal eliminative structuralism, see Hellmann (1989). On the affinity between some of Russell's views and modal eliminative structuralism, see Hellman (2004, 564). Of course, the standard objection to a view like Russell's is that Russellian logic includes the theory of classes, which is no longer considered to be obviously logical. For contemporary readers, then, this view just collapses into set theoretic realism.

So-called if-thenism is closely related to eliminative structuralism. Reck and Price (2000) read Russell in *POM* as a kind of if-thenist, as does Musgrave (1977). Gandon (2012) argues at length that Russell in *POM* is not an if-thenist about pure geometry. Unfortunately, again space considerations preclude the extended discussion that Gandon's claims merit.

classes: The class of all classes of relations that satisfy the axioms of descriptive geometry is contained in the class of terms that are ψ. Thus, the relational type is a class. Since Russell never suggests a structuralist interpretation of the theory of classes, this alternative still does not provide what the non-eliminative structuralist would want. In fact, this alternative is really a kind of set-theoretic realism.

Interestingly, in the parts of *POM* that were written first in late 1900, such as part III (on quantity), Russell suggests a third reading of relational types that has a stronger structuralist flavor. When writing these sections, Russell endorsed a novel program using "abstraction" principles. By "abstraction" principles, Russell means principles, such as Frege's famous "Hume's Principle" (Frege 1884, §63), that analyze equivalence relations (say, among classes) into identity claims about some new entities (say, cardinal numbers). Thus, cardinal numbers are defined by the biconditional *The number of Fs = the number of Gs iff the Fs and the Gs are equinumerous.* Similarly, directions are defined by the biconditional *The direction of l = the direction of m iff l and m are parallel lines.* He makes free use of abstraction principles in these parts of *POM*. For instance, in §231 he defines the ordinal number ω by abstraction as the abstractum to which all progressions (which are themselves related by the equivalence relation of isomorphism) are related. Thus, when two collections of objects, classes, and relations both satisfy the same logically describable axiom system, they are related by an equivalence relation (*having the same relational type as*), and their common relational type is then defined by abstraction. At various places, he suggests that the entity defined by abstraction is "unanalyzable" and thus distinct from any class (see, e.g., §155 and §157 on magnitudes).[14] By spring 1901, Russell rejected definitions by abstraction (see §110, written in June 1901). However, if this program of late 1900 and very early 1901 had been carried out to completion, this would have been close to what non-eliminative structuralists would want. That is, a mathematical theory such as number theory would have as its object some abstract object, distinct from all concrete progressions and distinct from classes.

Just as in the case of his theory of relations and his pure/applied distinction, not only *could* Russell's notion of a relational type be employed in a structuralist account of mathematical objects, in fact it *was* used in precisely this way. In his review of *POM*, Cassirer emphasized Russell's project of identifying the various relational types that characterize the various branches of mathematics (Cassirer 1907, 5). Later, Cassirer systematically used Russell's logic of relations to identify

[14] Russell was not consistent on this point, even in late 1900: elsewhere Russell suggests that the abstracta picked out by definitions by abstraction are just classes of equivalent terms (see, e.g., §231).

Russell (1919, chap. 5) introduces what he calls a "relation-number," which is a class of "similar" (i.e., isomorphic) relations. This is clearly the descendant of *POM*'s relational type, now interpreted in this third way, where the equivalence relation is isomorphism and the abstracta are classes of isomorphic relations.

the relational type of some mathematical theories (e.g., Cassirer [1910] 1923, 37–39), before applying an act of abstraction to identify the "system of relations" (110), which constitutes the true object of pure mathematics. In fact, Cassirer's position is what one gets by taking the object *C* mentioned in §378, that is, *the relational type of all descriptive spaces*, considered not as a class, but as distinct kind of abstractum. Furthermore, Carnap self-consciously draws on Russell's notion of relational types in identifying structures in his structuralist "general axiomatics project" from the mid-1930s, and in his pre-*Syntax* period philosophy of mathematics. In many writings from these periods, Carnap follows Russell's procedure of axiomatizing a mathematical theory, removing all nonlogical vocabulary, and treating the resulting axioms as a definition of a higher-order propositional function that applies to tuples of objects, relations, etc. Indeed, Carnap at various points endorses all three of Russell's interpretations of relational types.[15]

2. Russell's Criticism of Dedekind's Structuralism

Although Russell's philosophy could furnish the raw materials for essential components of a worked out non-eliminative structuralism such as Cassirer's, Russell himself presented a sustained and multipronged attack on non-eliminative structuralism, in the form in which Dedekind had developed it. He returned to Dedekind's structuralism again and again in a series of writings, both published and unpublished, between 1898 and 1901.[16] In this section I present three groups of criticisms that Russell developed of Dedekind's non-eliminative structuralism in these years.

2.1. Russell's Earliest Criticisms: The Priority of Cardinals over Ordinals

Russell first read Dedekind's *Was sind und was sollen die Zahlen?* in April 1898.[17] Even on his first reading, Russell was alert to the non-eliminative structuralist aspect of Dedekind's work, and he found it untenable. In particular, from this

[15] See Schiemer's chapter on Carnap for details and references on the structuralist aspects of Carnap's "general axiomatics project" and his pre-*Syntax* philosophy of mathematics. Schiemer's chapter also clearly lays out Russell's influence on Carnap.

[16] Since many of Russell's criticisms can be adequately understood only in the context of the particular views and preoccupations he had at the time of their writing, I will discuss the chronology of Russell's criticisms of Dedekind's structuralism. However, I cannot here give a full defense of the chronology, nor can I give a complete account of Russell's rather complex history of reading and writing about Dedekind in this period. I hope to come back to these issues in more detail elsewhere.

[17] See "What Shall I Read?" (Russell [1891–1902] 1983).

first reading, he responded critically to the passage in *Was sind* where Dedekind presents his version of non-eliminative structuralism. The passage (§73) reads as follows:

> If in the consideration of a simply infinite system *N* set in order by a transformation φ we entirely neglect the special character of the elements; simply retaining their distinguishability and taking into account only the relations to one another in which they are placed by the order-setting transformation φ, then are these elements called *natural numbers* or *ordinal numbers* or simply *numbers*, and the base-element 1 is called the *base-number* of the *number-series* N. With reference to this freeing the elements from every other content (abstraction) we are justified in calling numbers a free creation of the human mind.

This act of "freeing the elements from every other content" is now often called "Dedekind abstraction." It purportedly allows one to move from a representation of a particular model of the Peano axioms to a new independent object—what we might call the "structure" shared by all models, or simply "the numbers."

From his earliest reading,[18] Russell highlighted two features of Dedekind's view. First, what Dedekind calls "natural numbers" or simply "the numbers" are finite ordinal numbers, not cardinals. Dedekind thus defines the finite ordinals independently of defining cardinal numbers, and in fact he defines the finite cardinals in terms of the ordinals. (That is, Dedekind shows that *there are n Fs* just in case the *F*s can all be paired off 1-1 with the ordinals from 1 to *n*. See Dedekind [1888] 1963, §161.) Second, Dedekind believes that the natural numbers are arrived at by what he calls "abstraction."

I'll say more about the second feature in the following two sections. Concerning the first feature, Russell argued in the following way.[19] To say of the *F*s and the *G*s that they have the same cardinal number requires only the notion of a "correlation," i.e., a 1-1, onto relation. Modifying Russell's terminology and symbols for readability, Russell suggests the following:

> The cardinal number of *F*s = the cardinal number of *G*s iff there is a 1-1, onto relation from the *F*s to the *G*s.

On the other hand, to say of *x* (under some relation *R*) and *y* (under some relation *R′*) that they have the same ordinal number requires *both* the notion of a

[18] These two features are highlighted in a long marginal comment Russell made in April 1898 in his copy of *Was sind* next to §73. This copy is available at the Russell archives at McMaster University.

[19] This criticism was articulated in a set of notes from October 1900 (available at McMaster: RA 230.030870), and written out in prose in §232 of *POM*, which was written in November 1900.

correlation and the notion of a "serial relation." Again modifying Russell's terminology and symbols for readability, Russell holds:

> The ordinal number of x = the ordinal number of y iff x is in the co-domain, but not the domain[20] of the serial relation R, and similarly for y and the serial relation R', and there is a correlation S from the field of R to the field of R', such that for all x', y', x'', y'', if $x'Sx''$ and $y'Sy''$, then $x'Ry'$ iff $x''R'y''$.

Neither of these definitions presupposes the other. Thus, Russell holds, the ordinals need not be defined using the notion of a cardinal number, nor do the cardinals need to be defined using the notion of an ordinal number (as Dedekind in essence does). Nevertheless, since under Russell's proposed analysis of *the cardinal number of Fs = the cardinal number of Gs* and *the ordinal number of x = the ordinal number of y*, the first proposition requires only the notion of a correlation, and the second requires that same notion and a further one (namely, of a serial relation), the notion of a cardinal number is simpler than that of an ordinal number. Thus, the cardinal numbers are prior to the ordinals, when ordered by conceptual complexity.

The question of the relative priority of the notion of an ordinal and of a cardinal has been a mainstay of philosophical reflection on structuralism since the very beginning. Cassirer highlighted and defended Dedekind's view that the ordinals are conceptually prior to the cardinals, criticizing Frege's and Russell's alternative view (Cassirer 1950, 59ff.). Dummett, in his wide-ranging, probing, and highly influential critical discussion of Frege and Dedekind in his *Frege: Philosophy of Mathematics*, also highlights the issue of the conceptual priority of ordinals and cardinals (1991, 53, 293). Dummett criticizes Dedekind and other structuralists, who hold that the natural numbers are intrinsically ordinal, and defends the Fregean and Russellian view that numbers are intrinsically cardinal.[21] Charles Parsons has defended structuralism against this objection (2008, §14, 73ff.), as have W.W. Tait (1996, §§VI–VII) and Reck (2013, 159). Given this later history, it is very noteworthy that from his very first reading of Dedekind's book, Russell isolated the core philosophical issue of the priority of the cardinal and ordinals as a potential objection to Dedekind's non-eliminative structuralist theory of the natural numbers.

[20] A term that is in the co-domain but not the domain of a relation is a referent but not a relatum of the relation, as (for instance) the number 4 is in the finite ordinals up to 4 related by the successor relation. It is the "last" term in the series.

[21] For Dummett, the structuralist view of the natural numbers as intrinsically ordinal violates what has come to be called "Frege's constraint," that the definition of a mathematical object (e.g., a natural number) should make its canonical application obvious (e.g., its role in giving the cardinality of things). This argument was in fact given explicitly by Russell (1919, 9–10): "We want our numbers not merely to verify mathematical formula, but to apply in the right way to common objects. We want to have ten fingers and two eyes and one nose . . . and this requires that our numbers should have a *definite* meaning, not merely that they should have certain formal properties."

2.2. *Principles of Mathematics*, Chapter 30

As we've seen, from his very first reading of *Was sind*, Russell saw clearly the philosophical significance of the non-eliminative structuralist view suggested by Dedekind in §73, and focused on two issues: the alleged priority of ordinal over cardinal notions, and the philosophical tenability of "Dedekind abstraction." I discussed the first issue in the last section; in this section I turn to the second.

Russell addressed this second issue in earnest in a compressed and difficult-to-interpret passage that, though it was published in 1903 as chapter 30 ("Dedekind's Theory of Number") of *POM*, was actually written in November 1900. I believe that it is important to keep this date in mind, since the criticism of Dedekind abstraction in chapter 30 was written *before* Russell adopted his classic definition of cardinals as classes of equinumerous classes.[22]

In §241 of chapter 30, Russell quotes *Was sind*, §73, where Dedekind presents the natural numbers as abstractions from some simply infinite system. He objects as follows (I have numbered Russell's sentences to make later references easier, and italicized key phrases):

(1) Now it is impossible that this account should be quite correct. For it implies that the terms of all progressions other than the ordinals are complex, and that *the ordinals are elements in all such terms*, obtainable by abstraction. But this is plainly not the case. A progression can be formed of points or instants, or of transfinite ordinals, or of cardinals, in which, as we shall shortly see, the ordinals are not elements.

(2) Moreover it is impossible that the ordinals should be, as Dedekind suggests, nothing but the terms of such relations as constitute a progression. *If they are to be anything at all, they must be intrinsically something; they must differ from other entities as points from instants, or colours from sounds.*

(3) What Dedekind intended to indicate was probably a definition by means of *the principle of abstraction*, such as we attempted to give in the preceding chapter. But a definition so made always indicates some class of entities having (or being) a genuine nature of their own, and not logically dependent upon the manner in which they have been defined. The entities defined should be visible, at least to the mind's eye; *what the principle asserts is that, under certain conditions, there are such entities*, if only we knew where to look for them. But whether, when we have found them, they will be ordinals or cardinals, or even something quite different, is not to be decided off-hand.

[22] Russell adopted this definition sometime between March and June 1901. See Gregory Moore's introduction to Russell (1993, xxvii).

It will take a bit of unpacking to understand Russell's objections.[23] I will take the three objections in turn, starting with the second.

Objection (2) is directed against the metaphysical commitments that Russell finds in Dedekind's claim that the numbers "retain their distinguishability" despite having no "special character," standing only in relations to one another. Russell's objections draw on his own reflections on the metaphysics of relations. Since the time of his dissertation (in 1896), Russell had been preoccupied with an apparent paradox concerning points. Since each point is qualitatively indistinguishable from every other point, points must be distinguished by their relations to other points. If, for instance, there are two congruent triangles ABC and A′B′C′, A differs from A′ inasmuch as it stands in a certain relation to BC which A′ does not, and A′ in a certain relation to B′C′ that A does not. But what distinguishes BC from B′C′? A circularity or vicious regress threatens. Russell called this the "paradox of relativity": "a conception of difference without a difference of conception" (Russell [1898] 1983, 259; see Griffin 1991, 181ff., 317ff.; Galaugher 2013, 29ff.).

By 1900, Russell was keen to block this paradox. Russell's maneuver—which was articulated in a series of papers written in the summer of 1900, and incorporated into chapter 51 of *POM*, written in December 1900—was radical: though each point is qualitatively indistinguishable *to us*, he insisted that points are in fact all qualitatively different, even if we cannot detect these intrinsic properties.

> And more generally, two terms cannot be distinguished primitively by difference of relations to other terms; for difference of relation presupposes distinct terms, and cannot therefore be the reason why the two terms are distinct. Thus if there is any diversity at all, there must be immediate diversity, and this kind of diversity occurs between the various points of space. . . . As with people so with points: the impossibility of recognizing them must be attributed, not to the absence of individuality, but exclusively to our incapacity. (1900a, 255; cf. *POM* §428)

The supposed paradox of relativity concerning points in space, then, contravenes a principle that Russell believes holds generally: every term must have intrinsic properties peculiar to it, and no two terms can ever be distinguished by relational properties alone.

[23] There has been some discussion of Russell's criticisms of Dedekind in §241. Much of this literature, I believe, misinterprets Russell's meaning in various ways. For example, Shapiro (1997, 175), in a brief discussion, remarks only that Russell's objection "looks like Frege's Caesar problem." In fact, as I'll show, Russell's objections are quite different from the Caesar problem. See also Dummett (1991, 51–52); Tait (1996, §III); Hellman (2004, 570); Reck (2013, 145–147).

This reply to the paradox was surely fresh in his mind when he reread §73 of *Was sind* and formulated objection (2). He saw clearly that the structuralist view of the natural numbers, as intrinsically identical objects that differ only in their relational properties, was exactly like the paradoxical theory of points he rejected. Dedekind abstraction purports to take some particular progression, composed of terms with intrinsic properties, and form for us a new progression—*the* natural numbers, *the structure* common to all progressions—composed of terms that lack intrinsic properties. Russell rejects this move: "If they are to be anything at all, they must be intrinsically something."[24]

In objection (3), Russell argues that, even if Dedekind were correct in holding that the natural numbers are defined by "abstraction," it would not follow that the numbers have *only* the relational properties identified by this definition. This is because, on Russell's view, no definition (whether by abstraction, or otherwise) guarantees that the defined entities have *only* the properties that follow from the definition.

In formulating this objection, Russell interprets Dedekind abstraction in an idiosyncratic way: as an instance of what he calls definition by the "principle of abstraction." A definition by the "principle of abstraction" is a definition based on a principle, such as "Hume's Principle," that analyzes an equivalence relation into an identity claim about some new entities (see section 1.3).[25] In late 1900 and early 1901, Russell held that these definitions could be justified by a general principle, which he called the "principle of abstraction":

> This principle asserts that, whenever a relation, of which there are instances, has the two properties of being symmetrical and transitive, then the relation in question is not primitive, but is analyzable into sameness of relation to some other term; and that this common relation is such that there is only one term at most to which a given term can be so related, though many terms may be so related to a given term. (*POM*, §157)

[24] Although Russell does not point this out in *POM* §241, the paradox of relativity emerges in non-eliminativist structuralism in a more direct way. In symmetric structures, such as the integers together with addition, there is apparently no non-circular way to distinguish, say, −1 from 1. This paradox has been discussed in the contemporary literature on structuralism: e.g., Keränen (2001) and Parsons (2008, 107ff.). Contemporary philosophers have noted the affinity between this paradox and Kant's argument from incongruent counterparts; Russell had noted, a century earlier, an affinity between Kant's argument and the paradox of relativity (*POM*, §214n).

Of course, Euclidean 3-space is symmetric in uncountable ways, and so admits of uncountably many structure preserving nontrivial automorphisms. So the paradox discussed by Keränen and Parsons applies even more radically to space than to the integers. In this sense, this contemporary paradox is a special case of the more general paradox of relativity. Again, Russell's solution would be to deny the very possibility of objects with no distinguishing intrinsic properties.

[25] Russell in fact defines cardinal numbers in just this way in the first draft of "Logic of Relations" (Russell 1900b, §3, proposition 1.4).

According to this principle, the relation of equinumerosity (for example) between the class *F* and the class *G* is analyzable into a new relation, *is the number of*, that holds between both *F* and *G* and some new object, a cardinal number. Cardinal numbers are then defined as those objects to which equinumerous classes stand in the *is the number of* relation. Thus, when Russell was composing objection (3), he accepted definition by the "principle of abstraction" as an acceptable form of abstraction, and interpreted Dedekind abstraction accordingly.[26]

Russell's objection, then, is that though we define the numbers only in terms of the structural properties mentioned in the definition, it does not follow that the entities defined have *only* the properties that are mentioned in the definition. As Russell put it: "a definition so made always indicates some class of entities having (or being) a genuine nature of their own, and not logically dependent upon the manner in which they have been defined." Thus, though we make no mention of intrinsic properties in the definition, it does not follow that the defined entities themselves in fact lack intrinsic properties. A more pedestrian example will make this clear. If *A* and *B* are full siblings, then—in accordance with the principle of abstraction, since *is a full sibling with* is an equivalence relation—*A* and *B* stand in some common relation to some common third thing—in this case, a common set of parents. We can then define *the parents of A and B* by abstraction. But it surely does not follow that *A*'s and *B*'s parents have *only* the property of being parents—they are also intrinsically a certain height and weight. Each of them is an "an actual [person] with a tailor and a bank-account or a public-house," to repurpose a well-known Russellian passage (§56).

One possible reply to this objection would be to emphasize Dedekind's claim that the numbers are a "free creation of the human mind." On one possible interpretation of this phrase, Dedekind means that the mathematician, in performing Dedekind abstraction, *creates* a new set of objects.[27] These objects, plausibly, would fail to have nonstructural properties because the mathematician,

[26] Though Russell interprets Dedekind abstraction idiosyncratically as an instance of definition by the principle of abstraction, I do not believe that Russell's objection (3) depends on this interpretation. After all, Russell denies that definition by Dedekind abstraction picks out objects with only structural properties, not because of some specific feature of definition by the principle of abstraction, but because of a general feature of definitions in general: it never follows, from the fact that an object is defined as φ, that an object is only φ and therefore lacks properties that are not implied by the definition.

[27] A psychologistic reading of Dedekind is suggested by Dummett 1991; a non-psychologistic reading was first given by Cassirer (and by many others since: e.g., Reck 2013; Yap 2017). For Cassirer's non-psychologistic reading of Dedekind, see Reck's chapter on Cassirer.

Russell, at least in *POM* and earlier, does not read Dedekind psychologistically. (In this way, Russell's discussion of Dedekind's structuralism is both more sympathetic and more interesting than many later objections, e.g., by Dummett.) Russell's best reconstruction of Dedekind abstraction interprets it as definition from the principle of abstraction, which he took to be a candidate logical (not psychological) principle, motivated by a mind-independent metaphysical fact about equivalence. Indeed, none of the objections that are surveyed in this chapter depend on reading Dedekind in a psychologistic way.

in creating them, removed these intrinsic properties. Russell does not read Dedekind in this psychologistic way, and so does not formulate explicitly a response to this reply. However, it is clear that Russell would be deeply opposed to this way of thinking. We saw already, in his assertion that points do have intrinsic properties, even if they are indistinguishable *to us*, that Russell was deeply committed to the mind-independence of all entities, even mathematical entities. What is true is independent of the mind, both in its being, and in its being true. In the same vein, things do not come into being by being defined by us. The principle of abstraction does not bring new abstract objects into being. It is simply a true proposition about mind-independent reality: "what the principle asserts is that, under certain conditions, there are such entities." Furthermore, the defined entities are not under our control; it is emphatically not the case that they have only the properties that *we* give them.

Objection (1) draws on Russell's peculiar way of defending the "principle of abstraction." Russell in November 1900 motivated the principle on the grounds that it is an explication of the widespread philosophical intuition that equality and other relations akin to it (namely, equivalence relations) are "always constituted by possession of a common property" (§157). If two classes are equinumerous, they must have something in common (namely, the property of having *n* members); if two lines are parallel, the two lines must share some feature (namely, having such and such direction). However, Russell raises a worry about this defense. Plausibly, the intuition that equivalence relations are constituted by possession of a common property could be explicated in this way: for any relation *R* that is transitive, symmetrical, and non-empty,

$$(*)\ \exists S \text{ such that } \forall x, y\big(xRy \equiv \exists z(xSz \,\&\, ySz)\big).$$

This appears to be a perfectly correct explication of the intuition, where the "common property" is *being related by S to z*. However, on this explication, the right-hand side of the biconditional does not guarantee the transitivity of the relation *R*, for the following reason. Suppose A is equivalent under *R* to B, and A and B share property P, while B is equivalent under *R* to C, and B and C share property Q. It would therefore not follow that there is any property that A and C share. Thus, the fact that two equivalent terms share some property cannot be an *analysis* of what it is to stand in an equivalence relation, since *sharing a property*, in the sense of (*) guarantees only the symmetry, not the transitivity, of *R*.

Russell blocks this worry by insisting that, in the cases where we want to use a principle of abstraction to analyze an equivalence relation, the relation *S* is many-one: "In order that [the relation *R*] may be transitive, the relation [*S*] to the common property must be such that only one term at most can be the property of any given term" (§157). An example of a many-one relation is *x is the number of Fs*, which could be used to analyze by abstraction the relation *equinumerosity*; an example that is not many-one is *x is a parent of y*, which therefore could not be used to analyze by abstraction the relation *being a full sibling*. But what reason could be given, for a specific equivalence relation, that would guarantee that the relation that holds between the equivalent terms and the common property be many-one? Russell addresses this worry in the context of the relation *equality*, which he analyzes, through the principle of abstraction, in terms of abstract magnitudes: Two quantities (for instance, two material bodies A and B) are equal (say, in mass) if the body A *has* the magnitude *M* and body B *has* the magnitude *M*. Russell further claims that in this case, the abstractum (the magnitude; in our example, a magnitude of *M grams*) is an "element" of the concreta (the quantities; in our example, the two material bodies A and B) from which it can be abstracted (*POM*, §157). The relation between the quantity and the magnitude that it has is many-one, since, Russell argues, it is an "axiom" that only one magnitude can exist at a given spatiotemporal place. Thus, there cannot be *two* magnitudes of a given kind that both exist in the location where body A is located. That means that the troublesome case that I described in the previous paragraph cannot arise for equal quantities and their common properties, and transitivity is guaranteed after all.

The fact that a quantity has one and only one magnitude as its "element,"[28] then, explains why the principle of abstraction can be used to analyze the relation of equality, and magnitude can be defined by abstraction. Will the same be true in the case of the natural numbers, if they are defined by abstraction? We saw, in the case of objection (3), that Russell used his particular way of understanding definitions by abstraction to try to make sense of Dedekind's talk of "abstraction." I believe that this is true also of objection (1), and explains why he alleges that Dedekind's procedure can make sense only if ordinals are always "elements" of any terms arranged in a progression. Russell writes that Dedekind "implies that the terms of all progressions other than the ordinals are complex, and that the ordinals are *elements* in all such terms, obtainable by abstraction" (emphasis added).

Let me spell out in some more detail how Russell is interpreting Dedekind's "abstraction." Each progression, whether it be of natural numbers, points, or propositions, stands in an equivalence relation (namely, *being isomporphic to*) to every other progression. Similarly, each element in a progression stands in an equivalence relation to every corresponding element in some other progression. For example, 4, the fourth element of the natural number series, stands in an equivalence relation to D, the fourth element in the English alphabet: the

[28] By "element," Russell most likely here means what he calls a "part" in *POM*, chap. 16 ("Whole and Part").

relation is *having the same ordinal position in one's series as.*[29] Using the abstraction schema (*), the fact that 4 and D stand in the relation *R* (*having the same ordinal position in one's series as*) implies that there is some relation *S* between 4 and D and some abstract object *z*. The abstract object *z* is the "position" of 4 and D, and the relation *S* is the relation between 4 and the position that it occupies. But why think that the relation between 4 and its position is many-one?

Russell is probing what is plausibly a vulnerable commitment in Dedekind's picture: what guarantees that a definition by Dedekind abstraction will pick out a unique set of objects, *the* natural numbers?[30] For Russell, the only plausible reason is if the natural numbers are *elements* of all the objects that are ordered into progressions, just as (he claimed) magnitudes are elements in all quantities. Thus, Dedekind requires that "terms of all progressions other than the ordinals are complex, and that the ordinals are elements in all such terms." But, Russell alleges, this is plainly not the case. As Russell emphasizes strongly (§231), the position that a term occupies in a series is not intrinsic to the term itself, and there are infinitely many possible orderings of, say, the finite cardinals into a progression. In one series (1, 2, 3, 4, . . .) 4 is fourth, but in another (1, 3, 5, 7, 9, 2, 4, 6, 8, 10, . . .) 4 is seventh. So the cardinal number 4 must contain as an element both the ordinal 4 and the ordinal 7, and clearly an infinite number of other elements besides. But this is absurd.

2.3. *Principles of Mathematics*, Part II, Chapter 14

Russell returned to Dedekind's theory of the natural numbers seven months later, in June 1901, when he wrote Part II of *POM*, on cardinal numbers. In this part, he presents his classic definition of cardinal numbers as classes of equinumerous classes (§111), which he had developed sometime in March to June of 1901[31]—after he wrote the texts I discussed in sections 2.1 and 2.2 of this chapter. In *POM* Part II, Russell uses his definition of cardinal numbers to define (in chap. 14) the natural numbers (in essence, as classes of equinumerous finite classes). In defending this definition, he considers other definitions of the natural numbers by abstraction (§122). In this section, he poses the question: "Is any process of abstraction from all systems satisfying the five [Peano] axioms . . . logically possible?"

[29] More precisely: the series of numbers up to 4 is ordinally equivalent to the series of letters up to D. This is the notion of "likeness," which Russell defines in *POM*, §231.

[30] This objection is particularly pressing on psychologistic readings of Dedekind. Suppose I take some progression and freely create, by abstraction, a new system of objects, the numbers. Suppose you take the same progression and freely create a system: need it be the same system as the one I freely created? Or suppose I perform the act of abstraction a second time on the same progression: will I again get the same system of abstract objects? There needs to be some reason why the answer to these questions must be yes.

[31] See note 22.

He answers in the negative, giving a series of new objections to theories of abstraction such as Dedekind's.[32] In this section, I identify three such objections.

The first objection concerns the identity of the abstracta. Suppose Dedekind could identify the natural numbers as the unique elements of a progression that have merely structural properties. Even so, each of the progressions 0, 1, 2, . . . and 1, 2, 3, . . . satisfies Dedekind's definition of a progression, and each can make an equally good claim to be composed of elements with merely structural properties. So which progression is *the* numbers?[33] As Russell points out, if we consider the numbers with respect to their cardinal character, we can distinguish these two cases, but Dedekindian structuralists preclude this when they conceive of the numbers as having no features besides the structural features they have in virtue of being elements in a progression.[34] Perhaps, one might contend, the numbers are what one gets when one abstracts away the differences between the progression 0, 1, 2, . . . and the progression 1, 2, 3, ... But this is absurd, for then the products of that abstraction—the numbers themselves—would have to be distinct from the progressions from which they are abstracted: that is, they would have to be distinct from every progression of numbers.[35]

Russell considers, and rejects, one plausible escape from this objection. One might insist that the natural numbers are to be identified with neither 0, 1, 2, . . . nor 1, 2, 3, . . . since *the* natural numbers are that unique progression that has nothing but merely structural, and so no intrinsic, properties. Thus, the first element of the natural number progression is neither 0 nor 1, since it is not intrinsically anything other than the first element in the progression. But as we saw in objection (3) from section 2, Russell denies that possibility: "there is therefore no term of a class which has merely the properties defined by the class and no others" (§122). So there is no progression in the class of progressions that is merely a progression and nothing else.

[32] In §122, Russell specifically targets Peano's view that the natural numbers are defined by abstraction from what all progressions have in common. (On the use of definitions by abstraction in Peano and his school, see Mancosu 2016, chap. 2.2.1.) He clearly intended his criticisms to support his class-theoretic definition by undermining *every* definition of the natural numbers "by abstraction"—not just Peano's. Moreover, most of the objections leveled against Peano would, if valid, also apply to Dedekind's definition of the natural numbers by abstraction.

[33] This objection arises even on psychologistic readings of Dedekind. For suppose I create the numbers, and then pick them out ostensively as *the progression that I just created.* Still, each progression can also make an equally plausible claim to being *the progression that I just created*—since, if I create a new progression by abstraction and call it "the numbers," I would still be at a loss whether the first element is 0 or 1.

[34] One reply to this worry is to admit that the progression of numbers, defined by Dedekind abstraction from (0, N, S), cannot be identified with either series. However, when we bring in arithmetical operations and define the numbers by Dedekind abstraction from (0, N, S, +, ×), we expand the structure and definitively settle on one of the two alternatives. Russell does not consider this reply.

[35] *POM*, §122. More recently, this objection was directed against non-eliminative structuralists by Dummett (1991, 53). Parsons (2008, 76–78) provides a reply to Dummett, which to me at least is convincing. This objection is also articulated, and endorsed, though without reference to *POM* §122, in Hellman (2004, 572).

There is, however, one way that Russell identifies for Dedekind and other abstractionists to get around this objection: they could regard the symbols "0," "successor," and "number" as really variables:

> [One could] regard 0, number, and succession as a class of three ideas belonging to a certain class of trios defined by the five primitive propositions. It is very easy so to state the matter that the five primitive propositions become transformed into the nominal definition of a certain class of trios. There are then no longer any indefinables or indemonstrables in our theory, which has become a pure piece of Logic. But 0, number and succession become variables, since they are only determined as one of the class of trios. (§122)

This of course is the eliminative structuralism that we first encountered in section 1.3 in the context of Russell's discussion of relational types. On this view, a sentence of arithmetic is just a universally quantified conditional: for every x, class N, and relation S, if $\{x, N, S\}$ are an object, class, and relation that satisfy the axioms of arithmetic, then $\psi(x, N, S)$. This brings us to Russell's second objection: once this eliminative structuralist alternative is clearly articulated, Dedekind's non-eliminative structuralism becomes unmotivated. Dedekind insists that the intrinsic character of the numbers is irrelevant; but this insistence is satisfied by the eliminative procedure (whereby arithmetic is about all objects that form progressions, regardless of their intrinsic properties), just as much as it is satisfied by the non-eliminative procedure (whereby arithmetic is about some sui generis objects with no intrinsic properties).

Nevertheless, eliminative structuralism itself faces one last significant hurdle. Even if we construe the primitive symbols of arithmetic as variables, and treat every sentence of arithmetic as a claim about every class $\{x, N, S\}$ that satisfies the axioms of arithmetic, "nothing shows that there are such classes as the definition speaks of" (§123). Suppose the Dedekindian structuralist were able to evade objection (1) from section 2.2 of this chapter, by coming up with a principled reason why the relation S between the progression from which the numbers are abstracted and the numbers themselves is many-one. There is still a more fundamental worry, which even the eliminative structuralist must face. What justifies the claim that there is any relation S at all, or any abstract objects z? Surely, if definition by abstraction were creative, and the mathematician's act of abstraction produced the abstracta, these existence claims could be satisfied. But Russell rejects creative definitions.[36] Instead, Russell suggests that the existence claim can be justified only by explicitly constructing the numbers from classes. The class {0, N, successor}, defined in Russell's now well-known way in terms of

[36] Dedekind famously argued that his *Gedankenwelt* is an instance of a progression ([1888] 1963, §66). On Russell's reception of this argument, see Reck (2013, 147–149).

classes of equinumerous finite classes, proves the existence of trios that satisfy the Peano axioms. But, now, even eliminative structuralism is unmotivated. For once we've explicitly constructed in a class-theoretic way finite cardinals that satisfy the Peano axioms, the extra step of treating sentences of arithmetic in the eliminative structuralist way itself feels otiose. And the particular brand of logicism that Russell made famous in the published version of *POM*, and later in *Principia*, is left as the only plausible philosophy of mathematics.

This last objection to even eliminative structuralist is the very objection that Russell famously expressed, almost 20 years later, in his quip about theft and honest toil. The sentiments behind this quip have been well studied and elaborated in the century since it was written. Far less, unfortunately, has been devoted to the wealth of Russell's thinking that I have laid out in this chapter. I hope that this chapter has shown, though, that Russell's engagement with structuralist ideas was far deeper, more extensive, and more complex than a narrow focus on the virtues of honest toil would suggest.

Acknowledgments

This chapter owes a great deal to the Bertrand Russell Research Centre and the Bertrand Russell Archive at McMaster University, whose staff enabled my visit in the fall of 2015 to be so productive and enjoyable. I would especially like to thank Nicholas Griffin and Kenneth Blackwell for helping me to navigate and interpret Russell's manuscripts. I would also like to thank the philosophy department at McMaster, especially Sandra Lapointe and Richard T. W. Arthur, for hosting me as the 2015–16 Bertrand Russell Visiting Professor. Thanks also to Bahram Assadian and to audiences at the University of California, Riverside, and the Logic Seminar at the University of California, Irvine, for very helpful feedback and criticisms. Finally, I would like to heartily thank the editors for their sage advice, both philosophical and editorial.

References

Carnap, Rudolf. 1922. *Der Raum: Ein Beitrag zur Wissenschaftslehre*. Volume 56 of *Kant-Studien Ergänzungshefte*. Berlin: Reuther & Reichard.

Cassirer, Ernst. 1907. Kant und die moderne Mathematik. *Kant-Studien* 12, 1–40.

Cassirer, Ernst. [1910] 1923. *Substanzbegriff und Funktionsbegriff: Untersuchungen über die Grundfragen der Erkenntniskritik*. Berlin: Bruno Cassirer. Translated by William Curtis Swabey and Marie Collins Swabey in *Substance and Function and Einstein's Theory of Relativity*. Chicago: Open Court.

Cassirer, Ernst. [1921] 1923. *Zur Einstein'schen Relativitätstheorie*. Berlin: Bruno Cassirer Verlag. Translated by William Curtis Swabey and Marie Collins Swabey in *Substance and Function and Einstein's Theory of Relativity*. Chicago: Open Court.

Cassirer, Ernst. 1950. *The Problem of Knowledge: Philosophy, Science, and History since Hegel*. Translated by William H. Woglom and Charles W. Hendel. New Haven: Yale University Press.

Dedekind, Richard. [1888] 1963. *Was sind und was sollen die Zahlen?* Translated as The Nature and Meaning of Numbers. In *Essays on the Theory of Numbers*, translated by W. W. Beman, pp. 29–115. New York: Dover.

Dummett, Michael. 1991. *Frege: Philosophy of Mathematics*. Cambridge, MA: Harvard University Press.

Frege, Gottlob. 1884. *Die Grundlagen der Arithmetik: Eine logisch-mathematische Untersuchung über den Begriff der Zahl*. Breslau: Koebner.

Galaugher, Jolen. 2013. *Russell's Philosophy of Logical Analysis*. New York: Palgrave Macmillan.

Gandon, Sébastien. 2012. *Russell's Unknown Logicism: A Study in the History and Philosophy of Mathematics*. New York: Palgrave Macmillan.

Grassmann, Hermann. [1844] 1894. *Lineale Ausdehnungslehre, eine neuer Zweig der Mathematik*. In *Gesammelte mathematische und physikalischer Werke*, vol. 1, part 1, edited by Friedrich Engel. Leipzig: Teubner.

Griffin, Nicholas. 1991. *Russell's Idealist Apprenticeship*. Oxford: Clarendon Press.

Heis, Jeremy. 2011. Ernst Cassirer's Neo-Kantian Philosophy of Geometry. *British Journal for the History of Philosophy* 19, 759–794.

Hellman, Geoffrey. 1989. *Mathematics without Numbers*. Oxford: Oxford University Press.

Hellman, Geoffrey. 2004. Russell's Absolutism vs. (?) Structuralism. In *One Hundred Years of Russell's Paradox*, edited by G. Link, pp. 561–575. Berlin: de Gruyter.

Keränen, Jukka. 2001. The Identity Problem for Realist Structuralism. *Philosophia Mathematica* 3, 308–330.

Mancosu, Paolo. 2016. *Abstraction and Infinity*. New York: Oxford University Press.

Musgrave, A. 1977. Logicism Revisited. *British Journal of Philosophy of Science* 38, 99–127.

Parsons, Charles. 2008. *Mathematical Thought and Its Objects*. New York: Cambridge University Press.

Pieri, M. 1898. I principia della geometria di posizione composti in sistema logico deduttivo. *Memoria della R. Accademia della scienze di Torino* 48, 1–62.

Reck, Erich. 2003. Dedekind's Structuralism: An Interpretation and Partial Defense. *Synthese* 137, 369–419.

Reck, Erich. 2013. Frege or Dedekind? Towards a Reevaluation of their Legacies. In *The Historical Turn in Analytic Philosophy*, edited by Erich Reck, pp. 139–170. London: Palgrave Macmillan.

Reck, Erich, and Michael P. Price. 2000. Structures and Structuralism in Contemporary Philosophy of Mathematics. *Synthese* 125, 341–383.

Russell, Bertrand. 1891–1902] 1983. What Shall I Read? In *The Collected Papers of Bertrand Russell*, vol. 1, edited by Kenneth Blackwell, Andrew Brink, Nicholas Griffin, Richard A. Rempel, and John G. Slater, pp. 345–370. London: George Allen & Unwin.

Russell, Bertrand. 1897. *An Essay on the Foundations of Geometry*. Cambridge: Cambridge University Press.

Russell, Bertrand. [1898] 1983. On the Principles of Arithmetic. In *The Collected Papers of Bertrand Russell*, vol. 2, edited by Nicholas Griffin and Albert C. Lewis, pp. 245–260. London: Unwin Hyman.

Russell, Bertrand. 1900a. The Notion of Order and Absolute Position in Space and Time. In Russell 1993, pp. 234–258.

Russell, Bertrand. 1900b. The Logic of Relations with Some Applications to the Theory of Series. First draft. In Russell 1993, pp. 590–612.

Russell, Bertrand. 1901a. Recent Work on the Principles of Mathematics. In Russell 1993, pp. 363–379.

Russell, Bertrand. 1901b. *Principles of Mathematics, Part I. 1901 Draft*. In Russell 1993, pp. 181–208.

Russell, Bertrand. 1901c. The Logic of Relations with Some Applications to the Theory of Series. In Russell 1993, pp. 310–349.

Russell, Bertrand. 1902. Geometry, Non-Euclidean. In Russell 1993, pp. 470–504.

Russell, Bertrand. 1903. *Principles of Mathematics*. Cambridge: Cambridge University Press.

Russell, Bertrand. 1937. *Principles of Mathematics*. 2nd ed. Cambridge: Cambridge University Press.

Russell, Bertrand. 1919. *Introduction to Mathematical Philosophy*. New York: Macmillan.

Russell, Bertrand. 1993. *The Collected Papers of Bertrand Russell*. Vol. 3. Edited by Gregory H. Moore. New York: Routledge.

Russell, Bertrand. 2002. *The Selected Letters of Bertrand Russell*. Vol. 1: *The Private Years, 1884–1914*. Edited by Nicholas Griffin. 2nd ed. New York: Routledge.

Schiemer, Georg. 2018. Cassirer and the Structural Turn in Modern Geometry. *Journal for the History of Analytical Philosophy* 6(3), 163–193.

Shapiro, Stewart. 1997. *Philosophy of Mathematics: Structure and Ontology*. New York: Oxford University Press.

Tait, W. W. 1996. Frege versus Cantor and Dedekind: On the Concept of Number. In *Frege: Importance and Legacy*, edited by M. Schirn, pp. 70–113. Berlin: de Gruyter.

Whitehead, A. N. 1898. *A Treatise on Universal Algebra*. Cambridge: Cambridge University Press.

Yap, Audrey. 2017. Dedekind and Cassirer on Mathematical Concept Formation. *Philosophia Mathematica* 25(3), 369–389.

13

Cassirer's Reception of Dedekind and the Structuralist Transformation of Mathematics

Erich H. Reck

For much of the 20th century, Ernst Cassirer was seen as an intellectual historian, besides being the last member of Marburg Neo-Kantianism. More recently, he has been rediscovered as an original, substantive philosopher in his own right, perhaps even one of the great philosophers of the 20th century. This concerns his contributions to the philosophy of science (relativity theory, quantum mechanics, etc.), his mature, wide-ranging philosophy of symbolic forms (leading to a "philosophy of culture"), and the ways in which his views position him, in potentially fruitful ways, at the intersection of "analytic" and "continental" philosophy.[1] In addition, Cassirer was a keen observer of developments in pure mathematics, especially of their philosophical significance. There are two separable, though not unrelated, strands on that topic in his writings. The first concerns his reflections on revolutionary changes in geometry during the 19th century, culminating in David Hilbert's and Felix Klein's works. The second strand involves parallel changes in algebra, arithmetic, and set theory, where Evariste Galois, Richard Dedekind, and Georg Cantor played key roles.

In this essay the main focus will be on the second of the strands just mentioned, and in particular, on Cassirer's reception of Dedekind, which still deserves more attention.[2] As we will see, Cassirer was a perceptive reader of Dedekind, arguably still his subtlest philosophical interpreter. What he was concerned about with

[1] For the philosophy of physics, cf. Ryckman (2005) and French (2014), more generally also Ihmig (2001) and Part I of Friedman and Luft (2017); for the philosophy of culture, cf. Recki (2004), Luft (2015), and Part III of Friedman and Luft (2017); for Cassirer's position between the analytic and continental traditions, cf. Friedman (2000) and Part II of Friedman and Luft (2017). For more general discussions, see also Ferrari (2003) and Kreis (2010).

[2] Cf. Reck (2013), chapter 4 of Biagioli (2016), Yap (2017), and Heis (2017) for some recent discussions of the topic. For Cassirer's views on geometry, Klein's Erlangen program, and Hilbert's axiomatics, which have found more attention in the literature already, cf. Ihmig (1997, 1999), Mormann (2007), Heis (2011), most of Biagioli (2016), and Schiemer (2018). I will come back to the latter briefly later in this essay.

Erich H. Reck, *Cassirer's Reception of Dedekind and the Structuralist Transformation of Mathematics* In: *The Prehistory of Mathematical Structuralism*. Edited by: Erich H. Reck and Georg Schiemer, Oxford University Press (2020). © Oxford University Press.
DOI:10.1093/oso/9780190641221.003.0013

respect to mathematics in general was the introduction of "ideal elements", together with related, very significant expansions of its scope over time. This led to a reconsideration of its subject matter, including the rejection of the traditional view that mathematics is "the science of quantity and number". Cassirer's discussion of this topic often took place under the label of "concept formation" in science; and he identified a corresponding shift from "substance concepts" to "function concepts", seen as culminating in the 19th and early 20th centuries. What he arrived at with such considerations was, in later terminology, a structuralist conception of mathematical objects; and that conception was rooted in observations about mathematical methodology. Dedekind's work was decisive for Cassirer since he saw in it the clearest and most powerful example of the structuralist "unfolding" of mathematics, i.e., of the systematic, mature development of older structuralist "germs" in it.[3]

This essay will proceed as follows: first, an outline of Cassirer's overall perspective on mathematics will be provided. In the second section, we will turn to a brief summary of Dedekind's relevant contributions, one in which their crucial but also controversial structuralist dimension will be highlighted. Third, we will see how Cassirer's sympathetic reception of Dedekind's structuralism contrasts sharply with criticisms and dismissals by other philosophers, starting with Frege and Russell. This will lead to some historically grounded and philosophically significant observations about "existence," "determinateness," and "givenness" in modern mathematics. In the fourth section, several aspects of Cassirer's own views about structuralism, related to but also going beyond Dedekind, will be discussed. The latter will include a deeper motivation for structuralism than is usually provided today; some original views about the role of constructions in structuralist mathematics, together with its historical "unfolding"; and his insistence on the fact that the metaphysics and the methodology of mathematics, or of any science for that matter, should be viewed as inseparable. A brief conclusion will round off the essay.

1. Cassirer's Overall Perspective on Mathematics

Dedekind's contributions to mathematics play a prominent role in Cassirer's writings from early on. The first clear expressions of this fact occurs in his survey article "Kant und die moderne Mathematik" (1907), the second in his first systematical book, *Substanzbegriff und Funktionsbegriff* (1910). Dedekind remains an important reference point later on, e.g., in *Die Philosophie der*

[3] A second aspect of Dedekind's work important for Cassirer was his logicism. While not unrelated, I will leave it largely aside here; cf. Reck and Keller (forthcoming) for more.

Symbolischen Formen, vol. 3 (1929) and in *The Problem of Knowledge*, vol. 4 (1950).[4] The general context is Cassirer's discussion, in the relevant parts of these works, of the rise of modern mathematics and mathematical science—from Kepler's, Galilei's, and Descartes's innovative "mathematization" of nature, through the introduction of the integral and differential calculus by Leibniz, Newton, and their followers, to a number of developments in the 19th century.

With respect to the emergence of modern mathematics, there are two main strands one can distinguish: the gradual acceptance and systematization of various new geometries (projective, elliptic and hyperbolic, etc.), leading to David Hilbert's formal axiomatics and Felix Klein's "Erlangen program"; and the parallel expansion and diversification of algebra and arithmetic (complex numbers, Galois theory, Hamilton's quaternions, new conceptions of the real numbers, etc.), which brought with it the rise of set theory (including Cantor's transfinite numbers) and the replacement of Aristotelian logic by modern mathematical logic (Boole, Frege, Peano, Russell, and others). One noteworthy component of both strands is the introduction and systematic use of "ideal elements" in modern mathematics, such as points at infinity in projective geometry or, earlier, the complex numbers.

Cassirer was not the only philosopher surveying and analyzing these developments at the time. In fact, in this respect he followed his teachers in the Marburg School: Hermann Cohen and Paul Natorp (cf. Cohen 1883 and Natorp 1910). However, both Cohen and Natorp make the concept of the infinitesimal central to their accounts of science, while Cassirer shifts to a different perspective. He fully accepts the "arithmetization of analysis" by Cauchy, Bolzano, Weierstrass, Cantor, Dedekind, and others, with its replacement of infinitesimals by the familiar ε-δ treatment of limits. Unlike Cohen and Natorp, he also emphasizes that set theory and modern logic are natural next steps in this development, just as Hilbert's and Klein's approaches are with respect to unifying the new geometries. Moreover, Cassirer explicitly endorses Cantor's and Dedekind's emphasis on "mathematical freedom", i.e., the fact that modern mathematics has gone far beyond what is suggested in applications to nature and is exploring radically new "conceptual possibilities".[5]

What all these developments require, if we want to account for them systematically, is a novel conception of mathematics with respect to both its methodology and its subject matter. In Cassirer's own words:

[4] Cassirer mentions Dedekind in other writings too, including his early book on Leibniz (1902), his monumental series, *Das Erkenntnisproblem*, vols. 1–3 (1906–1910), and some of his later works, e.g., *An Essay on Man* (1944). But Cassirer (1907, 1910, 1929, 1950) will be the main sources of evidence for me, since they contain the most relevant and extensive discussions.

[5] For the idea of "mathematical freedom," cf. Tait (1996), for the exploration of new "conceptual possibilities," Stein (1988). (While in line with Cassirer's approach, neither of them mentions him.)

> Mathematics is no longer—as it was thought of for centuries—the science of quantity and number, but henceforth encompasses all contents for which complete law-like determinateness and continuous deductive interconnection is achievable. (Cassirer 1907, 40, my trans.)

It should be clear what is given up here, namely the view of mathematics as "the science of quantity and number", with its roots going back to Euclid. But what does Cassirer have in mind when he writes about "complete law-like determinateness" and "deductive interconnection"? Presumably these are meant to encompass the new developments in geometry, algebra, and arithmetic already mentioned. But how exactly; and what are some specific examples?

As we will see soon, it is Dedekind's treatment of the natural numbers and the real numbers that serves as the new paradigm for Cassirer here. It is primarily, although not exclusively, with those examples in mind that he writes:

> Here we encounter for the first time a general procedure that is of decisive significance for the whole formation of mathematical concepts. Wherever a *system of conditions* is given that can be realized in different contents [*das sich in verschiedenen Inhalten erfüllen kann*], we can hold on to the form of the system as an *invariant*, putting aside the difference in contents, and develop its laws deductively. (Cassirer [1910] 1923, 40, trans. modified)

As a relevant "system of conditions", consider Dedekind's characterization of the real numbers in terms of the concept of a continuous ordered field; and as two ways of "realizing" these conditions, take Dedekind's construction via the system of cuts on the rational numbers and Cantor's alternative construction via equivalent classes of Cauchy sequences (more on both later). The "invariant" to which we hold on in this case is "the real numbers"; and we "develop their laws deductively" based on Dedekind's definitions. This, then, is a paradigm of "law-like determinateness" and "logical interconnection".

Cassirer does not call the resulting "invariant", or the system of abstract objects thereby characterized, a "structure". But he comes close, e.g., when he writes:

> In this way we produce a new "objective" formation [*Gebilde*] whose structure [*Struktur*] is independent of all arbitrariness. But it would be uncritical naïveté to confuse the object thus arising with sensuously real and actual things. We

> cannot read off its "properties" empirically; nor do we need to, for it is revealed in all its determinateness as soon as we have grasped the relation from which it develops in all its purity. (1910, 40–41, trans. modified)

In the example just used, the "objective formation" is the system of real numbers as introduced by Dedekind—which "has", or alternatively "is", a certain structure. The fact that its "determinateness" is independent of empirical facts corresponds to the "mathematical freedom" Dedekind and Cantor emphasized. And the resulting "purity" has to do with the fact that all of this can be done in "pure logic" for both Dedekind and Cassirer. Finally, what is crucial about this conception of mathematics for Cassirer is that it is applicable equally to older, seemingly concrete parts of mathematics, such as elementary arithmetic or Euclidean geometry, and to novel, more abstract parts involving "ideal elements", e.g., complex numbers and points at infinity—both can now be understood as concerning (relational or functional) structures. Along such lines, pure mathematics in its entirety concerns "ideal" objects.

In Cassirer's 1910 book, *Substanzbegriff und Funktionsbegriff*, the conception of pure mathematics and mathematical science that results is characterized as involving "function concepts", as opposed to "substance concepts". Before examining further how he understands that distinction, let me complete my initial survey of Cassirer's perspective on mathematics throughout his career. While the focus in his 1910 book is on "scientific cognition", Cassirer broadens his point of view considerably during the 1920s and 1930s, by developing his wide-ranging philosophy of symbolic forms. Basically, a "symbolic form" is a way of "objectifying" various things, or better, a way of "constituting" both subjects and objects; and Cassirer now identifies several of them as integral parts of human culture.[6] The symbolic form at play in mathematical science, especially in its modern shape, remains a prime example (in some sense the most advanced example, although all are interdependent in the end); but there is also a variety of others, including mythical and religious thought, ordinary language and ordinary knowledge, art, history, law, technology, etc. (in an open-ended list).

According to Cassirer's mature position, human thought always involves symbolic processes, i.e., various ways of determining, constituting, and presenting things, be it in science or in other cultural spheres. The primary foil in this connection, i.e., the view to which he is fundamentally opposed, is a kind of naive realism according to which objects are simply "given" to subjects in experience, without any symbolic mediation or constitution (with nature already "cut at its

[6] What exactly a "symbolic form" is, or how Cassirer understands the underlying notion of "symbol," is a complex question. Roughly, a "symbolic form" is a system of signs, rules, and practices used to represent, and constitute in the first place, aspects of the world or of oneself. For more, cf. Cassirer (1923, 1927, 1929), Ferrari (2003), chapter 6, and much of Kreis (2010).

joints", as it were). Cassirer follows in Kant's footsteps in this respect, and more specifically, his "critical" approach to philosophy" (in the form adopted by the Marburg School).[7] According to how he develops that position further, from the 1920s on, his focus on the symbolic constitution of subjects and objects requires close attention to logical and methodological issues.

Cassirer calls the general perspective that results "logical idealism". With his original example of mathematics in the foreground (although the core points apply more generally), he characterizes it as follows:

> Logical idealism starts from an analysis of mathematical "objects" and seeks to apprehend the peculiar determinacy of these objects by explaining them through the peculiarity of the mathematical "method," mathematical concept formation, and the formulation of its problems. (Cassirer [1929] 1965, 405, trans. modified slightly)

Cassirer's paradigm in the case of pure mathematics, i.e., his main inspiration and illustration, remains Dedekind besides Cassirer 1929, cf. Cassirer 1950 and 1999. And it is to Dedekind's (methodological and metaphysical) structuralism that we now turn in more detail.

2. Dedekind's Structuralism and Its Critical Reception

The two texts by Dedekind on which Cassirer focuses, like most later philosophers of mathematics, are his 1872 essay, *Stetigkeit und irrationale Zahlen*, on the real numbers $\mathbb{R}$, and his 1888 essay, *Was sind und was sollen die Zahlen?*, on the natural numbers $\mathbb{N}$.[8] In both, Dedekind does exactly what we saw Cassirer highlight: he formulates "systems of conditions" that can be "realized in different contents"; and he considers a corresponding "objective formation", i.e., an abstract structure that is logically and fully determined by the system of conditions.

In the 1872 essay, the relevant "system of conditions"—which defines a (higher-order) concept—is that for a "continuous ordered field". Actually, Dedekind introduced the concept of a field (*Körper*) already earlier, in his writings on algebra and algebraic number theory.[9] What he adds now is the concept of continuity (*Stetigkeit*) (or line-completeness). Famously, the latter is defined in terms

[7] Kant's "Copernican Revolution" plays a central role here; cf. Keller (2015). With respect to my general understanding of Cassirer, and this point especially, I owe a big debt to Pierre Keller.

[8] Somewhat surprisingly, Cassirer does not comment on Dedekind's important contributions to algebra, algebraic number theory, etc., while he mentions some closely related works, e.g., by Galois and Hamilton. Cf. Reck (2016) for connections between all of Dedekind's contributions.

[9] Cf. the essay on Dedekind in the present volume, co-authored by José Ferreirós and me.

of Dedekind's concept of cut. Dedekind then considers the system of all cuts on the rational numbers $\mathbb{Q}$, endowed with a corresponding ordering and field operations (induced by those on $\mathbb{Q}$), and he shows that that system is a continuous ordered field. He is well aware that alternative such systems can be constructed too, most prominently that of all equivalence classes of Cauchy sequences on $\mathbb{Q}$, as Cantor and others had done. In other words, the "conditions" for being a continuous ordered field are "realized" by several systems. In a final step, Dedekind introduces "the real numbers" as a separate "pure" system corresponding to the system of cuts (isomorphic to but not identical with it); and he calls its introduction an act of "creation".[10]

Implicit in the procedure of Dedekind's 1872 essay, in the introduction of the system of cuts on $\mathbb{Q}$, are two assumptions: first, that we have the (infinite) system of all rational numbers available; second, that we can perform certain "logical" or set-theoretic constructions on it (essentially by forming the power-set of $\mathbb{Q}$). A main aim of Dedekind's 1888 essay, *Was sind und was sollen die Zahlen?*, is to provide a framework in which both of these assumptions can be justified further, i.e. a general theory of sets (*Systeme*) and functions (*Abbildungen*).[11] Within that framework, he then formulates another crucial "system of conditions", by defining the (higher-order) concept of a "simply infinite system". The latter depends, in turn, on several previously introduced concepts that are all "logical" (the concept of "infinity" and the more technical concept of "chain"). After that, he gives an argument that there are simply infinite systems (involving "thoughts", "thoughts of thoughts", etc.), parallel to his construction of the system of cuts on $\mathbb{Q}$ in 1872. And at that point, Dedekind adds a step not present in his earlier essay yet (although it can be supplemented retrospectively). Namely, he proves that any two simply infinite systems are isomorphic (his famous categoricity theorem). Finally, he uses both results to justify the "free creation"—via a process of "abstraction"—of a system that deserves to be called "the natural numbers".[12]

[10] In Dedekind's own words: "Whenever, then, we have to do with a cut (A_1, A_2) produced by no rational number, we create a new, an irrational number α, which we regard as completely defined by this cut (A_1, A_2); we shall say that the number α corresponds to this cut, or that it produces this cut" (Dedekind 1872, 15).

[11] The justification of the two assumptions mentioned remains implicit, however. Dedekind does not formulate basic laws for his set- and function-theoretic constructions; nor does he explicitly construct $\mathbb{Q}$ from $\mathbb{N}$, although he was familiar with how to do so. Cf. Reck (2003, 2016) for details.

[12] In Dedekind's own words again: "If in the consideration of a simply infinite system N set in order by a mapping φ, we entirely disregard the particular character of the elements, retaining merely their distinctness, and taking into account only the relations to one another in which they are placed by the order-setting mapping φ, then are these elements called *natural numbers* or *ordinal numbers* or simply *numbers*, and the base-element 1 is called the *base-number* of the *number-series* N. With reference to this freeing the elements from every other content (abstraction) we are justified in calling numbers a free creation of the human mind" (Dedekind 1888, 68).

The natural way to understand Dedekind's talk of "free creation" (although not an uncontroversial one) is the following: by a kind of "abstraction" we move from a previously constructed, relatively concrete system of objects (a particular continuous ordered field, a particular simply ordered system) to a new system that, while isomorphic, is distinct and more basic ("pure", more abstract, defined structurally). Understood as such, Dedekind's position amounts to a version of "non-eliminative structuralism" (in terminology introduced by Charles Parsons).[13] In the next section, I will provide further evidence that this is how Cassirer understands Dedekind, also that it is the position he adopts himself. It is striking, then, that most interpreters have had a very different, more critical reaction. Or rather, while almost all readers of Dedekind have accepted his technical contributions to the foundations of mathematics (his definitions of cut, continuity, infinity, simple infinity, his construction of the system of cuts, his categoricity theorem for simple infinities, etc.), his informal, more philosophical views about "abstraction" and "free creation", together with the resulting structuralism, have often been seen as problematic.

Bertrand Russell's critical reaction to Dedekind's structuralism is a good early illustration, one that was also highly influential. Basically, Russell could not make sense of objects introduced purely "relationally", like Dedekind's natural numbers, i.e., what Russell calls the finite "ordinals". As he puts it in his 1903 book, *The Principles of Mathematics*:

> IIt is impossible that the ordinals should be, as Dedekind suggests, nothing but the terms of such relations as constitute a progression. If they are to be anything at all, they must be intrinsically something. (Russell [1903] 1992, 249)

Russell assumes here, without further argument, that any object must have an "inner nature", one that goes beyond purely relational or structural properties (an assumption other philosophers have found plausible too). Hence, he finds the notion of an abstract or pure structure unintelligible. The other side of the coin is that he finds Dedekind's notion of "abstraction" unclear and unacceptable. In an attempt to be charitable, he concludes: "What Dedekind presents to us is not the numbers, but any progression [i.e., simply infinite system]: what he says is true of all progressions alike" (249). He then suggests using his own "principle of abstraction" instead, which amounts to the construction of the natural numbers in terms of equivalence classes of classes, as is well known. But this will

[13] Cf. Reck (2003) for further details.

not do for Dedekind's purposes, because Russell's form of "abstraction" does not lead to a system isomorphic to the original one.[14]

A second philosopher whose early criticisms of Dedekind were quite influential is Gottlob Frege. In Frege's *Grundgesetze der Arithmetik*, volume 2, also published in 1903, he considers Dedekind's theory of the real numbers. He thereby lumps Dedekind with several other thinkers (Stolz, Heine, Cantor, etc.) who talk about the mental "creation" of mathematical objects. In Frege's view, this is problematic for at least two reasons: First, it seems to lead to a subjectivist, perhaps even solipsistic position in the end. Second, it is in danger of being inconsistent; and this is especially so if the "creation" at issue is not backed up by explicit principles or basic laws (like the ones Frege formulates for his own approach). Frege is also critical of Dedekind's talk of set formation in his 1888 essay in terms of "mental" operations, since he sees that as problematically psychologistic too. Finally, Frege and Russell take the application of the natural numbers as cardinal numbers to be more basic than their ordinal use. Their definition as cardinal numbers, in the form proposed by both of them, thus appears more justifiable and appropriate.[15]

Frege's and Russell's criticisms of Dedekind's views, especially of his remarks about "abstraction" and "free creation", produced many echoes in later philosophy. A particularly explicit and stark example occurs in Michael Dummett's 1995 book, *Frege: Philosophy of Mathematics*. In that book, both Frege and Russell are appealed to as authorities, specifically with their arguments just mentioned, in support of Dummett's claim that Dedekind's position amounts to "mystical structuralism"—clearly a position not to be taken seriously. Finally, even after the re-emergence of a variety of structuralist positions in the philosophy of mathematics from the 1960s on, the corresponding authors (Paul Benacerraf, Michael Resnik, Stewart Shapiro, Geoffrey Hellman, and others) have remained suspicious of Dedekind's original, seemingly psychologistic ways of putting things, while appropriating him as a predecessor more generally. In other words, even in the writings of self-proclaimed structuralists, Frege's and Russell's early criticisms still reverberate strongly.[16] This is in striking contrast to Cassirer's sympathetic reception of Dedekind, which we consider next.

[14] While Russell is dismissive of Dedekind's structuralist position in his 1903 book, unpublished manuscripts show that he was more sympathetic originally and that his dismissive stance was the last of several stages through which he went; cf. the essay by Heis in this collection for details.

[15] For more details concerning Frege's reaction to Dedekind, cf. Reck (2019). As suggested in that article, it might be possible to defend Dedekind by formulating both "construction" and "abstraction" principles for him, although it is questionable if this can be done "purely logically."

[16] Cf. Reck (2013) for further details on Dedekind's reception, including by Frege, Russell, Dummett, and later structuralists. In mathematics, his writings were received more positively, e.g., by Ernst Schröder, David Hilbert, Ernst Zermelo, and Emmy Noether. But this was also not universal; and his remarks about "abstraction" and "free creation" were often simply ignored in that context.

3. Cassirer's Sympathetic Reception of Dedekind

As previously mentioned, Cassirer takes Dedekind's approach to the natural and real numbers to be essentially correct, even paradigmatic, already in 1907, only a few years after Frege's and Russell's criticisms. He also defends Dedekind explicitly against their criticisms, arguing that his approach is superior to theirs. Concerning the natural numbers, this defense includes taking an "ordinal" approach to be as basic as, and in some respects more fundamental than, a "cardinal" approach. This leads Cassirer to the following remark:

> [Dedekind showed that] in order to provide a foundation for the whole of arithmetic, it is sufficient to define the number series simply as the succession of elements related to each other by means of a certain order—thereby thinking of the individual finite numbers, not as "pluralities of units," but as characterized merely by the "position" they occupy within the whole series. (Cassirer 1907, 46)

The conception of natural numbers as "pluralities of units" is the traditional one traceable back to Euclid. It constitutes both a "cardinal" approach and a "substance-based" view, in Cassirer's terminology. As such, it is inferior to, and to be replaced by, Dedekind's "ordinal" and "function-based" conception, in which the natural numbers are treated simply as "positions" in a series.

With this characterization of natural numbers as "positions", we have arrived at Cassirer's own structuralism. In his 1910 book, he adds the following about it:

> It becomes evident that the system of numbers as pure ordinal numbers can be derived immediately and without circuitous route through the concept of class; since for this we need to assume nothing but the possibility of differentiating a sequence of pure thought constructions by different relations to a determinate base element, which serves as a starting point. The theory of the ordinal numbers thus represents the essential minimum that no logical deduction of the concept of number can avoid. (Cassirer [1910] 1923, 53, trans. modified)

It is, of course, Frege and Russell who define the natural numbers "through the concept of class". One reason for seeing Dedekind's ordinal conception as superior is that, instead of using such a "circuitous route", it brings out "the essential minimum" on which arithmetic relies. (Cassirer's point is confirmed by the possibility, and now standard practice, of developing arithmetic simply based on the Dedekind-Peano axioms.) With respect to the real numbers, he remarks along related lines:

> We thus see that, to get to the concept of irrational number, we do not need to consider the intuitive geometric relationships of magnitudes, but can reach this goal entirely within the arithmetic realm. A number, considered purely as part of a certain ordered system, consists of nothing more than a "position." ([1910] 1923, 49, my trans.)

In this case the traditional conception, thus Cassirer's foil, starts from an appeal to intuitively given geometric magnitudes, a conception widely shared well into the 19th century. That conception gets replaced by Dedekind's purely "arithmetic", or even "logical", approach in terms of cuts, continuity, etc.

What Dedekind has thus provided, as noted by Cassirer explicitly, is "the essential conceptual characterization" for both $\mathbb{N}$ and $\mathbb{R}$ (1907, 53); and in doing so, he has provided the "logical foundations of the pure concept of number" ([1910] 1923, 35). This goes significantly beyond Frege's and Russell's class-based constructions, in his opinion. Remember also Cassirer's use of the term "position" (in the two preceding passages, among others) to describe the resulting conception. This is more than 50 years before, in the 1960s, Paul Benacerraf reopened the debate about structuralism in English-speaking philosophy of mathematics; and it is more than 70 years before, from the 1980s on, Michael Resnik, Stewart Shapiro, and others started to use that term prominently to characterize non-eliminative structuralism.[17]

Earlier we encountered Russell's core objection to a non-eliminative structuralist conception, based on his assumption that numbers, like all objects, must be "intrinsically something". Cassirer takes up this point directly, as follows:

> If the ordinal numbers are to be anything, they must—so it seems—have an "inner" nature and character; they must be distinguished from other entities by some absolute "mark," in the same way that points are different from instants, or tones from colors. But this objection mistakes the real aim and tendency of Dedekind's formation of concepts. What is at issue is just this: that there is a system of ideal objects whose content is exhausted in their mutual relations. The "essence" of the numbers consists in nothing more than their positional value. (Cassirer [1910] 1923, 39, trans. modified)

In Cassirer's eyes, Russell's view that objects have to be distinguished by some "absolute mark" is unwarranted. But more than that, it shows Russell to hold on

[17] Cf. Reck and Price (2000) for references. One may wonder whether there was a direct influence in this connection. I am not aware of any references to Cassirer in published works by Benacerraf, Resnik, or Shapiro. But in Benacerraf's dissertation (on logicism), Cassirer's use of "position" in his 1910 book is quoted in a footnote (Benacerraf 1960, 162), as pointed out to me by Sean Walsh.

to an older, obsolete, "substance-based" view (despite his commendable introduction of a "logic of relations", which Cassirer praises and adopts himself). This is what makes Russell's position, and similarly Frege's, less adequate to modern mathematics than Dedekind's.[18]

Cassirer responds to the psychologism change against Dedekind—raised by Frege and, more vehemently, by neo-Fregeans like Dummett—as well. As Cassirer understands him, Dedekind's appeal to "abstraction" and "free creation" should not be interpreted along problematic psychologistic lines. In fact:

> [In Dedekind's works] abstraction has the effect of a liberation; it means logical concentration on the relational system, while rejecting all psychological accompaniments that may force themselves into the subjective stream of consciousness, which form no constitutive moment [*sachlich-konstitutives Moment*] of this system. ([1910] 1923, 39, trans. modified)

By taking Dedekind abstraction to involve "logical concentration on the relational system", Cassirer points to its logical and structural nature. This is what the critics, with their subjectivist interpretation of Dedekind's remarks about "thought", "abstraction", "free creation", etc., miss. It also reveals another respect in which his approach is superior, according to Cassirer's assessment.

Dedekind's talk of "free creation", in particular, is taken by his critics to imply that numbers exist as "mental entities" for him, i.e., in the subjective consciousness of people thinking about them. Cassirer rejects such a reading, as just noted.[19] Nor does he accept, however, that numbers exist "out there" in some crude realist sense. For him, both of those options misrepresent modern mathematics. What matters instead is "complete logical determinateness" (Cassirer 1907, 49), which he understands in a sense tied to mathematical methodology. In the case of introducing the real numbers by means of cuts, Cassirer clarifies this point as follows:

> The "existence" of an irrational number in Dedekind's sense is not intended to mean more than such determinateness: its "being" consists simply in its function of marking a possible division of the realm of rational numbers and thus of a "*position*." (Cassirer 1907, 49 n. 26, my trans.)

[18] As discussed in the essay by Jeremy Heis in the present volume, Russell made other noteworthy contributions to the rise of 20th-century structuralism, however.

[19] For more on Cassirer's defense of Dedekind against the psychologism charge, cf. Yap (2017).

In Cassirer's 1910 book, the point is explained further:

> The "things" referred to in this treatment are not posited as independent existences [*selbständige Existenzen*] present prior to any relation, but they gain their whole being [*Bestand*], insofar as it is of any concern for the arithmetician, first in and with the relations predicated of them. (Cassirer [1910] 1923, 36, trans. modified)

Similarly two pages later in the same text:

> The whole "being" of numbers rests, along these lines, upon the relations which they display within themselves, and not upon any relations to an outer objective reality [*gegenständliche Wirklichkeit*]. They need no foreign "basis" [*Substrat*], but mutually sustain and support each other insofar as the position of each in the system is clearly determined by the others. (38, trans. modified)

Cassirer's reference to what "concerns the arithmetician", i.e., what matters in terms of mathematical methodology, is significant here. So is his rejection of the view that any "outer objective reality" is involved, either mental or physical. Finally, noting that numbers "need no foreign basis, but mutually sustain and support each other" brings out another core aspect of a structuralist position.

For Cassirer, to ask further questions about the "objective reality" of numbers—ontological questions that go beyond their "logical determinateness"—would bring us back to the realist perspective to which he is fundamentally opposed. This has the following consequence: While the structuralist conception of mathematical objects that Cassirer attributes to Dedekind, and that he accepts himself, amounts to a non-eliminative position, it is not a realist position (in any traditional metaphysical sense); nor is it a form of subjective idealism, psychologism, or nominalism. Cassirer rejects all of these views explicitly. This distinguishes his approach right away from many current forms of structuralism, where the realism vs. nominalism opposition is central.[20] It also brings us back to the "logical idealism" he adopts instead. To quote the crucial passage one more time:

> Logical idealism starts from an analysis of mathematical "objects" and seeks to apprehend the peculiar determinacy of these objects by explaining them through the peculiarity of the mathematical "method," mathematical

[20] One exception is Charles Parsons's form of structuralism. Like Cassirer, Parsons is careful to separate the "non-eliminative" aspect of his position from any additional "realist" or "anti-realist" aspect. It is no coincidence that Parsons's perspective is also shaped strongly by Kant.

concept formation, and the formulation of its problems. (Cassirer [1929] 1965, 405, trans. modified slightly)

As we saw, the core of Cassirer's "logical idealism" is to account for mathematical "existence," "objects," etc., in terms of their "logical determinateness"; and the latter is tied closely to "mathematical method". Or to be more precise, it reflects the state of mathematical method at Cassirer's time, after the structuralist transformation of modern mathematics. This remark leads over to some further aspects of his position that deserve renewed attention.

4. Function Concepts, Constructions, and Unfoldings

In this section, three further aspects of Cassirer's "logical idealism" concerning mathematics will be highlighted, each of which goes beyond the current literature on structuralism in a noteworthy way. They involve, respectively, his notion of "function concept" and how it is situated historically; the important role Cassirer assigns to "constructions" in mathematics; and his argument that a structuralist conception constitutes the "unfolding" of "germs" present already in earlier stages of mathematics.

4.1. Function Concepts and Functional Thinking

As we saw, in his early works Cassirer characterizes the core difference between more traditional approaches to mathematical science and the novel structuralist perspective, most clearly represented in Dedekind's works, in terms of the distinction between "substance concepts" and "function concepts". What exactly that distinction amounts to is subtle, as it involves a number of ingredients that are never discussed in a fully clear, unified, and definitive way by him.[21] Nevertheless, some of what matters is clear enough. At its core, the crucial change is switching from an Aristotelian perspective on concept formation to a neo-Kantian perspective, both as understood by Cassirer.[22]

According to the position Cassirer ascribes to Aristotle (somewhat crudely, as one might add), concept formation proceeds as follows: We, as thinking subjects, encounter essentially independent objects in the natural world. We then ignore

[21] Cf. Heis (2014) for a helpful, but admittedly still partial, discussion of this topic. See also Kreis (2010), especially chapters 2–4.

[22] In recent discussions, certain forms of structuralism are described as "Aristotelian," as opposed to "Platonist," including some close to Dedekind. From Cassirer's point of view, the latter is rather problematic; i.e., it misrepresents "Dedekind abstraction" fundamentally.

various "marks" these objects have so as to distill out one or a few others, basically by "focusing on them selectively". A simple example would be to observe a red apple and to form the concept of "redness" simply by ignoring everything else about it. This is an illustration of the "substance concept" perspective, both in terms of the underlying realism and the particular conception of abstraction involved (a conception shared by various empiricist thinkers into the 19th century, e.g., J. S. Mill). "Function concepts", in contrast, should be thought of very differently. Not only do we not start with the assumption of fully formed subjects that are affected by independent objects; we also recognize that concept formation, especially in modern science, always involves a form of "constitution" and Kantian "synthesis". And crucially, the latter is based on a kind of "functional unity".

An illustration particularly relevant for present purposes is the difference between thinking of natural numbers as "multitudes of units" and Dedekind's approach to numbers. Along traditional lines, one assumes that some "heap" of objects is given to us directly. One then forms the idea of a corresponding "multitude of units" by ignoring all the differences between the objects in the heap except their numerical distinctness. We are led to Dedekind's alternative "function concept" once we recognize the following: Underlying any such supposedly basic, immediate procedure is a prior ability of "functionally relating" objects, including identifying and distinguishing them in the first place. But then, what is involved in forming a number cannot be as simple as just sketched; it must involve Kantian "synthesis". In fact, already the differentiation of a series of objects, one distinct from the next, does so.[23] And once we recognize that, we are led to thinking of the whole number series in terms of Dedekind's notion of a simple infinity. Rather than abstracting from the "marks" of a given heap of objects in a "subtractive" sense, the form of abstraction at play is more positive. In involves "logical concentration" on the functionally determined structure, here the natural number structure, just as Dedekind taught us. For Cassirer, this is a paradigmatic example of "functional unity".

As this brief sketch indicates, Dedekind's approach to the natural numbers is crucial for Cassirer not just by providing a novel conception of the natural numbers, but by being a model for something deeper and more general. Actually, Dedekind himself is aware of the depth at issue, as the following passage—quoted prominently and approvingly by Cassirer—indicates:

> If we trace closely what is done in counting a group or collection of things, we are led to consider the ability of the mind to relate things to things, to let one

[23] According to Cassirer's neo-Kantian perspective, basic "synthetic" activities include: identifying and differentiating, relating to one another, naming, etc. (see below for more). Here, as at related places in this essay, I am heavily indebted to conversations with Pierre Keller.

> thing correspond to another thing, or to represent one thing by another, an ability without which no thought is possible. (Dedekind 1888, 32)

We can understand this passage better if we relate it to another remark in the same text. As background, consider the following: How should we answer one of the two questions raised by Dedekind in the title of his 1888 essay, namely: "Was . . . sollen die Zahlen?" (What is the nature, or better, the point or role of numbers?) His own answer, formulated in the essay's preface, is this: "[Numbers] serve as a means of apprehending more easily and more sharply the difference of things" (31). As these passages indicate, Dedekind is reflectiong on our very ability to think; and for him that includes identifying and differentiating things, representing some by others, naming them, interrelating them in other ways, etc. The most basic role of numbers is to help us in this task, e.g., by arranging things in series: a first, a second, etc. This idea points right back to the concept of simple infinity. It also leads to Dedekind's answer to the second of his two title questions: "Was sind . . . die Zahlen?" (What are numbers, or what is their nature?) Namely, the natural numbers are the things obtained, via "Dedekind abstraction", from any simple infinity. And when suitably extended, such an approach leads to the negative, rational, real, and complex numbers as well, as illustrated most explicitly by his 1872 essay.

One striking thing about the passages by Dedekind just quoted, and about Cassirer's reception of them, is that the notion of function is made absolutely central. The notion of set is not as central; but it too plays a basic role, for both Dedekind and Cassirer (e.g., with respect to the domains and ranges of functions). Nor is the notion of relation quite as central, although it is again important (e.g., when considering the ordering relation on the rational numbers so as to form cuts). Why exactly is the notion of relation not as primary as that of function? The answer is, as Cassirer remarks briefly, that the idea of relation "can be traced back to the more fundamental idea of 'functionality'" (Cassirer 1907, 43); and likewise for the idea of set. In other words, using sets and relations, as we do in modern logic, involves "thinking functionally" in the end. In Frege's and Russell's new logic, with its emphasis on relations and sets or classes, we are moving toward this insight, but we do not quite reach it yet.

4.2. The Crucial Roles of Set-Theoretic Constructions

As just argued, what lies at the bottom of Dedekind's approach to mathematics, and Cassirer's reception of it, is "functional thinking"; and this is illustrated by the central role the successor function plays for the natural numbers. Nevertheless,

Dedekind employs sets in crucial ways too. Cassirer picks up on the latter point by emphasizing the role of constructions in modern mathematics more generally. In fact, with his strong emphasis on set-theoretic constructions Cassirer goes, at least in part, against a distinction made prominent by Hilbert and his followers, namely between the "genetic" and the "axiomatic" method. As often claimed by Hilbertians, mathematics in the late 19th and early 20th centuries involved the switch from a "genetic" to an "axiomatic" approach. For Cassirer such a contrast is spurious, since both sides remain crucial.

Cassirer position in this context can again be illustrated, and justified, in relation to Dedekind's work. Take Dedekind's treatment of the real numbers. It is true that the concept of a continuous ordered field does, in some sense or to some degree, provide the basis for that treatment. Along Hilbertian lines, it is then the axiom system by means of which that concept is defined that becomes crucial. But we should not forget about the construction of the system of cuts on the rational numbers. What is the point of that construction, i.e., which basic role or roles does it play? The first such role, explicitly acknowledged by Dedekind and noted by Hilbert as well, is to establish the (semantic) consistency of the concept of a continuous ordered field, or of the corresponding axiom system. But for both Dedekind and Cassirer there is more. The system of cuts also provides the basis for the "abstraction" by means of which "the real numbers" are introduced. This is the second basic role of the set-theoretic construction. A third role is this: it is only in terms of the cuts that we know how to operate with the real numbers, as is reflected in the fact that the ordering and the arithmetic operations on "the real numbers" are induced directly by those on the system of cuts.

The fact that set-theoretic constructions, like those of Dedekind cuts, play such crucial roles in modern mathematics has more general implications for Cassirer. Let me mention three of them briefly. First, it is in terms of constructing novel mathematical objects out of older ones that these new objects—including all the "ideal elements" characteristic of 19th-century mathematics—become intelligible and acceptable in the first place. This involves making it possible to operate with, say, the real numbers in terms of rational numbers. More basically, it is how we identify and differentiate them, i.e., it grounds their identity. A closely related second point is this: the constructions at issue establish connections between older and newer parts of mathematics. The newer parts are thus not separate and isolated, but integrated into mathematics as a whole from the start. In fact, it is this integration, or a network of corresponding links, that constitutes the unity of mathematics, as Cassirer notes. A third point concerns less mathematics itself than philosophy. For Cassirer, what the importance of such constructions establishes is that Kant was right with his claim that mathematics involves "the construction of concepts". Admittedly, Kant was focused too narrowly on traditional geometric constructions, while with Dedekind's works we see that it is set-theoretic constructions that are crucial for modern mathematics.[24]

4.3. The Historical Unfolding of Structuralist Aspects

I want to mention one more distinctive feature of both Cassirer's reception of Dedekind and of his own structuralist philosophy of mathematics. His juxtaposition of "substance" and "function concepts", as discussed above, may initially be taken to imply that he conceives of the history of mathematics as involving a radical discontinuity or rupture (the move from "substance" to "function concepts"). But this is not quite right. In fact, Cassirer wants to emphasize a corresponding continuity as well. Moreover, that continuity is not unrelated to some of the roles of constructions just sketched. As he writes:

> The new forms of negative, irrational and transfinite numbers are not added to the number system from without but grow out of the *continuous unfolding* of the fundamental logical function that was effective in the first beginnings of the system. (Cassirer [1910] 1923, 67, emphasis added)

The way in which mathematicians like Dedekind have gone from the natural numbers through the negative, rational, and real numbers all the way to the complex numbers by means of set-theoretic constructions is a main example of the "unfolding" Cassirer has in mind (in Dedekind 1854 already). But his conception of "unfolding", and of the corresponding continuity of mathematics, is both richer and subtler than that. Cassirer never spells out that conception clearly and fully in his writings, he only hints at it (including in unpublished manuscripts, e.g., Cassirer 1999). Here is what I take to be the core point: even very early forms of mathematics contain some "functional" aspects, i.e., aspects of the kind of "functional thinking" sketched previously, albeit not in pure forms yet. These aspects are refined and generalized over time, and they come to the fore in the 19th century, especially in works such as Dedekind's. Still, their "germs" go way back, to rudimentary and rather informal parts of mathematics, in fact even beyond what one would normally consider mathematics today.[25]

[24] For more on this point about Cassirer and Kant, cf. Reck and Keller (forthcoming).

[25] For Cassirer, ordinary and mystic ways of thinking are included here, e.g., in terms of the use of number words in magic (where other aspects overshadow the functional/structuralist ones, although they are present in very rudimentary ways). This is one way in which the various "symbolic forms," highlighted in his later writings, are interrelated and build on each other.

A basic illustration of this phenomenon is the following (cf. Heis 2017): consider the natural numbers in the traditional way, i.e., as involving "multitudes of units". Now think of adding two such numbers, e.g., 5 and 7. We can conceive of this as involving three steps: first we count five units, labeled by "1", "2", . . ., "5" ; then we add seven further units, labeled "6", "7", . . ., "12"; finally we record where this leads us, namely to the number 12. Note now that in the second step we treated the sixth unit "as a new 1", by bringing to bear its "position" in the number series. That is to say, we started to reiterate the successor operation with it (the relevant number of times). What we did, in other words, is to utilize an initial segment of the number series and its systematic, step-by-step extension. While obscured somewhat by thinking of numbers as "multitudes of units", this indicates that certain of the aspects distilled out by Dedekind are already at play in this context.

Cassirer's general point here is this: while often mixed together with more traditional and "impure" aspects—geometric, broadly intuitive, also sometimes formalist aspects— in earlier phases of mathematics, "functional" or structuralist aspects can be discerned in all of mathematics, even going back beyond Euclid. Once again, this establishes a unity or continuity for its historical development, across the supposed "substance" vs. "function concept" divide. Put differently, it is what allows us to speak of "mathematics" as one discipline, with a history from at least the ancient Greeks to Cassirer's time. By embedding it in this broad historical panorama, Cassirer has provided a rich historical background and motivation for structuralism in mathematics.

5. Summary and Concluding Remarks

This essay focused on Cassirer's reception of Dedekind's work, which he took to be paradigmatic for a shift from "substance" to "function concepts" in the mathematical sciences. With his sympathetic response to Dedekind's contributions, including defending his remarks about "abstraction" and "free creation", Cassirer went against the mostly critical, often dismissive reactions by other philosophers, both during his time and later, as the examples of Frege, Russell, and Dummett illustrated. And with his characterization of mathematical objects in terms of the notion of "position" in a structure, such as the natural number series, Cassirer anticipated the revival of structuralism in the philosophy of mathematics, by Benacerraf, Resnik, Shapiro, and others 50–70 years later. Both of these facts are remarkable, and a main goal of the present essay was to direct attention to them.

With his positive reception of a Dedekindian structuralism Cassirer did not just anticipate current structuralist positions, however. There are aspects to his approach that are genuinely original and make it distinctive. One example is his discussion—under the umbrella of "logical idealism"—of the specific form of "determinateness" operative in modern mathematics, which is closely related to the rejection of both realist and psychologistic views by him. Three other examples, discussed later in this essay, are the way in which Cassirer emphasizes, with Dedekind, the fundamental role of functional thinking; his emphasis on the roles played by constructions along Dedekindian lines; and the point that the historical development of mathematics, even across the substance/function divide, involves continuity in terms of the "unfolding" of structuralist "germs".

Overall, what Cassirer provides is a treatment of the structuralist transformation of modern mathematics that illuminates not only its logical and metaphysical aspects, but embeds it in a rich developmental and historical story. My summary of it could be enriched further by also covering his reflections on parallel developments in geometry. In the present essay, the focus was exclusively on the side of arithmetic. Both sides led Cassirer to essentially the same conclusions, however.[26] It should be acknowledged, finally, that there are limitations to Cassirer's discussion of structuralism in mathematics too, thus ways in which the current debates go beyond it. For example, he contributed little to its technical development, in the sense that he provided no formal reconstructions of core concepts and proved no new mathematical theorems. After all, he was not a mathematical logician. Then again, with respect to the philosophical and historical dimensions, his treatment deserves to be reconsidered today.

Acknowledgments

This essay has been in the making for over 10 years. An early version was presented at HOPOS 2008, in Vancouver, Canada; subsequent versions at the University of Turin, Italy, in 2010; at the University of Paris 7, Laboratoire SPHERE, France, in 2013; and at the University of Vienna, Austria, in 2015. I would like to thank all the audience members at these events for their comments, and their organizers for giving me a chance to present the material. Beyond that, I am indebted to Jeremy Heis for many conversations about Cassirer's philosophy of mathematics and, even more, to Pierre Keller, who pointed me toward Cassirer in the first place and who continued to influence my views on him substantially. Finally, I am grateful to Georg Schiemer for comments on the penultimate version of this chapter. All the remaining inaccuracies and mistakes are, as usual, my own.

[26] Cf. again Heis (2011), Biagioli (2016), and Schiemer (2018), among others.

References

Benacerraf, Paul. 1960. Logicism: Some Considerations. Ph.D. thesis, Princeton University.

Biagioli, Francesca. 2016. *Space, Number, and Geometry from Helmholtz to Cassirer*. New York: Springer.

Cassirer, Ernst. 1907. Kant und die moderne Mathematik. *Kant Studien* 12, 1–49.

Cassirer, Ernst. [1910] 1923. *Substanzbegriff und Funktionsbegriff*. Berlin: Bruno Cassirer. Translated by William Curtis Swabey and Marie Collins Swabey in *Substance and Function and Einstein's Theory of Relativity*. Chicago: Open Court.

Cassirer, Ernst. 1923. Der Begriff der symbolischen Form im Aufbau der Geisteswissenschaften. In *Vorträge der Bibliothek Warburg*. Leipzig: Teubner, pp. 11-39. English trans., The Concept of Symbolic Form in the Construction of the Human Sciences. In Cassirer 2013, pp. 72–100.

Cassirer, Ernst. 1927. Das Symbolproblem und seine Stellung im System der Philosophie. *Zeitschrift für Ästhetik und allgemeine Kunstwissenschaft* 21, 191–208. English trans., The Problem of the Symbol and Its Place in the System of Philosophy. In Cassirer 2013, pp. 254–271.

Cassirer, Ernst. [1929] 1965. *Die Philosophie der Symbolischen Formen*. Vol. 3. Bruno Cassirer: Berlin. English trans., *The Philosophy of Symbolic Forms*. Vol. 3. New Haven: Yale University Press.

Cassirer, Ernst. 1950. *The Problem of Knowledge: Philosophy, Science, and History since Hegel*. New Haven: Yale University Press.

Cassirer, Ernst. 1999. *Ziele und Wege der Wirklichkeitserkenntnis* (*Nachgelassene Manuskripte und Texte, Band 2*). Hamburg: Meiner.

Cassirer, Ernst. 2013. *The Warburg Years (1919–1933): Essays on Language, Art, Myth, and Technology*. New Haven: Yale University Press.

Cohen, Hermann. [1883] 2013. *Das Prinzip der Infinitesimal-Methode und seine Geschichte*. Vienna: Verlag Turia.

Dedekind, Richard. 1854. Über die Einführung neuer Funktionen in der Mathematik. In Dedekind 1930–32, vol. 3, pp. 428–438.

Dedekind, Richard. 1872. *Stetigkeit und irrationale Zahlen*. In Dedekind 1930–32, vol. 3, pp. 315–334. English trans., *Continuity and Irrational Numbers*. In Dedekind 1963, pp. 1–22.

Dedekind, Richard. 1888. *Was sind und was sollen die Zahlen?* Reprinted in Dedekind 1930-32, vol. 3, pp. 335–391. English trans., *The Nature and Meaning of Numbers*. In Dedekind 1963, pp. 29–115.

Dedekind, Richard. 1930–32. *Gesammelte Mathematische Werke*. Vols. 1–3. Edited by R. Fricke et al. Braunschweig: Vieweg.

Dedekind, Richard. 1963. *Essays on the Theory of Numbers*. Edited and trans. by W. W. Beman. New York: Dover. Originally by Chicago: Open Court, 1901.

Dummett, Michael. 1995. *Frege: Philosophy of Mathematics*. Cambridge, MA: Harvard University Press.

Ferrari, Massimo. 2003. *Ernst Cassirer: Stationen einer philosophischen Biographie*. Hamburg: Meiner.

Frege, Gottlob. 1903. *Grundgesetze der Arithmetik*. Vol. 2. English trans., *Basic Laws of Arithmetic*, vols. 1–2. Edited and translated by P. Ebert and M. Rossberg. New York: Oxford University Press, 2013.

French, Steven. 2014. *The Structure of the World: Metaphysics and Representation*. New York: Oxford University Press.

Friedman, Michael. 2000. *A Parting of the Ways: Carnap, Cassirer, and Heidegger*. Open Court: Chicago.

Friedman, J. T., and S. Luft, eds. 2017. *The Philosophy of Ernst Cassirer: A Novel Assessment*. Berlin: de Gruyter.

Heis, Jeremy. 2011. Ernst Cassirer's Neo-Kantian Philosophy of Geometry. *British Journal for the History of Philosophy* 19, 759–794.

Heis, Jeremy. 2014. Ernst Cassirer's *Substanzbegriff und Funktionsbegriff*. *Journal of the International Society for the History of the Philosophy of Science (HOPOS)* 4, 241–270.

Heis, Jeremy. 2017. Arithmetic and Number in the Philosophy of Symbolic Forms. In Friedman & Luft 2017, pp. 123–140.

Ihmig, K.-I. 1997. *Cassirers Invariantentheorie der Erfahrung und seine Rezeption des Erlanger Programms*. Hamburg: Meiner.

Ihmig, K.-I. 1999. Ernst Cassirer and the Structural Conception of Objects in Modern Science: The Importance of the "Erlanger Program". *Science in Context* 12, 513–529.

Ihmig, K.-I. 2001. *Grundzüge einer Philosophie der Wissenschaften bei Ernst Cassirer*. Darmstadt: Wissenschaftliche Buchgesellschaft.

Keller, Pierre. 2015. Cassirer's Retrieval of Kant's Copernican Revolution in Semiotics. In Friedman and Luft 2017, pp. 259–288.

Kreis, Guido. 2010. *Cassirer und die Formen des Geistes*. Berlin: Suhrkamp.

Luft, Sebastian. 2015. *The Space of Culture: Towards a Neo-Kantian Philosophy of Culture (Cohen, Natorp & Cassirer)*. New York: Oxford University Press.

Mormann, Thomas. 2007. Idealization in Cassirer's Philosophy of Mathematics. *Philosophia Mathematica* 16, 151–181.

Natorp, Paul. 1910. *Die logischen Grundlagen der exakten Wissenschaften*. Berlin: Teubner.

Reck, Erich. 2003. Dedekind's Structuralism: An Interpretation and Partial Defense. *Synthese* 137, 369–419.

Reck, Erich. 2013. Frege or Dedekind? Towards a Reevaluation of their Legacies. In *The Historical Turn in Analytic Philosophy*, edited by Erich Reck,. Palgrave Macmillan: London, pp. 139-170.

Reck, Erich. 2016. Dedekind's Contributions to the Foundations of Mathematics. In *Stanford Encyclopedia of Philosophy*, edited by Edward N. Zalta, pp. 1 32. Original version online in 2008.

Reck, Erich. 2019. Frege's Relation to Dedekind: *Basic Laws* and Beyond. In *Essays on Frege's Basic Laws of Arithmetic*, edited by P. Ebert and M. Rossberg. New York: Oxford University Press, pp. 264–284.

Reck, Erich, and Pierre Keller. Forthcoming. From Dedekind to Cassirer: Logicism and the Kantian Heritage. In *Kant's Philosophy of Mathematics*, vol. 2, *Reception and Developments after Kant*, edited by O. Rechter and C. Posy. New York: Cambridge University Press.

Reck, Erich, and Michael Price. 2000. Structures and Structuralism in Contemporary Philosophy of Mathematics. *Synthese* 125, 341–383.

Recki, Birgit. 2004. *Kultur als Praxis: Eine Einführung in Ernst Cassirers Philosophie der symbolischen Formen*. Berlin: Akademie Verlag.

Russell, Bertrand. [1903] 1992. *The Principles of Mathematics*. London: Routledge.

Ryckman, Thomas. 2005. *The Reign of Relativity: Philosophy in Physics, 1915–1925*. New York: Oxford University Press.

Schiemer, Georg. 2018. Cassirer and the Structural Turn in Modern Geometry. *Journal for the History of Analytic Philosophy* 6, 163–193.

Stein, Howard. 1988. *Logos*, Logic, *Logistiké*: Some Philosophical Remarks on Nineteenth-Century Transformations of Mathematics. In *History and Philosophy of Modern Mathematics*, edited by William Aspray and Philip Kitcher. Minneapolis: University of Minnesota Press, pp. 238–259.

Tait, W. W. 1996. Frege versus Cantor and Dedekind: On the Concept of Number. In *Frege: Importance and Legacy*, edited by M. Schirn. Berlin: de Gruyter, pp. 70–113.

Yap, Audrey. 2017. Dedekind, Cassirer, and Mathematical Concept Formation. *Philosophica Mathematica* 25, 369–389.

14

Methodological Frames: Paul Bernays, Mathematical Structuralism, and Proof Theory

Wilfried Sieg

Mathematical structuralism is deeply connected with Hilbert's and Bernays's proof theory and its programmatic aim to ensure the consistency of all of mathematics. That goal was to be reached on the sole basis of finitist mathematics, a distinguished, elementary part of mathematics. Gödel's second incompleteness theorem forced a step from *absolute finitist* to *relative constructivist* proof-theoretic reductions. The mathematical step was accompanied by philosophical arguments for the special nature of the grounding constructivist frameworks.

Against this background, I examine Bernays's reflections on proof-theoretic reductions of *mathematical structures* to *methodological frames* via *projections*. However, these reflections—from the mid-1930s to the late 1950s and beyond—are focused on narrowly arithmetic features of frames. Drawing on our broadened metamathematical experience, I propose a more general characterization of frames that has ontological and epistemological significance; it is rooted in the internal structure of mathematical objects that are uniquely generated by inductive (and always deterministic) processes.

The characterization is given in terms of *accessibility*: domains of objects are accessible if their elements are inductively generated, and principles for such domains are accessible if they are grounded in our understanding of the generating processes. The accessible principles of inductive proof and recursive definition determine the generated domains uniquely up to a canonical isomorphism. The determinism of the inductive generation allows us to refer to the mathematical objects of an accessible domain, and the canonicity of the isomorphism justifies at the same time an "indifference to identification." Thus is ensured the intersubjective meaning of mathematical claims concerning accessible domains.

Wilfried Sieg, *Methodological Frames* In: *The Prehistory of Mathematical Structuralism*. Edited by: Erich H. Reck and Georg Schiemer, Oxford University Press (2020). © Oxford University Press.
DOI: 10.1093/oso/9780190641221.003.0014

1. Describing the Context

Paul Bernays viewed mathematics as *the science of idealized structures*.[1] His perspective highlights the methodological changes that expanded, indeed *transformed* the subject during the 19th century. In his (1930), Bernays pointed to three related features characterizing this transformation: (1) the advancement of the *concept of set*, (2) the emergence of *existential* or *structural axiomatics*, and (3) the evolution of a close *connection between mathematics and logic*. He saw these developments as confronting the philosophy of mathematics with novel insights and new problems. In this early essay, Bernays took on the task of situating proof theory within the philosophy of mathematics and, in particular, clarifying the character of *mathematical cognition* (*mathematische Erkenntnis*).

More than 50 years later, Howard Stein observed in his (1988) that the 19th-century transformation of mathematics revealed a capacity of the human mind. He also asserted that this capacity had been discovered already in ancient Greece between the 6th and 4th centuries B.C. Stein emphasized that its *re*discovery teaches us something new about its nature and claimed that what has been learned "constitutes one of the greatest advances in philosophy." However, he did not explicitly formulate the "something new that has been learned" and, thus, did not clarify the dramatic philosophical advance. If we want to grasp this advance, we must deepen our understanding of the mind's mathematical capacity or, even more broadly, its capacities as they come to light in mathematics and its uses.

Taking a step toward deepening our understanding, section 2 begins by discussing the character of the 19th-century transformation as it is revealed in *existential axiomatics* and various foundational frames for it. I prefer to call existential axiomatics *structural* since it is the form of mathematical structuralism that evolved from Dedekind's work and is fully expressed in Bourbaki's *Éléments de mathématique*. Section 3 introduces Bernays's restricted *methodological frames* and his idea of viewing the formalization of axiomatic systems as a means of uniformly *projecting* them into such restricted frames.[2] This builds on

[1] Without taking on the task of interpreting "idealized," I consider for the purpose of this chapter "idealized" only to mean that structural definitions are obtained by a special kind of abstraction emphasized by Lotze; see (Sieg and Morris 2018, 32–34) and the remark by Bernays quoted in note 3. This "Begriffsbildung" is for me the core of the 19th-century *transformation* of mathematics; it is exemplified in Dedekind's *Was sind und was sollen die Zahlen?* My (2016) indicates the more philosophical side of the transition from Kant through Dedekind to Hilbert and beyond.

[2] Charles Parsons (2008) analyzes Bernays' "anti-foundationalism" and "structuralism." He focuses on Bernays's "later philosophy" and compares his structuralism to the *philosophical* structuralism of modern analytic philosophy, whereas I emphasize the continuity in his foundational reflections and connect his structuralism to the *mathematical* structuralism that originated in the 19th-century transformation of mathematics; see also notes 1 and 3. An informative survey of different forms of structuralism is found in (Reck and Price 2000). Finally, contemporary *scientific structuralism* as advocated by Suppes and many others is rooted in the mathematical structuralism as it emerged in the second half of the 19th century with deep connections, in particular, to Gauss,

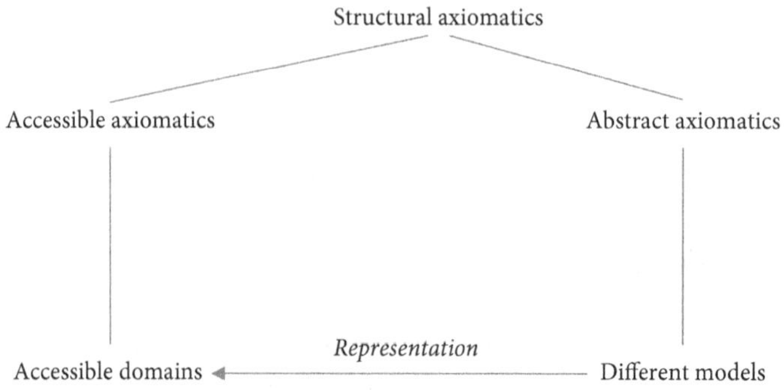

Figure 1 Accessible and abstract axiomatics

a philosophically significant distinction between two kinds of models for structural axiomatic theories, namely, those whose domains just satisfy broad structural conditions and those whose domains are in addition inductively generated. Bernays made this distinction in an elementary form for extensions of Hilbert's consistency program. Given our broader proof-theoretic experience, I generalize in section 4 "inductive generation" and introduce "accessible domains." These considerations lead to an informative and principled distinction between "abstract" and "accessible" axiomatics, both kinds falling under structural axiomatics. The diagram of Figure 1 reflects that distinction.[3]

The preceding incorporates, however, also Hilbert's perspective that analysis and geometry, for example, can be *represented* in set theory. That means that the different models of the axiomatic theories can be viewed as defined in subdomains of Zermelo's set theory, when the latter are viewed as accessible domains. Hilbert's perspective is discussed in section 2.

The elements of *accessible domains* have an internal structure grounding the principles of the structural axioms, but also ensuring that the domains are *canonically* isomorphic. I highlight cognitive aspects that make accessible domains

Riemann, Dedekind, and Hilbert. These connections, evident also in the work of Hertz, deserve a separate investigation. A first step was taken in a talk Aeyaz Kayani and I gave at the 2016 HOPOS meeting in Minneapolis; the talk was entitled *Roots of Suppes' Scientific Structuralism.*

[3] The diagram respects the distinctions made in (Bernays 1970) under the influence of Gonseth. Bernays asserts there, "Mathematical idealization is especially accentuated by the axiomatic treatment of theories." In German: "Die mathematische Idealisierung kommt insbesondere zur Geltung durch die axiomatische Behandlung von Theorien" (181). He continues, "As one knows, one has to distinguish two different kinds of axiomatics." Bernays follows Gonseth in calling the one "axiomatisation schématisante" and the other "axiomatisation structurante." That distinction is "parallel" to the one I am making between "generative structural definitions" (*accessible axiomatics*) and "abstract structural definitions" (*abstract axiomatics*). It would be of real interest to examine the philosophical aspects of their "axiomatisation schématisante" and compare them to those of accessible axiomatics.

suitable to serve as the core of *methodological frames* and as the basis for important relative consistency proofs. Finally, I formulate in section 5 a particular way in which we can investigate, I hope very fruitfully, "the mind's capacities as they come to light in mathematics and its uses."

2. Structural Axiomatics and Frames

There is a form of axiomatization in mathematics that is not tied to theories of modern mathematical logic with their formal languages and logical calculi; I am thinking of the axioms for *abstract* concepts like that of a *group*, *field*, or *topological space*. The axioms really are just characteristics (*Merkmale*) of *structural definitions*. These structural definitions stand in a venerable tradition that goes back, in particular, to Dedekind's work in algebraic number theory, but also to his essay *Was sind und was sollen die Zahlen?* (*WZ*). In this 1888 essay, Dedekind discards natural numbers as abstract objects and introduces instead the concept of a *simply infinite system* via a structural definition. If one reads from this perspective his 1872 essay *Stetigkeit und irrationale Zahlen* (*SZ*), then one can see that Dedekind defines there the structural notion of a *complete ordered field*.

The concept of a complete ordered field, with a different way of formulating topological completeness, is also defined in Hilbert's 1900 essay *Über den Zahlbegriff*. The axioms of his *Grundlagen der Geometrie* are yet another example of a structural definition, namely, that of a *Euclidean space*. As the final example of such a definition, consider Zermelo's 1908 axioms for set theory: they give a structural definition of the concept *Mengenbereich over a set of urelements*. Zermelo leads in three steps to the axioms (262-263): (1) "Set theory is concerned with a *domain* B of individuals which we shall call simply *objects* and among which are the *sets*." (2) "Certain *fundamental relations* of the form $a \in b$ obtain between the objects of the domain B." (3) "The fundamental relations of our domain B, now, are subject to the following *axioms*, or *postulates*." These steps are typical for the definition of structural notions and parallel almost verbatim Hilbert's steps in the papers just mentioned; they are then followed by the successive introduction and detailed discussion of the axioms. To re-emphasize, the above axiom systems are not formal theories, but structural definitions in the Dedekindian mold.

When commenting on *SZ* for the third volume of Dedekind's *Gesammelte Abhandlungen*, Emmy Noether attributed to Dedekind an *axiomatic conception* (*axiomatische Auffassung*). Three different points informed her judgment. Here are the first two: Dedekind structurally defined the concept of a complete ordered field and proved that the system of all cuts of rational numbers constitutes an instance of that definition. For the third point she referred to an 1876 letter to Lipschitz in which Dedekind expressed his view on the systematic and quite formal development of analysis or any other mathematical theory:

> All technical expressions [can be] replaced by arbitrary, newly invented (up to now meaningless) words; the edifice must not collapse, if it is correctly constructed, and I claim, for example, that my theory of real numbers withstands this test. (Dedekind 1932, 479)

These are indeed the *three crucial elements* of the modern axiomatic method as Noether and others practiced it in the 1920s. It is incisively described in Helmut Hasse's talk *Die moderne algebraische Methode* (1930); the talk addressed a general mathematical audience and suggested an expansion of the "algebraic method" to other parts of mathematics. In characterizing the algebraic method, Hasse emphasized the three aspects Noether pointed to—generalizing, of course, her first two points to other structural definitions.

The axiomatic method, when conceived of as structural, requires an intelligible and philosophically distinguished *methodological frame*, what Bernays calls "methodischer Rahmen." For Dedekind, as emphasized in the preface to the first edition of *WZ*, that was *logic* with a broad contemporaneous understanding; the same holds for the early Hilbert and Zermelo. This logical frame allowed novel *metamathematical* investigations. The central ones could be carried out due to the fact that a *form of semantics* was available: *model* is any system that "falls under" a structural concept or that "satisfies" its characteristic conditions.[4] Dedekind introduced *mappings* (*Abbildungen*) to relate different models in structure-preserving ways.[5] Within this frame, carefully exposed in *WZ*, he proved the concept of a simply infinite system to be *categorical* and argued for the *proof-theoretic equivalence* of any two models.[6]

Hilbert was a master in using models to give independence and relative consistency proofs. Among other things, his investigations show in the most striking way the irrelevance of the "nature" of the objects making up a system that falls under a structural definition.[7] Hilbert's beautiful geometric model of the

[4] This pre-Tarskian semantics was sustained from Dedekind through Hilbert and Ackermann to Gödel in his thesis (1929); it is still used in contemporary mathematical practice.

[5] For Dedekind, mappings form a distinct second category of mathematical entities; they are understood as being given by laws. Sieg and Schlimm (2014) analyze the evolution of the notion of mapping and its use for such metamathematical purposes.

[6] The concept of *proof-theoretic equivalence* was introduced in (Sieg and Morris 2018, section B.2) in order to illuminate section 134 of *WZ* and Dedekind's deeply connected description of the science of numbers in section 73.

[7] John Burgess coined the apt phrase "indifference to identification." In his letter to Frege, written on December 29, 1899, Hilbert asserted that "any theory is only a framework [*Fachwerk*] or a schema of concepts together with the necessary relations between them." The basic elements (*Grundelemente*), he continued, "can be thought in arbitrary ways."

arithmetic concept *Archimedean ordered field* makes that point quite directly and convincingly; see (Hilbert 1899, secs. 13 and 15). However, Dedekind had articulated in *WZ*, and even more explicitly in his letter to Keferstein (Dedekind 1890), a crucial foundational demand for his frame, namely, to give a *logical existence proof* (*logischer Existenzbeweis*) of a model of the concept of a simply infinite system.[8] Dedekind asserted that such a proof was needed to guarantee that the newly introduced concept did not contain an internal contradiction. Hilbert formulated this demand, from the very beginning of his axiomatic investigations, in a quasi-syntactic way and required that no contradiction can be obtained in finitely many logical steps. (It is only *quasi*-syntactic, as no logical steps were explicitly presented.)

The methodological frame was also seen as deeply significant for the representation of mathematical practice. In Dedekind's *WZ*, the representation of elementary number theory was at stake and was achieved through the justification of both the principle of proof by induction and that of definition by recursion.[9] Hilbert dealt with geometry and analysis around the turn of the century. In lectures from 1920, *Probleme der mathematischen Logik*, he expressed the representational strategy with respect to Zermelo's set theory:

> Set theory encompasses all mathematical theories (like number theory, analysis, geometry) in the following sense: the relations that hold between the objects of one of these mathematical disciplines are represented in a completely corresponding way by relations that obtain [between objects] in a subdomain of Zermelo's set theory. (330)

Only a short time later, this representation is refined proof-theoretically, shifting from semantic model to syntactic reduction; that is the core of my discussion in section 4. Coming back to Dedekind's logical frame, we can observe that the development of his theory of systems and of mappings is quite principled: the part concerning systems uses *full comprehension* and the *extensionality principle*, whereas the part concerning mappings uses, for example, closure under composition, and inversion (for bijections). This framework is used to introduce chains (of systems) as a central concept and to develop elementary set theory up

[8] The proof Dedekind gave is problematic, but not because of any "psychologistic" aspects. Frege viewed it as essentially correct; see (Frege 1969, 147–148). For Bernays the real reason for its being problematic is "the idea of a closed totality of all logical objects that can be thought at all" (Bernays 1930, 47).

[9] A similar remark can be made about Dedekind's *SZ*, where he sketches in section 6 the beginning steps of analysis. In section 7 he establishes, in a quite dramatic way, the equivalence of his continuity principle to a theorem of analysis, namely, that every bounded, monotonically increasing sequence has a limit.

to the Cantor-Bernstein theorem.[10] Zermelo's system Z can be understood as a reconceptualization of Dedekind's logical frame: the contradictory comprehension principle is replaced by the restricted separation principle and the latter is supplemented by suitable set existence principles, e.g., the power set and union axioms and the axiom of infinity. It should be noted that mappings are no longer considered as belonging to a separate category of mathematical entities but are rather defined as sets.

Zermelo's system Z developed into ZF during the next 20 years and was adopted as the framework for structural axiomatics. This way of looking at mathematics from a conceptual point of view was clearly articulated by Bourbaki. In their programmatic *The Architecture of Mathematics* from 1950, the role of *principal structures* (*structures mères*) is brought out, and their role in making mathematics intelligible is emphasized. Bourbaki clarifies (1950, 225–226) "what is to be understood, in general, by a mathematical structure":

> The common character of the *different concepts* [my emphasis] designated by this generic name [mathematical structure], is that they can be applied to sets of elements whose nature has not been specified; to define a structure, one takes as given one or several relations, into which these elements enter (in the case of groups, this was the relation $z = x \tau y$ between three arbitrary elements); then one postulates that the given relation or relations, satisfy certain conditions (which are explicitly stated and which are the axioms of the structure under consideration).

The striking parallelism of this description with Hilbert's and Zermelo's formulations should be obvious. Indeed, Hilbert had expressed that perspective in his letter to Frege as follows: "Well, it is surely obvious that every theory is only scaffolding of concepts or a schema of concepts together with their necessary relations to each other, and the basic elements can be thought in arbitrary ways" (Frege 1980, 13). For Bourbaki the expression "this system of mathematical objects has the structure of . . ." is synonymous with "this system of mathematical objects falls under the concept of . . . " Bourbaki concludes this passage on structures-in-general as follows:[11]

[10] The Cantor-Bernstein theorem is not actually formulated in *WZ*. However, Theorem 63—a theorem that is neither proved nor needed for the further development in *WZ*—is used in a contemporaneous manuscript to prove the Cantor-Bernstein theorem; see (Sieg and Walsh 2017).

[11] For a detailed understanding of Bourbaki's notion of "structure" one has, of course, to consult their mathematical exposition in their "Théorie des ensembles" and, additionally, study the very informative papers (Dieudonné 1939), (Cartan 1942), and (Bourbaki 1949): they lay bare their methodological considerations and sympathies.

> To set up the axiomatic theory of a given structure . . . amounts to the deduction of the logical consequences of the axioms of the structure, excluding every other hypothesis on the elements under consideration (in particular every hypothesis as to their own nature).

Again, one should notice the parallelism to Dedekind, Hilbert, and Zermelo. The pure structuralism exemplified by Bourbaki is also formulated in Bernays (1955, 109):

> Not only did Euclidean geometry lose its distinguished position and thus its role as the evident theory of space, but now also the arithmetic theory of magnitudes appears just as the theory of one structure among others. The dominant viewpoint is now one of a general formal theory of structures.[12]

The papers mentioned in note 11 that precede the programmatic (1950) show Bourbaki as being in the direct and deeply "formalist" tradition of Hilbert but refusing to take on the methodological challenge of his foundational program. And what a challenge it was, or turned out to be.

The consistency problem was for Hilbert, as I mentioned already, a quasi-syntactic one. However, all the proof ideas concerning the consistency of the arithmetic of real numbers—indicated in lectures or publications from this early period—are of a semantic kind. In his Heidelberg talk of 1904 Hilbert gave for the first time a "direct" syntactic consistency proof, but it was given for a woefully weak system, a purely equational theory for natural numbers without any logical principles. Impressed by Poincaré's well-known criticism of his proof, Hilbert gave up on the syntactic approach until around 1920, when he returned to it after he had taken, what prima facie seems to be a very roundabout path or a genuine detour.

In 1913, the group around Hilbert started a systematic study of *Principia Mathematica* (*PM*) that ultimately resulted in the lectures *Prinzipien der Mathematik*. These lectures were given by Hilbert in the winter term of 1917–18 and written up by Bernays; they are the real, exquisite beginning of *mathematical logic* and literally provide most of the content in Hilbert and Ackermann's influential book (1928). The possibility of formally developing parts of mathematics, in particular number theory and analysis, made it reasonable to reconsider the syntactic approach to consistency. Such formalizations are indeed the

[12] Here is the German text: "Nicht nur, daß die Euklidische Geometrie ihre ausgezeichnete Stellung und damit ihre Rolle als evidente Raumlehre verlor, auch die arithmetische Größenlehre erscheint jetzt mehr nur als die Lehre von einer Struktur unter anderen. Der beherrschende Gesichtspunkt ist jetzt der einer allgemeinen formalen Strukturlehre."

basis for the *uniform projection* of (the mathematical development of) structural definitions into domains of special mathematical objects. The suggested connection to consistency and the special character of these objects must be clarified. Before doing so in the next section, I will let Hilbert speak one more time about his conception of mathematics at this point.[13]

In lectures from the winter term of 1919 (*Natur und mathematisches Erkennen*), Hilbert wanted to support the claim that "the formation of concepts in mathematics is constantly guided by intuition and experience, so that on the whole mathematics is a non-arbitrary, unified structure." Having presented a construction of the continuum and an investigation of non-Archimedean extensions of the rational numbers, he formulated this general point:

> The different mathematical disciplines are consequently necessary parts in the construction of a systematic development of thought; this development begins with simple, natural questions and proceeds on a path that is essentially traced out by compelling internal reasons. There is no question of arbitrariness. Mathematics is not like a game that determines the tasks by arbitrary invented rules, but rather a conceptual system of internal necessity that can only be thus and not otherwise. (Hilbert 1919, 19)

I quoted this passage to make it crystal clear that formalization is a *tool* for Hilbert; this tool allowed him to reconsider the consistency problem in a truly syntactic way. However, it took a while before features of this tool would inspire the particular methodological distinctions of proof theory and would be used in the pursuit of its reductive aims.

3. Formalizability and Reductive Projections

In the 1917–18 lectures, Hilbert and Bernays transformed a part of the system of *PM* with the axiom of reducibility into a *tool for formalizing analysis*. Having proved the least-upper-bound principle in this system of second-order logic, their final comment was,

[13] The development to strict formalization of mathematical practice and the emergence of formal axiomatics is discussed in my other contribution to this volume, namely, "The Ways of Hilbert's Axiomatics: Structural and Formal." The appendix contains additional information about Hilbert's and Bernays' "formalism."

> Thus it is clear that the introduction of the axiom of reducibility is the appropriate means to turn the ramified calculus into a system out of which the foundations for higher mathematics can be developed. (Hilbert 1917–18, 214)

The core methodological question was, Does this system provide a *logicist foundation* for mathematics? If it did, a philosophically satisfying reduction of mathematics to logic would have been obtained. In his talk of September 1917 at the Zurich meeting of the Swiss Mathematical Society, Hilbert reiterated Dedekind's view that mathematics is part of logic. The fundamental work of Frege and Russell bolstered that view, and Hilbert remarked:

> But since the examination of consistency is a task that cannot be avoided, it appears necessary to axiomatize logic itself and prove that number theory and set theory are only parts of logic.
>
> This method was prepared long ago (not least by Frege's profound investigations); it has been most successfully explained by the acute mathematician and logician Russell. One could regard the completion of this magnificent Russellian enterprise of the *axiomatization of logic* as the crowning achievement of the work of axiomatization as a whole. (Hilbert 1918, 1113)

To help him reach this crowning achievement, Hilbert asked Bernays to become his assistant for the foundations of mathematics—at this very meeting in Zurich. Bernays accepted Hilbert's offer and returned to Göttingen, his alma mater, for the following winter semester. From the very beginning, there was a productive collaboration between Hilbert and Bernays that led to an immediate and significant outcome, namely, the 1917–18 lectures *Prinzipien der Mathematik* I just discussed.

Addressing the methodological question of section 3, Hilbert and Bernays analyzed *PM* in subsequent lectures and examined the nature of the axiom of reducibility. They concluded that its acceptance amounted to using structural axiomatics with its existential presupposition in a different guise, applied to the system of predicates concerning individuals. Thus, Russell's approach did not resolve the foundational problem.[14] Bernays articulated in his (1922b) the issue of assuming the existence of a model for any structural notion as follows:

> In the assumption of such a system with particular structural properties lies something transcendental, so to speak, for mathematics, and the question

[14] Their quite compelling arguments were exposed in the lectures (Hilbert 1920, 361–362) and are quite carefully reviewed in (Bernays 1930, 49–50). The evolution of Hilbert's thought in the period from 1917 to 1922 is discussed in my (1999); see also (Ferreirós 2009).

> arises which principled position with respect to it should be taken. (Bernays 1922b, 10)

An intuitive grasp of the completed sequence of natural numbers or even of the manifold of real numbers should not be excluded outright, Bernays asserted. Alluding to contemporaneous tendencies in the exact sciences, he suggested a different strategy, namely, to see "whether it is not possible to give a foundation of these transcendental assumptions in such a way that only primitive intuitive knowledge is used" (Bernays 1922b, 11).

Bernays's programmatic suggestion is brought to life through the idea of *projecting* structural definitions into a constructive domain and examining the image from a constructivist standpoint: the *formalization* of the structural notion was seen as the means of projecting. In Bernays's still pre-Gödel essay of 1930 one finds the remark:

> At this point, the investigation of mathematical proofs by means of the logical calculus is brought to bear in a decisive way. This [investigation] has shown that the concept formations and the inference patterns used in the theories of analysis and set theory are reducible to a limited number of processes and rules; in that way we succeed in totally formalizing these theories within the frame of a precisely delimited symbolism. (Bernays 1930, 57)[15]

Note that the *total* formalization with restricted processes and rules is at stake, not the syntactic completeness of the formal theory used to capture the structural concept. At this point, normative considerations as to the effectiveness of formal theories entered; after all, it *should be decidable* by a finite procedure whether a given syntactic configuration constitutes a formal proof or not. The total and effective formalizability underlies Hilbert's view that the consistency problem for formal theories is a constructive one. Hilbert and Bernays saw the evolving *formal axiomatics* as applying in identical ways to different parts of mathematics. The significance of this fact is expressed even in *Grundlagen der Mathematik I*:

> Formal axiomatics, too, requires for the checking of deductions and the proof of consistency certain evidences, but with the crucial difference (when compared to contentual axiomatics) that this evidence does not rest on a special epistemological relation to the particular domain, but rather is one and the same for any

[15] Here is the German text: "Hier kommt nun die Untersuchung der mathematischen Beweise mit Hilfe des logischen Kalküls entscheidend zur Geltung. Diese hat gezeigt, daß die Begriffsbildungen und Schlußweisen, die in den Theorien der Analysis und der Mengenlehre angewandt werden, auf eine begrenzte Anzahl von Prozessen und Regeln zurückführbar sind, so daß es gelingt, diese Theorien im Rahmen einer genau abgegrenzten Symbolik restlos zu formalisieren."

> axiomatics; this evidence is the primitive manner of recognizing truths that is a prerequisite for any theoretical investigation whatsoever. (Hilbert and Bernays 1934, 2)

This remark provides the reason for the *uniform* character of the projections' images in a single finitist frame.

Bernays explicitly introduced the image of *projection* in the early 1920s. The appendix to this chapter, "Transition to Hilbert's Proof Theory in 1922," describes the related pre-finitist considerations in (Bernays 1922a) and the use of projections there. As late as 1970, Bernays wrote:

> Taking the deductive structure of a formalized theory as an object of investigation, the (structural axiomatic) theory is *projected* as it were into the number-theoretic domain. (Bernays 1970, 186)

The result of this projection will usually be different from the structure intended by the theory. Nevertheless, the projection has an important point:

> The number-theoretic structure can serve to recognize the consistency of the theory from a standpoint that is more elementary than the assumption of the intended structure. (Bernays, 1970, 186)

The emphasis on *number-theoretic* structures is an artifact of the developments in the wake of Gödel's (1931), namely, the arithmetization of metamathematics. Initially, Hilbert and Bernays viewed the exclusive focus on natural numbers in the foundational discussion as a "methodological prejudice."[16] In their proof-theoretic studies during the 1920s, they operated with what they thought of as broader classes of mathematical objects, namely, finite syntactic configurations like formulae and derivations, and accepted induction and recursion principles for them. The methodological situation is diagrammatically depicted in Figure 2, making clear the reductive role of the projection: it avoids the role of models and their representation, creating, rather, an image in the finitist domain. Hilbert and Bernays never precisely characterized the "finitist domain" and did not offer a rigorous delimitation of finitist mathematics, though the image was to be investigated from the *finitist standpoint*.[17]

[16] The remark is quoted in full and its context is analyzed in (Sieg 1999, 117–118). In his (1970, 188), Bernays calls "the arithmetizing monism in mathematics an arbitrary thesis." In his (1937, 81) he emphasizes that the "total elimination of geometric intuition" might be viewed as "unsatisfactory and artificial." He claims there, "The reduction of the continuous to the discrete succeeds indeed only in an approximate sense."

[17] The term "finite Mathematik" was seemingly a familiar one at this point in early 1922; it had been used in (Bernstein 1919) as covering any "constructive" tendency whatsoever.

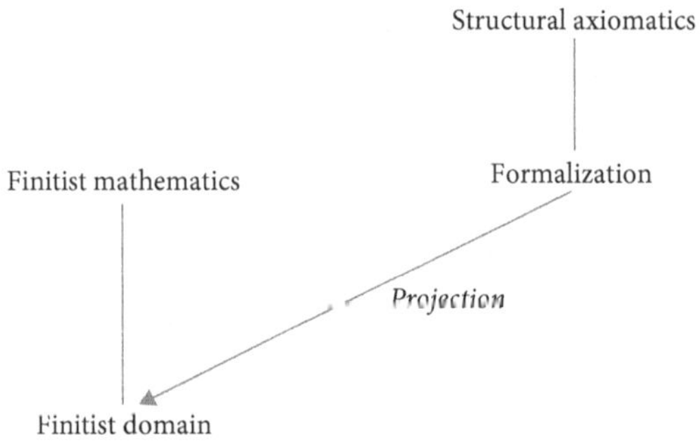

Figure 2. Projection into the finitist domain

As an exemplification of this methodological schema, consider the structural concept of a *complete ordered field* as formalized in second-order number theory; in the finitist domain one has to represent only the elementary formalism, not the infinite objects of its models.

In line with the inspiration from science for the proof-theoretic enterprise, Bernays emphasized in (1922b) the significance of what we now call the *reflection principle*. That principle is equivalent to the consistency of a formal theory T and states that the provability in T of a finitistically meaningful statement implies its finitist correctness; see section 4.[18] This refined metamathematical approach to the consistency problem was successfully realized in early 1922 for the quantifier-free system of *primitive recursive arithmetic*, which is a theory of definite mathematical interest. In the lectures (Hilbert 1921–22), one finds explicitly the beginning of Hilbert's proof theory and his finitist program. Attempting to extend this approach to theories with quantifiers, Hilbert's *Ansatz* from late 1922 replaced quantifiers by epsilon terms and investigated the resulting proofs by the substitution method; that approach was successfully taken up by Ackermann in his thesis (1924), though not in as sweeping a way as it was at first believed. Hilbert's address to the Bologna Congress in 1928 was a bold political act expressing his deep commitment to the international mathematical community, but it was also a remarkable scientific statement: the evolution of mathematical logic is described with great lucidity; the state of proof theory is presented, albeit mistakenly, as including consistency proofs for full elementary number theory by Ackermann and von Neumann; important metamathematical problems are formulated, in particular, the consistency problem for analysis, the

[18] See also (Hilbert 1928, 474) and (Bernays 1930, 55; 1937, 80; 1938, 153).

syntactic completeness problem for number theory, and the semantic completeness problem for first-order logic.

Gödel gave in his thesis (1929) a positive solution of the last problem; in his attempt to address the first problem, he discovered in August 1930 the syntactic incompleteness of familiar theories like *PM*, *ZF*, and von Neumann's set theory. A few months later he proved his second incompleteness theorem, which was viewed by some as radically undermining Hilbert's finitist consistency program.[19] That program, Gödel noted, had been attractive to mathematicians and to philosophers alike; in his 1938 lecture at Zilsel's, he wrote:

> If the original Hilbert program could have been carried out, that would have been without any doubt of enormous epistemological value. The following requirements would both have been satisfied: (A) Mathematics would have been reduced to a very small part of itself. . . . (B) Everything would really have been reduced to a concrete basis, on which everyone must be able to agree. (Gödel 1938, 113)

Gödel explored in this lecture a variety of extensions of finitist mathematics: from transfinite induction used in Gentzen's 1936 proof of the consistency of arithmetic to his own system of computable functionals of finite type that led eventually to the *Dialectica interpretation*; see (Sieg and Parsons 1995).

At this point, when thinking from our contemporary perspective about *extensions* of the constructive basis for Hilbert's program, it is important to examine which structural notions need to be reduced and to reflect on the domains to which they are to be reduced. After all, the simplicity of the universal finitist basis has been lost, but there may be other bases, "on which everyone must be able to agree." In the same year in which Gödel made his remarks at Zilsel's, Bernays contributed a paper to *Les Entretiens de Zürich*, entitled *Über die aktuelle Methodenfrage der Hilbertschen Beweistheorie*; the paper was published in French three years later (Bernays 1941). Bernays addressed the same question Gödel had asked at Zilsel's: How can one extend the finitist standpoint? Both examined Gentzen's 1936 consistency proof of elementary number theory via transfinite induction up to the first epsilon number, and both asserted that this principle went beyond finitist mathematics.

Gödel referred to the French publication of this paper in a letter to Bernays of January 16, 1942. He writes with obvious surprise:

[19] For the developments that arose out of Gödel's Königsberg remarks and his (1931), see (Sieg 2011); von Neumann had independently discovered the second incompleteness theorem already in November 1930 as we know from his correspondence with Gödel that was published in volume 5 of Gödel's *Collected Works* (2003b).

> I read your article in the *Entretiens de Zürich* from the year 1938 with great interest; only what you say on p. 152, lines 8–11 is not comprehensible to me. Wouldn't that be tantamount to giving up the formalist standpoint? (Gödel 2003a, 133)

Gödel points to the last sentence of a paragraph in which Bernays answered his own question: What is the methodological restriction of proof theory, if it is not the restriction to the elementary evidence of the finitist standpoint? Bernays wrote (and I translate from the German original of his *Entretiens* contribution (1938, 16)):

> One can respond [to this question] that the general nature of the methodological restriction remains in principle exactly the same. However, if we want to keep open the possibility of extending the methodological frame, then we must avoid using the concepts of evidence and certainty in a sense that is too absolute.[20]

The paragraph ends with the sentence Gödel had pointed to: "In this way we gain, on the other hand, the fundamental advantage of not being forced to view the usual methods of analysis as unjustified or dubious" (Bernays 1938, 16).[21] Bernays agrees in his response to Gödel's letter that this perspective is not that of strict formalism, but he also emphasizes that he has never taken a formalist position.[22] Positively, he argues:

> It does not seem appropriate to posit in an absolute sense one methodological standpoint per se as evident and the standpoints differing from it as dubious or as only technically justified. That sort of opposition is also not at all necessary . . . as long as one decides to distinguish between different layers and kinds of evidence. (In Gödel 2003a, 139)

Bernays then points out that the certainty of a *thought system* (*Gedankensystem*) is not given from the beginning but is acquired through a kind of *intellectual*

[20] Here is the German text: "Hierauf ist zu erwidern, dass die Tendenz der methodischen Beschränkung grundsätzlich dieselbe bleibt, nur dass wir—wenn wir uns die Möglichkeit von Erweiterungen des methodischen Rahmens offen halten wollen—vermeiden müssen, die Begriffe der Evidenz und der Sicherheit in einem zu absoluten Sinne zu gebrauchen."

[21] Here is the German text: "Damit gewinnen wir andrerseits den grundsätzlichen Vorteil, dass wir nicht genötigt sind, die üblichen Methoden der Analysis als ungerechtfertigt oder bedenklich zu problematisieren."

[22] Bernays points to his (1930) and his essay "Sur le platonisme dans les mathématiques" as exemplary essays in which he took exception from such a perspective. Clearly, he had taken already in his 1922 papers such a "non-formalist" position.

experience (*geistige Erfahrung*). That observation pertains also to analysis. Nevertheless, he emphasizes, "that does not prevent one from contrasting the methods of analysis with an approach of more elementary evidence and of a more specifically arithmetic character" (In Gödel 2003a, 139).

Having articulated an open perspective that allows distinguishing between different layers and kinds of evidence, Bernays insists on the methodological significance of syntactic consistency proofs:

> The task of establishing the inner harmony of analysis from such a standpoint of more elementary evidence as a syntactic necessity by formalizing the inferences of analysis, that task gains in this way its methodological significance.[23] (In Gödel 2003a, 138)

What standpoint of "more elementary" evidence can be taken? How is an approach based on "a more specifically arithmetic character" to be understood? In subsequent papers Bernays made some general suggestions, which point in a direction that can be given more weight and significance by exploiting our more extended experience with proof-theoretic investigations.

4. Accessible Objects and Principles

To indicate the core metamathematical and methodological issues, I will discuss three examples of relevant proof-theoretic work. However, before giving these examples, I briefly recall the context as described at the beginning of the previous section: structural definitions are to be projected, via their associated formal development, into a "constructive" domain; their images are to be investigated from a "constructive" standpoint with the goal of establishing the consistency of the structural definition. Indeed, the methods for consistency proofs in the pursuit of variants of Hilbert's Program have been required to be "constructive," i.e., processes should be effective, mathematical objects should be inductively generated, and proofs should shun the law of the excluded middle. Bernays highlighted these features in his (1954), as the metamathematical investigations must be embedded in a suitable methodological frame. To be suitable for the programmatic proof-theoretic aims, such a frame must satisfy the constructivity requirements just listed, in particular, the crucial condition on mathematical

[23] Here is the German text: "Die Aufgabe, die innere Einstimmigkeit der Analysis von einem solchen Standpunkt elementarerer Evidenz an Hand der Formalisierung der Schlussweisen der Analysis als eine syntaktische Notwendigkeit zu erweisen, erhält damit ihre methodische Bedeutsamkeit."

objects: "The objects (making up the intended model of the theory) are not taken from the domain as being already given but are rather constituted by generative processes" (1954, 12). The nature of the objects is as irrelevant for Bernays as it was for Dedekind, but the generative processes give them a unique *internal structure*. This internal structure is independent of the completed totality of all the generated objects. Keep this observation in mind when I discuss now three paradigmatic proof-theoretic studies.

The first proof-theoretic study is important for two reasons: (1) it showed that Hilbert's program could be pursued from an extended constructive standpoint, and (2) it exemplified an important shift, as the formalization of the broader constructive principles could be used to prove rigorously formulated *relative* consistency results. As to (1), John von Neumann and Jacques Herbrand believed that Gödel's results spelled the definite impossibility of the program for strong formal theories like analysis or even full number theory. When writing his (1931), Herbrand knew Gödel's results well and proved finitistically the consistency of fragments of first-order number theory (PA), when the induction principle is restricted to quantifier-free formulae. Gödel viewed Herbrand's theorem, even in December 1933, as the most far-reaching result in the pursuit of Hilbert's finitist program (Gödel 1933a, 52). What changed the general approach to the consistency problem was the metamathematical fact proved in this first study: Gödel and Gentzen independently established in 1932 the consistency of full elementary number theory (PA) *relative* to its intuitionist variant (HA).[24] According to Bernays (1967) and the historical record, finitist and intuitionist mathematics had been viewed as co-extensional up to the discovery of the reduction of (PA) to (HA). This result showed that intuitionist mathematics is a proper constructive extension of finitist mathematics.

One can view this result as having been obtained by a projection of the concept *simply infinite system* through (PA) into a subdomain of intuitionist mathematics. The arithmetic principles governing the relevant subdomain are those of (PA) and are joined with intuitionist logic; the resulting formal theory is Heyting Arithmetic (HA). Notice, first of all, that (PA) is adequate for the formalization of ordinary number theory. Observe, second, that derivations in (PA) are syntactically translated into proofs in (HA). Indeed, the translation and the metamathematical argument, showing that the translation yields HA-proofs, can be carried out in (HA). The resulting HA-proofs are, finally, recognized from an intuitionist standpoint as being "correct."[25]

24 As to the sequence of these discoveries see (Sieg 2011, 178).

25 *Correct* is to be understood in this context in two different ways. In the formal metamathematical argument, one establishes the partial reflection principle for (PA) within the constructive theory (HA) for a certain class of arithmetic statements. In the overall methodological considerations, one recognizes the proofs in (HA) as "fully correct" from the intuitionist standpoint. For a more detailed discussion, see my (1984) or its republication in (Sieg 2013, 250–252).

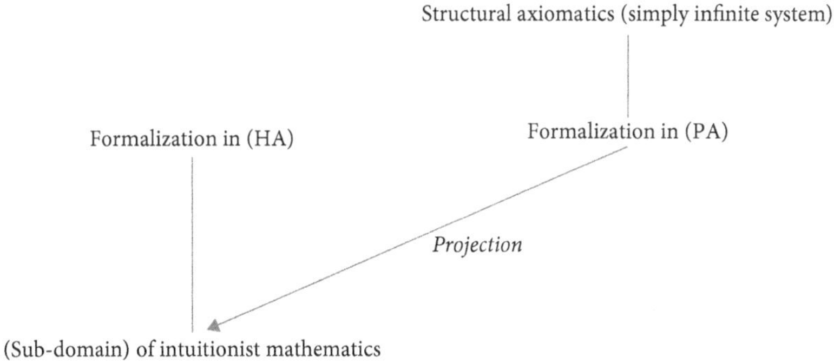

Figure 3. Projection into a sub-domain of intuitionist mathematics

Here we have a direct and essentially logical reduction. Figure 3 points again to formalization as the means of projection, but it incorporates the formalized principles needed for the *relative* consistency proof. Thus, the diagram has three central, distinct components: (i) an articulation of the abstract axiomatic theory as a formal one, (ii) the identification of the syntactic objects of the formal theory with elements of a suitable domain, and (iii) the precise formulation of constructive principles concerning that domain. The Gödel-Gentzen result is special in that both the classical and constructive theory have the same mathematical principles; it is "only" the underlying logic that is different.

Significantly later, results were obtained for the notion of a *complete ordered field*. That concept is also categorical, but I should emphasize that categoricity does not guarantee accessibility: Cauchy sequences, Dedekind cuts, and Hilbert's "Strecken" (of his geometric model) all constitute complete ordered fields that are isomorphic, but not canonically so. In the second study, the classical theory is a subsystem of analysis (i.e., of second-order arithmetic) with the comprehension principle for arithmetic formulae only; the system is denoted by $(ACA)_0$. It can be shown to be conservative over (PA) and, as (PA) is relative consistent to (HA), it is consistent relative to (HA).[26] Despite the fact that this subsystem of analysis is proof-theoretically not stronger than (HA), it is adequate for a significant part of mathematical practice: Weyl's development of classical analysis in *Das Kontinuum* can be formalized in $(ACA)_0$; see (Feferman 1988) and also (Takeuiti 1978). All of this is reflected through an easy modification of the

[26] A proof-theoretic argument for the conservative extension result is given in (Feferman and Sieg 1981, 112). It can be established in (HA); that fact is important for the proof of the partial reflection principle. Many reductions of classical to constructive theories are found in that paper. Significant reductive results are presented in (Rathjen and Sieg 2018) for a much-extended range of theories.

diagram in Figure 3: "simply infinite system" is replaced by "complete ordered field" and (PA) by $(ACA)_0$.

The third study is even more illuminating as to the broad methodological issues, but it is also mathematically more complex. We aim again for a projection of the notion of a *complete ordered field*, but this time through the classical and impredicative subsystem of analysis $(\Pi^1_1\text{-CA})_0$ into the domain of the finite constructive number classes whose principles are formalized in the intuitionist theory $ID(O)_{<\omega}$. $(\Pi^1_1\text{-CA})_0$ has the comprehension principle for Π^1_1-formulae, whereas $ID(O)_{<\omega}$ expands (HA) by closure and minimality principles for the O_n; these principles are formulated subsequently, once we have stated the generative clauses for the number classes. We first notice that $(\Pi^1_1\text{-CA})_0$ is adequate for the formalization of mathematical analysis. In Supplement IV of Hilbert and Bernays (1939), analysis is developed in second-order arithmetic. A careful examination of their development shows that the comprehension principle is used only for Π^1_1-formulae. Second, the reduction is obtained in two steps. In Feferman (1970), $(\Pi^1_1\text{-CA})_0$ is shown to be proof-theoretically equivalent to the classical theory of finitely iterated inductive definitions $ID_{<\omega}$. The first step is then followed by the reduction of the classical theory $ID_{<\omega}$ to intuitionist $ID(O)_{<\omega}$; this second step was taken in my 1977 thesis and involves crucially transformations of infinitary proof figures that are identified with elements of the constructive number classes. The transformed infinitary proofs of a subclass of arithmetic statements are recognized in $ID(O)_{<\omega}$ as correct. These considerations are reflected in a modification of the diagram in Figure 3: "simply infinite system" is again replaced by "complete ordered field," (PA) by $(\Pi^1_1\text{-CA})_0$, and (HA) by $ID(O)_{<\omega}$. The published presentation of this second step is Sieg (1981), but a sketch is given in (Sieg 2013, 254–256).

A summary discussion of the crucial aspects of these investigations can be given with the help of the three diagrams in Figures 1–3. The first diagram simply reflects my distinction between accessible and abstract axiomatics. The second diagram indicates the perspective for the finitist investigation of the images of structural axiomatic theories; the images have been obtained through the formalization of the theories. The third diagram adds two significant new components. The image of the projected abstract notion is no longer found in the finitist domain, but rather in that of intuitionist mathematics; that is the first new component. The second new component is the formal articulation of the theory in which the metamathematical investigations proceed, here (HA) and $ID(O)_{<\omega}$. An appropriately generalized diagram is found in Figure 4.

What general features should be required of methodological frames, so that they are suitable for extensions of Hilbert's constructivist program? Bernays reflected already in his (1938) on constraints for frames and took an *arithmetical perspective in the strict sense* as central:

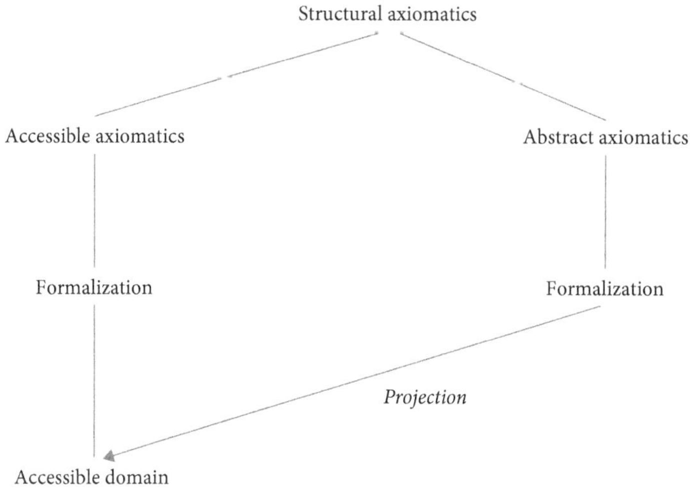

Figure 4. Reductions to accessible domains

> [Accordingly,] arithmetical is the representation [*Vorstellung*] of a figure that is composed of discrete parts, in which the parts themselves are considered *either only* in their relation to the whole figure *or* according to certain coarser distinctive features that have been specially singled out; arithmetical is also the representation of a formal process that is performed with such a figure and that is considered only with regard to the change that it causes. (Bernays 1983)

These considerations underlie the requirement that methodological frames must have domains of objects that are constituted through generative *arithmetical* processes that are then captured through the adopted principles. Bernays called this special form of structural axiomatics *sharpened axiomatics* (*verschärfte Axiomatik*).

Thus, the crucial question is, which procedures can be viewed as generative (arithmetical) ones? Elementary inductive definitions of syntactic notions, like formula or proof, were clearly viewed in that light from the very beginning. Due to Aczel's (1977) we have an extremely general way of generating mathematical objects that goes far beyond the arithmetical generation of Bernays. Aczel's ways allow, of course, the generation of natural numbers, elementary syntactic objects, but they also yield constructive ordinals and even the elements in segments of the cumulative hierarchy of sets.[27] I focus on just natural numbers **N** and

[27] The latter case has its roots in Zermelo's investigation of *Mengenbereiche* in his (1930); their quasi-categoricity ensures that Zermelo's work on *Mengenbereiche* is for sets what Dedekind's work on *einfach unendliche Systeme* is for natural numbers.

constructive ordinals **O**, the second constructive number class. **N** is generated from some element *0* using an injective successor operation *s* and two rules: *0* is in **N** and if *n* is in **N**, then *s(n)* is also in **N**. The second constructive number class is also generated with the help of two rules, namely, *0* is in **O** and if *e* is (the Gödel number of) a recursive function enumerating elements of **O**, then *e* is also in **O**.[28] The closure and minimality principles for domains are standard for **N** and can be articulated for **O** as follows:

$$(\forall x)\ (\mathrm{A}(\mathbf{O},x) \rightarrow \mathbf{O}(x))$$

and

$$(\forall x)\ (\mathrm{A}(F,x) \rightarrow F(x)) \rightarrow (\forall x)\ (\mathbf{O}(x) \rightarrow F(x))$$

The first formula expresses that **O** is *closed* under the generating clauses, whereas the second formula schema says (*F* being any formula in the language of HA expanded by the unary predicate **O**) that **O** is *minimal* among all predicates that are closed under the generating clauses. The latter is the principle of *proof by induction* for **O**. The intuitionist theory ID(**O**) is the extension of (HA) by the two preceding principles.[29]

N and **O** are examples of *i.d. classes* that obey not only the *principle of proof by induction* but also the *principle of definition by recursion*, because they are deterministic.[30] The deterministic i.d. classes are the *accessible domains*, and the associated accessible principles support canonical isomorphisms between any two such classes. They are centrally positioned in the final diagram of Figure 4 that combines and generalizes the diagrams from Figures 1 and 3.

The methodological point of projections and the resulting structural reductions is to coordinate and bring into harmony two crucial aspects of mathematical experience: the *conceptual* one involving abstract notions that have many different models, and the *constructive* one concerning accessible domains that are characterized uniquely up to a canonical isomorphism. The first aspect provides mathematical explanations that rest on conceptual understanding, whereas the second aspect facilitates thinking about mathematical objects and fundamental principles that are grounded in the inductive generation of those

28 The antecedents of these generating clauses can be expressed by a formula that is arithmetic in **O**. Their disjunction is abbreviated by A(**O**, *e*).

29 The intuitionist theory ID$(\mathbf{O})_{<\omega}$ is a similar expansion (HA) with principles for the finite constructive number classes $\mathbf{O}_n$; the latter are obtained by iterating the definition of **O**, but allowing in the second generating clause also branching over already obtained number classes $\mathbf{O}_k$ with *k* less than *n*.

30 An i.d. class is deterministic if the generating operations are injective. Consequently, all of its elements have an associated unique construction tree that is of course well-founded.

objects. *Reductive projections* are the crucial means for joining those aspects guaranteeing the coherence of abstract concepts. The philosophical significance of consistency proofs is to be assessed in terms of the objective underpinnings of the frames to which reductions are achieved. It is precisely here that the various accessible domains play a distinctive role and offer, through a comparison of their generating operations, a scale for assessing relative consistency proofs. This remains an open field for penetrating philosophical investigation and concrete mathematical work.[31]

In this open field, questions are being pursued that transcend traditional issues in the philosophy of mathematics and that are based on one common insight: mathematics *systematically* investigates *concepts* that are structurally defined. Which concepts are to be considered, which logical means are to be used for the development of their theories, and which methodological frames should be considered—these questions have been controversial. From this perspective, the controversy between "classical" and "constructive" mathematics can be transformed into two probing questions, (1) *what is characteristic of and possibly problematic in* classical mathematics and (2) *what is characteristic of and taken for granted as convincing in* constructive mathematics. Answers to these questions have hardly been advanced by "ideological" discussions. Some argue as if an exclusive alternative between *Platonism* (taken to be required for classical mathematics) and *intuitionism* (taken to be required for constructive mathematics) had emerged from sustained foundational work over the last 150 years or so; others argue as if that work were deeply misguided and had no bearing on our understanding of mathematics. Both attitudes prevent us from turning attention to two crucial and more specific tasks, namely, on the one hand, to understand the role of abstract structural concepts in mathematical practice and, on the other hand, to clarify the function of accessibility notions in philosophical analysis. These tasks have fundamentally to do with mathematical cognition; some fruitful directions for explorations are discussed in the next section, which also happens to be the last one.

[31] As to more up-to-date work in proof theory concerning proof-theoretic reductions, see the contribution to the *Stanford Encyclopedia of Philosophy* Rathjen and I wrote (Rathjen and Sieg 2018). The volumes (Kahle and Rathjen 2015) and (Jäger and Sieg 2017) are also rich sources for contemporary work in proof theory. To obtain an "abstract" grasp of accessible domains, I have been interested in their category-theoretic characterization for quite a while see (Sieg 2002, 372–373). Patrick Walsh worked on this very issue in his dissertation (Walsh 2019).

5. Exploring Cognitive Capacities

Reconnecting with Stein's remarks on capacities of the mathematical mind, I am led back to the 19th century and to Dedekind. In his *Habilitationsrede* from 1854 Dedekind remarks on different ways of conceiving the object of a science and asserts that this difference "finds its expression in the different forms, the different systems in which one seeks to frame its conception" (429).[32] The need to frame the conception of a science arises from the fact that our intellectual powers are imperfect. "Their limitation leads us to frame the object of a science in different forms, and introducing a concept means formulating a hypothesis on the inner nature of the science." How well the concept captures this inner nature is determined by its usefulness for the development of the science; in mathematics that mainly means its usefulness for constructing proofs. Dedekind put the theories from his foundational essays to this test by showing that they allow the direct, *stepwise* development of number theory and analysis by means of our *Treppenverstand* using exclusively the *characteristic conditions* (*Merkmale*) of the structural definition of the relevant notion as starting points. *Creating concepts* and *deriving theorems* are consequently the tools to overcome, at least partially, the limitations of our intellectual powers.[33]

The theme of such *specifically human* understanding is sounded also in a remark from Bernays (1954, 18): "Though for differently built beings there might be a different kind of evidence, it is nevertheless our concern to find out what evidence is for us." Bernays emphasized, as mentioned already, that evidence is acquired through intellectual experience and experimentation in an almost Dedekindian spirit. In 1946, he wrote, for example:

> In this way we recognize the necessity of something like intelligence or reason that should not be regarded as a container of [items of] a priori knowledge, but as a mental activity that reacts to given situations with the formation of experimentally applied categories. (Bernays 1946, 91)

Intellectual experimentation of this kind *in part* supports the creation of concepts that define abstract structures or characterize accessible domains; *in part* it is supported through the illuminating use of these concepts in proofs of significant theorems of mathematical practice. These aspects of the mind are central, if we

[32] Here is the German original: "Diese Verschiedenheit der Auffassung des Gegenstandes einer Wissenschaft findet ihren Ausdruck in den verschiedenen Formen, den verschiedenen Systemen, in welche man sie einzurahmen sucht."

[33] This is discussed in detail in (Sieg and Morris 2018). The functional role of concepts or, in Bourbaki's terminology, of structures is emphasized by Heinzmann and Petitot in their contribution to this volume.

want to grasp the subtle connection between reasoning and understanding in mathematics, as well as the role of *leading ideas* in guiding the construction of proofs and of *concepts* in providing explanations.[34]

How can we explore these issues in a systematic and yet open way? The investigation of proofs and their conceptual contexts is central for such research. In a way, I am arguing for an *expansion of proof theory* to consider informal mathematical proofs as objects of theoretical study; formal representations of proofs and their metamathematical investigation are important, but in the end—for our purposes—subservient to the examination of what Hilbert called "the notion of the specifically mathematical proof" (1918). Even for Gentzen in (1936, 499), "The objects of proof theory shall be the *proofs* carried out in mathematics proper." Hilbert had made already an additional claim concerning the general philosophical significance of formalized mathematics that "is carried out according to certain definite rules, in which the *technique of our thinking* is expressed":

> These rules form a closed system that can be discovered and definitively stated. The fundamental idea of my proof theory is none other than to describe the activity of our understanding, to make a protocol of the rules according to which our thinking actually proceeds. . . . If any totality of observations and phenomena deserves to be made the object of a serious and thorough investigation, it is this one. (Hilbert 1927, 475)

A good start for such an investigation is a thorough *computer-based formal reconstruction* of parts of the rich body of mathematical knowledge that *is systematic*, but that is also *structured for human intelligibility and discovery*.[35] In order to expand formal methods by *heuristics* (leading ideas) and to carry out proof search experiments, we must isolate truly creative elements in proofs and implement them. Thus, we will come closer to an understanding of the technique of our thinking, be it mechanical or non-mechanical. In a radio broadcast of 1951, Turing remarked: "The whole thinking process is still rather mysterious to us, but I believe that the attempt to make a thinking machine will help us greatly in finding out how we think ourselves" (Turing [1951] 2004, 486). It is no less mysterious more than 75 years later, but we have now powerful computational

[34] In my paper *Gödel's Philosophical Challenge (to Turing)* (2013a) I explore the ways in which Gödel and Turing, in quite different ways, try to overcome the limitations of particular formal theories. Turing appeals to "initiative" and varied mathematical experience, whereas Gödel seeks a deeper understanding of abstract concepts, in particular, that of "set."

[35] See my paper with Patrick Walsh on *natural formalization* (2017), but also our discussion of Gowers's "human-centered automatic theorem-proving" in (Gowers 2016) and (Ganesalingam and Gowers 2013, 2017).

and sophisticated logical tools as well as a broad methodological perspective for exploring human mathematical cognition. I am convinced that such explorations will illuminate "one of the greatest advances in philosophy."

Appendix: Transition to Hilbert's Proof Theory in 1922

Hilbert's consistency issue had been raised in a "model theoretic" form already by Dedekind. To guarantee that the concept of a simply infinite system does not contain internal contradictions, Dedekind proved the "logical existence" of a system falling under this concept. In the Second Problem of his Paris talk of 1900, Hilbert formulated the goal of ensuring the "mathematical existence" of a structurally defined concept by giving a consistency proof. In (Hilbert 1905), a *direct* syntactic consistency proof was given for a purely equational system of arithmetic. It took the integration of mathematical and logical investigations (as described in sections 2 and 3) to be able to resume such "proof theoretic" investigations in the early 1920s.

Bernays's contribution (1922a) to the issue of *Die Naturwissenschaften* that celebrated Hilbert's 60th birthday was fully aligned with Hilbert's conception of structural axiomatics. His sketch of how to address the consistency problem is based on talks Hilbert had given in Copenhagen and Hamburg during the first half of 1921; they were published as (Hilbert 1922).[36] The transitional features of Hilbert's paper are also reflected in Bernays's considerations.[37] For the axiomatic treatment of geometry, Bernays formulated matters as follows (1922a, 96):

> The spatial relationships are, so to speak, projected into the mathematical-abstract sphere; in this sphere, the structure of their connection presents itself as an object of pure mathematical thinking and is being investigated with the sole focus on logical relations.[38]

[36] The three aspects of 19th-century developments he pointed out in his (1930), and which I discussed at the very beginning of this paper, are already present here in (Bernays 1922a). As to the philosophical significance of this new kind of axiomatics, he emphasized that (1) it involves an "Abgehen vom Apriorismus" (95) and (2) mathematics, so understood and developed, is an "allgemeine Formenlehre" (99).

[37] The development of Hilbert's foundational investigations in this critical period between the 1917–18 lectures and the 1921–22 lectures is described in (Sieg 1999). All the relevant sources are, of course, available now in (Ewald and Sieg 2013).

[38] Here is the German text: "Die räumlichen Verhältnisse werden gleichsam in die Sphäre des Mathematisch-Abstrakten projiziert, in welcher die Struktur ihres Zusammenhanges sich als ein Objekt des rein mathematischen Denkens darstellt und einer Forschungsweise unterzogen wird, die nur auf die logischen Beziehungen gerichtet ist."

How is such an *investigation* to be realized? The structural axiomatic treatment provides the basis for the exclusive focus on logical relations. Any mathematical proof is taken to be "a concrete object all of whose parts can be surveyed; it must be possible, at least in principle, to communicate it [the proof] completely from beginning to end" (97). That a proof does or does not end in a contradiction is "a concretely checkable property." At exactly this point, the logical calculus of "Peano, Frege, and Russell" comes in: these three logicians expanded the calculus in such a way "that the thought-inferences of mathematical proofs can be completely reproduced by symbolic operations" (98). A joint formal development of mathematics and logic is thus ensured, but there is no sense yet of the theoretical means needed for metamathematical investigations. Bernays only writes that, in principle, it is possible to obtain consistency proofs for analysis through "elementary, ostensively certain considerations." Hilbert (1922) thought that one would not have to appeal to any principle of induction, thus sidestepping Poincaré's objection to his earlier syntactic consistency proof.

This brief appendix is simply to point out that Bernays, in his first paper on foundational matters, is fully aligned with Hilbert and uses the representation of mathematical proofs in formalisms as a tool for their investigation, not as a way for characterizing mathematics as a formal game.

Acknowledgments

The perspective on foundational problems I expressed in this chapter is deeply shaped by my intellectual experience as a student of mathematics in Berlin: I was fascinated by *structuralist mathematics* as taught by Karl Peter Grotemeyer, learned the elements of category theory, and read a lot of Bourbaki; at the same time, I was affected by Paul Lorenzen and his philosophically critical attitude toward the foundations of that very mathematics.

It was only later, after having studied mathematical logic in Münster under Dieter Rödding and worked in proof theory at Stanford with Solomon Feferman, that I started to appreciate the balanced perspective of Paul Bernays: his characterization of mathematics as *the science of idealized structures* and the philosophically significant role proof theory was assigned in his scheme of things.

That position was alluded to at the end of my first reflective essay (1984) and became topical and connected to Bernays in (1990). I formulated matters more pointedly in two seminars at the University of Bologna on April 11 and 12, 2007, under the title *Reductive Structuralism: Joining Aspects of Mathematical Experience*. In June 2015, I gave a talk at the University of Vienna under the title *Reductive Structuralism*; the present chapter is an elaboration of those talks.

The translations in this paper are mine, unless quoted from a particular source. I want to thank Erich Reck and Georg Schiemer, who read earlier drafts and made many helpful suggestions. Critical remarks of Patrick Walsh prompted me to rethink and rewrite the central section 4.

References

Ackermann, W. 1924. Begründung des "tertium non datur" mittels der Hilbertschen Theorie der Widerspruchsfreiheit. *Mathematische Annalen* 93, 1–36.

Aczel, P. 1977. An Introduction to Inductive Definitions. In Barwise 1977, pp. 739–782.

Barwise, J., ed. 1977. *Handbook of Mathematical Logic*. Amsterdam: North-Holland.

Bernays, P. 1922a. Hilberts Bedeutung für die Philosophie der Mathematik. *Die Naturwissenschaften* 4, 93–99.

Bernays, P. 1922b. Über Hilberts Gedanken zur Grundlegung der Mathematik. *Jahresbericht der DMV* 31, 10–19.

Bernays, P. 1930. Die Philosophie der Mathematik und die Hilbertsche Beweistheorie. Reprinted in Bernays 1976, pp. 17–61.

Bernays, P. 1938. Über die aktuelle Methodenfrage der Hilbertschen Beweistheorie. Unpublished manuscript from the Bernays Nachlass. it was presented at Les entretiens de Zürich in December 1938 and published in French as Bernays 1941.

Bernays, P. 1941. Sur les questions méthodologiques actuelles de la théorie Hilbertienne de la démonstration. In *Les entretiens de Zürich sur les fondements et la méthode des sciences mathématiques*, edited by F. Gonseth, pp. 144–152. Discussion, pp. 153–161. Zürich: Leemann.

Bernays, P. 1946. Gesichtspunkte zum Problem der Evidenz. Reprinted in Bernays 1976, pp. 85–91.

Bernays, P. 1954. Zur Beurteilung der Situation in der beweistheoretischen Forschung. *Revue internationale de philosophie* 8, 9–13. Discussion, pp. 15–21.

Bernays, P. 1967. Hilbert, David. In *Encyclopedia of Philosophy*, edited by P. Edwards, vol. 3, pp. 496–504. New York: Macmillan.

Bernays, P. 1970. Die schematische Korrespondenz und die idealisierten Strukturen. Reprinted in Bernays 1976, pp. 176–188.

Bernays, P. 1976. *Abhandlungen zur Philosophie der Mathematik*. Darmstadt: Wissenschaftliche Buchgesellschaft.

Bernstein, F. 1919. Die Mengenlehre Georg Cantors und der Finitismus. *Jahresbericht der DMV* 28, 63–78.

Bourbaki, N. 1949. Foundations of Mathematics for the Working Mathematician. *Journal of Symbolic Logic* 14, 1–8.

Bourbaki, N. 1950. The Architecture of Mathematics. *Mathematical Monthly* 57, 221–232.

Buchholz, W., S. Feferman, W. Pohlers, and W. Sieg. 1981. *Iterated Inductive Definitions and Subsystems of Analysis: Recent Proof-Theoretical Studies*. Vol. 897 of Lecture Notes in Mathematics. New York: Springer.

Cartan, H. 1943. Sur the fondement logique des mathématiques. *Revue Scientifique* 81, 3–11.

Dedekind, R. 1854. Über die Einführung neuer Funktionen in der Mathematik, Habilitationsvortrag. In Dedekind 1932, pp. 428–438. Translated in Ewald 1996, pp. 754–762.

Dedekind, R. 1872. Stetigkeit und irrationale Zahlen. Vieweg. Reprinted in Dedekind 1932, pp. 315–324. Translated in Ewald 1996, pp. 765–779.

Dedekind, R. 1888. Was sind und was sollen die Zahlen? Vieweg. Reprinted in Dedekind 1932, pp. 335–391. Translated in Ewald 1996, pp. 787–833.

Dedekind, R. 1890. Letter to H. Keferstein, Cod. Ms. Dedekind III, I, IV. Printed in Sinaceur 1974, pp. 270–278. Translated in van Heijenoort 1967, pp. 98–103.

Dedekind, R. 1932. *Gesammelte mathematische Werke*. Vol. 3. Edited by R. Fricke, E. Noether, and Ö. Ore. Braunschweig: Vieweg.

Dieudonné, J. 1939. Les méthodes axiomatiques modernes et les fondements des mathématiques. *Revue Scientifique* 77, 224–232.

Ewald, W. B., ed. 1996. *From Kant to Hilbert: A Source Book in the Foundations of Mathematics*. 2 vols. New York: Oxford University Press.

Ewald, W,. and W. Sieg, eds. 2013. *David Hilbert's Lectures on the Foundations of Arithmetic and Logic, 1917–1933*. New York: Springer.

Feferman, S. 1988. Weyl Vindicated: "Das Kontinuum" 70 Years Later. In *Temi e prospettive della logica e della filosophia della scienza contemporanee*, vol. 1, *Logica*, pp. 59–93. Bologna: Cooperativa Libraria Universitaria Editrice Bologna.

Feferman S., and W. Sieg. 1981. Proof-Theoretic Equivalences between Classical and Constructive Theories for Analysis. In Buchholz et al. 1981, pp. 78–142.

Frege, G. 1969. *Nachgelassene Schriften und wissenschaftlicher Briefwechsel*. Edited by H. Hermes, F. Kambartel, and F. Kaulbach. Hamburg: Meiner Verlag.

Ferreirós, J. 2009. Hilbert, Logicism, and Mathematical Existence. *Synthese* 170, 33–70.

Ganesalingam, M., and W. T. Gowers. 2013. A Fully Automatic Problem Solver with Human-Style Output. arXiv: 1309.4501.

Ganesalingam, M., and W. T. Gowers. 2017. A Fully Automatic Problem Solver with Human-Style Output. *Journal of Automated Reasoning* 58, 253–291.

Gentzen, G. [1933] 1964. Über das Verhältnis zwischen intuitionistischer und klassischer Arithmetik. *Archiv für mathematische Logik und Grundlagenforschung* 16, 119–132.

Gentzen, G. 1936. Die Widerspruchsfreiheit der reinen Zahlentheorie. *Mathematische Annalen* 112, 493–565.

Gödel, K. 1929. Über die Vollständigkeit des Logikkalküls. Dissertation, Vienna. In Gödel 1986, pp. 60–101.

Gödel, K. 1931. Über formal unentscheidbare Sätze der Principia Mathematica und verwandter Systeme I. In Gödel 1986, pp. 126–195.

Gödel, K. 1933. Zur intuitionistischen Arithmetik und Zahlentheorie. In Gödel 1986, pp. 286–295.

Gödel, K. 1933a. The Present Situation in the Foundations of Mathematics. In Gödel 1995, pp. 36–53.

Gödel, K. 1938. Vortrag bei Zilsel. In Gödel 1995, pp. 85–113.

Gödel, K. 1946. Remarks before the Princeton Bicentennial Conference on Problems in Mathematics. In Gödel 1990, pp. 150–153.

Gödel, K. 1986. *Collected Works*. Vol. 1, *Publications 1929–36*. Edited by S. Feferman et al. New York: Oxford University Press.

Gödel, K. 1995. *Collected Works*. Vol. 3, *Unpublished Essays and Lectures*. New York: Oxford University Press.

Gödel, K. 2003a. *Collected Works*. Vol. 4, *Correspondence A-G*. Edited by S. Feferman et al. New York: Oxford University Press.

Gödel, K. 2003b. *Collected Works*. Vol. 5, *Correspondence H–Z*. New York: Oxford University Press.

Gowers, W. T. 2016. Interview with A. Diaz-Lopez. *Notices of the AMS* 63, 1026–1028.

Hallett M., and U. Majer, eds. 2004. *David Hilbert's Lectures on the Foundations of Geometry, 1891–1902*. New York: Springer.

Hasse, H. 1930. Die moderne algebraische Methode. *Jahresbericht der DMV* 39, 22–34.

Herbrand, J. 1931. Sur la non-contradiction de l'arithmétique. *Crelles Journal für die reine und angewandte Mathematik* 166, 1–8.

Hilbert, D. 1899. Grundlagen der Geometrie. In *Festschrift zur Feier der Enthüllung des Gauss-Weber-Denkmals*, pp. 1–92. Göttingen: Teubner.

Hilbert, D. 1900. Über den Zahlbegriff. *Jahresbericht der DMV* 8, 180–183.

Hilbert, D. 1900a. Mathematische Probleme. *Nachrichten der Königlichen Gesellschaft der Wissenschaften zu Göttingen*, 253–297.

Hilbert, D. 1905. Über die Grundlagen der Logik und der Arithmetik. *In Verhandlungen des Dritten Internationalen Mathematiker-Kongresses*, pp. 174–185. Göttingen: Teubner.

Hilbert, D. 1917–18. Prinzipien der Mathematik. Lecture notes by P. Bernays. In Ewald and Sieg 2013, pp. 59–214.

Hilbert, D. 1921–22. Grundlagen der Mathematik. Lecture notes by P. Bernays. In Ewald and Sieg 2013, pp. 431–519.

Hilbert, D. 1927. Die Grundlagen der Mathematik. *Abhandlungen aus dem mathematischen Seminar der Hamburgischen Universität* 6, 65–85. Translated in van Heijenoort 1967, pp. 464–479.

Hilbert, D. 1928. Probleme der Grundlegung der Mathematik. *Mathematische Annalen* 102, 1–9.

Hilbert, D. 1930. Naturerkennen und Logik. *Die Naturwissenschaften* 18, 959–963.

Hilbert, D., and W. Ackermann. 1928. *Grundzüge der theoretischen Logik*. Berlin: Springer. Reprinted in Ewald and Sieg 2013, pp. 806–915.

Hilbert, D., and P. Bernays. 1934. *Grundlagen der Mathematik*. Vol. 1. Berlin: Springer.

Hilbert, D., and P. Bernays. 1939. *Grundlagen der Mathematik*, Vol. 2. Berlin: Springer. 2nd ed., 1970.

Jäger, G., and W. Sieg, eds. 2017. *Feferman on Foundations: Logic, Mathematics, Philosophy*. New York: Springer.

Kahle, R., and M. Rathjen, eds. 2015. *Gentzen's Centenary: The Quest for Consistency*. New York: Springer.

Parsons, C. D. 2008. Paul Bernays' Later Philosophy of Mathematics. In *Logic Colloquium 2005*, edited by C. Dimitracopoulos, L. Newelski, D. Normann, and J. R. Steel, pp. 129–150. Lecture Notes in Logic 28. New York: Association for Symbolic Logic and Cambridge University Press.

Rathjen, M., and W. Sieg. 2018. Proof Theory. *Stanford Encyclopedia of Philosophy*.

Reck, E., and M. Price. 2000. Structures and Structuralism in Contemporary Philosophy of Mathematics. *Synthese* 125, 341–383.

Sieg, W. 1977. Trees in Metamathematics (Theories of Inductive Definitions and Subsystems of Analysis). Dissertation, Stanford University.

Sieg, W. 1981. Inductive Definitions, Constructive Ordinals, and Normal Derivations. In Buchholz et al. 1981, pp. 143–187.

Sieg, W. 1984. Foundations for Analysis and Proof Theory. *Synthese* 60, 156–200. Reprinted in Sieg 2013, pp. 229–261.

Sieg, W. 1990. Relative Consistency and Accessible Domains. *Synthese* 84, 259–297. Reprinted in Sieg 2013, pp. 299–326.

Sieg, W. 1999. Hilbert's Programs: 1917–1922. *BSL* 5, 1–44. Reprinted in Sieg 2013, pp. 91–127.

Sieg, W. 2002. Beyond Hilbert's Reach? In *Reading Natural Philosophy: Essays in the History and Philosophy of Science and Mathematics*, edited by D. Malament, pp. 345–375. Chicago: Open Court. Reprinted in Sieg 2013, pp. 345–375.

Sieg, W. 2011. In the Shadow of Incompleteness: Hilbert and Gentzen. In *Epistemology versus Ontology: Essays on the Philosophy and Foundations of Mathematics in Honour of Per Martin-Löf*, edited by P. Dybjer, S. Lindström, E. Palmgren, and G. Sundholm, pp. 155–192. New York: Springer. Reprinted in Sieg 2013, pp. 155–192.

Sieg, W. 2013. *Hilbert's Programs and Beyond*. New York: Oxford University Press.

Sieg, W. 2013a. Gödel's Philosophical Challenge (to Turing). In *Computability: Turing, Gödel, Church, and Beyond*, edited by B. J. Copeland, C. J. Posy, and O. Shagrir, pp. 183–202. Cambridge, MA: MIT Press.

Sieg, W. 2014. The Ways of Hilbert's Axiomatics: Structural and Formal. *Perspectives on Science* 22(1), 133–157.

Sieg, W. 2016. On Tait on Kant and Finitism. *Journal of Philosophy* 112(5–6), 274–285.

Sieg, W., and D. Schlimm. 2005. Dedekind's Analysis of Number: Systems and Axioms. *Synthese* 147, 121–170. Reprinted in Sieg 2013, pp. 35–72.

Sieg, W., and D. Schlimm. 2014. Dedekind's Abstract Concepts: Models and Mappings. *Philosophia Mathematica* 10.1093/philmat/nku021, 2014, pp. 1–26.

Sieg, W., and R. Morris. 2018. Dedekind's Structuralism: Creating Concepts and Deriving Theorems. In *Logic, Philosophy of Mathematics, and Their History: Essays in Honor of W. W. Tait*, edited by E. Reck. London: College Publication, pp. 251–301.

Sieg, W., and C. D. Parsons. 1995. Introductory note to Gödel 1938. In Gödel 1995, pp. 62–84. Reprinted in Sieg 2013, pp. 193–213.

Sieg, W., and P. Walsh. 2017. Natural Formalization: Deriving the Cantor-Bernstein Theorem in ZF. Manuscript.

Stein, H. 1988. *Logos*, Logic, and *Logistiké*: Some Philosophical Remarks on Nineteenth Century Transformation of Mathematics. In *History and philosophy of modern mathematics*, edited by W. Aspray and P. Kitcher, pp. 238–259. Minneapolis: University of Minnesota Press.

Takeuti, G. 1978. *Two Applications of Logic to Mathematics*. Publications of the Mathematical Society of Japan 13. Tokyo: Iwanami Shoten Publishers; Princeton, NJ: Princeton University Press.

Turing, A. M. 1936. On Computable Numbers, with an Application to the Entscheidungsproblem. *Proceedings of the London Mathematical Society* 42, 230–265.

Turing, A. M. [1951] 2004. Can Digital Computers Think? Radio broadcast. In *The Essential Turing: The Ideas That Gave Birth to the Computer Age*, edited by J. Copeland, pp. 482–486. New York: Oxford University Press.

Turing, A. M. 1954. Solvable and Unsolvable Problems. *Science News* 31, 7–23.

Walsh, P. 2019. Categorical Characterization of Accessible Domains. Dissertation, Department of Philosophy, Carnegie Mellon University.

Weyl, H. 1918. *Das Kontinuum*. Leipzig: Verlag von Veit.

Zermelo, E. 1908. Untersuchungen über die Grundlagen der Mengenlehre I. *Mathematische Annalen* 65, 261–281. Translated in Ebbinghaus et al. (2010), pp. 189-229

Zermelo, E. 1930. Über Grenzzahlen und Mengenbereiche. *Fundamenta Mathematicae* 16, 29–47. Translated in Ebbinghaus et al. (2010), pp. 401-429

15

Carnap's Structuralist Thesis

Georg Schiemer

1. Introduction

Rudolf Carnap's philosophy of mathematics of the 1920s and 1930s is usually identified with his work on Fregean or Russellian logicism and with the principle of logical tolerance first formulated in his *Logical Syntax of Language* (Carnap 1934).[1] However, recent scholarly work has shown that Carnap also made significant contributions to the logical analysis of modern axiomatics and its (meta-)theory, in particular in his unpublished manuscript *Untersuchungen zur allgemeinen Axiomatik*, written between 1927 and 1929. While the early metalogical work presented there has been investigated in detail (e.g., Awodey and Carus 2001; Reck 2007), no closer attention has so far been paid to the structuralist account of mathematics underlying Carnap's "general axiomatics" project.

This chapter investigates Carnap's mathematical structuralism in his work on formal axiomatics as well as in related contributions from the time. As will be shown, his account is based on a genuinely structuralist assumption, namely that axiomatic theories describe abstract structures or the structural properties of the objects in their domains. A central motivation for his work in the 1920s and early 1930s was to give a logical analysis and explication of this structural content of theories. I will dub this assumption Carnap's *structuralist thesis*.

The aim in the present chapter is twofold: first, to show that Carnap, in his various contributions to the philosophy of mathematics from the time, proposed different ways to characterize the notion of mathematical structure. Three approaches will be analyzed in detail here. According to the first one, structure is what can be specified axiomatically, that is, in terms of "implicit definitions" expressed in a formal axiom system. Second, mathematical structures are also characterized in Carnap's work in terms of "logical constructions," more specifically, in terms of explicit definitions in a purely logical type-theoretical language.

[1] See, e.g., Carnap (1930, 1931), as well as Bohnert (1975) for a detailed study of Carnap's account of logicism. Compare, e.g., Friedman (1999) and Wagner (2009) for surveys of Carnap's contributions in his *Logical Syntax*.

Georg Schiemer, *Carnap's Structuralist Thesis* In: *The Prehistory of Mathematical Structuralism*. Edited by: Erich H. Reck and Georg Schiemer, Oxford University Press (2020). © Oxford University Press.
DOI:10.1093/oso/9780190641221.003.0015

Finally, again in the context of his general axiomatics project, Carnap proposes a way to think about the structures shared by isomorphic models of a given theory in terms of the notion of structural abstraction. Thus, so-called model structures are explicitly specified in *Untersuchungen* as isomorphism types that can be specified by means of "definitions by abstraction." The chapter will survey Carnap's different approaches to characterize the structuralist thesis and point out several connections between them.

The second aim is to re-evaluate Carnap's early contributions to the philosophy of mathematics in light of current work on mathematical structuralism. Specifically, I will discuss two connections between his approaches to characterize mathematical structures and present philosophical debates. The first point of contact concerns his attempt to specify structures in terms of definitions by abstraction, or equivalently, by abstraction principles. The general idea here is to specify an identity criterion for structures based on the notion of isomorphism between mathematical systems that instantiate these structures. As we will see, different versions of this type of structural abstraction have also been introduced in recent work on structuralism.[2]

Another point of contact with the present debate concerns the notion of "structural properties" of mathematical objects. Informally speaking, structural properties are characterized as those properties not involving the intrinsic nature of objects, but rather their interrelations with other objects in a given system. In Carnap's work from the late 1920s, one can find two suggestions how to specify such properties, namely (i) in terms of the notion of logical definability and (ii) in terms of the notion of invariance under isomorphic transformations of a given system. As will be shown, a similar *duality* between two ways to think about structural properties is also discussed in contemporary work on structural mathematics.

The chapter is organized as follows: section 2 will provide a brief overview of Carnap's work on the philosophy of mathematics before the publication of his *Logical Syntax*. Section 3 will then focus on Carnap's structuralist account of mathematics, in particular on three ways to characterize the structuralist thesis, namely in terms of axiomatic definitions (section 3.1), logical constructions (section 3.2), and definitions by abstraction (section 3.3). Given this, section 4 will then compare Carnap's position with modern mathematical structuralism. The comparison will focus on the notion of structure abstraction (section 4.1) and the duality between definability-and invariance-based approaches to thinking about structures (section 4.2). Section 5 will contain a brief summary.

2 See, in particular, Linnebo and Pettigrew (2014), Leach-Krouse (2017), and Reck (2018).

2. Pre-*Syntax* Philosophy of Mathematics

Carnap made central contributions to the philosophy of mathematics throughout his intellectual career. For the purpose of the present chapter, it makes sense to distinguish between two phases in his engagement with modern mathematics, namely a "structuralist" phase in his work from the 1920s and early 1930s and the subsequent turn to a "syntactic" period leading to the publication of his *Logical Syntax of Language* in 1934.[3] Our focus here will be limited to Carnap's pre-*Syntax* work on the philosophy of mathematics.[4]

His research on mathematics from this period mainly focuses on three areas that, on closer inspection, are connected with each other in interesting ways. Carnap's most well-known work is foundational in character and concerns a Fregean or Russellian logicism, that is, the reduction of mathematics to higher-order logic. Carnap was a central proponent of the logicist program and published a number of articles on the topic.[5] Logicism is described in these works as based on two main assumptions, namely (i) that all mathematical primitive terms can be explicitly defined in a purely logical language and (ii) that all mathematical axioms (such as the axioms of Peano arithmetic) can be derived from purely logical principles (see, e.g., Carnap 1931).

While Carnap does not specify in detail the logical system to be used for such a logicist reduction in his work, he indicates at several places that it should be a simplified version of Russell and Whitehead's type theory (henceforth TT) first presented in *Principia Mathematica* (Russell and Whitehead 1962). Logicism is thus understood as the general project of interpreting mathematical theories in a logical type theory. More specifically, according to Carnap, the logicist thesis consists of several interpretability results according to which the language of a given mathematical theory (such as Peano arithmetic) can be translated into a purely logical language such that all mathematical axioms and theorems become deducible from certain definitions and the logical principles of TT alone.[6]

[3] Compare Awodey (2017) for a similar distinction in Carnap's work on the philosophy of mathematics. See Awodey and Carus (2007) for a study of Carnap's transition to a purely syntactical approach in the philosophy of logic and mathematics. Compare the articles contained in Part 2 of Wagner (2009) for more detailed discussions of Carnap's account of mathematics in his *Logical Syntax*.

[4] This should not suggest that that Carnap has made no interesting contributions on the topic in *Logical Syntax* or in later work. The present focus on Carnap's early work is due to the fact that his structuralist understanding of mathematics is formulated in most detail here. See, for instance, Goldfarb and Ricketts (1992) for a more general discussion of Carnap's philosophy of mathematics. Compare also Awodey (2007) for a study of Carnap's post-syntactic philosophy of mathematics and logic developed in his later work on semantics.

[5] See, in particular, Carnap (1930, 1931), and also Carnap (1929).

[6] Compare Carnap on this understanding of the logicist thesis: "Every provable mathematical statement can be translated into a statement that consists only of logical primitive signs and that is provable in logic" (1931, 95).

Given this type-theoretic version of classical logicism, it should be noted that Carnap's approach differed in several respects from earlier accounts of the logicist thesis. Most importantly, logicism was not developed by him in strong opposition to other foundational programs such as Hilbert's formalism. Rather, at least from 1931 onward, his work on the foundations of mathematics can be seen as the attempt to "reconcile" Frege's logicism with the emphasis on the formal axiomatic method in the Hilbert school.[7]

A second field of Carnap's research on mathematics concerns the foundations of geometry and the nature of space. While this work precedes his contributions to type-theoretic logicism by a number of years, one can nevertheless identify several interesting thematic connections or continuities between the two fields. One such connection will be discussed in detail in section 3.2. It concerns Carnap's use of logical or set-theoretic constructions (also of central importance for the logicist program) in the representation of geometrical structures such as the structure of topological or projective space.

Carnap's central contribution in this respect is *Der Raum: Ein Beitrag zur Wissenschaftslehre* of 1922, a monograph based on his 1920 dissertation written under the supervision of the neo-Kantian Bruno Bauch (Carnap 1922). Carnap's aim in the book is to settle the long-term debate on the nature of space by distinguishing between three types of geometrical space, namely *formal, intuitive*, and *physical* space, and by studying their respective interrelations. These different notions of space can be investigated by different types of geometrical theories: formal space presents an abstract "order-configuration" whose properties can be specified in terms of a formal axiomatic theory. Intuitive space, in turn, is described by geometrical principles grounded in some form of a priori intuition (or a Husserlian *Wesenserschauung*). Physical space is described in applied or physical geometry, based on conventions concerning its metrical properties.[8]

This novel philosophical analysis of geometrical space was clearly motivated by several fundamental developments in 19th-century geometry as well as by a long-standing debate on the status of geometrical axioms. The immediate mathematical background of Carnap's book includes Grassmann's *Ausdehnungslehre*, Riemann's theory of formal manifolds presented in his *Habilitationsschrift*, Klein's and Sophus Lie's algebraic study of different geometries in terms of transformation groups, and Hilbert's axiomatization of Euclidean geometry presented in his *Grundlagen der Geometrie* of 1899, to name only some. On the philosophical side, Carnap's book engages with work on the reception of the Kantian

[7] See Awodey and Carus (2001), Reck (2004), and Schiemer (2012b) on Carnap's attempted synthesis of the different foundational approaches in mathematics.

[8] See Carus (2007) for a detailed discussion of the neo-Kantian background of Carnap (1922). See Friedman (1999) and Mormann (2007) for different analyses of Carnap's philosophy of geometry, in particular, on his account of the role of conventions in the book.

account of geometrical knowledge in works by Natorp and Cassirer, Poincare's geometrical conventionalism, as well as with contributions by Helmholtz and others on the status of geometrical axioms.

Carnap's philosophical investigation of the nature of space and its axiomatic description in 1922 is closely connected to a third area of research, namely his subsequent work on formal axiomatics. The axiomatic method in mathematics is investigated in detail in several publications from the late 1920s and early 1930s. A main contribution is the second part of his logic textbook *Abriss der Logistik* (Carnap 1928) entitled "Applied Logicistic." Carnap discusses here the logic of axiomatic definitions as well as the formalization of different axiomatic theories (including arithmetic, set theory, projective geometry, and topology).

A second important source is Carnap's already-mentioned *Untersuchungen zur allgemeinen Axiomatik*.[9] In this unpublished manuscript, he develops a general study of the methodology of axiomatic mathematics and a logical explication of several metatheoretic concepts. This includes different notions of (relative) consistency, independence, and completeness of axioms or axiom systems that were discussed informally in preceding mathematical work. Carnap's immediate mathematical background comprises work by Hilbert, the Italian "Peanists," the American postulate theorists, as well as Richard Dedekind's proto-axiomatic study of arithmetic (Dedekind 1888). Moreover, regarding the study of different completeness properties of axiom systems, Carnap frequently refers to Fraenkel's influential *Einleitung in die Mengenlehre* (1928) as an important background for his own more systematic contributions.[10]

Given these thematic fields in Carnap's early philosophy of mathematics, one comment concerning his general structuralist thesis is in order here. His analysis of the nature of formal geometry in 1922 and, more importantly, of general axiomatics from the late 1920s clearly shows that Carnap was not only a "foundationalist," but also an early proponent of a version of philosophical structuralism. Interestingly, his structuralism was not an isolated position at that time, but shared by several other prominent philosophers, including Russell, Cassirer, and Quine.[11] What clearly distinguishes Carnap's account from that of his contemporaries is that the structuralist thesis for him was not just an informal position regarding the nature of mathematics. On the contrary, a central motivation

[9] Related articles written by Carnap on modern axiomatics are Carnap (1929, 1934), and Carnap and Bachmann (1936).

[10] See Awodey and Carus (2001) and Schiemer, Zach, and Reck (2017) for surveys of Carnap's early metatheoretic work. Compare, in particular, Awodey and Reck (2002) for a detailed study of the development of metatheoretic notions in 19th- and early 20th-century mathematics.

[11] See the articles on these philosophers in the present volume for detailed studies of their respective structuralist accounts of mathematics.

underlying his work was to characterize in logical terms the *structural content* of formal theories. So what precisely is Carnap's mathematical structuralism?

3. Three Structuralist Ideas

Carnap's work on the philosophy of mathematics from the 1920s and 1930s contains three distinct but interrelated proposals on how to characterize the structuralist thesis, that is, how to specify the structural content of mathematics:

(i) *Structures via axiomatic definitions*: a mathematical structure is what can be defined in terms of an axiom system.
(ii) *Structures via logical constructions*: a mathematical structure is what is logically constructible in terms of explicit definitions in a purely logical language.
(iii) *Structures via definitions by abstraction*: a mathematical structure is what can be specified in terms of definitions by abstraction (or by abstraction principles).

In the following section, I will give a more detailed discussion of these approaches as well as of Carnap's understanding of their relations. Moreover, I will also discuss how the different methods of thinking about mathematical structure are connected to his generalized logicism.

3.1. Formal Axiomatics

Carnap's early writings on the philosophy of mathematics are strongly motivated by the development of modern axiomatics in work by Hilbert, Dedekind, the Peanists, and the American postulate theorists (among others).[12] What characterizes their contributions is a novel conception of the nature of mathematical theories. Axiomatized theories were no longer understood descriptively, that is, as organizing our knowledge about a pre-theoretically given system such as physical space or the natural numbers. Rather, they came to be understood prescriptively, as definitions of abstract mathematical structures.[13]

[12] Compare, e.g., Torretti (1978), Grattan-Guinness (2000), and Gray (2008) for historical accounts of the development of modern axiomatics.

[13] See Schlimm (2013) for a more detailed discussion of this development and the distinction between a *descriptive* and *prescriptive* account of axiomatic theories.

Interestingly, this new account of the axiomatic method was applied not only in the case of algebraic theories such as the theory of groups, but also to theories traditionally viewed as descriptive in character. Hilbert's axiom system for Euclidean geometry in his *Grundlagen der Geometrie* (1899) is a case in point here. Compare Paul Bernays's apt characterization of the abstract character of Hilbert's approach:

> A main feature of Hilbert's axiomatization of geometry is that the axiomatic method is presented and practiced in the spirit of the abstract conception of mathematics that arose at the end of the nineteenth century and which has generally been adopted in modern mathematics. It consists in abstracting from the intuitive meaning of the terms . . . and in understanding the assertions (theorems) of the axiomatized theory in a hypothetical sense, that is, as holding true for any interpretation . . . for which the axioms are satisfied. Thus, an axiom system is regarded not as a system of statements about a subject matter but as a system of conditions for what might be called a relational structure. (Bernays 1967, 497)

Two issues are particularly noteworthy about Bernays's account of the "abstract conception of mathematics" characteristic of modern axiomatics. (As we will see, both issues also play a significant role in Carnap's own work on the topic.) The first one is a methodological point: the meaning of primitive mathematical terms is not supposed to be specified independently of the axiomatic theory, for instance, by reference to some form of empirical or a priori intuition. Instead, their meaning is determined solely through their occurrence in the axioms in terms of *implicit definitions*. Second, this change is related to a new understanding of the very subject matter of an axiomatic theory. As is highlighted by Bernays, the axiomatic approach of Hilbert is characterized by the assumption that relational structures form the real content of mathematical theories.[14]

The idea that axiomatic theories deal with abstract structures also forms a central assumption in Carnap's work on the philosophy of mathematics. One of his earliest works on the topic, *Der Raum* of 1922, already contains a specification of this structuralist account of theories. As mentioned previously, Carnap distinguishes here between three different concepts of space, namely "formal," "intuitive," and "physical space." The former type of space is the one investigated in pure or formal geometry. It is characterized by Carnap in the introduction to the book in terms of the concept of an "order system" (*Ordnungsgefüge*):

[14] Compare Torretti (1978) as well as the articles on Hilbert, Bernays, Dedekind, and Cassirer contained in the present volume for more detailed accounts of this structuralist understanding of modern axiomatics.

> Formal space is a general order-system of a certain kind. By "general order-system" we mean a system of relations—not between certain objects of a sensible or nonsensible domain, but between entirely indeterminate relata about which we only need to know that one kind of link entails a different kind of link in the same domain. So formal space deals not with the figures usually considered spatial, such as triangles or circles, but with meaningless relata whose place may be taken by an enormous variety of things (numbers, colors, degrees of kinship, judgments, people, etc.). (Carnap 1922, 5–6)

Notice the emphasis on the purely relational character of a formal space and on the fact that the nature of the primitive elements is irrelevant for its geometrical study. In fact, as Carnap points out, these objects are left "indeterminate" in the sense that only their interrelations to other objects are specified by the theory in question.

The first section of the book contains a closer specification of the characteristic properties of a formal space. It is here that the background of Carnap's understanding, namely modern axiomatics in the spirit of Hilbert's work, becomes most explicit. Compare the following remark on the role of axiomatic definitions:

> Only relations among the elements . . . are specified by the axioms. . . . Theorems are then derived from the axioms with no regard whatever for the intuitive meaning of these elements and relations. . . . If we think of all the theorems as put into this more general form, then instead of geometry proper (that of points, lines, and planes) we have a "pure theory of relations" or "theory of orders," i.e., a theory of indefinite objects and of the equally indefinite relations holding among them. (Carnap 1922, 7–8)

An axiom system (such as Hilbert's axiomatization of Euclidean geometry) is described here as a "pure theory of relations," that is, roughly as a formal theory in the modern sense of the term. The primitive terms of a theory are not interpreted but understood schematically. Axioms and theorems derived from the former are, in turn, not assertoric statements about a concrete space, but reinterpretable relative to different systems of the specified structure.

Interestingly, formal space itself is identified by Carnap with this abstract structure shared by the different models of the theory in question. Compare again Carnap on this characterization of the subject matter of formal geometrical theories:

> The object of this discipline is not space, i.e., the system of points, lines, and planes determined by *geometrical* axioms (which we call "intuitive space" to distinguish it), but a "relational or order system" [*Beziehungs- oder*

> *Ordnungsgefüge*] determined by the *formal* axioms. As this represents the formal design of the spatial system, and turns into the spatial system again when spatial elements are substituted for indeterminate relata, it too will be called "space": "*formal space*." (Carnap 1922, 8)

Notice that, in 1922, Carnap does not yet use the term "structure" to label such abstract forms or order systems. This use of terminology changes in the course of the 1920s, however, and Carnap eventually comes to introduce the notion of structure in his work on axiomatics. An early instance of this can be found in Carnap's lecture notes for a course entitled "Philosophy of Space; Foundations of Geometry" held at the mathematical department of the University of Vienna in 1928 and 1930 as well as in Prague in 1932.[15] The subject matter of a formal axiom system of pure geometry is sketched here as follows:

> The AS [axiom system] is about undetermined objects. It determines only a relational structure between them. . . .
>
> Implicit definition: but more precisely: definition of a class of systems of objects, that is the shared "structure" of these systems. . . .
>
> An AS determines (defines) one (or several) structure[s] of a relational system, the "theorems" [*Lehrsatze*] determine structural properties of that system that follow from this definition, the AS; therefore analytic. (RC 089-62-02)

These brief comments highlight Carnap's general conception of axiomatics at the time: a theory can define one or several abstract structures shared by different relational systems satisfying the axiom system in question. How is the notion of an axiomatically defined structure understood here?

This issue as well as the method of implicit definition is first addressed in closer detail in Carnap's "Eigentliche und uneigentliche Begriffe" of 1927 as well as in his logic textbook *Abriss der Logistik* of 1929. The article contains a number of interesting observations regarding the axiomatic method, in particular on so-called definitions through axiom systems. Carnap illustrates this type of definition based on the example of a theory of basic arithmetic.[16] According to him, this theory can either be understood as describing the properties of the intended

[15] See documents RC 089-62-02 of the Rudolf Carnap Papers at the Archive of Scientific Philosophy (Hillman Library, University of Pittsburgh).

[16] The axiom system presented here is based on Russell's theory of arithmetical progressions presented in Russell (1919). Compare section 15.3.3 for a more detailed discussion of the theory. A second paradigmatic example discussed in the text is again Hilbert's axiomatization of Euclidean geometry in Hilbert (1899).

or standard model of the natural numbers. Alternatively, it can also be viewed as a formal theory in the following sense:

> We take the words "number" and "successor" as new terms that have not yet been given a meaning, and we stipulate that they are to refer to those concepts with the character specified by the AS. Thus here the AS makes no initial assumptions, but rather only through it is a class determined, which will then be called "the numbers," and a relation, which will be called "successor." In contrast to the determination of a concept by explicit definition, as discussed earlier, here the new concepts are not connected to old ones, but are specified by the formal characteristics they inherently possess; hence the terminology "implicit definition" for the determination of a concept by an AS. (Carnap 1927, 360–361)

As Carnap makes clear, this account of the implicit definition of primitive terms implies that theories so construed can be interpreted relative to different models. As will be shown in section 3.3, Carnap developed a detailed account of the model theory of axiomatic theories in his manuscript *Untersuchungen zur allgemeinen Axiomatik*, also written around the same time.

More important in the present context is how these models are related to the general structure defined by an axiom system. In the case of elementary arithmetic, this relation between the possible interpretations of the axiom system (including the standard or intended model) and their shared structure is described as follows:

> The first model, the sequence of cardinal numbers, is that for the sake of which the AS was set up. As we see, however, the AS, and therefore the implicit definition it expresses, applies not only to that case, but also to infinitely many others, namely all those that agree with it with respect to the specified formal properties, i.e., the structure. In the theory of relations, the sequences with these properties are called "progressions." . . . The implicit definition of the sequence of numbers therefore does not uniquely determine the number sequence, but only the unique class of all progressions. (Carnap 1927, 362)

Given this model-theoretic account of axiom systems and their interpretations, what does Carnap mean by the structure of a theory? In addressing this issue, his distinction between "improper" and "proper" concepts plays a central role. Briefly put, an axiom system provides an implicit definition of several improper concepts whose meaning remains indeterminate. In the case of arithmetic, these are the concepts expressed by the primitive terms "natural number," "zero," and "successor" respectively. In addition, an axiom system can also be understood as an explicit definition of a proper concept whose meaning is in turn fully determined by the definition.

According to Carnap, this "explicit concept" of an axiom system closely corresponds to the class of models or realizations satisfying the axioms in question. In fact, in the case of elementary arithmetic discussed in his 1927 article, the relevant explicit concept (i.e., the "Peano number concept") is simply defined as the "class" of all arithmetical progressions (see Carnap 1927, 368). This insight that an axiom system not only provides an implicit definition of its primitive terms, but also an explicit definition of a higher-level mathematical concept, was not new at the time. In fact, it is likely that Carnap adopted the idea from his teacher Frege and the latter's critical discussion of Hilbert's work.[17]

Carnap's reformulation of the Fregean understanding of axiom systems as definitions of higher-level concepts is left informal in the 1927 article. This changes in Carnap's *Abriss der Logistik* of 1929, where the topic is taken up again. In Part II of the book, titled "Applied Logistic," Carnap gives a type-theoretic explication of the notion of axiomatic theories and their content. Roughly put, axiom systems are formalized here in a language of simple type theory in the following way: the primitive terms of a theory are expressed by free variables (of a given order and type) $X_1, \ldots, X_n$. Axioms and theorems are expressed as propositional functions $\Phi(X_1, \ldots, X_n)$, that is, as open formulas in the modern sense of the term.[18]

Given this formalization of mathematical theories, Carnap reiterates the point that an axiom system not only provides an implicit definition of the primitive terms occurring in the axioms, but also an explicit definition of a higher-order concept, the "explicit concept" of an axiom system. He gives the following formal account of the notion in the *Abriss*:

> For instance, if $x, y, \ldots \alpha, \beta, \ldots P, Q, \ldots$ are the primitive variables of the *AS* and if we name the conjunction of axioms (that is a propositional function) $\mathrm{AS}(x, y, \ldots \alpha, \beta, \ldots P, Q, \ldots)$, then the definition of the explicit concept of this *AS* is
>
> $$\hat{x}, \hat{y}, \ldots \hat{\alpha}, \beta, \ldots \hat{P}, \hat{Q}, \{AS\,(x, y, \ldots \alpha, \beta, \ldots P, Q, \ldots)\}$$ (Carnap 1929, 72)

[17] Frege's view of formal axiom systems as definitions of higher-level concepts is first expressed in his famous exchange with Hilbert. It is also presented in Frege's lecture "Logic in Mathematics" presented in Jena in 1914. Compare Carnap's notes of the lecture as well as Gottfried Gabriel's introduction, both published in Awodey and Reck (2004).

[18] This convention to express primitive mathematical terms as variables and axioms as propositional functions has a rich mathematical prehistory and is discussed more extensively in Carnap (1927).

The formula in this passage stands for a class of n-tuples of possible interpretations of the primitive variables of a given axiom system *AS*. Put in modern terms, an explicit concept is thus understood purely extensionally here, as determined by the class of models defined by the theory.[19] Carnap's notion of the explicit concept of an axiom system can thus be reconstructed in modern terms as a genuinely model-theoretic notion, namely as the model class of a given theory. Regarding the previous example of elementary arithmetic, Carnap holds that

> The explicit concept of Peano's AS of the numbers, e.g., is the class of number sequences that satisfy the AS; this is the logical concept *prog* (class of the progressions). (Carnap 1929, 72)

The central point to note here is that this notion of explicit concepts can be understood as Carnap's first attempt of a formal specification of the informal notion of "structure" (or "order system") used previously to describe the subject matter of a theory. To put it in Howard Stein's words, "A Fregean 'second-level concept' simply is the concept of a species of structure" (1988, 254).

3.2. Logical Construction

A significant part of Carnap's pre-*Syntax* work on the philosophy of mathematics was dedicated to foundational issues, in particular, to the further articulation of Frege's and Russell's logicist program. In the relevant publications on this topic, Carnap's understanding of concept formation in mathematics seems to be at odds with his structuralist thesis.[20] In particular, he states a strong preference here for the "logical construction" of mathematical concepts based on explicit definitions compared to the mere "postulation" of them in terms of axiomatic conditions. This clearly echoes Russell's preceding discussion of the genetic and the axiomatic method and his well-known remark on "theft over honest toil" in Russell (1919).

Logicism for Carnap too is based on a constructivist account of mathematics that distinguishes it from the axiomatic tradition of Hilbert and Dedekind. Compare Carnap on this general difference in his discussion of impredicative definitions:

[19] One should add here that, strictly speaking, the explicit concept of a theory cannot be identified with its class of models. Rather, what Carnap seems to suggest here is more of a "methodological identification" in the sense that one can study the one by studying the other. I would like to thank Erich Reck for emphasizing this point to me.

[20] See, e.g., Carnap (1930) and Carnap (1931).

> The essential point of this method of introducing the real numbers is that they are *not postulated but constructed*. The logicist does not establish the existence of structures that have the properties of the real numbers by laying down axioms or postulates; rather, through explicit definitions, he produces logical constructions that have, by virtue of these definitions, the usual properties of the real numbers. As there are no "creative definitions," definition is not creation but only name-giving to something whose existence has already been established. . . . This "constructivist method" forms part of the very texture of logicism. (Carnap 1931, 94)

The logicist approach to the formation of concepts in analysis (as well as in other mathematical fields) stated here is clearly incompatible with Hilbert's understanding of axiom systems as implicit definitions of the primitive terms of a theory.[21] How did Carnap address the apparent conflict between the two foundational approaches, namely logicism and formal axiomatics?

Interestingly, the two traditions are usually not treated separately in his work. In fact, Carnap's writings from the time have been described as a systematic attempt to "reconcile" Frege's logicist constructivism with Hilbert's structuralist understanding of mathematics.[22] One approach relevant here has to do with Carnap's own characterization of the structuralist thesis. According to him, mathematical structures can be specified not only through axiomatic definitions, but also as those entities characterizable in *purely* logical terms. Thus, a principled way to think about mathematical structures for Carnap is to say that structure is what is logically definable in higher-order logic (where higher-order logic is usually taken to be a system of simple type theory).

A closer look at his writings from the 1920s helps to see how this "logicist" account of the structuralist thesis and its relation to the axiomatic approach were understood by him. A first formulation of the former approach can be found already in *Der Raum*. In the first chapter of the book and based on the discussion of Hilbert's axiomatic approach, Carnap introduces a second way to specify a formal space (understood again as an abstract "order system"):

> The construction of formal space can also be undertaken by a different path, however, not just by the above way of setting up certain axioms about classes and relations: by deriving (ordered) series and, as a special case, continuous

[21] Compare again Carnap on the constructivism underlying Frege's logicism: "A concept may not be introduced axiomatically but must be constructed from undefined, primitive concepts step by step through explicit definitions" (Carnap 1931, 105).

[22] Compare Awodey and Carus (2001), Reck (2004), and Schiemer (2012a) for more detailed discussions of this point.

> series from *formal logic*, the general theory of classes and relations. (Carnap 1922, 8)

This logical construction of formal space is specified as follows: based on work by Russell (in particular Russell 1903), Carnap first introduces the notion of order relations and order systems, so-called *series*. Special types of such order systems are series of the natural numbers, that is, arithmetical progressions in the sense specified in the previous section as well as continuous series of the real numbers. Given the latter, Carnap argues, one can set-theoretically construct continuous series of higher levels, that is, sets of ordered tuples of real numbers. A formal space (of n dimensions) is then defined as a "continuous series of n-th level (a series of series)" (Carnap 1922, 14). Put in modern terms, this is a manifold of n-ary tuples of real numbers.

Given this general notion of a formal space—also called a topological space R_{nt} here—one can construct other spaces such as projective space or different metrical spaces by imposing "more restrictive conditions on the order relations in these series" (Carnap 1922, 14). Now, Carnap does not specify in detail how these restrictive conditions are to be understood. It becomes clear from his remarks, however, that they should not be identified with axiomatic conditions.[23] More important to note here is that each of the resulting spaces remain formal in the sense specified above. Compare again Carnap on this point:

> We are here still dealing with merely formal relations, without any assumptions about what sort of objects have these relations to each other. The different *R*'s are therefore also called systems of order-relations (systems of ordinal relations), briefly, order-systems. (Carnap 1922, 17)

Given this set-theoretical construction of formal spaces as manifolds of real numbers, two points of commentary are in order here. The first point concerns the relation between Carnap's logicist account of geometrical structures and the axiomatic approach discussed in the previous section. How precisely does the specification of structure in terms of entities characterizable in purely logical terms correspond to the one in terms of axiomatic definitions?

[23] In fact, in an interesting passage, Carnap mentions the axiomatic method as an alternative approach to the specification of such a formal space: "Now, it has emerged that the resulting order-structures (e.g., R_{3p}), if they are to be investigated on their own (i.e., without reference to R_{3t} or R_{nt}), are simpler to construct if they are presented directly as structures of certain simple relations whose formal properties are given—rather than taking the circuitous route by way of continuous series of the first, and then of the third level subject to certain limiting conditions" (Carnap 1922, 15). See Mormann (2007) and Carus (2007) for more detailed discussions of Carnap's approach.

Interestingly, we saw that at least in Carnap (1922), Carnap viewed the two approaches as essentially equivalent ways to think about formal space. More specifically, as pointed out by Friedman in his editorial notes in Carnap (2019), the axiomatic approach gives implicit definitions of the primitive terms of a theory, whereas the logicist approach consists in "*explicitly* defining a model for such an axiom system within . . . set theory." One could therefore think of the connection between the axiomatic and the logicist approach in the following way: a formal space, conceived of as an "order system," is treated here as a concrete model of an axiomatic theory that is representable in set theory.[24] It thus forms a particular instance falling under the higher-level "explicit concept" defined by the theory.

It should be noted however that, strictly speaking, Carnap does not identify the subject matter of a formal geometry with a particular order system (conceived as a set-theoretic model of the theory) in 1922. Given the notion of number series, Carnap introduces the notion of a similarity between such systems. This corresponds roughly to the modern notion of an isomorphism between two ordered sets.[25] An "order type" of a particular series is then defined as the concept holding of all series similar to it. In the case of progressions, this is the order type ω; in case of continuous number series, this is order type λ. Compare again Carnap on this point:

> To express more briefly what holds for these mutually similar series, we assert it of a single formal representative of them that we construct for this purpose. . . . Strictly speaking, this representative of the progressions is nothing other than their concept (in our sense of the word). (Carnap 1922, 13)

Applied to Carnap's account of formal spaces sketched previously, it follows that a (topological, projective, or metrical) space should not be understood as a particular order system. Rather, it presents an order type, that is, a higher-level similarity concept or, put in purely extensional terms, a similarity class of such a system. Thus, both in Carnap's axiomatic approach and in the set-theoretic approach, mathematical structures are identified with higher-level concepts. We will return to his conception of structures as similarity (or isomorphism) types in the next section.

Turning to the second point, one immediate consequence of Carnap's approach in 1922 is that formal geometry itself becomes a part of logic or set theory. This fact was clearly intended and led him to formulate a *generalized* logicism

[24] The latter approach, to think about structure in terms of logically definable models, can be found also in subsequent work by Carnap, in particular in his *Untersuchungen* manuscript. See Schiemer (2012b) for a closer discussion of this point.

[25] Compare section 15.3.3 for a closer discussion of Russell's notion of the similarity of relations and Carnap's later generalization of it.

not limited to number theory and analysis.[26] The view that formal space (as the subject matter of pure geometry) is essentially a logical concept is expressed at different stages in his work on the foundations of geometry. An early formulation of the idea is contained in his dissertation manuscript of 1920, which formed the basis for *Der Raum*:

> An [abstract space] is a logical system of relations among indefinite elements. It says: in case certain relations, specified purely formally, hold among the elements of a set, then certain theories hold for this system. (Unpublished manuscript, Quoted from Carus 2007, 110)

Compare also a related remark concerning the status of pure geometry in Carnap's lectures notes of 1928:

> (Mathematical) geometry is essentially relation theory (theory of relations, of structures, of order systems) a branch of formal logic, therefore analytic. (RC 089-62-02)[27]

Pure geometry forms a part of logic because its subject matter, namely abstract space, can be represented in terms of sets of real number tuples that, given Frege's thesis, are effectively reducible to arithmetical and thus to purely logical notions.

A different but related account of the logical nature of geometry can be identified in Carnap's subsequent work on axiomatics. Returning again to his *Abriss der Logistik*, we saw that an axiom system not only gives an implicit definition of its primitive terms, but also an explicit definition of a higher-level concept applying to all models of the theory in question. Carnap discusses a number of mathematical examples to illustrate this Fregean account, including Peano arithmetic, Zermelo-Freankel set theory, projective geometry, and topology (among others).

For instance, Carnap presents the following formalization of Hausdorff's neighborhood axioms for topological spaces: the theory describes one primitive binary relation, namely $\{\alpha U x\}$ standing for "α is a neighborhood set of x." The class of points is defined as the range of relation U, that is, as $\mathbf{pu} := Ran(\mathrm{U})$.

[26] This geometrical logicism, i.e., the fact that pure space is constructable in pure logic, essentially goes back to Russell's extensive discussion of different geometries in his *Principles of Mathematics* (1903). Carnap frequently refers to this book, as well as to Russell and Whitehead's *Principia Mathematica* as the primary sources for his own discussion of formal space. See, in particular, Gandon (2009) and Gandon (2012) for further details on Russell's approach.

[27] I leave open the issue here how the concept of analyticity used here was understood by Carnap in his pre-syntactical work.

The theory of neighborhoods is given by the following axioms (in slightly modernized form):

> Ax1a: $Dom(\mathbf{U}) \subset \wp(\mathbf{pu})$ (Neighborhoods are classes of points.)
>
> Ax1b: $\mathbf{U} \subset Kon(\in)$ (A point belongs to each of its neighborhoods.)
>
> Ax2: $\forall \alpha, \beta, x(\alpha \mathbf{U} x \wedge \beta \mathbf{U} x \rightarrow \exists y(y\mathbf{U}x \wedge y \subset \alpha \cap \beta))$ (The intersection of two neighborhoods of a point contains a neighborhood.)
>
> Ax3: $\forall \alpha, y(\alpha \in Dom(\mathbf{U}) \wedge y \in \alpha \rightarrow \exists \gamma(\gamma \mathbf{U} y \wedge \gamma \subset \alpha))$ (For every point of a neighborhood α, a subclass of α is also a neighborhood.)
>
> Ax4: $\forall x, y(x, y \in \mathbf{pu} \wedge x \neq y \rightarrow \exists \alpha, \beta(\alpha \mathbf{U} x \wedge \beta \mathbf{U} y \wedge \alpha \cap \beta = \varnothing))$ (For two distinct points, there exist two corresponding neighborhoods with no points in common.)

Given this axiomatization, it seems natural to say that the explicit concept "*hausd*" represents the structure defined by Axioms 1–5, i.e., the structure shared by all concrete models satisfying the theory. Moreover, given the fact that in Carnap's formalization of the theory, the only primitive term, U (standing for the neighborhood sets), is symbolized as a relation variable, it follows that the concept *hausd* turn out to be purely logical in character. Compare Carnap on this point:

> The explicit concept of a geometrical AS . . . presents the logical concept of the relevant type of space (e.g., the concept "projective space"). In this sense geometry can also be represented as a branch of logistic itself (as arithmetic) instead of being a case of application of logistics to a nonlogical domain. (Carnap 1929, 72)

Concerning the specific example of Hausdorff topology, he goes on to add:

> The explicit concept of the AS is the class of the "Hausdorffian neighborhood systems" (*hausd*), a purely logical concept. (Carnap 1929, 76)

These passages illustrate Carnap's attempt to reconcile the logicist's emphasis on explicit definitions with structural axiomatics. The resulting version of the logicist thesis does not amount to the claim that the individual models of an axiomatic theory are logically constructible. Rather, Carnap adopts the Fregean strategy to represent the structural content of a mathematical theory in terms of a *higher-level* concept defined by the theory's axioms. Since such explicit concepts of theories (such as hausd) are definable in a language of *pure* type theory, it follows that the represented mathematical content is also purely logical, and the axiomatic theory thus "a branch of logic."

3.3. Model Structures

An important characteristic of modern axiomatics is the new focus on metatheoretic properties of theories and their interpretations. As a consequence of this "metatheoretic turn" at the end of the 19th and early 20th century, axiom systems themselves became an object of (meta)mathematical investigation. Moreover, mathematicians working in geometry, number theory, and other disciplines started to investigate systematically the content of theories in terms of structure-preserving mappings between their models.[28]

This metatheoretic approach in modern axiomatics is usually characterized today in structuralist terms, that is, by referring to the structures or structural properties defined by an axiom system. More specifically, it is usually held that one can investigate the logical structure of a given theory not only by deriving theorems, but also by analyzing how particular axioms contribute to the specification of this content, how the structure is changed if particular axioms are added or omitted from the system, and so on.[29]

Interestingly, a similar approach to expressing the metatheoretic properties of theories in structuralist terms can be identified in Carnap's work on general axiomatics from the late 1920s. We saw in section 3.1 that Carnap, from his *Der Raum* onward, defended the view that an axiom system defines a structure (or an "order system") that in turn can be instantiated by different "formal models" or physical "realizations." While he does not discuss models and their properties in published work in closer detail, the model-theoretic account of theories is developed in his project on "general axiomatics," in particular, in his *Untersuchungen zur allgemeinen Axiomatik*. The manuscript contains a detailed discussion of the logical formalization of axiomatic theories that is similar to the account presented in *Abriss der Logistik* (see again section 3.1 for details). In addition, Carnap's manuscript also contains a logical explication of several genuinely metatheoretical concepts (such as the notions of logical consequence, truth in

[28] This line of research includes Dedekind's categoricity result for arithmetic in *Was sind und was sollen die Zahlen* (1888), Hilbert's consistency and independence proofs in his *Grundlagen* (1899), as well as the formulation of different notions of completeness in subsequent work by the postulate theorists. See Awodey and Reck (2002) for a rich study of early metatheoretic work in modern axiomatics. Compare also the articles on Hilbert and Dedekind in the present volume for further details.

[29] Compare Hintikka for a characterization of this general approach: "An axiom system is also calculated to serve also as an object for a metatheoretical study. . . . For the purpose of reaching such a metatheoretical overview, it is crucial to grasp the logical structure of the theory in question, in the sense of seeing what the different independent assumptions of the theory are, of seeing which theorems depend on which of these basic assumptions and so on. For this purpose, the axiomatic method is eminently appropriate" (Hintikka 2011, 72–73).

a model, etc.), as well as several metatheorems on the relation between different notions of completeness.[30]

A central concept defined in this context is that of a "model isomorphism," that is, a mapping between two models of a given theory that preserves their relational structure. The isomorphism relation (or, in Carnap's terms, the "isomorphism correlation") between two models is defined roughly in the modern sense as a bijective function between the respective individual domains that induces correlations between the higher-order domains and thus preserves the relations in the models.[31] Based on this notion, Carnap specifies several completeness properties that turn out to be crucial for the understanding of the "logical structure" of theories, including the notions of *non-forkability* and *monomorphicity* (or, in modern terminology, of semantic completeness and categoricity).[32]

How does Carnap specify the structural content of axiomatic theories in *Untersuchungen*? In contrast to previous work, he argues here that an axiom system does not only define an "explicit concept" (conceived of as the class of its models), but possibly also several more fine-grained structures, so-called *model structures* (conceived of as subclasses of its model class). Roughly put, a model structure is the structure shared by isomorphic models of a given theory. As Carnap points out, such structures are to be identified with the classes of isomorphic models:

> In logistic, one tends to define structures, including also the cardinalities, in terms of isomorphism classes. (Carnap 2000, 72)

In the related article (Carnap and Bachmann 1936), a more detailed specification of model structures in terms of the notion of a "complete isomorphism" is given:

> Since the complete isomorphism between n-place models (i.e., sequences with n members) is a $2n$-ary equivalence relation, n-place relations can be defined over the field of this relation . . . such that the n-place relations have the following properties: for each model there exists exactly one such relation which is satisfied by the constituents of the model and is satisfied by the constituents of two different models if and only if the models are completely isomorphic. The

[30] See Carnap (2000). Several of the concepts introduced here were published later on in Carnap and Bachmann (1936). Compare Awodey and Carus (2001), Reck (2004), and Schiemer, Zach, and Reck (2017) for further details on Carnap's axiomatics project.

[31] Carnap's definition of "model isomorphism" is actually more complex than this since it takes into account a mapping between "inhomogeneous models," that is, models with relations of different types and orders. See Carnap (2000) and also Carnap and Bachmann (1936) for further details. See also Carnap (1929) for a simplified definition.

[32] Compare, in particular, Awodey and Carus (2001) and Schiemer, Zach, and Reck (2017) for assessments of Carnap's early metatheory and of the limitations of his approach.

> relations so determined we will call *structures* and will say that model M_1 *has* structure S_1 if "$S_1(M_1)$" is analytic. (Carnap and Bachmann 1981, 74)

Structures are specified here as unary relations that hold between any two models of a given theory in case there exists an isomorphism between them. In an attached footnote to the passage, Carnap goes on to add that structures in this sense are relations introduced by a "definition through formation of abstraction classes" or simply by "definition through abstraction" (Carnap and Bachmann 1936, 171).[33]

Translated into modern terminology, the idea expressed here is to treat structures as particular equivalence classes, namely as "isomorphism classes" of models. Let K be the class of models defined by a theory T. Let $K/_{\cong}$ be the partition of class K induced by a suitable isomorphism relation $\cong$ between the objects in this class. For a model $M \in K$, the relevant model structure is simply the isomorphism class $[M]_{\cong} := \{N \mid N \cong M\}$. Each model structure of T is a cell of the partition of K induced by $\cong$. Moreover, given that $K/_{\cong}$ forms a partition, for any two different model structures we have the following two results: (i) $[M]_{\cong} \cap [N]_{\cong} = \varnothing$ and (ii) $\bigcup_{M \in K} [M]_{\cong} = K/_{\cong}$.[34]

Given this approach, two further points of commentary are in order here. First, it should be noted that Carnap's approach to thinking about mathematical structures in terms of definitions by abstraction was not new, but fairly conventional at the time. In fact, in his *Abriss* and in other publications, Carnap refers to Frege's famous definition of cardinal numbers in terms of an abstraction principle as well as to work by Couturat and Weyl for further details on the method. Concerning the notion of mathematical structure, Carnap's central background is Russell's logical work on the general theory of relations. In fact, the notion of "model structures" outlined in *Untersuchungen* present a straightforward generalization of the notion of "relational structures" previously introduced by Russell in his *Introduction to Mathematical Philosophy* (Russell 1919).[35]

In chapter 6 of his book of 1919, Russell first defines what he calls a "similarity relation" between relations: two n-ary relations R, S are similar if there exists a monotone, that is, a structure-preserving function $f: R \rightarrow S$ such that $x_1, \ldots, x_n \in R$ iff $f(x_1), \ldots, f(x_n) \in S$ (Russell 1919, 52–55). The "relation-number" of a given relation is then defined as "the class of all those relations that are similar to the given relation" (Russell 1919, 56). Based on this, Russell then introduces the

[33] This notion of structure based on the method of definition by abstraction but restricted to a single relation is discussed also in Carnap's *Abriss*. See section 15.4.1 for further details.

[34] A natural way to think about the kind of structural abstraction from isomorphic models underlying this approach is in terms of abstraction principles. I will return to this point in section 15.4.1.

[35] See the article by Heis in the present volume for a more detailed study of Russell's structuralist views.

notion of "structure" in the sense that two similar relations "have the same structure." More explicitly, he holds that

> two relations have the same structure when they have likeness, *i.e.* when they have the same relation-number. Thus what we defined as the "relation-number" is the very same thing as is obscurely intended by the word "structure"—a word which, important as it is, is never (so far as we know) defined in precise terms by those who use it. (Russell 1919, 61)

This passage shows how strongly Carnap's account of structures in his general axiomatics project is influenced by Russell's preceding ideas. In particular, in *Untersuchungen* and also in Carnap and Bachmann (1936), Russell's notion of similarity is generalized to apply also to "non-homogenous" relations as well as to models understood as ordered sequences of such relations. Similarly, the Russellian account of structures as "relation numbers," that is, as similarity classes of relations, is adopted in Carnap's work to apply also to formal models of different arities and of more complex type levels.

The second point to emphasize here is that Carnap's motivation for the introduction of model structures was clearly metatheoretic in spirit. Talk of such structures allowed him to develop a more refined account of the subject matter of axiomatic theories than in previous work. More specifically, instead of identifying the structural content of a theory with a single "explicit concept," Carnap proposes a classification of axiomatic theories here based on the number and type of model structures they describe. In the completed first part of *Untersuchungen* (published as Carnap 2000), he introduces the notion of the "structure number" of theories as the number of isomorphism classes they describe. Categorical theories such as second-order Peano arithmetic have number 1; noncategorical theories such as group theory or Hausdorff topology have structure numbers greater than 1.

In the projected but unfinished second part of the manuscript (RC 081-01-01 to 081-0133) as well as in Carnap and Bachmann (1936), a further specification of the structural content of theories is given based on the notion of so-called extremal structures.[36] The fundamental idea here is that the content of a theory is not only determined by the number of its isomorphism classes of models, but also by possible relations between them. A central notion introduced by Carnap for the study of such interrelations between structures is that of a "proper

[36] In the following, I refer mainly to the published results in Carnap and Bachmann (1936). For a closer discussion of the differences between the 1936 paper and the existing notes on Part 2 of *Untersuchungen* see Schiemer (2013).

structure extension" (or "proper substructure"). Carnap proposes the following definition of this notion in his article with Friedrich Bachmann of 1936:

> We call a structure S a proper substructure of a second structure T, if S and T are distinct and every model having the structure S is isomorphic to a proper part of every model having the structure T. (Carnap and Bachmann 1936, 175)

Put differently, given a theory T and two model structures S, T described by it, we say that S is a proper substructure of T, in symbols $S \sqsubset T$, if and only if (i) $S \neq T$ and (ii) for every model M with structure S and for every model N with structure T, there exists a mapping that *embeds* (in the model-theoretic sense of the term) M into N. Notice that the relata of this substructure-relation are the structures (conceived as isomorphism classes) themselves and not the models instantiating them.

Based on this notion of substructure, defined in terms of isomorphisms and embeddings between models, Carnap suggests an ordering of the class of model structures of a given axiomatic theory in terms of their extremal structures. The extremal structures consist of "initial structures," "end structures," and "isolated structures," defined in the following way. Given the class of structures defined by a theory T, we say that

1. S is an "*initial structure*" iff there exists no T of theory T such that $T \sqsubset S$;
2. S is an "*end structure*" iff there is no T of theory T such that $S \sqsubset T$;
3. S is an "*isolated structure*" iff there is no T of theory T such that $S \sqsubset T$ or $S \sqsubset T$.[37]

Put less formally, the initial structures (taken together with the isolated structures) represent the structures of minimal models of the theory in question, that is, of models that do not contain isomorphic copies of other models as submodels. Similarly, end structures (taken together with the isolated structures) represent structures of maximal models, that is, models not embeddable in other models. Isolated structures stand for models without any embeddings to other, non-isomorphic models.

This framework of extremal structures was explicitly introduced by Carnap to further analyze the structural content of axiomatic theories.[38] In particular, according to him, each theory can be assigned a "structure diagram," that is, a

[37] Compare Carnap's slightly different definition of these extremal structures in terms of the domain and range of the substructure relation (Carnap and Bachmann 1936, 176).

[38] We refer the reader to Schiemer (2012a) for a closer discussion of this theory of extremal structures and the limitations of Carnap's approach.

(possibly infinite) directed graph where the nodes represent model structures defined by the theory and the edges represent the proper substructure relation (and thus the embedding properties between models of different structures). Such a structure diagram of a theory can thus be viewed as a graphical representation of its structural content.

To see how Carnap thought about this structural content in terms of his newly introduced terminology, let us briefly look at one of his mathematical examples discussed in this context, namely the theory of elementary arithmetic. This is essentially a version of Russell's theory of arithmetical progressions with a single primitive relation *R(x,y)* (standing for a successor relation) and based on four axioms:

*b*1 $\forall x \forall y \big(R(x,y) \rightarrow \exists z (R(y,z))\big)$

*b*2 $\forall x \forall y \forall z \Big(\big(R(x,y) \wedge R(x,z) \rightarrow y = z\big) \wedge \big(R(x,y) \wedge R(z,y) \rightarrow x = z\big)\Big)$

*b*3 $\exists! x \big(x \in Dom(R) \wedge x \notin Ran(R)\big)$

*b*4 $Min_S\,\big(b1 - b3; R\big)$ (Carnap and Bachmann 1936, 179)

Axiom *b*1 states that relation *R* is endless. Axiom *b*2 states that *R* is an injective function. Axiom *b*3 states that there exists a base element in the progression. Axiom 4 is a so-called minimal axiom similar in effect to an induction axiom. It effectively imposes that all models satisfying axioms *b*1–*b*3 belong to minimal structures in the sense specified. What is particularly interesting about Carnap's discussion of this mathematical axiom system is the way in which he characterizes its structural content by analyzing the corresponding structure diagrams of its subtheories. Consider the two graphs in Figure 1, presenting the possible structures of models satisfying axiom systems *b*1–*b*2 and *b*1–*b*3 respectively.

The structures described by subtheory *b*1–*b*2(*R*) include the intended natural number structure, i.e., all isomorphic models of the form of a "progression" as well as infinitely many cycles of order 1 up to infinity. These structures, as well as the possible combinations of them, are presented by the nodes in the diagram on the right-hand side.

By adding axiom *b*3 to the system, the structural content is significantly restricted. In particular, as is illustrated in the diagram on the left-hand side, adding *b*3 to the base theory will have the effect that all structures of isolated cycles will be eliminated. The model class of the theory now contains the models of the intended structure *P* (i.e., progressions) as well as unintended

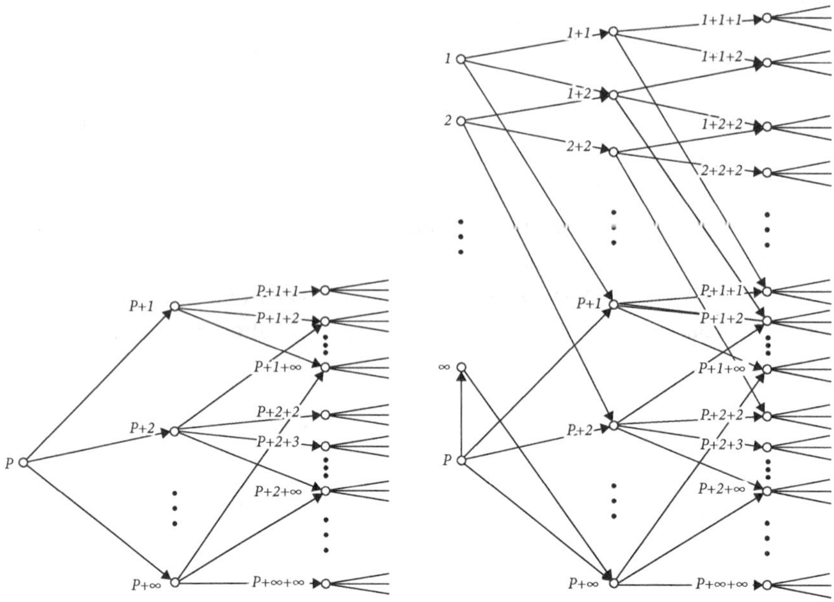

Figure 1. Structure diagrams of theories b1-b3(R) and b1-b2(R).

models consisting of combinations of progressions and cycles. Adding the minimal axiom *b*4 finally has the effect that all further unintended models are ruled out and the only remaining structure defined by the theory is that of an arithmetical progression *P*. In other words, adding axiom *b*4 to the system *b*1–*b*3 will render the resulting theory categorical or, in Carnap's own terminology, monomorphic.

4. Points of Contact with Modern Structuralism

The previous section has shown that one can identify several proposals in Carnap's early philosophy of mathematics on how to characterize the structuralist thesis. Interestingly, not only does his general structuralism connect his work with that of several of his contemporaries, including Russell, Husserl, Cassirer, and Quine, but one can also find several parallels between Carnap's early views on the structural nature of mathematical theories and contemporary structuralism. In this section, we will focus on two specific points of contact with the present philosophical debate.

4.1. Structural Abstraction

Carnap's treatment of model structures in his work on general axiomatics is based on the notion of abstraction. Specifically, we saw that the structure of a model of a given theory was identified with its isomorphism type, that is, with the class of models isomorphic to it. The main philosophical background for his approach was clearly Russell's work, in particular, the extensive treatment of abstraction principles in Russell (1903) and the subsequent discussion in Russell (1919). Interestingly, Carnap's abstraction-based approach is also closely connected to much more recent debates on mathematical structuralism.

Present research on the topic is based on a general distinction between two ways to think about the nature of mathematical structures. According to "eliminativist" structuralists, the mathematicians' talk about abstract structures should be understood merely as an abbreviation for generalizing over all models of a given theory. In contrast, "non-eliminative" structuralists such as Parsons, Shapiro, and others are realists about mathematical structures. For them, abstract entities such as the structure of the natural numbers exist in addition to the particular (set-theoretic) systems satisfying a theory.[39]

In the literature on non-eliminative structuralism, a further distinction is usually made between forms of *ante rem* and *in re* structuralism.[40] Briefly put, *ante rem* structuralists hold that abstract structures are bona fide objects that exist independently of their instantiating systems. Thus, the structure of the natural numbers exists irrespectively of whether there are particular number systems satisfying the axioms of Peano arithmetic. In contrast, *in re* structuralists usually argue that such higher-order entities are conceptually or ontologically dependent on their instantiating systems. Thus, according to this position, the natural number structure shared by all models of second-order Peano arithmetic exists only insofar as there are concrete models of the theory that instantiate the structure.

Carnap's own account of model structures outlined in *Untersuchungen* can be understood as an early formulation of *in re* structuralism about mathematics. In particular, his method of introducing structures by definitions by abstraction, i.e., by taking equivalence classes of isomorphic models, can be considered as one way to specify the conceptual dependency between structures and particular systems. Structures—conceived of as isomorphism types or classes—exist only if there are models of the axiomatic theory in question.[41] Comparable accounts

[39] See Reck and Price (2000) for an overview of the different accounts of mathematical structuralism.

[40] See, in particular, Shapiro (1997) on this distinction.

[41] One should add here that in Carnap's understanding of structures as isomorphism classes, the conceptual dependency between structures and systems is only given under the assumption that the classes in question are non-empty.

of such an abstraction-based structuralism can also be found in the current literature on the topic. Linnebo and Pettigrew have recently introduced a version of non-eliminative structuralism based on Fregean abstraction principles that determines this kind of abstraction from concrete systems to pure abstract structures (Linnebo and Pettigrew 2014).[42] The motivating idea underlying their approach is described as follows:

> A pure structure is the result of some operation of abstraction on a class of systems that are pairwise isomorphic. (Linnebo and Pettigrew 2014, 270)

Pure structures such as the structure of the natural numbers or of complete ordered fields can be introduced by abstracting away all nonessential or nonstructural properties of the objects in such systems. Such properties are identified here with properties not shared by isomorphic systems. The corresponding principle of structural abstraction has the form

$$[S]=[S'] \Leftrightarrow S \cong S' \qquad \text{(SA)}$$

where S, S' represent relational systems of the same signature, $\cong$ symbolizes the isomorphism relation between such systems, and $[S],[S']$ express the structures of S and S' respectively. The principle (SA) specifies an identity condition for abstract structures: for any two systems of a given signature, one can say that they share the same abstract structure just in case they are isomorphic.[43]

From a methodological point of view, this abstraction-based account of structuralism (as developed by Linnebo, Pettigrew, and Reck) is clearly similar to Carnap's position from the late 1920s. Mathematical structures are specified here and there as general forms shared by isomorphic models or systems. Moreover, even though Carnap does not explicitly introduce a structural abstraction principle of the form of (SA) in his work on axiomatics, a similar principle can be found in his *Abriss der Logistik* of 1929. In §22, in the context of his discussion of relations, the structure (or relation number) of a relation is identified with the "class of its isomorphic relations." Theorem L 22-24 then states an abstraction principle very similar to the one given above (Carnap 1929, 90):

$$P\, Smor\, Q \,.\equiv.\, Nr'\, P = Nr' Q$$

[42] A related account of mathematical structuralism based on a notion of "Dedekind-Cantor abstraction" has recently been developed in Reck (2018).

[43] Linnebo and Pettigrew also formulate structural abstraction principles for positions and relations in such abstract structures. See Linnebo and Pettigrew (2014) for further details.

where P, Q are relations of a given type and order, Smor stands for the isomorphism relation between them and $Nr'P$, $Nr'Q$ stand for the structure of P and Q respectively.[44]

Despite the obvious similarity between Carnap's and the contemporary accounts, there are also important differences concerning the very notion of structural abstraction. With respect to abstraction principles such as (SA), this relates to the question how the abstraction operator used on the left-hand side of the equivalence statement is understood. Such operators are usually treated as functions from a domain consisting of relational systems to a codomain of abstract structures. How can the codomain of the structural abstraction operator be understood?

In addressing this question, it is interesting to compare recent contributions to abstraction-based structuralism with different uses of abstraction principles (and definitions by abstraction) in 19th- and early 20th-century mathematics. In his recent study of this topic, Mancosu has shown that one can distinguish between at least three ways in which the operator in abstraction principles was understood in mathematics (Mancosu 2016). The values of abstraction functions were either taken to be (i) (canonical) representatives of the equivalence cells determined by an equivalence relation between mathematical objects or (ii) the equivalence classes themselves. Alternatively, the values of a given abstraction operator were also sometimes thought of (iii) as newly introduced *abstracta*, that is, as a type of "new object not coinciding with the equivalence class or one of its representatives" (Mancosu 2016, 87).

Mancosu's taxonomy of the possible values of abstraction functions corresponds closely to the different ways in which structural abstraction is described in the literature on structuralism. We saw that in Carnap's case, structures of models are identified with their isomorphism types.[45] Similar versions of this understanding of mathematical structures as equivalence classes can also be found in the more recent literature. Compare, for instance, how the nature of abstract structures in the case of basic arithmetic is described by Benacerraf in his influential article of 1965:

[44] Carnap refers to Russell (1919) for further discussion of the notion of structure in this section. Compare Heis's article in the present volume for a detailed discussion of similar structural abstraction principles in Russell's work.

[45] It should be noted here that a corresponding structural abstraction principle of the form (SA) can lead to inconsistency in case the structures on the right-hand side of the biconditional can also be inserted as models on the left-hand side. This fact is related to the Burali-Forti Paradox and has been discussed in the (neo-)logicist literature and in philosophy of mathematics more generally. See, in particular, Linnebo and Pettigrew (2014) on this point. Notice that this danger of yielding an inconsistent account of structural abstraction is excluded in Carnap's type-theoretic framework given the fact that model structures are required to be of a higher type than their instantiating models.

> If we identify an abstract structure with a system of relations (in intension, of course, or else with the set of all relations in extension isomorphic to a given system of relations), we get arithmetic elaborating the properties of . . . all systems of objects (that is, *concrete* structures) exhibiting that abstract structure. (Benacerraf 1965, 70)

While Benacerraf does not address the issue of structural abstraction from systems to abstract structures here, he explicitly mentions the possibility of identifying such structures as isomorphism classes of a given system.[46]

A different view of structural abstraction is presented in a recent paper by Leach-Krouse (2017). Leach-Krouse discusses different "structural" abstraction principles for models of axiomatic theories in the context of a neo-logicist approach to mathematics. The principles introduced here are similar in logical form to the structural abstraction principles already mentioned. However, the abstraction operators are understood neither in Carnap's nor in Linnebo and Pettigrew's sense.[47] Instead, Leach-Krouse's account follows an "approach to abstraction favored by Georg Cantor and Richard Dedekind, on which abstraction serves to introduce the isomorphism type of a mathematical structure as a first-class citizen of the mathematical universe" (Leach-Krouse 2017, 3).

More specifically, given a finitely axiomatizable theory T expressed in a second-order language with signature $\Sigma = \{R_1, \ldots, R_n\}$, a structural abstraction principle A_T for T is characterized here as a second-order sentence of the form

$$(\forall \overline{X}_n)(\forall \overline{Y}_n)[\S_T(\overline{X}_n) = \S_T(\overline{Y}_n) \leftrightarrow \overline{X}_n E_T \overline{Y}_n] \qquad (A_T)$$

The (sequences of) variables $\overline{X_n}$ and $\overline{Y_n}$ present model variables in Carnap's understanding of the term, that is, ordered sequences of relation or function variables substituted for the primitive terms of the theory. The binary relation E_T presents an isomorphism relation between models of theory T. The terms $\S_T(\overline{X_n})$ and $\S_T(\overline{Y_n})$ present the structures of models $\overline{X_n}$ and $\overline{Y_n}$ respectively. Thus, in a sense comparable to (SA), this principle states that any two models of T

[46] In his 1965 article, Benacerraf does not mention Carnap as an early proponent of such an account of mathematical structures.

[47] Leach-Krouse explicitly mentions the possibility of identifying mathematical structures with isomorphism classes (Leach-Krouse 2017, 5–6).

that are isomorphic also share the same structure and vice versa (Leach-Krouse 2017, 9–10).

In contrast to Carnap's account of structural abstraction, the abstraction operator $\S_T$ does not give isomorphism classes as values here. Rather, $\S_T$ expresses a type-lowering function just as in the case of Hume's principle in the neo-logicist project. More precisely, it presents a function that assigns an object of the individual domain *dom* of the object language to each model of the theory T. The only constraint on the interpretation of $\S_T$ determined by the principle (A_T) is that the function will assign the same individual to isomorphic systems. Thus, unlike in Carnap's account, the structure of a given model is specified here in terms of "first-order representatives" from the domain of the object language.[48]

A third possible approach to structural abstraction is presented in Linnebo and Pettigrew (2014) as well as in Reck (2018). In both accounts, the abstraction operator in (SA) gives as values pure structures of relational systems (of a given mathematical signature) that are thought of neither as equivalence classes nor as first-order representatives, but rather as newly introduced abstracta or "*sui generis* objects" (Linnebo and Pettigrew 2014, 274). More specifically, structures are themselves structured systems consisting of a domain of pure positions (or placeholders) and pure relations that can be exemplified by concrete set-theoretic systems. I cannot enter here into a closer discussion of the different approaches to structural abstraction or their philosophical implications.[49] Instead, let us turn to a second point of contact between Carnap's early structuralism and the modern debate.

4.2. Invariance and Definability

The notion of structural properties of (objects in) mathematical systems plays a central role in modern structuralism. In fact, structuralism is often characterized by reference to this notion: it is the thesis that mathematical theories investigate only the structural or relational properties of the objects in their respective domains.[50] According to this view, mathematical systems such as groups or

[48] Leach-Krouse's approach to structural abstraction seems similar to Mancosu's first strategy of thinking of the values of a mathematical abstraction operator in terms of representatives of a given equivalence class. Notice, however, that in Leach-Krause's account it is not required that the structures conceived of as first-order representatives form elements of the relevant isomorphism classes they stand for.

[49] In this respect, it might be interesting to give a closer discussion of possible connections between an abstraction-based structuralism and debates on the metaontology and logic of abstraction principles in neo-logicism. See, e.g., Linnebo (2018).

[50] Compare, for instance, Hellman on this point: "On a structuralist view, . . . the mathematician claims knowledge of structural relationships on the basis of proofs from assumptions that are frequently taken as *stipulative of the sort of structure(s) one means to be investigating*" (Hellman 1989, 5).

number systems are usually specified axiomatically, that is, in terms of implicit definitions. The task of the mathematician is then to investigate the "structural relationships" between the objects in such systems based on deductive proofs.

As we saw, this account of modern axiomatics closely corresponds to Carnap's view. Compare again the passage from his lecture notes on geometry (already quoted in section 3.1):

> An AS determines (defines) one (or several) structure[s] of a relational system, the "theorems" [*Lehrsätze*] determine structural properties of that system that follow from this definition, the AS; therefore analytic. (RC 089-62-02)

In his work on type theory and general axiomatics, Carnap proposes two ways in which this notion of structural properties can be made logically precise. The first approach— presented in Carnap (1929) and Carnap (2000)— is to specify structural properties of relations (and henceforth also of models of axiomatics theories) in terms of the notion of invariance under isomorphic transformations. Carnap gives the following definition in his *Untersuchungen* manuscript:

Definition 1.7.1. *The property fP of relations is called a "structural property" if, in case it applies to a relation P, it also applies to any other relation isomorphic to P. . . .*

$$(P,Q)\left[\left(fP \,\&\, Ism(Q,P)\right) \rightarrow fQ\right]$$

> The structural properties are so to speak the invariants under isomorphic transformation. They are of central importance for axiomatics. (Carnap 2000, 74)

The structural properties of a relation are thus those properties left invariant or preserved under suitable isomorphisms. Typical examples of such properties mentioned by him concern the arity and type of relations, the cardinality of their fields, as well as properties such as the reflexivity, symmetry, and transitivity of a binary relation.

In addition to this invariance-based account, Carnap proposed a second way to think about structural or "formal" properties in his monograph *Der Logische Aufbau der Welt* of 1928. In the first section of the book, the notion of a relational structure is characterized in terms of "the totality [*Inbegriff*] of its formal properties" (Carnap 1928, 13). Put differently, the structure of a given relation can be determined by considering all formal properties that apply to it. Formal properties, in turn, are specified in the *Aufbau* as follows:

> By formal properties of a relation, we mean those that can be formulated without reference to the meaning of the relation and the type of objects between which it holds. They are the subject of the theory of relations. The

> formal properties of relations can be defined exclusively with the aid of logistic symbols, i.e., ultimately with the aid of the few fundamental symbols which form the basis of logistics (symbolic logic). (Carnap 1928, 21)

Such properties are thus determined by means of the notion of logical definability: a property of a relation is formal just in case it is definable in a pure type-theoretic language.

Given Carnap's suggestions on how to explicate the notion of structural properties, two further remarks should be made here. First, Carnap was clearly one of the first philosophers to reflect on a general *duality* between two conceptually distinct ways to specify the structural content of mathematics. This is the use of invariance criteria on the one hand and the method of logical definability on the other hand. This duality has also been discussed in more recent work on logic and model theory. Compare, for instance, Hodges's characterization:

> In a sense, structure is whatever is preserved by automorphisms. One consequence . . . is that a model-theoretic structure implicitly carries with it all the features which are set-theoretically definable in terms of it, since these features are preserved under all automorphisms of the structure. There is a rival model-theoretic slogan: structure is whatever is definable. Surprisingly, this slogan points in the same direction as the previous one. (Hodges 1997, 93)

The general observation stated here is clearly in line with Carnap's two attempts to specify the structural properties of mathematical relations.[51]

Second, Carnap's approach to structural properties is also closely related to recent work on mathematical structuralism. In particular, one can find both ways to think about such properties, namely in terms of isomorphism invariance and logical definability, also in the present literature. For instance, a definability-based account of structural properties of positions in abstract structures is discussed in detail in work on non-eliminative structuralism, e.g., in Keränen (2001) and Shapiro (2008). An invariance-based account of structural properties of such positions is presented in Linnebo and Pettigrew's (2014) work on structural abstraction.[52]

Moreover, the notion also plays a crucial role in several of the systematic debates in these fields, for instance, on identity criteria for positions in abstract structures (e.g., in work by Shapiro, Keränen, and Leitgeb). The central bone

[51] An important difference from Hodges's account concerns the logical framework in which structural properties are specified. Whereas Hodges's book is about first-order model theory, Carnap's focus is on the definability of properties in a higher-order language of logical type theory.

[52] See Korbmacher and Schiemer (2018) for a more systematic comparison of the two definitions of structural properties.

of contention here concerns the question whether structurally indiscernible positions in a given structure—that is, positions that share the same structural properties—should be identified. A related Leibnizian principle of structural indiscernibility is usually formulated as follows:

For all structural properties P and all objects a, b in the domain of a structure S:

$$(P(a) \Leftrightarrow P(b)) \Leftrightarrow a =_S b$$

The objects (conceived of as pure positions) in a structure are thus identified if there exists no structural property that allows one to discriminate between them. In this case, the objects can be said to play the same role in a given structure.[53]

Interestingly, Carnap developed a similar account of structuralist identity conditions in his work from the late 1920s. In the *Aufbau*, he first states the idea of a purely "structural description" of an object in a domain in terms of its formal properties. His example of the graphical representation of the European-Asian railway network is used as an illustration of how one can, in principle, discriminate between different objects (that is, train stations) by considering only such properties (Carnap 1928, 17–19).[54] Carnap adds that in the hypothetical case that two objects share exactly the same formal properties, they have to be "treated as identical in the strict sense" of the term (Carnap 1928, 19). A similar account of structural identity is expressed in his work on general axiomatics. The notes of the fragmented second part of the *Untersuchungen* contain a section titled "Reduction of the Primitive Concepts" (RC 081-01-12). Here Carnap addresses the question which objects of a given relation are identifiable purely in terms of the relation. He holds that "an R-element x is describable through R if there exists a formal property with respect to R that only applies to x and to no other R-element" (RC 081-01-12/1). Carnap's specification of this approach is based on the further distinction between two properties of pairs of elements of a relation, which he calls "homotopical" and "heterotopical." Roughly put, two objects x, x' are homotopical with respect to a relation R if there exists an automorphism $f: R \cong R$ such that $f(x) = x'$. Objects that are not homotopical with any other object in R are called heterotopical R-elements.[55]

[53] See Keränen (2001) and Shapiro (2008). Compare also Leitgeb and Ladyman (2008) for a critical discussion of such a "structuralist" identity criterion.

[54] In his concrete example, these are graph-theoretic properties of the nodes in the unlabeled graph representing the structure of the railway system.

[55] According to Carnap, systems consisting only of heterotopical objects are called heterotopical systems. This concept corresponds closely to the modern notion of rigid systems, i.e., systems like the natural number systems whose automorphism class contains only the trivial automorphism. Compare Leitgeb and Ladyman (2008).

The relevant result stated in Carnap's *Untersuchungen* is that it is precisely the heterotopical objects in a relation (or a model) that can be identified in terms of formal properties. In contrast, in models consisting only of pairwise homotopical objects, such a discrimination of individuals is not possible given that there are often infinitely many non-trivial automorphisms of the model. Carnap's observation is obviously connected to the modern debate on the principle of identity of structurally indiscernible objects. In particular, it has been pointed out by Keränen (among others) that adopting such a principle will force structuralists to identify objects in nonrigid structures that can be mapped to each other by nontrivial automorphisms.

The second point to be mentioned here concerns the notion of the structural identity of relations or relational systems. In *Untersuchungen*, structural properties are defined for models of a given axiomatic theory in terms of the notion of isomorphisms. As Carnap points out in the passage cited earlier, such properties present the "invariants" under isomorphic transformations. A point not discussed in his 1928 manuscript, but briefly addressed in the *Abriss*, is whether one can formulate a structural property for a given relation (or a system of relations) that allows one to discriminate it from all other non-isomorphic relations.

This directly relates to the question whether, for a given system, one can identify a complete invariant or, in Carnap's terminology, a complete "structure characteristic" of it. Put in modern terms, an invariant is simply a function f that assigns the same value to isomorphic systems, that is, for any two systems R, S, one has $f(R) = f(S) \Leftrightarrow R \cong S$. An invariant for a given type of systems is complete if it also allows one to discriminate between any two non-isomorphic systems.[56] Compare Carnap's characterization of such complete invariants in his *Abriss*:

> The task of presenting a "structure characteristic" . . . is to present a procedure by which one can assign a formula expression (for instance one consisting of numbers) to the given relations . . . in a way that two relations are assigned the same characteristic if and only if they are isomorphic. (Carnap 1929, 55)

Carnap made a rough suggestion in *Abriss* on how to formulate such a complete invariant for finite relations based on their graph-theoretical representations and the corresponding adjacency matrices. Unfortunately, he did not further develop the ideas sketched there (see Carnap 1929, §22e). The relevant point for us to

[56] Notice that the operators in the structural abstraction principle (SA) discussed in the previous section present complete invariants in this sense.

note is that Carnap's work already contains several of the key ideas—particularly, on the identity criteria for objects and systems—that are prominently discussed in contemporary debates on structuralism.

5. Conclusion

This chapter surveyed Carnap's contributions to a structuralist account of mathematics from the 1920s and early 1930s. As several other chapters in the present volume show, his early structuralism was by no means an isolated position but shared by several other philosophers working at the time. Carnap's contemporaries Ernst Cassirer, Bertrand Russell, and also Edmund Husserl can be mentioned in this respect. Characteristic of their respective work is the fact that it is based on a close philosophical reflection of several methodological developments in 19th- and early 20th-century mathematics.

As we saw, this also holds of Carnap's pre-*Syntax* philosophy of mathematics. In his contributions from the period in question, one can identify three ways to characterize the thesis that mathematical theories are about abstract structures. The first method concerns axiomatic definitions which, according to Carnap, can be both understood as implicit definitions of the primitive terms of a given theory as well as explicit definitions of its class of models. The second method is based on the notion of logical constructions, specified by him in terms of explicit definitions in a logical type theory. Finally, Carnap's work on general axiomatics depends crucially on the notion of model structures, characterized as isomorphism classes of models, specifiable in terms of definitions by abstraction.

The study of Carnap's logical analysis of these different approaches allowed us to highlight several aspects of his early structuralism. First, Carnap took the different ways to characterize structures to be essentially equivalent. In particular, it is clear from the discussion given in *Der Raum* and in later writings that he understood Hilbert's axiomatic approach and Russell's genetic approach as two alternative ways to characterize the structural content of a theory. Second, it was shown that there are close connections in Carnap's work between a structuralist account of mathematics and his understanding of the logicist thesis. More specifically, his proposal to treat the content of mathematical theories in terms of explicit concepts has direct ramifications for a *generalized* logicism: it allows one to treat also non-arithmetical theories as reducible to logic and directly motivates an "if-thenist" reconstruction of mathematical theorems.

Finally, I presented two points of contact between Carnap's early philosophy of mathematics and recent debates on structuralism. The first concerns the role of structural abstraction principles in the formulation of versions of *in re* structuralism. Carnap, closely following Russell in this respect, proposed to think of structures of relational systems in terms of equivalence classes. Alternative ways to treat the operators in structural abstraction principles have been developed in work by Linnebo, Pettigrew, Reck, and Leach-Krouse. The second point of contact with modern work concerns Carnap's suggestions on how to explicate the notion of structural properties, namely in terms of the notions of definability and invariance. As we saw, this proposal connects his early contributions to structuralism with debates on adequate structuralist identity conditions for positions in mathematical structures.

The focus of this chapter was on Carnap's early contributions to the philosophy of mathematics. It would be interesting to give a comparison between the structuralist thesis concerning mathematical knowledge developed there and Carnap's more general scientific structuralism in his later work on the logic of science. Specifically, the present literature on Carnap still lacks a closer analysis of how his early contributions to general axiomatics are related to his mature work on logical theory reconstruction, for instance, on the *ramsification* of theories. A comparative study of Carnap's structuralist ideas from different periods of his intellectual career will have to be developed elsewhere.

Acknowledgments

This project has received funding from the European Research Council (ERC) under the European Union's Horizon 2020 research and innovation program (grant agreement No. 715222). Earlier versions of this chapter were presented at the conference "Early Carnap in Historical Context" at the University of Konstanz in 2017 as well as at the Sixth Annual Conference of the Society for the Study of the History of Analytical Philosophy at the University of Calgary in 2017. I would like to thank the respective audiences for their comments as well as Francesca Biagioli, John Wigglesworth, Christian Damböck, André Carus, Andrea Sereni, Christoph Limbeck-Lillienau, Hannes Leitgeb, Erich Reck, and Steve Awodey for helpful discussions on the topics addressed in this chapter.

References

Awodey, S. 2007. Carnap's Quest for Analyticity: The Studies in Semantics. In *The Cambridge Companion to Carnap*, edited by M. Friedman and R. Creath, pp. 226–247. Cambridge: Cambridge University Press.

Awodey, S. 2017. Carnap and the Invariance of Logical Truth. *Synthese*, special issue: Carnap on Logic (194), 67–78.

Awodey, S., and A. Carus. 2001. Carnap, Completeness, and Categoricity: The *Gabelbarkeitssatz* of 1928. *Erkenntnis* 54, 145–172.

Awodey, S., and A. Carus. 2007. Carnap's Dream: Godel, Wittgenstein, and *Logical Syntax*. *Synthese* 159(1), 23–45.

Awodey, S., and E. Reck. 2002. Completeness and Categoricity, Part 1: 19th Century Axiomatics to 20th Century Metalogic. *History and Philosophy of Logic* 23, 1–30.

Awodey, S., and E. Reck. 2004. *Frege's Lectures on Logic: Carnap's Student Notes 1910–1914*. Chicago: Open Court.

Benacerraf, P. 1965. What Numbers Could Not Be. *Philosophical Review* 74(1), 47–73.

Bernays, P. 1967. Hilbert, David. In *Encyclopedia of Philosophy*, edited by P. Edwards, vol. 3, pp. 496–405. New York: Macmillan.

Bohnert, H. G. 1975. Carnap's Logicism. In *Rudolf Carnap, Logical Empiricist*, edited by J. Hintikka, pp. 183–216. Dordrecht: Reidel.

Carnap, R. 1922. *Der Raum: Ein Beitrag zur Wissenschaftslehre*. Vol. 56 of *Kant-Studien Ergänzungshefte*. Berlin: Reuther & Reichard. English translation in Carnap (2019), 22–208

Carnap, R. 1927. Eigentliche und uneigentliche Begriffe. *Symposion* 1, 355–374.

Carnap, R. 1928. *Der logische Aufbau der Welt*. Springer: Berlin.

Carnap, R. 1929. *Abriss der Logistik*. Vienna: Springer.

Carnap, R. 1930. Die alte und die neue Logik. *Erkenntnis* 1, 12–26.

Carnap, R. 1931. Die logizistische Grundlegung der Mathematik. *Erkenntnis* 2, 91–105.

Carnap, R. 1934. *Logische Syntax der Sprache*. Vienna: Springer. translated as *The Logical Syntax of Language*. London: Open Court, 2002.

Carnap, R. 2000. *Untersuchungen zur allgemeinen Axiomatik*. Darmstadt: Wissenschaftliche Buchgesellschaft.

Carnap, R. 2019. *The Collected Works of Rudolf Carnap*. Vol. 1, *Early Writings, 1922–1927*. Edited by A. W. Carus et al. Oxford: Oxford University Press.

Carnap, R., and F. Bachmann. 1936. Uber Extremalaxiome. *Erkenntnis* 6, 166–188.

Carnap, R., and F. Bachmann. 1981. On Extremal Axioms. *History and Philosophy of Logic* 2, 67–85. Translation of Carnap and Bachmann 1936.

Carus, A. 2007. *Carnap and Twentieth-Century Thought*. Cambridge: Cambridge University Press.

Dedekind, R. 1888. *Was sind und was sollen die Zahlen?* Braunschweig: Vieweg.

Friedman, M. 1999. *Reconsidering Logical Positivism*. Cambridge: Cambridge University Press.

Gandon, S. 2009. Toward a Topic-Specific Logicism? Russell's Theory of Geometry in the *Principles of Mathematics*. *Philosophia Mathematica* 17(1), 35–72.

Gandon, S. 2012. *Russell's Unknown Logicism: A Study in the History and Philosophy of Mathematics*. New York: Palgrave Macmillan.

Goldfarb, W., and T. Ricketts. 1992. Carnap and the Philosophy of Mathematics. In *Science and Subjectivity. The Vienna Circle and 20th Century Philosophy*, edited by D. Bell and W. Vossenkuhl, pp. 61–78. Berlin: Akademieverlag.

Grattan-Guinness, I. 2000. *The Search for Mathematical Roots, 1870–1940: Logics, Set Theories and the Foundations of Mathematics from Cantor through Russell to Gödel*. Princeton, NJ: Princeton University Press.

Gray, J. 2008. *Plato's Ghost: The Modernist Transformation of Mathematics*. Princeton, NJ: Princeton University Press.

Hellman, G. 1989. *Mathematics without Numbers: Towards a Modal-Structural Interpretation*. Oxford: Oxford University Press.

Hilbert, D. 1899. *Grundlagen der Geometrie*. Leipzig: Teubner.

Hintikka, J. 2011. What Is the Axiomatic Method? *Synthese* 183(1), 69–85.

Hodges, W. 1997. *A Shorter Model Theory*. Cambridge: Cambridge University Press.

Keränen, J. 2001. The Identity Problem for Realist Structuralism. *Philosophia Mathematica* 9(3), 308–330.

Korbmacher, J., and G. Schiemer. 2018. What Are Structural Properties? *Philosophia Mathematica* 3(26), 295–323.

Leach-Krouse, G. 2017. Structural-Abstraction Principles. *Philosophia Mathematica* 25(1), 45–72.

Leitgeb, H., and L. Ladyman. 2008. Criteria of Identity and Structuralist Ontology. *Philosophia Mathematica* 16(3), 388–396.

Linnebo, O. 2018. *Thin Objects: An Abstractionist Account*. Oxford: Oxford University Press.

Linnebo, O., and R. Pettigrew. 2014. Two Types of Abstraction for Structuralism. *Philosophical Quarterly* 64(255), 267–283.

Mancosu, P. 2016. *Abstraction and Infinity*. Oxford: Oxford University Press.

Mormann, T. 2007. Geometrical Leitmotifs in Carnap's Early Philosophy. In *Cambridge Companion to Carnap*, edited by R. Creath and M. Friedman, pp. 57–75. Cambridge: Cambridge University Press.

Reck, E. 2004. From Frege and Russell to Carnap: Logic and Logicism in the 1920s. In *Carnap Brought Home: The View from Jena*, edited by S. Awodey and C. Klein, pp. 151–180. Chicago: Open Court.

Reck, E. 2007. Carnap and Modern Logic. In *The Cambridge Companion to Carnap*, edited by M. Friedman and R. Creath, pp. 176–199. Cambridge: Cambridge University Press.

Reck, E. 2018. On Reconstructing Dedekind Abstraction Logically. In *Logic, Philosophy of Mathematics, and Their History: Essays in Honor of W.W. Tait*, edited by E. Reck, pp. 113–137. London: College Publications.

Reck, E., and M. Price. 2000. Structures and Structuralism in Contemporary Philosophy of Mathematics. *Synthese* 125(3), 341–383.

Russell, B. 1903. *The Principles of Mathematics*. Cambridge: Cambridge University Press.

Russell, B. 1919. *Introduction to Mathematical Philosophy*. London: George Allen & Unwin.

Russell, B., and A. N. Whitehead. 1962. *Principia Mathematica*. 3 vols. Cambridge: Cambridge University Press.

Schiemer, G. 2012a. Carnap on Extremal Axioms, "Completeness of the Models," and Categoricity. *Review of Symbolic Logic* 5(4), 613–641.

Schiemer, G. 2012b. Carnap's *Untersuchungen*: Logicism, Formal Axiomatics, and Metatheory. In *Rudolf Carnap and the Legacy of Logical Empiricism*, edited by R. Creath. Berlin: Springer, 13–36.

Schiemer, G. 2013. Carnap's Early Semantics. *Erkenntnis* 78(3), 487–522.

Schiemer, G., R. Zach, and E. Reck. 2017. Carnap's Early Metatheory: Scope and Limits. *Erkenntnis* 194(1), 33–65.

Schlimm, D. 2013. Axioms in Mathematical Practice. *Philosophia Mathematica* 21(1), 37–92.

Shapiro, S. 1997. *Philosophy of Mathematics: Structure and Ontology*. New York: Oxford University Press.

Shapiro, S. 2008. Identity, Indiscernibility, and Ante Rem Structuralism: The Tale of i and –i. *Philosophia Mathematica* 3(16), 285–309.

Stein, H. 1988. *Logos*, Logic, and *Logistiké*: Some Philosophical Remarks on Nineteenth-Century Transformation of Mathematics. In *History and Philosophy of Modern Mathematics*, volume 11, edited by W. Aspray and P. Kitcher, pp. 238–259. Minneapolis: University of Minnesota Press.

Torretti, R. 1978. *Philosophy of Geometry from Riemann to Poincaré*. Boston: Reidel.

Wagner, P., ed. 2009. *Carnap's Logical Syntax of Language*. New York: Palgrave Macmillan.

16

Explication as Elimination: W. V. Quine and Mathematical Structuralism

Sean Morris

W. V. Quine has long been recognized as an important influence on the development of mathematical structuralism. Stewart Shapiro, for instance, uses the following remark from Quine's "Ontological Relativity" as the epigraph to his own structuralist treatise, *Philosophy of Mathematics: Structure and Ontology* (1997):[1]

> Expressions are known only by their laws, the laws of concatenation theory, so that any constructs obeying those laws . . . are *ipso facto* eligible as explications of expression. Numbers in turn are known only by their laws, the laws of arithmetic, so that any constructs obeying those laws—certain sets, for instance—are eligible in turn as explications of number. Sets in turn are known only by their laws, the laws of set theory. (Russell 1919, 44)

This statement is certainly clear in its structuralist commitments, but, taken out of context, its overall philosophical aims are far less clear. Many commentators have simply taken Quine to be part of that tradition of mathematical structuralism associated with Paul Benacerraf, stemming from his classic article "What Numbers Could Not Be" ([1965] 1983).[2] Here, Benacerraf argues that since structuralism about the natural numbers opens the way to a variety of mutually incompatible theories of what the numbers are, the conclusion to draw is the numbers are not objects at all. In one way or another, modern structuralists typically aim to respond to Benacerraf's challenge, to in some sense answer the question, what then are the numbers? We might also take Quine to be attempting to answer this question, but this seems potentially contrary to his naturalism if the question and answer are construed in a robustly metaphysical way. Furthermore, assimilating Quine to this tradition ignores that the beginnings of his structuralism

[1] In addition to Shapiro's work, see Parsons (1990, 2004); and Resnik (1997).

[2] Quine cites Benacerraf's paper in "Ontological Relativity," agreeing with the idea that arithmetic is all there is to the numbers, but also remarking that Benacerraf's "conclusions differ in some ways from those I shall come to" (1969b, 45 n. 9).

Sean Morris, *Explication as Elimination* In: *The Prehistory of Mathematical Structuralism*. Edited by: Erich H. Reck and Georg Schiemer, Oxford University Press (2020). © Oxford University Press.
DOI:10.1093/oso/9780190641221.003.0016

can be found already in his earliest work, his 1932 dissertation, "The Logic of Sequences," some 30 years before Benacerraf's article. I will argue instead then that Quine's structuralism is much better situated and understood within the context of an early form of structuralism, specifically the structuralism Russell put forward as part of his program for scientific philosophy. While there is much diversity among the views of the early structuralists (as there is also among contemporary structuralists), which include also Dedekind and Carnap, one thing that unites them is the rejection of a more metaphysical view of mathematics and of structures more generally. They all put forward views of mathematics that, in a sense, answer only to mathematics itself. The basic idea here is that all an account of mathematical objects requires is that the entities—whatever they are—that serve as these objects satisfy the relevant postulates and theorems. Here we can see how Quine's early work in the foundations of mathematics leads in a natural way to the more general naturalism of his later philosophy.

In what follows, I will look at the development and motives for Quine's particular brand of mathematical structuralism. I will argue that Quine, unlike many contemporary mathematical structuralists, does not appeal to structuralism as a way of accounting for what the numbers *really* are. Instead, he denies the very conception of analysis that gives rise to such philosophical projects, that is, a conception of analysis that aims to divulge some deeper hidden extra-scientific metaphysical reality.[3] In this way, I see Quine's philosophy as firmly rooted in the tradition of scientific philosophy and its critical attitude toward more metaphysical varieties of philosophizing. The tendency to treat Quine's philosophy as part of the contemporary analytic scene, I think, misconstrues the radical nature of his views and its deep connections to the tradition of scientific philosophy, starting with Russell and running through to Carnap and, then, culminating in Quine. The structure of this chapter is as follows. In section 1, I provide a brief account of Russell's structuralism with a particular emphasis on its anti-metaphysical motivations. Here, I focus on Russell's work of the 1910s as this is where his own particular version of scientific philosophy emerges most clearly.[4] It is also the period in which he wrote his *Introduction to Mathematical Philosophy*, probably the text that Quine most often cites as inspiring his own commitment to structuralism. In section 2, I present the beginnings of Quine's structuralism, arguing that it emerged from his early and careful engagement with Russell's work in the foundations of mathematics. In section 3, I move to Quine's mature view. As we saw already, Quine's structuralism is often traced to his 1969 "Ontological Relativity." But I turn instead to his 1960 *Word and Object*,

[3] Where science, for Quine especially, includes mathematics.

[4] I think there are also structuralist aspects in Russell's earlier work as well, though they may differ in key ways from the view put forward in the 1910s.

as it is here that we find his most detailed discussion of his structuralism during this period. To use the taxonomy of Reck and Price, Quine's structuralism here is of the relativist variety: there are many models that satisfy the structural properties of mathematical objects, and the relativist structuralist simply chooses one of these models as, for example, the natural numbers. A different model could have been chosen, but so long as the choice is a consistent one, no conflict arises. For most relativist structuralists, Quine included, set theory provides the model (Reck and Price 2000, sec. 4).[5] Finally, in section 4, I argue that Quine's appeal to structuralism largely stands apart from the concerns of contemporary structuralism stemming from Benacerraf and his challenge that numbers are not objects.

1. Scientific Philosophy and the Russellian Background

Since I aim to situate Quine's structuralism in the context of Russell's program for scientific philosophy, let me begin by very briefly characterizing the tradition of scientific philosophy as it began to emerge in the second half of the 19th century.[6] The terminology "scientific philosophy" began to appear in the literature in the mid-1800s in reaction to the very speculative metaphysics of post-Kantian idealism and its attempts to distinguish the methods and aims of philosophy from those of the sciences. Alan Richardson emphasizes two aspects, in particular, to characterize this movement: first, a critical attitude toward metaphysics, sometimes extending to philosophy as a whole; and second, a cooperative spirit between philosophy and the sciences (1997, 426–427). This latter feature arose largely in reaction to philosophy's attempts during the 19th century to distinguish itself from the sciences by following artistic or religious models for philosophizing. This latter aspect is apparent from the start in Russell's work in the philosophy of mathematics. He frequently appeals to the results of mathematicians such as Peano and Cantor and urges that philosophers engaging in the philosophy of mathematics study the most up-to-date foundational work on the subject. Similarly, I think this characterization would be uncontroversial for Quine's work as well, taking naturalism as the central tenet of his philosophy. Indeed, I think most contemporary analytic philosophers would grant that philosophy should be done in cooperation with the latest results of science. The

[5] I think this is the best characterization of Quine's position, and I think it is also how Quine understands Russell's position in the 1910s. Reck and Price do point out a problem for relativist structuralism in that it seems that the objects of the basic theory, in most cases sets, are treated differently than the other objects of mathematics. Unlike, say, the numbers, the sets are not eliminated in favor of some other structure. I am not sure that Quine would feel the force of this objection. As we will see in section 16.4, Quine thinks that all objects—sets, numbers, atoms, tables, chairs, etc.—are given only by their structural properties.

[6] On the tradition of scientific philosophy, Friedman (2012) and Richardson (1997).

former characterization, however, is one that distinguishes the earlier tradition of scientific philosophy from much of contemporary analytic philosophy. And so it is this one that I will focus on throughout this chapter.

In the mid-1910s Russell explicitly put forward his program for scientific philosophy, urging that philosophy take up a scientific methodology so as to yield to philosophy that kind of progress already found in the sciences. He envisioned groups of independent researchers, each focusing on their own specialized research so that philosophy might proceed piecemeal. He diagnosed philosophy's floundering as rooted in its striving for a single grand system of the world. Russell proposed instead that

> The essence of philosophy . . . is analysis, not synthesis. To build up systems of the world, like Heine's German professor who knit together fragments of life and made an intelligible system out of them, is not, I believe, any more feasible than the discovery of the philosopher's stone. What is feasible is the understanding of general forms, and the division of traditional problems into a number of separate and less baffling questions. "Divide and conquer" is the maxim of success here as elsewhere. (2004b, 87)

Along with the rejection of such grand systematizing came also a skepticism toward more metaphysical approaches to philosophy.[7] For example, Russell considers the common-sense belief in the existence of permanent, rigid bodies such as tables, chairs, stones, and such as "a piece of audacious metaphysical theorizing; objects are not continually present to sensation, and it may be doubted whether they are there when they are not seen or felt" (1993, 107).[8] Elsewhere, he compares this assumption as akin to a Kantian *Ding an sich*.[9] He thinks such assumptions introduce unnecessary doubt into philosophy[10] and instead,

[7] This emerges as a definite theme in Russell's work of the 1910s. Russell's desire to emphasize this new focus perhaps also explains his retitling of his 1901 "Recent Work on the Principles of Mathematics" to "Mathematics and the Metaphysicians" (2004a) for its 1918 reprinting. There are many ways that metaphysics might be characterized. In this chapter I will focus on the idea that metaphysics divulges some sort of hidden reality that is in some way more real than the reality described by the natural sciences or, in this case, mathematics. I should add that Russell himself leaves open the possibility of metaphysics from within scientific philosophy (see, for example, 2004c, 127). I am emphasizing the strand of his thought that rejects what he refers to as "traditional metaphysics." Similarly, I am emphasizing the anti-metaphysical strand of Quine's thought. But Quine's view parallels Russell in also leaving open a scientifically acceptable metaphysics. Certainly, his aim of "limning the most general traits of reality" (1960, 161) has a metaphysical ring to it. Indeed, I think Quine's philosophy could be accurately described as the naturalizing of metaphysics. But we might then wonder whether this is metaphysics in any sense that a more traditional metaphysical philosopher would accept.

[8] *Our Knowledge of the External World* is scattered with remarks such as this, as are his other works from this period. For other examples see pp. 111–12, 134.

[9] See, for example, Russell 1993, 92. We will later see that Quine follows Russell in his motivations for structuralism here.

[10] See for example, Russell 1993, 134.

recommends trying to find constructions out of less dubious entities. He sums up this view as his "supreme maxim in scientific philosophizing": "*Wherever possible, logical constructions are to be substituted for inferred entities*" (2004c, 121).

Still, we might wonder how we can be assured that we have given an appropriate logical construction to serve the role of the desired object. To this point, Russell responds,

> Given a set of propositions nominally dealing with the supposed inferred entities, we observe the properties which are required of the supposed entities in order to make these propositions true. By dint of a little logical ingenuity, we then construct some logical function of less hypothetical entities which has the requisite properties. This constructed function we substitute for the supposed inferred entities, and thereby obtain a new and less doubtful interpretation of the body of propositions in question. (2004c, 122)

This is just the sort of structuralism about mathematics that he would go on to describe in his 1919 *Introduction to Mathematical Philosophy*, a text Quine read and often cites as inspiring his own structuralism.[11] In this later work, Russell presents Peano's axioms (this is "the body of propositions in question," in this case) for arithmetic and observes that any progression will satisfy them and also that any series satisfying the axioms is a progression. In this way, these axioms define the class of progressions.[12] Hence, any progression can be taken to do the work of the natural numbers in pure mathematics.[13] We simply identify the first object of the progression with zero, the second with one, the third with two, and so on. But since any progression will do, the members of the progression will not necessarily be the numbers as we ordinarily think of them. Russell says that they may be points in space, moments in time, or any other such infinite collection of objects: "Each different progression will give rise to a different interpretation of all the propositions of traditional pure mathematics; all of these possible interpretations will be equally true" (1919, 8–9). Russell later makes clear the philosophical import of this structuralism in explaining that similar constructions can also be carried out for geometry. He observes here that from a mathematical standpoint all questions about the "intrinsic nature" of geometric objects, such

[11] Quine cites this as one of the books that most influenced his philosophical direction (2008a, 328). Russell also makes this point with regard to mathematics in (1993, 209–210), another text that Quine read.

[12] Throughout this chapter, I use "class" and "set" interchangeably since this fits with Russell's and Quine's typical usage. I am not drawing the common distinction between sets and (proper) classes. Nor are Russell and Quine.

[13] Russell adds the condition that any such a progression should also be suited to applications of mathematics (1919, 9). Quine shows how this condition can easily be met by any progression (1960, 262–263).

as points, lines, and planes, can be put aside. Since points need be nothing more than what makes their axioms true, there is nothing further that needs to be said about them. All that a point requires is that "it has to be something that as nearly as possible satisfies our axioms, but it does not have to be 'very small' or 'without parts.' Whether or not it is those things is a matter of indifference, so long as it satisfies the axioms" (1919, 59).

Russell concludes his discussion by generalizing this account not only to the rest of mathematics but also to the rest of science, remarking: "This is only an illustration of the general principle that what matters in mathematics, and to a very great extent in physical science, is not the intrinsic nature of our terms, but the logical nature of their interrelations" (1919, 59).[14] By emphasizing the importance of structure here, Russell makes clear the form that his critical attitude toward more metaphysical approaches to philosophy takes. From his scientific standpoint, questions about the intrinsic nature of objects are dismissed. All that science demands of an object is that it satisfy the axioms or postulates of the relevant science. There is no deeper, mysterious essence about objects to be discovered; their structural properties are enough to meet the demands of scientific philosophizing.

2. The Beginnings of Quine's Structuralism

In his autobiography, Quine makes explicit Russell's influence, stating that he inspired Quine to philosophy and referring to *Principia Mathematica* as "the crowning glory" of his undergraduate honors reading (1985, 59). Quine's own 1932 dissertation, "The Logic of Sequences," was a reworking of roughly the first 400 pages of *Principia* and was written under the direction of Russell's coauthor, Alfred North Whitehead.[15] Here, too, Quine draws a further important connection to Russell, remarking on the philosophical significance of their respective works: "Outwardly my dissertation was mathematical, but it was philosophical in conception; for it aspired, like *Principia*, to comprehend the foundations of logic and mathematics and hence the abstract structure of all science" (1985, 85). Here we see already that Quine had absorbed Russell's point that what mattered most in mathematics and science was structure. In this section, I will sketch out

[14] Similarly, he makes this point concluding the previously quoted passage from "The Relation of Sense Data to Physics, remarking of his logical constructions: "This method, so fruitful in the philosophy of mathematics, will be found equally applicable in the philosophy of physics, where, I do not doubt, it would have been applied long ago but for the fact that all who have studied this subject hitherto have been completely ignorant of mathematical logic" (Russell 2004c, 122). As we will see in the final section, Quine, too, extends his structuralism to all of science.

[15] For background on the dissertation see Quine (1985, 84–86), as well as his preface to the version of the dissertation published in 1990 and also Dreben (1990).

the beginnings of Quine's structuralism and argue that it was motivated by the same critical attitude toward metaphysics that we saw in Russell. In this early period, however, Quine's views develop not with a focus on numbers but more generally on the nature of propositions.

Explicit philosophical discussions are largely absent from Quine's dissertation, but we do get some sense of the purpose that Quine's structuralism would later serve.[16] In particular, he places little weight on the intuitiveness of his system; there is no attempt to discover what the entities of mathematics *really* are. What matters is that the system be convenient for the mathematical work at hand. This emerges immediately in Quine's own account of propositions. To this end, he introduces the primitive operation of predication, which he describes as the binary operation upon function and sequence, yielding what he calls a *proposition*, expressed notationally by juxtaposing the two operands, φ and X, to get φX. This, Quine says, is all there is to a proposition: "Such is the manner in which propositions emerge in the present system. A proposition is for us a construct, a complex, wrought from a function and a sequence by the undefined operation of predication" (1990, 38).

Still, Quine recognizes that we might ask for more; we might reasonably think that a proposition is not just a formal construct:

> But, it may be asked, what sort of thing is this product of predication? From the official standpoint of our system, it is to be answered only that it is whatever predication yields; and predication is primitive. Unofficially, we may say that by a proposition we mean exactly what one ordinarily means by the term; and, from this standpoint, we may describe predication as that operation upon function and sequence which renders that latter argumental to the former and produces a proposition. In the terms of the present system, thus, the proposition is logically subsequent to the function and argument sequence which enter it. This treatment, however, is quite independent of metaphysical and epistemological considerations. It is altogether indifferent to the present system if function and argument be construed as abstractions which are, in some philosophical sense, subsequent to the proposition from which they are abstracted; just as it is irrelevant that, from a psychological standpoint, propositions are pretty certainly prior chronologically to function and sequences. *Nor, indeed, are we even concerned with maintaining that propositions are, in any absolute sense, logically subsequent to functions and sequences—mainly, perhaps, because we have little conception of what possible meaning such a statement might have.* The point is merely that is has proved convenient in the present system to frame

[16] Still, much Quinean philosophy can be pulled from the dissertation. For more on this topic, see Dreben (1990); and Morris (2015).

our primitives in such a way that, for us, the proposition emerges as complex. (1990, 38–39; my emphasis).

Here we see again Quine's technical approach to more traditionally philosophical concerns, but he also highlights that there may be a number of philosophical concerns about propositions that he does not address. Quine, however, does not see this as a deficiency but rather as a benefit. Philosophical controversies over propositions, such as the ones he points out, are irrelevant to the technical development of propositions in his system. Engaging in these controversies, then, would only lead to the kind of stagnation that scientific philosophy had sought to avoid. Indeed, Quine's remark that his account is "independent of metaphysical and epistemological considerations" might be aimed at Russell himself. It is just these sorts of concerns—about the function-argument analysis of propositions, whether to take the components or the propositions as prior, and most generally, how to account for the very unity of a proposition at all—that leave much unresolved on the philosophical side of Russell's account.[17] Quine's conclusion, unlike Russell's, is simply that we have no firm ground to stand upon to even know exactly what question we are asking in these cases. As we see in the italicized sentence, Quine simply rejects that there is any absolute sense of what a proposition is; there is no question to ask about what propositions *really* are, aside from the account that Quine's logical system provides.

There is one further remark to note in this passage. He observes, "Unofficially, we may say that by a proposition we mean exactly what one ordinarily means by the term." Here, I think Quine gives the first hint that something like the structuralism we saw already in Russell will be conducive to Quine's own philosophical position. Indeed, in recognizing the difficulties Russell had with propositions, Quine sees structuralism as a solution, or better, a dissolution of the whole problem. Exactly what we mean ordinarily by a proposition is far from clear, but what is important in understanding Quine's view here is that he thinks there is some agreed-upon meaning or role that we ascribe to propositions.[18] And any technical entity that satisfies this role has equal claim to being a proposition.

Despite the apparent success of his account of propositions over previous accounts, we might still wonder why we should take them to be sequences. Here Quine brings us back to his emerging structuralism. In his 1934 *A System of Logistic*, the published version of his dissertation, he observes that Whitehead had emphasized the non-assertiveness of propositions, meaning that only in

[17] See in particular secs. 480–483 of *Principles of Mathematics* (1937), Russell's appendix on Frege's views. For useful commentary see also Hylton (1990, 336–338, 342–350), also his (2005); as well as Ricketts (2001), along with his (2002).

[18] This seems to be precisely what he later rejects about them, or more specifically, about the notion of analyticity, in his "Two Dogmas of Empiricism" (1980, 25), first published in 1951.

making a judgment does the proposition assert something true or false. For example, the proposition that this book is red does not assert that the book is in fact red. This only comes when a judgment is made. Quine then adds of his own account of propositions that "the doctrine of propositions as sequences stands in striking agreement with Whitehead's point of view; it presents a definite technical entity fulfilling just the demands which he makes of a proposition" (1934, 33). Significant here is not so much Quine's agreement with Whitehead but rather Quine's remark that he has provided a definite technical entity that fulfills the role that we expect propositions to play. The kind of ordinary meaning of propositions that he had in mind earlier is now made somewhat more precise. Propositions are those sorts of things that are potentially true or false, that serve as the postulates and theorems of a logical system, that can be manipulated in accordance with the rules of the system, etc. Again, we see Quine already in his earliest philosophy leaning toward the sort of structuralism found in Russell's *Introduction to Mathematical Philosophy*. Quine is not merely adopting the sort of formalistic or technical approach characteristic of much mathematical work. Rather, he takes such an approach to have philosophical consequences when embodied in a kind of structuralism. Here he eliminates traditional philosophical worries specifically over the true nature, or essence of, propositions. There is no deeper question to be asked about them than what role it is that they ordinarily play. If we can find a sufficiently clear technical entity that satisfies this role, there can be no further demand to make, aside from pragmatic concerns over whether that particular entity best suits the particular task at hand. While the paradigm case for such an account is no doubt the sort that Russell introduced with the numbers, Quine's account of propositions is very much in this same spirit. As we will see, this is the kind of clarificatory work that he would later identify as a paradigm of philosophical analysis (1960, sec. 53).

After the mid-1930s all positive talk of propositions drops out of Quine's work.[19] Perhaps foreshadowing his later attack on the analytic/synthetic distinction, he came to realize that talk of propositions lacked contexts that were "clear and precise enough to be useful" (1980, 25). Furthermore, he may have come to see that his technical replacements could be rendered less controversial by simply calling them what they were—sequences, sentences, or what have you. Still, this early work on propositions is important in setting up Quine for the sort of structuralism he would adopt in his mature philosophy. After all, there are many entities—numbers among them—crucial to science and in need of very much the kind of analysis Quine had offered for propositions in this early period.

[19] Quine's other significant work on propositions, also from 1934, is his "Ontological Remarks on the Propositional Calculus" (1976b). The discussion here is complimentary to both the dissertation and *A System of Logistic*.

3. Quine's Mature View

In light of more recent structuralist approaches to mathematics, which tend to respond directly in one way or another to Benacerraf's challenge, I hope in this and the next section to give us a better sense of what Quine's structuralism is both meant and not meant to do. From the perspective of contemporary structuralism, Quine's discussion may perhaps appear simplistic or inadequate. He never addresses many of the worries that we see in current discussions. I take it that this is intentional on his part as his structuralism is largely meant to deny certain kinds of philosophical worries. Quine is not, for example, trying to answer the question of what the numbers *really* are or more generally, what structures *really* are.[20] Rather, as we will see, he aims to dissolve rather than solve philosophical puzzles such as this one.

While Quine's claim in "Ontological Relativity" that the numbers are known only by their laws is perhaps his most explicit statement of a kind of mathematical structuralism, his most sustained discussion of his view occurs in *Word and Object*. We saw in section 1 that Russell's structuralism arose out of his urging of the analytic method as the right way to pursue scientific philosophy. Quine continues with this approach, adopting in section 53, "The Ordered Pair as a Philosophical Paradigm," a kind of structuralism as a general method for philosophical analysis. Here he describes a common situation where we have a term that is in some sense defective but that is also very useful to our theorizing. We must then somehow make sense of it, preserving its utility while removing its defectiveness. Quine looks to the ordered pair as a particularly clear case of just this phenomenon. Typically, we find this device in mathematics where it allows us to assimilate relations to classes by treating the relations as classes of ordered pairs (1960, 257). Its defectiveness readily appears when we try to give an account of what an ordered pair is. Referring to Peirce's nearly impenetrable account in terms of a mental diagram, Quine recommends instead, "We do better to face the fact that 'ordered pair' is (pending added conventions) a defective noun, not at home in all the questions and answers in which we are accustomed to imbed terms at their full-fledged best" (WO, 257–258). He then explains that mathematicians take the single postulate

(1) If $\langle x, y\rangle = \langle z, w\rangle$ then $x = z$ and $y = w$,

[20] I take it that many contemporary structuralists would agree on the first question, but it does seem that the discussion then just shifts the worries that arose around numbers to worries about the structures themselves.

to govern all uses required of the ordered pair. So we want a single object that will do the work of two and that satisfies this condition. The solutions, Quine observes, are many, with Kuratowski's rendering of $\langle x, y \rangle$ as $\{\{x\},\{x, y\}\}$ being among the most common. But Norbert Wiener's $\{\{x\},\{y,\varnothing\}\}$ serves the purpose equally well. It is straightforward to show that either of these classes satisfies postulate (1) (1960, 258–259).[21] This, Quine declares, is precisely what a philosophical analysis should do:

> This construction is paradigmatic of what we are most typically up to when in a philosophical spirit we offer an "analysis" or "explication" of some hitherto inadequately formulated "idea" or expression. We do not claim synonymy. We do not claim to make clear and explicit what the user of the unclear expression had unconsciously in mind all along. We do not expose hidden meanings, as the words 'analysis' and 'explication' would suggest; we supply lacks. We fix on the particular functions of the unclear expression that make it worth troubling about, and then devise a substitute, clear and couched in terms to our liking, that fills those functions. Beyond those conditions of partial agreement, dictated by our interests and purposes, any traits of the explicans come under the head of "don't-cares." (1960, 258–259)[22]

The analysis of the ordered pair is unusual only in that the condition of partial agreement can be made so explicitly and simply. Other cases of analysis will not be so straightforward, but on Quine's account, this is still ultimately what any such analysis is meant to accomplish.[23]

There is then no answer to which of these analyses of the ordered pair is the correct one. Any object satisfying (1) has equal right to being the ordered pair, and this, Quine says, is the general situation with any analysis, or explication. For

> *explication is elimination*. We have, to begin with, an expression or form of expression that is somehow troublesome. It behaves partly like a term but not enough so, or it is vague in ways that bother us, or it puts kinks in a theory or encourages one or another confusion. But it also serves certain purposes

[21] See, for example, Enderton (1977, 35–36). There are plenty of other equally good analyses of the ordered pair; Quine (1960) gives further examples on p. 260.

[22] "Explication" is of course Carnap's terminology; see, for example, *Meaning and Necessity* (1956, 7–8). Part of what I hoped to have shown in section 16.2 was that Quine had this notion already in place prior to any serious engagement with Carnap's work. A more general conclusion, which I have not argued for in this chapter, is that Quine's and Carnap's shared philosophical aims can be traced back to the common influence of Russell.

[23] Here is at least part of his rejection of the analytic-synthetic distinction. Quine just does not think that we have any idea of what the conditions of partial agreement should be for the analysis of analyticity.

> that are not to be abandoned. Then we find a way of accomplishing those same purposes through other channels, using other and less troublesome forms of expression. The old perplexities are resolved. (1960, 260)

In the end, the question of what an ordered pair is is dissolved when this troublesome notion is replaced by some clearer notion. And now to bring us more directly back to structuralism about the numbers, Quine goes on to say exactly this of Frege's analysis of numbers as well, citing Russell's *Principles of Mathematics* as his source. Here Quine presents the more typically philosophical question "What is a number?" and—just as Wiener and Kuratowski did for ordered pairs—we have Frege replacing these somewhat mysterious entities with the better-understood classes. On this account, for each number *n*, we identify it with the class of all *n*-membered classes (the seeming circularity here can be paraphrased away). Quine then observes that to object that classes have different properties from numbers is to make no objection at all. It is just to misunderstand the point of explication:[24]

> Nothing needs be said in rebuttal of those critics, from Peano onward, who have rejected Frege's version because there are things about classes of classes that we have not been prone to say about numbers. Nothing, indeed, is more logical than to say that if numbers and classes of classes have different properties then numbers are not classes of classes; but what is overlooked is the point of explication. (1960, 262, footnote omitted).

Furthermore, again like the ordered pair, this is just one of many ways of explicating numbers. Von Neumann and Zermelo offered other analyses. None are equivalent but all serve perfectly well as the numbers. Quine concludes that, as with the ordered pair, we can provide a condition that any explication of number must satisfy. Such a condition is provided by the notion of a progression, and any objects satisfying it will serve perfectly well as the numbers.

4. Quine and Modern Structuralism

I have been describing the development of Quine's structuralism, but let me now come back to the more general point I wanted to make about Quine's place in the history of analytic philosophy. I began with Russell so as to emphasize his influence on Quine's structuralism, and in particular, the critical attitude toward

[24] Russell does this as well (1919, 18–19).

metaphysics, characteristic of the scientific tradition of philosophy. We saw this with regard to propositions, where Quine showed how propositions could be rendered in terms of the sequences of his logical system. There was no worry here about whether these are really what propositions are. Sequences of a certain sort turned out to fulfill just the role required of propositions in his system. Quine's point was that there was no further demand to be made of them. Here, I stressed that this was a decidedly philosophical view on Quine's part. We see it now fully developed in his later work. What is to be emphasized here is again his rejection of certain philosophical questions—by eliminating problematic entities in favor of some that are better understood, Quine not so much solves as dissolves philosophical questions (1960, 260). The importance of elimination here cannot be stressed enough for properly understanding the significance and purpose of the remark with which we began this chapter, that "numbers . . . are known only by their laws, the laws of arithmetic" (1969b, 44). It is precisely on this point that I think Quine's position can be distinguished from much of what goes on in contemporary mathematical structuralism. Let me try to explain why.

Most of the contemporary discussion of mathematical structuralism has been set by Benacerraf's "What Numbers Could Not Be" ([1965] 1983). In it, Benacerraf famously concludes that the numbers cannot be objects (290). Since numbers, unlike other sorts of objects, have their requisite properties only in relation to the other numbers, we cannot give an account of any particular number short of characterizing the entire abstract structure of arithmetic. As he explains:

> The pointlessness of trying to determine which objects the numbers are thus derives directly from the pointlessness of asking the question of any individual number. For arithmetical purposes the properties of numbers which do not stem from the relations they bear to one another in virtue of being arranged in a progression are of no consequence whatsoever. But it would be only these properties that would single out a number as this object or that.
>
> Therefore, numbers are not objects at all, because in giving the properties . . . of numbers you merely characterize an *abstract structure*—and the distinction lies in the fact that the "elements" of the structure have no properties other than those relating them to other "elements" of the same structure. . . .
>
> Arithmetic is therefore the science that elaborates the abstract structure that all progressions have in common merely in virtue of being progressions. It is not a science concerned with particular objects—the numbers. The search for which independently identifiable particular objects the numbers really are (sets? Julius Caesars?) is a misguided one. ([1965] 1983, 291)

Benacerraf's remarks here illustrate how far Quine's view is from the concerns of much of contemporary structuralism. The discussion here tends to attempt a direct response to Benacerraf's conclusion. Participants in the dialogue either accept it and try to work out more precisely what it means for the numbers to not be objects; or they reject it and try to show how despite being recognized only by their structural properties, numbers still have a claim to being objects of a rather special sort.[25] For Quine, this entire discussion assumes too much from the start, resting on the uncritical assumption that we have some conception of an object ready to hand within which we can make sense of these two options. I will not be able to treat fully Quine's views on ontology and objecthood here, but let me try to give some better indication of how I think Quine sees the matter.[26]

Whereas Benacerraf assumes at the outset that the notion of an object is well understood and that the numbers are not instances of it, Quine does not. For Quine, we cannot assume as given that we know what will be among the objects and what will not. He takes ontology itself as a theoretical undertaking, one to be worked out in accord with the best science of our day. So as to where to draw the boundary between object and non-object, Quine responds,

> It is a wrong question; there is no limit to draw. Bodies are assumed, yes; they are the things, first and foremost. Beyond them there is a succession of dwindling analogies. Various expressions come to be used in ways more or less parallel to the use of the terms for bodies, and it is felt that corresponding objects are more or less posited, *pari passu*; but there is no purpose in trying to mark an ontological limit to the dwindling parallelism. (1981b, 9)

So our paradigm for an object might be bodies, that is, ordinary physical objects, but beyond this, there are just "dwindling analogies." We cannot simply rely on the notion of an object as given to us as fully understood. But then what are we to do about ontological questions? Should they just be rejected wholesale in the spirit of Carnap? No, as Quine continues:

[25] For the former view, I have in mind an eliminative structuralist such as Geoffrey Hellman. For his view see, for example, his (1989). I will not discuss his views further as, with their reliance on modal notions, I think they are far from anything that Quine would be willing to accept. For the latter view, I have in mind philosophers such as Michael Resnik or Stewart Shapiro. Reck and Price identify Resnik and Shapiro as both being "pattern structuralists"; that is, they are both committed to some version of the view that mathematics investigates patterns, and these are in themselves real objects. Shapiro's pattern structuralism is the more robust of the two, identifying the numbers with a sort of universal pattern (he calls his own view *ante rem* structuralism). Resnik also claims that numbers are patterns, but takes Quine's doctrine of ontological relativity more seriously and so does not identify the numbers with any one pattern. I am brushing over many subtleties in their views, but see Reck and Price (2000, sec. 7) for a more detailed summary.

[26] For a more detailed account of Quine's views, see Hylton (2004) and on abstract objects specifically see his (2007, 258–259).

> My point is not that ordinary language is slipshod, slipshod though it be. We must recognize this grading off for what it is, and recognize that a fenced ontology is just not implicit in ordinary language. The idea of a boundary between being and nonbeing is a philosophical idea, an idea of technical science in a broad sense. Scientists and philosophers seek a comprehensive system of the world, and one that is oriented to reference even more squarely and utterly than ordinary language. Ontological concern is not a correction of a lay thought and practice; it is foreign to the lay culture, though an outgrowth of it. (1981b, 9)

Contrary to Benacerraf, then, Quine thinks that the notion of an object itself and what it is to be ontologically committed to it stands in need of philosophical explication. Without providing some explicit criteria here, we cannot say whether or not numbers are to be counted among the objects. First of all, Quine tells us that for something to count as an object, we must have identity criteria for it, as summed up in his oft-quoted slogan, "No entity without identity" (1969c, 23; 1981a, 102). This tells what might be acceptable as an object, but it does not yet tell us if we are in fact committed to the existence of some particular object.[27] For example, surely we know the identity criteria for numbers, but the question here is whether we are in fact committed to the existence of numbers as objects. Clearly, we do talk of numbers as if they are objects, making claims such as "There is a number that is the successor of zero." But as we saw Quine point out, ordinary language is not a sure guide to ontological commitment.

Accepting that we cannot just read off of our everyday language what objects there are, Quine proposes a technical substitute. Using first-order quantification theory, Quine recommends that we regiment our scientific theory and then simply read off its ontological commitments by way of the universal and existential quantifiers, understood respectively as "for all objects *x*" and "there exists an object *x*." His solution to this quandary about objects is nicely summed up in another of his familiar slogans: "To be is to be the value of a variable" (1939, 708). Given this account, we now have a clear sense of what it means for an object to exist or not. So, for Quine, unlike Benacerraf, the numbers have every right to be considered objects alongside our ordinary physical objects so long as we are willing to countenance both as values of variables. Of course, we might reject Quine's criterion for ontological commitment, a possibility that he is well aware of. He welcomes other proposals, but to the extent that they do not capture the locution "there exists an object *x*," he sees them as giving no intelligible account of ontological commitment.[28]

[27] This criterion is closely tied into how Quine sees reification setting in. For a much more complete account of Quine's views here, again see Hylton (2004).

[28] See, for example, Quine's "Existence and Quantification," where he compares his objectual quantification with substitutional quantification (1969a, 103–108).

Benacerraf was driven to reject numbers as objects because of what he saw as some of their rather odd characteristics, chief among them that many different structures would do the work of the numbers. Again, this is clearly something that Quine is well aware of, noting many times, following Russell, that any progression will do. And here we see also the importance of elimination in Quine's account. To say that the numbers are some progression, for example, von Neumann's set-theoretic account, raises the question of why the numbers are this progression and not, for example, the one given by Frege or by Zermelo. On Quine's account of explication, we do not make such an identity. We have eliminated some apparent objects, not well understood, and replaced them with objects that are in some sense better understood. Out of habit or convenience, we refer to these sets as the numbers, but they are in the end just sets. These sets preserve whatever we found useful about numbers while pushing off any other features of the old numbers as "don't cares." There were of course other options for our explication, but as Quine observes, "Any progression will serve as a version of number so long as and only so long as we stick to one and the same progression. Arithmetic is, in this sense, all there is to number: there is no saying absolutely what the numbers are; there is only arithmetic" (1969b, 45).[29] The choice may be guided by certain pragmatic concerns. So in some other context we are equally free to choose a different analysis or explication, better suited to whatever that particular context requires (1960, 263). Here again we see the importance of not losing sight of Quine's point about explication being elimination. Whatever explication, or analysis, of numbers we opt for is all that is left of the numbers. There is no further independent question about whether we have correctly identified *the* numbers. The numbers have been eliminated in favor of some progression that has whatever features made the numbers worth explicating in the first place.[30] A failure to appreciate this aspect of Quine's account leads to the sort of worry Benacerraf identifies—which of the various progressions are the numbers really? For Quine, we might say, this is just a metaphysical pseudo-question (2008b, 401, 405).

[29] Note that the wording here is very much like the wording in the passage from *The Logic of Sequences* saying that there is no absolute sense of propositions short of some particular system (1990, 39).

[30] Again, Reck and Price place Quine's structuralism under the heading of relativist structuralism. The idea here is that there may be many structures that will serve as the natural numbers, and what we do is just pick one of them and stick with it. Reck and Price raise as a central question for relativist structuralism what we are to do about the basic level, the sets. Should not these also be treated in some structuralist way? It seems to me that Quine does have in mind also treating sets along structuralist lines. For example, in the quotation with which we began this chapter he remarks that it is not just numbers that are known by their laws but also sets. Indeed, as we will see, Quine thinks that in a sense all there is to any sort of object is its place in a theory. In this sense, I think Resnik correctly identifies Quine's position as "structuralism all the way down" (1997, 266).

So, on Quine's account the numbers are objects, but there is another line of thought that also treats the numbers as objects but that still seems at odds with Quine's account. Many prominent contemporary mathematical structuralists, chief among them Michael Resnik and Stewart Shapiro, agree with Quine that numbers are objects, but they also think there is something to what motivates Benacerraf to his conclusion: the numbers do somehow seem different from other sorts of objects; they do not seem to be objects in any ordinary sense. With this thought in mind, each in his own way tries to work out how the numbers might still be objects of a sort. Putting aside much detail, both embrace what Charles Parsons has identified as the incompleteness of mathematical objects.[31] In short, mathematical objects are incomplete in the sense that there are certain questions that we cannot answer about them since, as Benacerraf observed, they are given only by their relations within the entire structure of mathematics. So we seem to be at a loss about what the intrinsic nature of each number is; again, whether the numbers are in fact these sets or those.[32] Whereas Benacerraf indicated this as a problem for treating numbers as objects, Resnik and Shapiro just take this as characteristic of the particular kind of objects that the numbers are.[33]

We have already had a hint of Quine's response to this sort of worry about mathematical objects. His appeal to the quantifiers not only tells us what objects there are but is also univocal—Quine has no modes of being; there is only a single all-purpose notion of existence, applying to all objects indiscriminately.[34] This could be taken as a weakness of Quine's account; perhaps we would be better off recognizing somehow that the numbers, while still objects (contra Benacerraf), are unique in being identifiable only by their role in the structure of arithmetic as a whole. Quine surely recognizes differences among abstract objects, such as numbers, and the more ordinary concrete objects. In particular, he notes that we can learn terms for visible concrete objects by ostension, whereas this is not possible for terms for abstract objects (though more accurately he says that this difference is better reflected in the distinction between observation and theoretical terms). This, however, is an epistemological difference, rather than one reflecting a difference in kind among the objects themselves (1998, 402; 1981b, 16).

[31] Resnik explicitly adopts Parsons's terminology. Shapiro does not but attributes the appropriate characteristics to the numbers for them to be incomplete in Parsons's sense. See MacBride (2005) for a much fuller elaboration of this issue. While generally against, as we will see, characterizing mathematical objects, and abstract objects generally, as existing in some way differently from how concrete objects exist, Quine does not object to Parsons's technical work on incomplete existence. He just thinks the resulting theory not open to ontological assessment (1998, 400).

[32] See MacBride for this characterization (2005, 564).

[33] It should be noted that both Resnik and Shapiro describe themselves as Quineans of a sort. It may be that in light of their attempts to respond more directly to Benacerraf's challenge, Quine might have re-evaluated his own view on the matter. I will not undertake this task here on Quine's behalf, though I think it a worthwhile undertaking on the whole.

[34] Hylton emphasizes this point; see his (2007, 258; see also 303).

But still, what about the seemingly unique structural aspect of numbers? Here, too, Quine would be unconvinced, for this does not seem to be a unique feature of numbers after all, as shown by his doctrine of ontological relativity. He illustrates this most straightforwardly with what he calls proxy functions, where such a function maps our old objects onto some new objects (1969b, 55–61; 1981b, 19). For example, we might have the function *f* taking each object to its spatiotemporal complement *f*(*x*). With the predicates and terms appropriately adjusted, evidential support for the old and new theories remains the same, and so they are empirically indistinguishable. Here we have a version of what Quine calls his "global ontological structuralism" (2008b, 405):

> Structure is what matters to a theory, and not the choice of its objects. F.P. Ramsey urged this point fifty years ago, arguing along other lines, and in a vague way it had been a persistent theme also in Russell's *Analysis of Mind*. But Ramsey and Russell were talking only of what they called theoretical objects, as opposed to observable objects.
>
> I extend the doctrine to objects generally, for I see all objects as theoretical. (1981b, 20)

As he sums up his point, "Save the structure and you save all" (2008b, 405).

The point I wish to draw from this last discussion is that Quine will not be tempted to describe mathematical objects as incomplete. His global structuralism shows that there is nothing unique about the structural aspects of mathematical objects; much the same can be said of concrete objects. For Quine, in a sense, either all objects are incomplete or none are. No special trait of mathematical objects is picked out by their apparent incompleteness. As Quine describes his own view: "My own line is a yet more sweeping structuralism, applying to concrete and abstract objects indiscriminately" (2008b, 402). Resnik is, I think, then correct in describing Quine's view as "structuralism all the way down" (1997, 266). Resnik, however, wishes to contain his own structuralism so that it applies only to mathematical objects:

> By contrast, mathematical structuralism, including my own, finds its roots in the philosophical remarks of Dedekind, Hilbert, Poincaré, and Russell, and Paul Benacerraf's provocative thoughts on the multiple reduction of arithmetic to set theory. It takes the thesis that mathematical objects are incomplete ("known only by their laws") as a datum and tries to explain it, and consequently it does not go as far as Quine's. (1997, 267)

Shapiro, in his own way, joins Resnik in this view. Now, I am not claiming that Quine would reject any of the technical work that Resnik and Shapiro have contributed toward a mathematical theory of structures. What worries Quine are the motivations—that there is a desire on the part of structuralists such as Resnik and Shapiro to preserve some special status for mathematics (not unlike Carnap's attempt to declare mathematics analytic).[35] We see this here in Resnik's remark that he takes the incompleteness of mathematical objects as a datum to be explained by structuralism. This is precisely the kind of assumption that Quine's doctrine of ontological relativity, and his associated structuralism, denies. He describes his own global structuralism as coming from his naturalism—that is, from science itself—and its rejection of "the transcendental question of the reality of the external world—the question whether or in how far our science measures up to the *Ding an sich*" (1981b, 22).[36] He does not begin by assuming that mathematical objects are unique in some way. Ontological relativity shows mathematical objects no more, and no less, incomplete than ordinary concrete objects are.

We might, however, think such an unorthodox view to be in tension with Quine's professed realism.[37] He thinks not:

> Naturalism itself is what saves the situation. Naturalism looks only to natural science, however fallible, for an account of what there is and what what there is does. Science ventures its tentative answers in man-made language, but we can ask no better. The very notion of object . . . is indeed as parochially human as the parts of speech; to ask what reality is *really* like, however, apart from human categories, is self-stultifying. It is like asking how long the Nile really is, apart from parochial matters of miles or meters. Positivists were right in branding such metaphysics as meaningless. (2008b, 405)

Naturalism allows no deeper insight into reality than what will tolerate Quine's doctrine of ontological relativity. His global structuralism, then, just tells us what can be coherently said of objects unless we allow for some form of mystical

[35] The situation is much like that with regard to the analytic/synthetic distinction. Quine saw no flaws in Carnap's technical work. It was the underlying philosophical motivations that worried him: "In recent classical philosophy the usual gesture toward explaining 'analytic' amounts to something like this: a statement is analytic if it is true by virtue solely of the meanings of words and independently of matters of fact. It can be objected, in a somewhat formalistic and unsympathetic spirit, that the boundary which this definition draws is vague or that the definiens is as much in need of clarification as the definiendum. This is an easy level of polemic in philosophy, and no serious philosophical effort is proof against it. But misgivings over the notion of analyticity are warranted also at a deeper level, where a sincere attempt has been made to guess the unspoken *Weltanschauung* from which the motivation and plausibility of a division of statements into analytic and synthetic arise" (1976a, 138).

[36] Recall this is one of the ways that Russell described the aim of his structuralism.

[37] For more on the radical nature of Quine's views here, see again Hylton (2004, especially sections IV and V).

insight into the *true* nature of reality. Here, I have been describing Quine in terms that may seem more appropriate to a discussion of Carnap, and in this passage, we see Quine himself doing so. While I do want to stress, much more than is usually done, the significant continuities between Quine and Carnap, especially as part of a tradition of scientific philosophy stemming from Russell, I do not want to abolish the differences. And nor does Quine, as he then explains. Where the positivists went wrong was in trying to deny ontological questions altogether (2008b, 405). Still, Quine's own countenancing of such ontological questions, and in particular, his structuralism, is not a return to a more traditional brand of metaphysical theorizing, as he concludes:

> My global structuralism should not . . . be seen as a structuralist ontology. To see it thus would be to rise above naturalism and revert to the sin of transcendental metaphysics. My tentative ontology continues to consist of quarks and their compounds, also classes of such things, classes of such classes, and so on, pending evidence to the contrary. My global structuralism is a naturalistic thesis about the mundane human activity, within our world of quarks, of devising theories of quarks and the like in the light of physical impacts on our physical surfaces. (2008b, 406)

And here Quine brings us back to Russell. What matters most in the ontology of mathematics, and in the sciences more generally, is not the intrinsic nature of the objects but rather their structural relations to one another.

References

Benacerraf, Paul. [1965] 1983. "What Numbers Could Not Be." In *Philosophy of Mathematics: Selected Readings*, 2nd ed., edited by Paul Benacerraf and Hilary Putnam, pp. 272–294. New York: Cambridge University Press.

Carnap, Rudolf. 1956. *Meaning and Necessity*. 2nd ed. Chicago: University of Chicago Press.

Dreben, Burton. 1990. "Quine." In *Perspectives on Quine*. Robert B. Barrett and Roger Gibson, pp. 82–95. Cambridge, MA: Basil Blackwell.

Enderton, Herbert B. 1977. *Elements of Set Theory*. San Diego: Academic Press.

Friedman, Michael. 2012. "Scientific Philosophy from Helmholtz to Carnap and Quine." In *Rudolf Carnap and the Legacy of Logical Empiricism*, ed. Richard Creath, pp. 1–11. *Vienna Circle Institute Yearbook* 16.

Hellman, Geoffrey. 1989. *Mathematics without Numbers: Towards a Modal-Structural Interpretation*. Oxford: Clarendon Press.

Hylton, Peter. 1990. *Russell, Idealism, and the Emergence of Analytic Philosophy*. Oxford: Clarendon Press.
Hylton, Peter. 2004. "Quine on Reference and Ontology." In *The Cambridge Companion to Quine*, edited by Roger F. Gibson, pp. 115–150. New York: Cambridge University Press.
Hylton, Peter. 2005. "Frege and Russell." In *Propositions, Functions, and Analysis: Selected Essays on Russell's Philosophy*, pp. 153–184. Oxford: Clarendon Press.
Hylton, Peter. 2007. *Quine*. New York: Routledge.
MacBride, Fraser. 2005. "Structuralism Revisited." In *The Oxford Handbook of Philosophy of Mathematics and Logic*, edited by Stewart Shapiro, pp. 563–589. New York: Oxford University Press.
Morris, Sean. 2015. "Quine, Russell, and Naturalism: From a Logical Point of View." *Journal of the History of Philosophy* 53(1), 133–155.
Parsons, Charles. 1990. "The Structuralist View of Mathematical Objects." *Synthese* 84, 303–346.
Parsons, Charles. 2004. "Structuralism and Metaphysics." *Philosophical Quarterly* 54(214), 56–77.
Quine, W. V. 1934. *A System of Logistic*. Cambridge, MA: Harvard University Press.
Quine, W. V. 1939. "Designation and Existence." Journal of Philosophy 36(26), pp. 701–709.
Quine, W. V. 1960. *Word and Object*. Cambridge, MA: MIT Press.
Quine, W. V. 1969a. "Existence and Quantification." In *Ontological Relativity and Other Essays*, pp. 91–113. New York: Columbia University Press.
Quine, W. V. 1969b. "Ontological Relativity." In *Ontological Relativity and Other Essays*, pp. 26–68. New York: Columbia University Press.
Quine, W. V. 1969c. "Speaking on Objects." In *Ontological Relativity and Other Essays*, pp. 1–25. New York: Columbia University Press.
Quine, W. V. 1976a. "Mr. Strawson on Logical Theory." In *The Ways of Paradox and Other Essays*, rev. ed., pp. 137–157. Cambridge, MA: Harvard University Press.
Quine, W. V. 1976b. "Ontological Remarks on the Propositional Calculus." In *The Ways of Paradox and Other Essays*, rev. ed., pp. 265–272. Cambridge, MA: Harvard University Press.
Quine, W. V. 1980. "Two Dogmas of Empiricism." In *From a Logical Point of View: Nine Logico-Philosophical Essays*, 2nd ed., pp. 20–46. Cambridge, MA: Harvard University Press.
Quine, W. V. 1981a. "On the Individuation of Attributes." In *Theories and Things*, pp. 100–112. Cambridge, MA: Harvard University Press.
Quine, W. V. 1981b. "Things and Their Place in Theories." In *Theories and Things*, pp. 1–23. Cambridge, MA: Harvard University Press.
Quine, W. V. 1985. *The Time of My Life*. Cambridge, MA: MIT Press.
Quine, W. V. 1990. *The Logic of Sequences: A Generalization of "Principia Mathematica"*. New York: Garland Publishing.
Quine, W. V. 1998. "Reply to Parsons." In *The Philosophy of W. V. Quine*, expanded ed., edited by Edwin Hahn and Paul A. Schilpp, pp. 396–404. Chicago: Open Court.
Quine, W. V. 2008a. "Books That Mattered to Me." In *Quine in Dialogue*, edited by Dagfinn Føllesdal and Douglas B. Quine, pp. 328–329. Cambridge, MA: Harvard University Press.
Quine, W. V. 2008b. "Structure and Nature." In *Confessions of a Confirmed Extensionalist and Other Essays*, pp. 401–406. Cambridge, MA: Harvard University Press.

Reck, Erich H., and Michael P. Price. 2000. "Structures and Structuralism in Contemporary Philosophy of Mathematics." *Synthese* 125(3), 341–383.

Resnik, Michael D. 1997. *Mathematics as a Science of Patterns*. Oxford: Clarendon Press.

Richardson, Alan. 1997. "Toward a History of Scientific Philosophy." *Perspectives on Science* 5(3), 418–451.

Ricketts, Thomas. 2001. "Truth and Propositional Unity in Early Russell." In *Future Pasts: The Analytic Tradition in Twentieth-Century Philosophy*, edited by Juliet Floyd and Sanford Shieh, pp. 101–122. New York: Oxford University Press.

Ricketts, Thomas. 2002. "Wittgenstein against Frege and Russell." In *From Frege to Wittgenstein: Perspectives on Early Analytic Philosophy*, edited by Erich H. Reck, pp. 227–251. New York: Oxford University Press.

Russell, Bertrand. 1919. *Introduction to Mathematical Philosophy*. London: George Allen and Unwin.

Russell, Bertrand. 1937. *Principles of Mathematics*. 2nd ed. London: George Allen and Unwin.

Russell, Bertrand. 1993. *Our Knowledge of the External World as a Field for Scientific Method in Philosophy*. New York: Routledge.

Russell, Bertrand. 2004a. "Mathematics and the Metaphysicians." In *Mysticism and Logic*, pp. 57–74. Mineola, NY: Dover.

Russell, Bertrand. 2004b. "On Scientific Method in Philosophy." In *Mysticism and Logic*, pp. 75–96. Mineola, NY: Dover.

Russell, Bertrand. 2004c. "The Relation of Sense Data to Physics." In *Mysticism and Logic*, pp. 113–141. Mineola, NY: Dover.

Shapiro, Stewart. 1997. *Philosophy of Mathematics: Structure and Ontology*. New York: Oxford University Press.

Whitehead, Alfred North, and Bertrand Russell. 1960. *Principia Mathematica*. 3 vols. 2nd ed. Cambridge: Cambridge University Press.

Name Index

For the benefit of digital users, indexed terms that span two pages (e.g., 52–53) may, on occasion, appear on only one of those pages

Abel, Niels, 16, 65–66, 66n14, 187
Aczel, Peter, 188n4, 371–72
Aristotle, 29, 342–43
Awodey, Steve, 3–4, 132, 136, 138, 179–81, 217, 268–69

Bachmann, Friedrich, 403–4
Benacerraf, Paul, 1–2, 5–6, 8, 11, 54–55, 131, 273, 337, 339, 339n17, 347, 409–10, 410n46, 421–23, 421n2, 430, 433-437, 437n33, 438
Bernays, Paul, 11, 14, 143, 155n9, 157–58, 157n10, 158–160, 228, 352–354, 354n3, 356, 357n8, 359–60, 360n13, 361, 361n14, 362–63, 363n16, 364–65, 364n18, 366–68, 366n22, 370–371, 374, 376–77, 389, 389n14
Boole, George, 15, 76, 243, 248n13, 252, 255–56, 306–7, 331
Bolyai, János, 63, 245
Bourbaki, Nicolas, 5, 11, 13, 16, 43, 50–51, 72n31, 73, 83, 187–93, 188n4, 189n6, 191n10, 193n13, 195–98, 197n19, 198n20, 199–203, 200n24, 201n25, 204, 206, 208, 210–11, 217, 217n4, 225, 232–33, 281n9, 353, 358, 358n11, 359, 374n33
Brianchon, Charles Julien, 109*f*, 109
Burgess, John, 6, 23, 356n7

Cantor, Georg, 69, 78n46, 82, 149, 154n8, 160n13, 168, 245, 252, 252n21, 254–55, 329, 331–333, 335, 337, 358, 358n10, 408n42, 410, 423
Cantù, Paola, 11, 275n3
Carnap, Rudolf, 7, 14–15, 217, 310, 310n12, 314, 314n15, 383–85, 383n1, 385n3, 385n6, 386–88, 386–87n10, 389–93, 391n15, 393n18, 394, 394n20, 395, 395n21, 396, 396n23, 397–99, 397n25, 398n27, 400–2, 401n32, 402n33, 403–8, 403–04n38, 407n41, 409, 409n44, 410–11, 410n46, 412–13, 413n51, 414–17, 414n55, 422, 431n22, 434, 439–40, 439n35
Cartan, Henri, 188–89, 189n6, 191-92, 192n12, 195, 201
Carter, Jessica, 13, 182–83
Cassirer, Ernst, 11, 14, 22, 31n13, 34–35, 35n19, 36, 45, 303, 307, 307n4, 308, 308n6, 310, 310n12, 313–14, 316, 320n27, 329–31, 329n2, 330n3, 331n5, 332–34, 333–34n8, 336–39, 339n17, 340–43, 340n19, 341n20, 342n22, 343n23, 344–46, 346n24, 346n25, 347–48, 386–88, 389n14, 416
Cauchy, A.-L., 61, 256, 331
Cayley, Arthur, 108, 118n26, 123n30, 124n32, 169, 245n8
Chevalley, Claude, 189, 189n6, 207
Chihara, Charles, 3
Connes, Alain, 13, 188–89, 203, 207, 209
Corry, Leo, 188n4, 197–98, 207, 211

Dedekind, Richard, 5, 7–8, 11–14, 23–24, 35n19, 59–61, 63–64, 64n12, 65–72, 66–67n16, 69n19, 70n23, 72n30, 73–76, 74n33–35, 75n37, 75n38, 76–77n40, 77n42, 77n44, 78n46, 78–83, 79–80n50, 80n52, 81n56, 82n58, 90–91, 128, 142, 144–50, 152, 152–53n6, 154n8, 160, 167–68, 172–177, 175n4, 177n5, 178, 180–81, 181n7, 183–84, 203, 203n29, 204, 206, 208, 220–21, 248n14, 252, 252–53n23, 255, 282, 282n11-12, 284n17, 303, 303n1, 304, 304n2, 305, 307, 307n4, 308, 314–15, 314n16, 316–18, 318n23, 319–21, 320n27, 322–24, 323n30, 324n34, 325, 329–31, 330n3, 331n4, 332–35, 334n8-9, 334–35n11, 336–37, 337n14–16, 338–40, 340n19, 341–42, 342n22, 343–48, 353, 353n1, 353–54n2, 355–56, 356n4-6, 357, 357n8-9, 358–59, 361, 36 8, 371n27, 374, 376, 387–88, 389n14, 394, 400n28, 408n42, 410, 422, 438

DeMorgan, Augustus, 306
Dieudonné, Jean, 189, 189n6, 190n9, 192–93, 193n13, 194, 204, 211
Dirichlet, P.G.L., 12, 60–62, 64–65, 67, 69, 75–76, 79, 82
Dummett, Michael, 316, 316n21, 318n23, 320n27, 324n35, 337, 337n16, 340, 347

Ellenberg, Samuel, 135, 189, 216–17, 220, 223–24, 226–29
Euclid, 28, 75n37, 90–91, 144, 246n11, 259–60, 259n28, 263, 266, 332, 338, 347

Feferman, Solomon, 8, 23, 49, 51n30, 53–54, 54n32, 228
Ferreirós, José, 12, 51n30, 90, 246n10, 334n9
Folina, Janet, 14, 16, 202n27
Fraenkel, Abraham, 178, 387
Frege, Gottlob, 23, 77, 77n44, 79, 79n49, 93n8, 99n16, 100, 100n18, 142–44, 146n2, 146, 150–52, 154, 181, 247, 276, 306, 313, 316, 316n21, 318n23, 330–31, 337, 337n14, 337n16, 338–40, 344, 347, 356–57n8, 358, 361, 377, 386, 393, 393n17, 394–95, 395n21, 398–99, 402, 428n17, 432, 436

Galois, Évariste, 65–69, 66n14, 66n15, 69n20, 108, 220, 275n5, 329, 334n8
Gauss, Carl Friedrich, 12, 16, 60, 60n4, 61, 65, 71, 71n27, 73n32, 220, 245, 274, 275n4, 353–54n2
Gentzen, Gerhard, 157n10, 159–60, 365, 368, 375
Gergonne, Joseph Diez, 92, 108–10, 111n13
Gödel, Kurt, 3, 157n10, 158, 160n13, 190n9, 202, 352, 356n4, 362–63, 365, 365n19, 366, 368, 375n34
Gordan, Paul, 167–70
Grassmann, Hermann, 5, 11–12, 15–16, 21–26, 25n3, 26–27n5, 27–28, 27n6, 29–33, 31n12–13, 33n14, 34–37, 34n16, 35n18, 37n20, 37n22, 38–42, 40–42n26, 43–46, 43–47n28, 47–52, 51n29–30, 53–55, 119n27, 275n3, 308n7, 386
Grothendieck, Alexander, 188, 189, 189n7, 196, 208–9, 223, 225, 228

Hamilton, W.R., 74, 74n33, 76, 256, 331, 334n8
Hankel, Hermann, 15, 22, 39, 41, 45, 187, 218n7, 218–19, 274–275
Hasse, Helmut, 208, 220, 356
Hausdorff, Felix, 398–99, 403
Heinzmann, Gerhard, 13, 191n11, 197n17, 281n9, 297n36, 374n33
Heis, Jeremy, 14, 337n14, 340n18, 348, 402n35, 409n44
Hellman, Geoffrey, 1–8, 14, 47, 273, 337, 411n50, 434n25
Herbrand, Jacques, 368
Hesse, Ludwig Otto, 107, 108, 115n21, 115–17, 116n23, 119, 121, 126, 128, 133, 138
Hilbert, David, 10–14, 23, 60, 72, 72n30, 73, 77, 77n40, 83, 91, 93, 93n8, 100–101, 142–46, 146n3, 147–53, 154–60, 154n8, 155n9, 168–71, 188, 190–92, 195–196, 202, 204, 211, 215, 218–219, 258, 284–86, 293, 329, 331, 337n16, 345, 352, 353n1, 354–57, 356n4, 356n7, 358–61, 360n13, 361n14, 362–65, 367–70, 375–77, 386–89, 389n14, 390, 391n16, 393, 393n17, 394–95, 400n28, 416, 438
Hodges, Wilfrid, 413, 413n51
Hurwitz, Adolf, 72, 154n8
Husserl, Edmund, 16, 77, 217–18, 386, 406

Isaacson, Dan, 8, 23, 49–52

Jordan, Camille, 108, 121n29

Kant, Immanuel, 31, 243, 245–47, 256, 256n25, 260, 265–66, 289, 289n24, 290, 295n32, 296n33, 297, 319n24, 330, 334, 334n7, 341n20, 345, 346n24, 353n1
Keferstein, Hans, 80, 147, 357
Keränen, Jukka, 52, 319n24, 413, 415
Klein, Felix, 5, 12–14, 16, 22, 40n24, 47–48, 106–8, 107n5, 108n6, 110–13, 112n15, 113n16, 114n18, 115n20, 117–19, 118n24–26, 119n27, 120–29, 121–25n33, 127n37, 128n39, 130, 130n42, 131–34, 134n54, 135–38, 137n60,170, 245n8, 308, 308n5, 329, 329n2, 331, 386
Kronecker, Leopold, 61, 68–70, 72n30, 79, 96n12, 99, 99n17, 143,147n4, 154n8, 157n10, 172–75, 217–18, 282n12
Kummer, Ernst, 71–72, 71n28, 78, 172, 172n2–3, 173–75
Kuratowski, Kazimierz, 431–32

Landry, Elaine, 3, 181, 184
Lautman, Albert, 16, 211n33
Lawvere, William, 46–47, 217, 226–27, 227n13

Leach-Krouse, Graham, 410n47, 410, 411n48, 417
Leibniz, Gottfried Wilhelm, 16, 21, 28, 33n15, 131, 210, 305, 331, 331n4
Leitgeb, Hannes, 413
Lie, Sophus, 108, 120n28, 123n31, 125, 171, 194, 287, 386
Linnebo, Øystein, 7–8, 408, 408n43, 410, 416–417
Lipschitz, Rudolf, 64, 78n46, 356
Lobachevsky, Nikolai Ivanovich, 245
Lotze, Hermann, 152–53n6, 353n1

Mac Lane, Saunders, 5, 11, 13, 72n31, 83n59, 135, 160, 168, 179–81, 189n7, 215–16, 215–18n5, 217–27, 226–27n12, 228–34
Mancosu, Paolo, 409, 411n48
Marquis, Jean-Pierre, 3–4, 130n42, 132–33, 133n51, 138
McLarty, Colin, 3–4, 13, 179, 181, 189n7, 281n9
Mill, John Stuart, 247–48, 256n25, 343
Monge, Gaspard, 108
Morris, Sean, 15

Neumann, John von, 1, 6, 82n58, 226, 276, 364–65, 365n19, 368, 432, 436
Noether, Emmy, 5, 13, 72, 72n31, 83, 166–77, 175–77n5, 178–82, 181n7, 183–85, 215, 216n3, 219–22, 281n9, 337n16, 355–56

Parsons, Charles, 1–8, 15, 23, 49, 52–53, 52n31, 54, 316, 319n24, 336, 341n20, 353–54n2, 407, 437, 437n31
Pascal, Blaise, 109*f*, 109
Pasch, Moritz, 12, 88–89, 91–102, 92n5, 97n13, 98n15, 99n17, 108, 110
Peano, Guiseppe, 9–10, 23–24, 43, 43n27, 248n14, 324n32, 331, 377, 394, 423, 432
Peirce, C.S., 11, 13, 73, 77, 241–63, 241–45n8, 247–48n14, 249–56n25, 259n28, 261–63n37, 264–65n42, 264–70, 266–67n45, 306, 430
Petitot, Jean, 13, 281n9, 374n33
Plato, 88, 210, 246n11
Plücker, Julius, 107–08, 111–15, 111–12n15, 114n17–n18, 116–17, 119, 121, 125–27, 127n37, 133–34, 138
Poincaré, Henri, 11, 14, 60, 60n2, 153, 155, 190–91, 202, 202n27, 275–76, 278–80, 279n8, 281–84, 282n11, 283n13, 284n16, 285–91, 289–90n24, 291n26, 292–95, 294n30, 295n32, 296–300, 296n33, 298n37, 300n40, 307–8, 310, 359, 377, 386–87, 438
Poncelet, Jean-Victor, 92–93, 108–11, 110n11, 111n13, 114–115, 117, 126n35, 129

Quine, W.V.O., 14–15, 215n2, 216, 387, 406, 421–24, 421–22n3, 424n7, 424n9, 425–31, 425–27n16, 429n19, 431n21–n23,432–34, 434–35n28, 435–40, 439n35, 439n37

Ramsey, Frank Plumton 438
Reck, Erich, 12, 14, 23, 35n19, 89n2, 89–90, 128, 128n40, 275n4, 282n11, 284n18, 307n4, 320n27, 394n19, 408, 416–17, 422–23, 423n5, 434n25, 436n30
Resnik, Michael, 1–3, 5, 8, 15, 273, 337, 339, 339n17, 347, 434n25, 436n30, 437, 437n31, 437n33, 438–39
Riemann, Bernhard, 9–10, 12, 15–16, 47–48, 60–64, 63n9, 64n11–n12, 65, 82, 119n27, 195, 204–205, 205–6n30, 230, 244n4, 244–46, 246n10, 386–87
Russell, Bertrand, 2, 7–8, 11, 14–15, 23–24, 35n19, 80–82, 143–44, 154, 303–11, 303n1, 306n3, 308–15n18, 312–17, 316n21, 317n22, 318–23n29, 318–26, 324n32, 324n34, 330–331, 336–38, 337n14, 337n16, 339–40, 340n18, 344, 347, 361, 377, 385, 387–88, 391n16, 394, 396, 397n25, 398n26, 402–3, 402n35, 405–407, 409n44, 416–17, 421–27, 422–23n5, 424–26n14, 428–30, 428n17, 431n22, 432–33, 436, 438, 439n36, 439–40

Scharlau, Wilfried, 66–67, 70
Schiemer, Georg, 12, 15, 23, 89n2, 183–84, 308, 310n12, 314n15
Schleiermacher, Friedrich, 21, 37, 37n20, 37n22, 40, 45–46
Schlimm, Dirk, 12
Schröder, Ernst, 77, 77n40, 306–7, 337n16
Shapiro, Stewart, 1–5, 7–8, 15, 23, 43n27, 49–52, 51n29, 131, 273, 277–78, 278n7, 279, 284, 287–88, 297, 299–300, 308n6, 337, 339, 339n17, 347, 407, 413–14, 421, 421n1, 434n25, 437, 437n31, 437n33, 438–39
Sieg, Wilfried, 12, 14, 91
Stein, Howard, 353, 374, 394

Thom, René, 195–96

van der Waerden, B.L., 65, 72, 166, 167–68, 178, 200
von Neumann, John, 82n58, 364–65, 365n19, 368, 432, 436

Weierstrass, Karl, 16, 63, 68–69, 76, 99n17, 154n8, 217–219, 230, 255, 331
Weil, André, 13, 189, 189n5, 189n6, 192, 203–210
Weyl, Hermann, 156, 167–68, 177–78, 215, 218–19, 218n5, 218n7, 232n18, 369, 402
Whitehead, Alfred North, 144, 215n2, 308n7, 385, 398n26, 426, 428–29
Wiener, Norbert, 431–32

Yap, Audrey, 13, 281n9

Zermelo, Ernst, 6, 76, 82, 82n58, 83, 226, 276, 337n16, 354–359, 371n27, 432, 436

Subject Index

For the benefit of digital users, indexed terms that span two pages (e.g., 52–53) may, on occasion, appear on only one of those pages

a priori, 68, 178, 245, 251, 285–90, 293–94, 296n33, 297–300, 300n39, 300n40, 374, 376n36, 386, 389
a posteriori, 204
abstraction
  Dedekind, 80–81, 83, 312, 315, 317, 319–23, 320n26, 320n27, 324–25, 336, 342n22, 344
  definition by, 133n52, 313, 313n14, 319, 322–24, 324n32, 325–26, 384, 388, 402, 402n33, 407–9, 416
  principle of, 7–8, 15, 317, 319, 320n26, 320–22, 320n27, 336–37 (*see also* abstraction principle)
  structural(ist), 7–8, 90, 128, 132–34, 383–84, 402n34, 408, 408n43, 409, 409n45, 410–11, 411n48, 413, 415n56, 416–17
abstraction principle, 8, 133, 313, 384, 388, 402, 402n34, 407–8, 408n43, 409, 409n45, 410, 415n56, 416–17. *See also* abstraction: principle of
accessibility, 352, 369, 373
algebra
  abstract, 22, 41, 43–44, 89, 167–68, 171–72, 185, 216–17
  (non-)commutative, 42, 168, 177, 178, 222
  linear, 22, 24, 26, 41, 43–44
  modern, 27, 41, 65, 70, 72–73, 91, 2
  universal, 41, 42, 44
algebraic *vs*. non-algebraic, 3, 51–52, 51n29
algebraic integers, 70–72, 167, 172, 175–78, 183–84. *See also* number: algebraic
algebraic number. *See* number: algebraic
algebraic topology, 193–94, 201n25
analysis
  arithmetization of, 9, 82, 331
  complex, 9
  rigorization of, 9, 255
  subsystem of, 369–70
anti-realism, 3, 276, 278, 288–89, 299
arithmetic
  consistency proof for (*see* arithmetic: consistency of)
  consistency of, 148, 162
  Dedekind-Peano (*see* arithmetic: Peano)
  foundations of, 11, 89, 102, 153n7, 269
  Peano, 15, 131, 385, 403, 407
  primitive recursive, 158, 364
automorphism, 3, 72, 83, 230–32, 230n15, 319n24, 413–15, 414n55
axiom(s)
  Dedekind-Peano, 2–3, 6–8, 79–80, 83, 251, 315, 325–26, 338, 385, 407, 425–26
  definition by, 93–94
  of induction, 248–49, 249n16, 249–50n17
  Peano (*see* Dedekind-Peano)
  structural, 168, 354–55
axiomatics, 23, 43, 156, 188, 190–91, 203, 210, 218, 284, 291, 353–54, 354*f*, 362–63, 364*f*, 371*f*, 376n36, 391, 398, 401n30, 407–8, 412
  abstract, 370, 371*f*
  formal, 12, 90, 142–44, 154, 331, 360n13, 362, 383, 386–87, 390–91, 395
  general, 314, 314n15, 383–84, 387–88, 400–1, 403, 407, 412, 414, 416–17
  modern, 383, 388, 388n12, 389, 389n14, 390, 400, 400n28, 412
  structural, 12, 143–44, 146, 150–51, 159–60, 178, 353, 358, 361, 369*f*, 371*f*, 376, 399
axiom system, 7–8, 110, 143, 145, 156, 160–62, 192, 227, 248, 285, 286, 313, 345, 355, 383, 387–93, 391n16, 394–95, 397–98, 400–1, 400n29, 405

calculus, 16, 34–35, 41–42, 61, 65, 74n34, 154–55, 154n8, 158–159, 188n4, 198n21, 243, 252, 255–57, 273–75, 248n13, 331, 361–62, 377, 429n19
Cantor-Bernstein theorem, 160n13, 358, 358n11

categoricity, 3, 10, 80–81, 83, 335–36, 369, 371n27, 400n28, 401
category, 46–47n28, 135–37, 180, 182, 200–1, 216, 220, 224–29, 230n15, 231, 231n16, 232–33
category of sets, 227, 233n19
category theory, 3–6, 8–9, 10, 12–13, 34, 47, 69n21, 81n56, 83, 107–8, 135–38, 135n56, 135n57, 137n60, 177, 179–85, 188–90, 188n2, 189, 189n7, 193, 200–1, 201n25, 217, 224–25, 227, 229–30, 232–33, 281n9, 373n31, 377
class
  equinumerous, 303, 317, 320, 323, 325–26
  equivalence, 7–10, 15, 25, 78n46, 183, 225, 332, 334–38, 402, 409, 411, 411n48, 416–17
  isomorphism, 7, 313n14, 401–4, 407–8, 407n41, 410–11, 410n47, 411n48, 416
  of models/structures, 48, 51–52, 51n29, 94, 189–90, 393–94, 394n19, 401, 404–7, 416
cognition, 77, 97, 292–93, 300, 333
  mathematical, 353, 373, 375–76
completeness, 10, 74, 145, 150, 153–54, 258, 334–35, 387, 400n28, 400–1
  axiom of (*or* completeness axiom), 144–46, 149
  semantic, 154n8, 401
  syntactic, 362, 364–65
  topological, 355
comprehension principle, 81–82, 357–58, 370
computation, 60n5, 218–19, 375–76
concept
  formation, 21–22, 34, 36–38, 37n20, 37n22, 40–41, 45–46, 51n30, 54, 152n5, 329–30, 332, 334, 339, 341–42, 360, 362, 394–95
  higher-order, 8, 334–35, 393
  substance vs. function concepts, 14, 329–30, 333, 342, 346–47
consistency, 3, 10, 12, 31, 80, 96, 146–49, 154–55, 155n9, 156–60, 157n10, 162, 201, 258, 285, 345, 352, 359–65, 368, 376, 387
  program (Hilbert), 353–54, 365
  proof, 143, 148, 149, 151, 154–55, 157–58, 160, 162, 190n9, 292, 354–55, 359, 364–65, 367–69, 372–73, 376–77, 400n28
construction
  logical, 76, 335, 383–84, 388, 394–96, 424–25, 426n14
  of proof, 159, 198, 374–75
  vs. postulation, 48, 50, 52
  set-theoretic, 75–76, 83, 133, 335, 335n11, 344–46, 386, 396, 416
constructivism, 45–46, 53–55, 156–57, 278, 288–89, 298–99, 395, 395n21
continuum, 76, 145–46, 149, 241n1, 253–55, 254n24, 282–83, 285n20, 289, 289–90n24, 293–94, 297, 300, 310, 360
convention (conventionalism), 202, 284–87, 293, 295, 310, 386–87, 386n8, 393n18, 430
coordinates, 9, 24, 110–13, 112n14, 113n16, 114–17, 119, 120–21, 123, 127, 231n16, 372–73. *See also* coordinate system

decomposition, 36–37, 71, 172–73, 175–77, 181, 203
Dedekind-Peano arithmetic. *See* arithmetic: Peano
Dedekind-Peano natural number. *See* number: natural
deduction, 14, 23, 37n22, 46, 77–78, 91–96, 149–50, 159, 191, 193–94, 196, 266, 280, 309, 311, 332, 338, 359, 362–63, 411–12
definition
  explicit, 94, 383–84, 388, 392–95, 395n21, 398–99, 416
  implicit, 43n27, 92–94, 98–99, 284–86, 383–84, 389, 391–93, 395, 397–98, 411–12, 416
  impredicative, 394
  inductive, 226, 370, 371–72
  recursive, 81, 157–58, 352
  structural, 145, 147–48, 152, 168, 176, 303, 353n1, 354n3, 355–57, 359–60, 362, 367–68, 374, 414
determinateness, 33, 37–38, 50–52, 330, 332–33, 340–42, 347–48
diagrammatic reasoning, 13, 241–42, 246–47, 254, 259–60, 263–66, 265n40, 267, 269
domain
  accessible, 352, 354–55, 354*f*, 371*f*, 372–73, 373n31, 374–75
  constructive, 362, 365, 367–70
duality, 12, 92–93, 95–96, 107–13, 109n9, 110n11, 111n13, 114–16, 129, 132–34, 138, 207n31, 384, 413
  principle of, 12, 107–13, 138
  projective, 92–93, 96, 107–10, 109n8, 111–14

Éléments (Bourbaki), 188–90, 189n6, 194–96, 197n19, 201, 201n25, 211, 225, 353
Elements (Euclid), 259–60, 263
empiricism, 88–89, 92–97, 100n18, 101–2, 247–48, 342–43
epistemology, 49, 54, 88–89, 217, 273, 284n17, 287, 289, 295, 296n13, 297–300, 352, 362–63, 365, 427–28, 437

equivalence, 8, 10, 15, 25, 28, 78n46, 127, 127n38, 132, 135n55, 137n61, 138, 183, 225, 293, 313, 313n14, 320n27, 320–22, 335–37, 357, 407, 409, 411, 411n48, 417
  proof-theoretic, 356, 356n6, 359
  structural, 12, 107–8, 135–36, 138, 401–2, 409
Erlangen program, 12, 47, 106–8, 111, 117–18, 120–21, 128n39, 130n42, 136–37, 245n8, 329n2, 331
existence. *See* Platonism
extension theory (*Ausdehnungslehre*), 22, 24, 26n4, 31, 31n12, 43n27, 33*t*

fictionalism, 278
field, 9–10, 65–70, 89–91, 173, 199–200, 204–9, 216–17
  continuous, 334–35
  finite, 68
  (complete) ordered, 74, 143, 355–56, 369–70, 408
finitism, 155n9
formal operation, 29–30, 38, 40, 49, 50, 52–53, 211
formalism, 92, 100n18, 156–58, 291, 360n13, 366, 377, 428–29
  Hilbert's, 156–58, 190–91, 200–1, 291, 359, 386
formalization, 101, 143, 153–54, 187, 195–96, 283–84, 353, 359–60, 362, 362n13, 368–70, 375n35, 393, 398–99, 400
foundation of mathematics, 8–10, 61, 74, 92, 94, 99n17, 101, 142, 146, 154, 217n4, 268–69, 336, 386, 421–23
  categorical, 180, 227, 233n19
foundational program. *See* foundation of mathematics
function, 7–10, 22, 34–36, 50, 61–68, 79, 83, 178, 203–4, 217, 225–26, 253, 312, 333, 335, 389, 393, 410–11, 425–28
  concept, 342–43
functor, 69n21, 136–37, 137n59, 193, 200–1, 216–17, 220, 224–25, 227, 231, 233

Galois theory, 9, 12, 59, 65–67, 66n15, 69n20, 69–70, 72, 78, 83, 182, 331
geometry
  algebraic, 13, 189, 194, 203, 206, 209, 216, 228
  applied, 245, 269, 307–8, 308n7, 310
  descriptive, 311–13
  differential, 216–17, 229
  Euclidean, 63, 91, 118n26, 121, 129, 218, 285–86, 307–10, 333, 359, 386–87, 389
  foundations of, 145, 386, 391, 397–98
  mathematical, 257, 285–87, 398
  metric, 123, 125, 245, 307–8, 309n10
  19th-century, 12, 14, 106–7, 106n3, 110, 118, 138, 386–87
  non-Euclidean, 9, 41, 43–44, 63, 107n55, 118n26, 124n32, 242, 244–45, 273–74, 307–10
  physical, 257, 386
  projective, 44, 47–48, 91–94, 96, 101–2, 107–10, 114, 116–18, 121, 124–26, 129, 135, 138, 206, 244–45, 254–55, 307–8, 331, 387, 398
  pure, 191, 245–46, 257, 307–8, 308n7, 310, 312n13, 391, 398
group
  abelian/commutative, 29, 49, 200
  additive, 29, 34–6
  axioms, 67, 89, 145, 150, 151
  cohomology, 216
  finite, 67, 193
  Galois, 66, 69–70, 83, 182, 183
  homomorphism, 221–24
  Lie, 171, 194
  permutation, 66, 108, 182, 256–57
  theory, 3, 9, 51n29, 60, 65, 106–7, 129, 232–33, 279
  topological, 200–1
  transformation, 12, 106–8, 118, 121–28, 130–36, 202, 386–87 (*see also* transformation: group of)

Heyting arithmetic, 368
hierarchy of sets, 196, 371–72
  cumulative, 371–72
homomorphism, 10, 78, 81, 83, 177, 180–81, 219, 221–23, 222n8, 224–25, 228, 233
  algebra, 231
  group, 221, 224
Hume's principle, 313, 319, 411

ideal, 9, 72, 77–78, 82–83, 172, 173–77, 180–81, 184, 287, 300
  agent, 97
  divisors, 9, 174–76
  element, 9, 329–31, 333, 345–46
  idealism, 218, 334, 341–42, 347–48, 423–24
  number, 71, 173–75
  object, 339
  prime factors, 172–74
  subject, 97
  theory, 64, 68–70, 72, 77–78, 82, 167–68, 171–72, 175n4, 176, 178, 218–19

identity
  geometrical, 129–30
  morphism, 228, 230
  structural, 106–07, 130–31, 138, 414–17
identity criterion, 51–53, 130n42, 131, 384, 414n53
incompleteness
  of mathematical objects, 437–39
  syntactic, 365
  theorem, 158–59, 160n13, 352, 365, 365n19
independence, 24, 39, 118, 129, 151, 160, 258, 277–78, 278n7, 288, 297, 300, 306n3, 320–21, 356–57, 387, 400n28
indifference to identification, 352, 356n7
induction, 26, 34, 79–81, 143, 155, 157–58, 249–50, 249–50n17, 290–93, 291n26, 294, 363
  axiom of, 248–49, 249n16, 405
  empirical, 291
  hypothesis, 250
  mathematical induction, 80
  numerical induction, 34
  pessimistic meta-induction, 294–95
  principle of, 250–51, 290, 357, 368, 372, 377
  transfinite, 365
inductively defined class, 352–54, 367–68
intuition, 21–23, 39, 43, 46, 76, 155n9, 203, 211, 218, 259–60, 280, 283–85, 285n20, 286, 288–94, 297–98, 300, 300n40, 321, 360, 368, 368n25
  a priori, 298–300, 386, 389
  geometric, 76, 190–91, 205, 285n20, 293, 298, 363n16
  mathematical, 46, 275–76, 288–89, 294, 297
  spatial, 39
intuitionism, 157n10, 373
invariance, 122, 129–30, 132n47, 171, 182, 202, 245n8, 384, 412–13, 417
invariant, 10, 12, 23, 47, 54, 65–66, 68, 107–8, 117–18, 122–25, 123n31, 127–132, 134–38, 167–71, 179, 181–84, 194, 206, 233n19, 332, 412, 415
  complete, 415n56, 415–16
  isomorphism, 226–27, 233–34, 233n19
isomorphism, 7, 9–10, 75, 75n38, 78, 78n46, 81, 83, 125–26, 126n36, 127–28, 132n47, 133–34, 134n53, 137, 137n59, 176–77, 180–81, 183, 200–1, 219, 221–24, 226–30, 232–34, 233n19, 383–84, 397, 401–2, 401n31, 403–4, 407–8, 407n41, 409, 410n47, 410–11, 411n48, 412–13, 415
  canonical, 352, 372
  natural, 216–17, 220
  relation, 313, 313n14, 401, 408
  type, 397, 409–10

law
  formal, 29–30n10, 30, 41–2, 48,
  laws of concatenation, 421
  logical, 31, 31n12, 150
  of arithmetic, 77–78, 252, 421, 433
  of composition, 67, 192, 199–200
  of reciprocity, 117
  of progression, 31, 31n13, 46
  of thought/thinking, 31n12, 33, 33n14, 76, 82–83
logic
  first-order, 3–4, 92–93, 364–65
  intuitionist, 368
  mathematical, 154–55, 159–60, 331, 348, 355, 359–60, 364–65, 377, 426–27
  modal, 3–4
  modern, 8–9, 12, 331, 344
  of relations, 14, 243, 246–47, 258–61, 304–7, 310, 310n12, 311–14, 319n25, 339–40
  quantificational, 306
  second-order, 360
  type-theoretic (*see* type theory)
logicism, 73, 76–77, 82–83, 279, 290–91, 303, 325–26, 330n3, 339n17, 383, 383n1, 385–86, 388, 394–95, 395n21, 397–98, 398n26, 416

magnitude
  extensive, 24–27, 26–27n5, 29–30, 33*t*, 34, 36, 38, 41, 43–44, 47–48, 52–54
  intensive, 33*t*
manifold, 16, 46, 63, 63n9, 119–27, 119n27, 127n37, 128–30, 130n43, 131–34, 134*f*, 135, 137–38, 200–1, 203, 229, 230n15, 231, 231n16, 303–4, 362, 386–87, 396
mathematical understanding, 159, 279–281
mathematics
  applied, 40–42, 46, 54, 234, 242, 269, 285–86, 305, 307, 309–10
  computational, 60n5, 218
  conceptual, 12, 34, 35n19, 38–40, 60–62, 65, 68–69, 72, 78, 168, 171, 174–75, 339 (*see also* structuralism: conceptual)
  constructive, 63, 155–56, 167, 363n17, 367–68, 373

finitist, 155, 155n9, 156–58, 191, 202, 352, 363–66, 368, 370
intuitionist, 144, 368–70
modern, 8–9, 128, 190–91, 210, 215, 330–31, 389
pure, 32, 46, 61, 73, 101, 128, 157, 181, 245–47, 304–5, 309–10, 312, 333, 425–26
mapping, 4, 22, 35n19, 36, 79, 79n47, 81, 81n56, 82, 107–8, 110, 114–17, 119, 122, 124–28, 127n38, 129, 133–38, 146n2, 152n5, 152, 199, 221, 225, 335n12, 356, 356n6, 357–58, 400–1, 401n31, 404
structure-preserving, 116–17, 134–36, 138
metamathematics, 191, 202, 363
metaphysics, 15, 246n11, 299–300, 318, 330, 423–24, 424n7, 426–27, 432–33, 439–40
metatheory, 48, 197, 401n32
methodological frame, 14, 352–57, 366–68, 370–71, 373
methodology, 1, 4–5, 10, 12–13, 15, 22–23n2, 23, 43, 59, 63, 65, 69, 72, 82–83, 89, 91, 128, 166–68, 170, 179, 211, 220, 273–74, 281, 289-90n24, 329–31, 340, 341, 387, 424
model
isomorphic, 3, 7, 80–81, 142n2, 233n19, 336–37, 384, 401–2, 402n34, 404–5, 407–8, 410–11, 415
non-standard, 92
model theory, 10, 48, 228, 232–33, 392, 413, 413n51

naturalism, 421–22, 423, 439–40
neo-Kantianism, 329
neo-logicism, 411n49
nominalism, 187, 278, 298n37, 341
number
algebraic, 68, 70–71, 75n37, 173, 203, 205*t*, 206, 208
cardinal, 147n4, 252–54, 291, 303, 313, 315–17, 323, 337, 392, 402
complex, 3, 9, 53, 74n33, 76, 79, 171–73, 203, 218–19, 229n14, 230–31, 230n15, 233, 257, 268, 275n4, 331, 333, 344, 346
cyclotomic integer, 172, 172n2
imaginary, 73, 256–58
irrational, 74–75, 143, 147–49, 147n4, 283, 334, 335n10, 336, 339–40, 355
natural, 1–3, 6–8, 25–27, 43, 53, 60–81, 96–97, 100–1, 147–50, 152–53, 166–67, 181, 217, 226–27, 248–50, 273, 276–78, 293, 304, 315–16, 319, 322–24, 334, 338, 343, 355, 363, 371–72, 392–93, 405, 418-7, 421–22
ordinal, 249–50n17, 292, 313, 315–16, 335n12, 338–39
rational, 27, 44, 73–76, 146n2, 147n4, 175, 205*t*, 332, 334–35, 340, 344, 355–56, 360
real, 26–27, 44, 68, 74–76, 78, 81, 143, 144, 146–49, 282–83, 303, 332–35, 337–38, 345–46, 395–96
number theory, 12, 22, 39, 59, 60–61, 67, 96, 100–1, 143, 154–59, 216, 231, 310, 357, 361, 364–65, 368, 397–98
algebraic, 12, 64–65, 70–71, 78, 334–35, 334n8, 355
analytic, 61, 193

object
abstract, 2–3, 6, 276, 313, 320–21, 322–23, 323n30, 332, 355, 434n26, 437
concrete, 54, 377, 437–39
geometrical, 114, 117, 127, 135
physical, 15, 88–89, 97, 218, 274–75, 434
syntactic, 157–58, 369
ontology, 49, 54–55, 135, 231–32, 297–98, 300, 433–35
metaontology, 411n49
structuralist, 232–34, 440
order system *(Ordnungsgefüge)*, 389–91, 395–97, 400–1

Peano arithmetic. *See* arithmetic: Peano
permutations, 33*t*, 66, 72, 121, 121n29
group of, 65–66, 108, 182, 255–57
phenomenology, 217
platonism, 51n30, 187, 276–78, 288, 299–300, 373
position in a structure, 2, 3, 6, 45, 106, 304–5, 322–23, 338–40, 347, 414
postulate. *See also* axiom
primitive recursive arithmetic. *See* arithmetic: primitive recursive
Principia Mathematica, 144, 154, 159, 215–16, 3 26, 359, 385, 398n26, 426–27
principle of abstraction. *See* abstraction: principle of
progression, 8, 305–6, 310, 313, 317, 319, 322–25, 336–37, 392–94, 396–97, 405–6, 425–26, 432–33, 436
law, of, 31, 46
projection, 121, 126, 129–30, 352, 359–60, 363, 368–69, 370, 372–73

proof
  theory of, 10, 12, 14, 142, 155, 352–53, 360, 363–66, 375, 377
property
  categorical, 136–37
  formal, 286, 316n21, 392, 396n23, 412–15
  geometric, 122, 124, 130n42, 150
  intrinsic, 6, 9, 106n1, 166, 267–68, 303, 316n21, 321n25, 336, 304, 425–26
  non-structural, 6, 90, 232, 234, 320–21, 408
  structural, 4, 10, 48, 53, 106, 130–31, 183, 230, 232, 269, 310, 324, 361–62, 384, 400, 411–15, 422–23, 433–34
provability, 156, 364
psychologism, 77, 80n52, 280, 320n27, 323n30, 324n33, 337, 340–41, 340n19, 348, 357n8

quantification theory. *See* logic

realism, 2, 3, 15, 51, 276, 278, 278n7, 288–89, 294–99, 294n30, 295n32, 296n33, 312n13, 313, 333–34, 340–43, 347–48, 407, 439
  structural, 51, 277–78, 288–89, 294–97
realist. *See* realism
reasoning
  algebraic, 137–38, 231
  diagrammatic, 13, 241–43, 245, 247, 254, 259–67, 267n45, 269
  geometric, 309
  logical, 279
  mathematical, 191, 192, 258–59, 268–69, 290–93, 305–6, 374–75
  methods/styles of, 106–7, 128n41, 148–49, 153, 158–59
relation
  congruence, 60, 133
  equivalence, 183, 313n14, 319–23, 401, 409
  form of a, 241
  fundamental, 30, 210, 355
  incidence, 109–10, 109n9, 112–13, 129
  logic of, 14, 243, 247, 258–59, 261, 304–7, 310, 312–13, 319n25, 339–40
  mutual (Hilbert), 148–49
  one-to-one, 253–54
  reciprocal, 28
  system of (*see* system: relational)
relational type. *See* type, relational
representation, 61–63, 68, 77, 111, 113–14, 119, 125, 128n41, 132–33, 135–36, 159, 167–68, 177–78, 187, 207, 222, 255, 262n34, 263, 265–67, 315, 357–58, 363, 371, 375, 377, 386, 404–5, 414, 415–16

semantic(s), 146–48, 155, 166–67, 356, 356n4, 357–59, 385n4
set-theoretic. *See* set theory
set theory, 1–4, 6–10, 23, 43, 47, 49, 72–78, 77–78n45, 79, 81–83, 149, 154, 156–59, 174–76, 183–84, 189–91, 196, 197n17, 200–1, 201n25, 217, 221–22, 225–28, 232n18, 242, 252–53, 254n24, 269, 312n13, 312–13, 329, 331, 335, 345–46, 354–55, 357–58, 361–62, 365, 387, 396–98, 407, 411, 413, 421, 423, 436, 438
space
  abstract, 43, 293, 305, 398
  element of space, 112–14, 114n18, 126–27n37, 131
  Euclidean, 1, 9–10, 63, 63n9, 111–14, 145, 152, 309–10, 319n24, 355, 359
  formal, 310n12, 310, 386, 390–91, 395–98, 396n23, 398n26
  intuitive, 118–19, 128, 289, 300n40, 386, 390–91
  metric, 9–10, 75n37
  n-dimensional, 8–9, 63, 63n9
  physical, 43–44, 244–46, 286–87, 294, 305, 307–10, 308n7, 386, 388–89
  projective, 110, 206, 386, 396, 399
  topological, 63n9, 179–80, 193, 200–1, 201n25, 202, 205–6, 224, 229, 355, 396, 398–99
  vector, 8–10, 24, 42–44, 51–53, 177, 201
structuralism
  about science/global, 294–95, 353–54n2, 417, 426n14, 438–40
  abstractionist, 407–8, 408n42, 409, 411n49
  ante rem, 2, 7, 14, 49, 55, 131, 132n49, 241, 277–78, 287–89, 297, 297n34, 299–300, 407, 434n25
  categorical, 4–5, 135–36, 136n58, 138, 180–81, 183–84, 241–42, 268–70
  category-theoretic (*see* categorical)
  concept, 8, 11–12, 23, 51, 51n30, 53–55
  eliminative, 2–4, 6–8, 11–12, 49–50, 180–81, 278, 297, 299, 304, 312, 312n13, 314, 325–26, 407, 434n25
  empiricist, 91–100
  in re, 7, 15–16, 49–50, 54–55, 132–33, 132n49, 138, 278, 407
  metaphysical, 4–5, 8, 11, 334
    (*see also* philosophical)

methodological, 5, 8, 11–16, 22, 22–23n2, 23, 43–47, 54, 59, 59n1, 81–83, 89–92, 91n4, 95–96, 101–2, 128n41, 166–68, 171, 178–79, 184, 241, 268–69, 273–75, 282, 284, 284n18, 286–88, 299–300, 334
modal, 2, 47, 180, 278, 312, 312n13, 434n25
non-eliminative, 2–4, 6–7, 12–14, 49, 52–53, 52n31, 54, 81n54, 106, 132, 138, 304, 304n2, 307–8, 312–15, 312n13, 316, 319n24, 325, 336, 339, 341n20, 407–8, 413
philosophical, 4–6, 8, 11, 22–23, 41, 48–49, 54–55, 128, 166–70, 175–76, 178, 184, 226, 230–32, 273–74, 277, 387 (*see also* metaphysical)
realist, 2, 276–78, 288–89, 294, 296–99, 340–41, 341n20, 347–48, 407
reductive, 14
relativist, 6–7, 423, 423n5, 436n30
set-theoretic, 6, 146
universalist, 7
structural(ist)
abstraction, 7, 80–81, 83, 90, 100, 128–29, 132–34, 152–53n6, 210, 282–86, 312, 313n14, 313–15, 317, 319–26, 320n26, 320n27, 323n30, 324n34, 335–37, 335n12, 337n16, 340, 342–45, 342n22, 347, 353n1, 383–84, 388, 402, 402n34, 407–11, 408–9n45, 413
approach/methodology/perspective, 4, 6–7, 45, 60, 72, 89–91, 100, 107, 131, 138, 166–76, 181–82, 210, 281, 283–84, 294–95, 307, 312–14, 341–42, 347–48, 383–84, 385n4, 389, 389n14, 395, 411n50, 421, 430
conception/view of mathematical objects, 2, 81, 83, 135–36, 183, 232, 276, 289, 303, 313–14, 316, 316n21, 319, 324, 329–30, 341, 415
conception/view of mathematics, 8–10, 92, 96, 132, 196, 268–70, 273–75, 284n17, 285–86, 310, 416
intuition, 288–89, 294
thesis, 131, 383–84, 387–88, 394–95, 406, 416–417
transformation of mathematics, 21, 41, 329–30, 342, 347–48
structures
abstract, 1–8, 41, 43, 48, 51, 89, 92, 94–96, 102, 106–9, 132–33, 135, 276–77, 281–83, 288–89, 296–98, 305–8, 310, 334, 336–37, 374–75, 383, 388–91, 407–10, 413–15, 426–27, 433
algebraic, 47, 137, 192, 196, 199–200, 205, 267–68
category-theoretic, 182–84, 188n2, 189–90, 225
deductive, 14, 17, 193–94, 332, 363
extremal (incl. minimal, maximal), 145, 370–72, 403–5, 404n37, 404n38
formal, 143, 156, 191
geometric, 130n43, 146n2, 273–74, 287, 293, 386, 396, 398, 400–1
model-theoretic, 50, 392, 394, 400–4, 407, 413
natural number, 1, 100, 276–78, 293, 300, 310, 343, 405, 407, 433
non-rigid, 3, 415
order, 192, 199–202, 205–9, 233n19, 396n23, 398
pure, 106, 336–37, 408, 411
set-theoretic, 50, 183–84, 197–98, 197n17
sui generis, 2, 200, 325, 411
topological, 196, 200, 205
system, 2–3, 9–10, 26n4, 43–44, 76–79, 89–91, 115–17, 144–50, 248–50, 276–79, 286, 332, 334–36
algebraic, 44, 205
axiom, 7–8, 12, 110, 143, 145, 156, 160–62, 169, 192, 227, 248, 285–86, 313, 345, 355, 383–84, 387–401, 405
coordinate, 24, 111–12, 112n14, 114–15, 117, 120–21, 126, 217
logical, 153, 267, 385, 398, 428–29, 433
number, 8–9, 46, 50, 52n31, 73–77, 78n45, 99, 100–1, 166, 168, 268, 277, 283, 407, 412, 414n55, 346
relational, 46, 59, 202, 250–51, 267–70, 313–14, 340, 391, 408–9, 411–12, 415–17
set theoretic (*see* set theory)
simply infinite, 79–81, 81n53, 91, 146–48, 150, 152, 226–27, 291n25, 292, 298, 300, 305, 315, 317, 335–36, 335n12, 355, 356–57, 368–70, 376

theory of forms
Grassmann's, 11, 27–28, 32, 54
Hankel's, 15, 21, 187, 218, 274
theory of groups. *See* group theory
topology, 3, 9–10, 51n29, 63, 192–94, 201, 201n25, 204–6, 216, 223, 232–33, 242, 245n7, 245, 255, 293, 387
Hausdorff, 399, 403
transfer principles, 12, 106–8, 115, 115n21, 117, 121, 124–26, 125n34, 126n36, 127–29, 133–35, 137–38
transformation, 46, 48, 54, 79n47, 110–11, 169, 315, 384, 412

theory of forms (*cont.*)
  continuous, 39, 46
  dual, 110–11, 117
  group of, 12, 106–7, 118–33, 130n42, 130n43, 138, 245n8, 386–87 (*see also* group: transformation)
  natural, 193, 216–17, 201, 224
translation, 46, 46–47n28, 125, 368
  between theorems, 110, 117, 126, 134
type
  isomorphism, 384, 407–10
  order, 290–91, 291n25, 397
type theory, 132, 160, 228, 385, 393, 395, 399, 412, 413n51, 416

unity of mathematics, 189, 192–94, 202, 210–11, 345–46